Triaxial Testing of Soils

Poul V. Lade

WILEY Blackwell

This edition first published 2016

Registered Office
John Wiley & Sons, Ltd, The Atrium, Southern Gate, Chichester, West Sussex, PO19 8SQ, United Kingdom.

Editorial Offices
9600 Garsington Road, Oxford, OX4 2DQ, United Kingdom.
The Atrium, Southern Gate, Chichester, West Sussex, PO19 8SQ, United Kingdom.

For details of our global editorial offices, for customer services and for information about how to apply for permission to reuse the copyright material in this book please see our website at www.wiley.com/wiley-blackwell.

Library of Congress Cataloging-in-Publication data applied for

ISBN: 9781119106623

A catalogue record for this book is available from the British Library.

Wiley also publishes its books in a variety of electronic formats. Some content that appears in print may not be available in electronic books.

Set in 10/12pt Palatino by SPi Global, Pondicherry, India

1 2016

Triaxial Testing of Soils

Contents

Preface

The triaxial test is almost always chosen for studies of new phenomena, because it is relatively simple and versatile. The triaxial test is the most suitable for such studies and it is required in geotechnical engineering for the purposes of design of specific projects and for studying and understanding the behavior of soils.

The first triaxial compression test apparatus, shown in Fig. P.1, was designed by von Kārmān (1910, 1911) for testing of rock cores. The scale may be deduced from the fact that the specimen is 4 cm in diameter (Vásárhelyi 2010). However, his paper was not noticed or it was forgotten by 1930 when Casagrande at Harvard University wrote a letter to Terzaghi at the Technical University in Vienna in which he describes his visit to the hydraulics laboratory in Berlin. Here he saw an apparatus for measuring the permeability of soil. Casagrande suggested that the cylindrical specimen in this apparatus could be loaded in the vertical (axial) direction to indicate its strength. Therefore, he was going to build a prototype, and Terzaghi proposed that he build one for him too. This appears to be the beginning of triaxial testing of soils in geotechnical engineering. The apparatus was immediately employed by Rendulic (Terzaghi and Rendulic 1934) for tests with and without membranes, the results of which played an important role in understanding the effective stress principle as well as the role of pore water pressure and consolidation on shear strength at a time when the effective stress principle was still being questioned (Skempton 1960; de Boer 2005).

Previous books on the developments of techniques for triaxial testing have been written by Bishop and Henkel (1957, 1962) and by Head (1986). The proceedings from a conference on Advanced Triaxial Testing of Soil and Rock (Donaghe *et al.* 1988) was published to summarize advances in this area. Other books have not appeared since then. To understand the present book, the reader is required to have a background in basic soil mechanics, some experience in soil mechanics laboratory testing and perhaps in foundation engineering.

In addition to triaxial testing of soils, the contents of the book may in part apply to more advanced tests and to the testing of hard soils – soft rocks. It is written for research workers, soil testing laboratories and consulting engineers. The emphasis is placed on what the soil specimen is exposed to and experiences rather than the esthetic appearance of the equipment. There will be considerable use of physics and mathematics to illustrate the arguments and discussions. With a few exceptions, references are made to easily accessible articles in the literature. Much of the book centers on how to obtain high quality experimental results, and the guiding concepts for this purpose have been expressed by the car industry in their slogans "Quality is Job One" (Ford Motor Company) and "Quality is never an accident, it is always the result of excellent workmanship" (Mercedes).

The book is organized in a logical sequence beginning with the principles of triaxial testing in Chapter 1, and the computations and presentations of test results in Chapter 2. The triaxial equipment is explained in Chapter 3, and instrumentation, measurements, and control is reviewed in Chapter 4. Preparation of triaxial specimens is presented in Chapter 5, and saturation of specimens is described in Chapter 6. The two testing stages in an experiment are made clear in Chapter 7: Consolidation and in Chapter 8: Shearing. Chapter 9 accounts for the corrections to the measurements, Chapter 10 informs about special tests and test conditions, and Chapter 11 puts the results from triaxial tests in perspective by reviewing results from

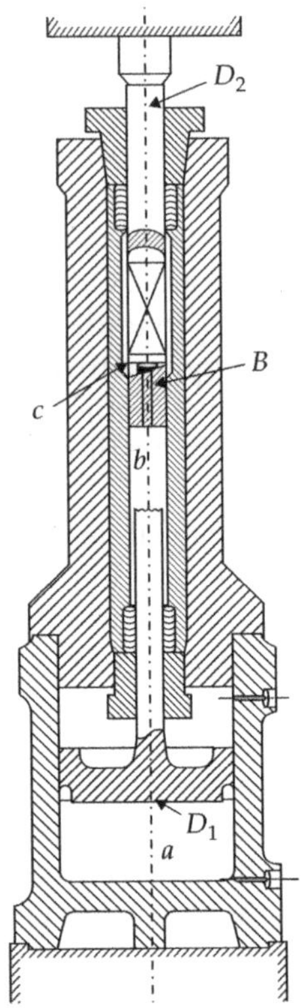

Figure P.1 Triaxial apparatus designed and constructed for testing of rock cores by von Kārmān (1910, 1911).

tests with three unequal principal stresses. Appendices are provided to explain special experimental techniques. Information on vendors for the various types of equipment may be obtained from the internet.

The author's background for writing this book consists of a career in laboratory experimentation at university level to study and model the behavior of soils. More specifically, he received an MS degree in 1967 from the Technical University of Denmark for which he wrote a thesis on the influence of the intermediate principal stress on the strength of sand and, in retrospect, ended up with the wrong conclusion on the basis of perfectly correct results. He received a PhD from the University of California at Berkeley in 1972 with a dissertation on "The Stress–Strain and Strength Characteristics of Cohesionless Soils," which included results from triaxial compression tests, true triaxial tests and torsion shear tests to indicate the effects of the intermediate principal stress on sand behavior, as well as a three-dimensional elasto-plastic constitutive model for the behavior of soils.

With his students, the author developed testing equipment, performed experiments and built constitutive models for the observed soil behavior while a professor at the University of California at Los Angeles (UCLA) (1972–1993), Johns Hopkins University (1993–1999), Aalborg University in Denmark (1999–2003), and the Catholic University of America in Washington, DC (2003–2015). Many of the experimental techniques developed over this range of years are explained in the present book.

Great appreciation is expressed to John F. Peters of the US Army Engineer Research and Development Center in Vicksburg, MS for his careful review of the manuscript and for his many comments. Special thanks go to Afshin Nabili for his invaluable assistance with drafting a large number of the figures and for modification of other diagrams for the book.

Poul V. Lade
October 2015

References

Bishop, A.W. and Henkel, D.J. (1957) *Measurement of Soil Properties in Triaxial Test*. Edward Arnold, London.

Bishop, A.W. and Henkel, D.J. (1962) *The Measurement of Soil Properties in the Triaxial Test*, 2nd edn. St. Martin's Press, New York, NY.

de Boer, R. (2005) *The Engineer and the Scandal.* Springer, Berlin.

Donaghe, R.T., Chaney, R.C., and Silver, M.L. (eds) (1988) *Advanced Triaxial Testing of Soil and Rock, ASTM STP 977*. ASTM, Philadelphia, PA.

Head, K.H. (1986) *Manual of Soil Laboratory Testing – Volume 3: Effective Stress Tests*. Pentech Press, London.

von Kārmān, T. (1910) *Magyar Mérnök és Építészegylet Közlönye*, **10**, 212–226.

von Kārmān, T. (1911) *Verhandlungen Deutsche Ingenieur*, **55**, 1749–1757.

Skempton, A.W. (1960) Terzaghi's discovery of effective stress. In: *From Theory to Practice in Soil Mechanics* (eds L. Bjerrum, A. Casagrande, R.B. Peck and A.W. Skempton), pp. 42–53. John Wiley and Sons, Ltd, London.

Terzaghi, K. and Rendulic, L. (1934) Die wirksame Flächenporosität des Betons. *Zeitschrift des Österreichischen Ingenieur- und Architekten Vereines*, **86**, 1–9.

Vásárhelyi, B. (2010) Tribute to the first triaxial test performed in 1910. *Acta Geology and Geophysics of Hungary*, **45**(2), 227–230.

About the Author

Poul V. Lade received his MS degree from the Technical University of Denmark in 1967 and he continued his studies at the University of California at Berkeley where he received his PhD in 1972. Subsequently his academic career began at the University of California at Los Angeles (UCLA) and he continued at Johns Hopkins University (1993–1999), Aalborg University in Denmark (1999–2003), and the Catholic University of America in Washington, DC (2003–2015).

His research interests include application of appropriate experimental methods to determine the three-dimensional stress–strain and strength behavior of soils and the development of constitutive models for frictional materials such as soils, concrete, and rock. He developed laboratory experimental apparatus to investigate monotonic loading and large three-dimensional stress reversals in plane strain, true triaxial and torsion shear equipment. This also included studies of effects of principal stress rotation, stability, instability and liquefaction of granular materials, and time effects. The constitutive models are based on elasticity and work-hardening, isotropic and kinematic plasticity theories.

He has written nearly 300 publications based on research performed with support from the National Science Foundation (NSF) and from the Air Force Office of Scientific Research (AFOSR). He was elected member of the Danish Academy of Technical Sciences (2001), and he was awarded Professor Ostenfeld's Gold Medal from the Technical University of Denmark (2001). He was inaugural editor of *Geomechanics and Engineering* and he has served on the editorial boards of eight international journals on geotechnical engineering.

1 Principles of Triaxial Testing

1.1 Purpose of triaxial tests

The purpose of performing triaxial tests is to determine the mechanical properties of the soil. It is assumed that the soil specimens to be tested are homogeneous and representative of the material in the field, and that the desired soil properties can in fact be obtained from the triaxial tests, either directly or by interpretation through some theory.

The mechanical properties most often sought from triaxial tests are stress–strain relations, volume change or pore pressure behavior, and shear strength of the soil. Included in the stress–strain behavior are also the compressibility and the value of the coefficient of earth pressure at rest, K_0. Other properties that may be obtained from the triaxial tests, which include time as a component, are the permeability, the coefficient of consolidation, and properties relating to time dependent behavior such as rate effects, creep, and stress relaxation.

It is important that the natural soil deposit or the fill from which soil samples have been taken in the field are sufficiently uniform that the soil samples possess the properties which are appropriate and representative of the soil mass in the field. It is therefore paramount that the geology at the site is well-known and understood. Even then, samples from uniform deposits may not "contain" properties that are representative of the field deposit. This may happen either (a) due to the change in effective stress state which is always associated with the sampling process or (b) due to mechanical disturbance from sampling, transportation, or handling in the laboratory. The stress–strain and strength properties of very sensitive clays which have been disturbed cannot be regenerated in the laboratory or otherwise obtained by interpretation of tests performed on inadequate specimens. The effects of sampling will briefly be discussed below in connection with choice of consolidation pressure in the triaxial test. The topic of sampling is otherwise outside the scope of the present treatment.

1.2 Concept of testing

The concept to be pursued in testing of soils is to simulate as closely as possible the process that goes on in the field. Because there is a large number of variables (e.g., density, water content, degree of saturation, overconsolidation ratio, loading conditions, stress paths) that influence the resulting soil behavior, the simplest and most direct way of obtaining information pertinent to the field conditions is to duplicate these as closely as possible.

Triaxial Testing of Soils, First Edition. Poul V. Lade.

However, because of limitations in equipment and because of practical limitations on the amount of testing that can be performed for each project, it is essential that:

1. The true field loading conditions (including the drainage conditions) are known.
2. The laboratory equipment can reproduce these conditions to a required degree of accuracy.
3. A reasonable estimate can be made of the significance of the differences between the field loading conditions and those that can be produced in the laboratory equipment.

It is clear that the triaxial test in many respects is incapable of simulating several important aspects of field loading conditions. For example, the effects of the intermediate principal stress, the effects of rotation of principal stresses, and the effects of partial drainage during loading in the field cannot be investigated on the basis of the triaxial test. The effects of such conditions require studies involving other types of equipment or analyses of boundary value problems, either by closed form solutions or solutions obtained by numerical techniques.

To provide some background for evaluation of the results of triaxial tests, other types of laboratory shear tests and typical results from such tests are presented in Chapter 11. The relations between the different types of tests are reviewed, and their advantages and limitations are discussed.

1.3 The triaxial test

The triaxial test is most often performed on a cylindrical specimen, as shown in Fig. 1.1(a). Principal stresses are applied to the specimen, as indicated in Fig. 1.1(b). First a confining pressure, σ_3, is applied to the specimen. This pressure acts all around and therefore on all planes in the specimen. Then an additional stress difference, σ_d, is applied in the axial direction. The stress applied externally to the specimen in the axial direction is

$$\sigma_1 = \sigma_d + \sigma_3 \tag{1.1}$$

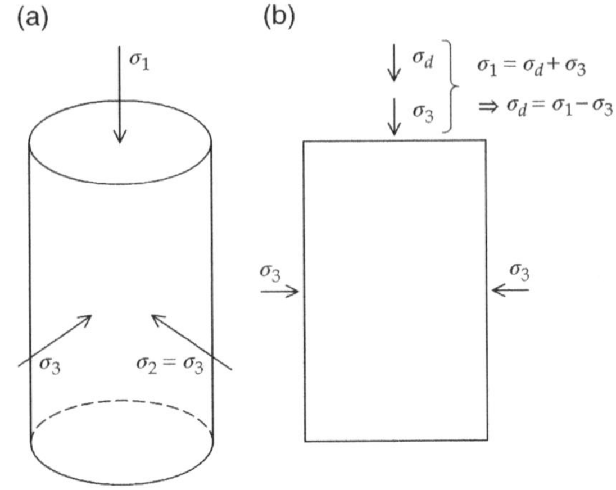

Figure 1.1 (a) Cylindrical specimen for triaxial testing and (b) stresses applied to a triaxial specimen.

and therefore

$$\sigma_d = \sigma_1 - \sigma_3 \tag{1.2}$$

In the general case, three principal stresses, σ_1, σ_2 and σ_3 may act on a soil element in the field. However, only two different principal stresses can be applied to the specimen in the conventional triaxial test. The intermediate principal stress, σ_2, can only have values as follows:

$$\sigma_2 = \sigma_3 : \text{Triaxial compression} \tag{1.3}$$

$$\sigma_2 = \sigma_1 : \text{Triaxial extension} \tag{1.4}$$

The condition of triaxial extension can be achieved by applying negative stress differences to the specimen. This merely produces a reduction in compression in the extension direction, but no tension occurs in the specimen. The state of stress applied to the specimen is in both cases axisymmetric. The triaxial compression test will be discussed in the following, while the triaxial extension test is discussed in Chapter 10.

The test is performed using triaxial apparatus, as seen in the schematic illustration in Fig. 1.2. The specimen is surrounded by a cap and a base and a membrane. This unit is placed in a triaxial cell in which the cell pressure can be applied. The cell pressure acts as a hydrostatic confinement for the specimen, and the pressure is therefore the same in all directions. In addition,

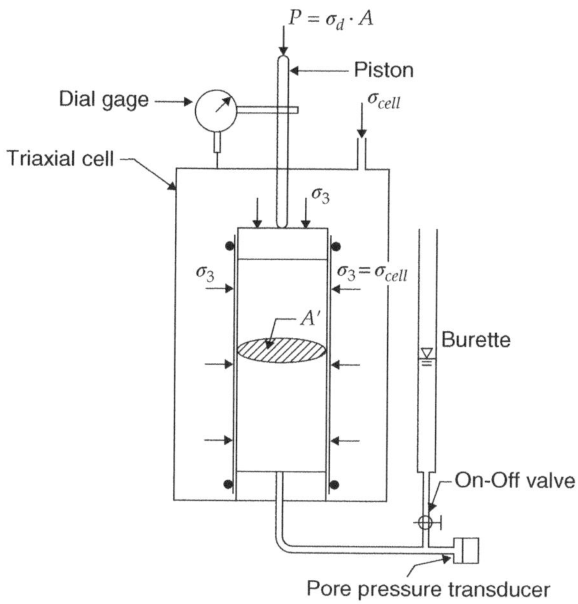

Figure 1.2 Schematic diagram of triaxial apparatus.

a deviator load can be applied through a piston that goes through the top of the cell and loads the specimen in the axial direction.

The vertical deformation of the specimen may be measured by a dial gage attached to the piston which travels the same vertical distance as the cap sitting on top of the specimen. Drainage lines are connected to the water saturated specimen through the base (or both the cap and the base) and connected to a burette outside the triaxial cell. This allows for measurements of the volume changes of the specimen during the test.

Alternatively, the connection to the burette can be shut off thereby preventing the specimen from changing volume. Instead the pore water pressure can be measured on a transducer connected to the drainage line.

The following quantities are measured in a typical triaxial test:

1. Confining pressure
2. Deviator load
3. Vertical (or axial) deformation
4. Volume change or pore water pressure

These measurements constitute the data base from which other quantities can be derived [e.g., stress difference ($\sigma_1 - \sigma_3$), axial strain ε_1, and volumetric strain ε_v].

1.4 Advantages and limitations

Whereas the triaxial test potentially can provide a substantial proportion of the mechanical properties required for a project, it has limitations, especially when special conditions are encountered and necessitates clarification based on experimentation.

The *advantages* of the triaxial test are:

1. Drainage can be controlled (on–off)
2. Volume change or pore pressure can be measured
3. Suction can be controlled in partially saturated soils
4. Measured deformations allow calculation of strains and moduli
5. A larger variety of stress and strain paths that occur in the field can be applied in the triaxial apparatus than in any other testing apparatus (e.g., initial anisotropic consolidation at any stress ratio including K_0, extension, active and passive shear).

The *limitations* of the triaxial test are:

1. Stress concentrations due to friction between specimen and end plates (cap and base) cause nonuniform strains and stresses and therefore nonuniform stress–strain, volume change, or pore pressure response.
2. Only axisymmetric stress conditions can be applied to the specimen, whereas most field problems involve plane strain or general three-dimensional conditions with rotation of principal stresses.
3. Triaxial tests cannot provide all necessary data required to characterize the behavior of an anisotropic or a cross-anisotropic soil deposit, as illustrated in Fig. 1.3.
4. Although the axisymmetric principal stress condition is limited, it is more difficult to apply proper shear stresses or tension to soil in relatively simple tests.

The first limitation listed above can be overcome by applying lubricated ends on the

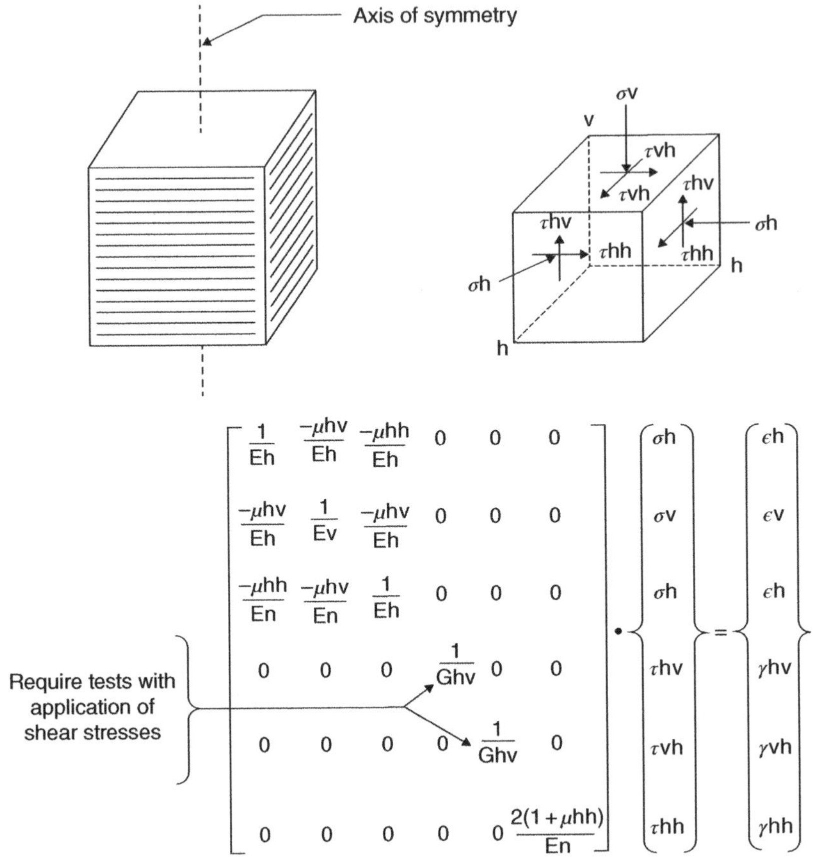

Figure 1.3 Cross-anisotropic soil requiring results from more than triaxial tests for full characterization.

specimen such that uniform strains and stresses and therefore correct soil response can be produced. This is discussed in Chapter 3. In addition to the limitations listed above, it should be mentioned that it may be easier to reproduce certain stress paths in other specialty equipment than in the triaxial apparatus (e.g., K_0-test).

Although the triaxial test is limited as explained under points 2 and 3 above, it does combine versatility with relative simplicity in concept and performance. Other equipment in which three unequal principal stresses can be applied or in which the principal stress directions can be rotated do not have the versatility or is more complicated to operate. Thus, other types of equipment have their own advantages and limitations. These other equipment types include plane strain, true triaxial, simple shear, directional shear, and torsion shear apparatus. All these pieces of equipment are, with the exception of the simple shear apparatus, employed mainly for research purposes. Their operational modes, capabilities and results are reviewed in Chapter 11.

1.5 Test stages – consolidation and shearing

Laboratory tests are made to simulate field loading conditions as close as possible. Most field conditions and the corresponding tests can be simplified to consist of two stages: consolidation and shearing.

1.5.1 Consolidation

In the first stage the initial condition of the soil is established in terms of effective stresses and stress history (including overconsolidation, if applicable). Thus, stresses are applied corresponding to those acting on the element of soil in the field due to weight of the overlying soil strata and other materials or structures that exist at the time the mechanical properties (stress–strain, strength, etc.) are sought. Sufficient time is allowed for complete consolidation to occur under the applied stresses. The condition in the field element has now been established in the triaxial specimen.

1.5.2 Shearing

In the second stage of the triaxial test an additional stress is applied to reach peak failure and beyond under relevant drainage conditions. The additional stress applied to the specimen should correspond as closely as possible to the change in stress on the field element due to some new change in the overall field loading situation. This change may consist of a vertical stress increase or decrease (e.g., due to addition of a structure or excavation of overlying soil strata) or of a horizontal stress increase or decrease (e.g., due to the same constructions causing the vertical stress changes). Any combination of vertical and horizontal stress changes may be simulated in the triaxial test. Examples of vertical and horizontal stress changes in the field are shown in Fig. 1.4.

Usually, it is desirable to know how much change in load the soil can sustain without failing and how much deformation will occur under normal working conditions. The test is therefore usually continued to find the strength of the soil under the appropriate loading conditions. The results are used with an appropriate factor of safety so that normal working stresses are always somewhat below the peak strength.

The stress–strain relations obtained from the triaxial tests provide the basis for determination of deformations in the field. This may be done in a simplified manner by closed-form solutions or it may be done by employing the results of the triaxial tests for calibration of a constitutive model used with a numerical method in finite element or finite difference computer programs.

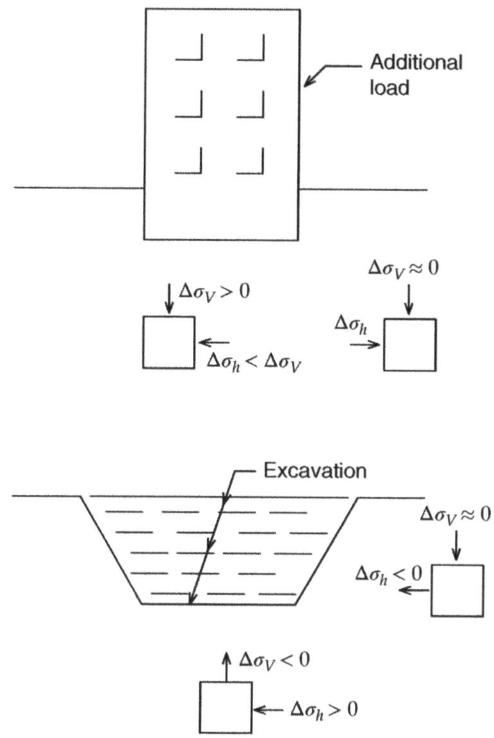

Figure 1.4 Examples of stress changes leading to failure in the field.

1.6 Types of tests

The drainage conditions in the field must be duplicated as well as possible in the laboratory tests. This may be done by appropriate drainage facilities or preventions as discussed above for the triaxial test. In most cases the field drainage conditions can be approximated by one of the following three types of tests:

1. Consolidated-drained test, called a CD-test, or just a drained test
2. Consolidated-undrained test, or a CU-test
3. Unconsolidated-undrained test, or a UU-test

These tests are described in ASTM Standards D7181 (2014), D4767 (2014), and D2850 (2014), respectively.

Which condition of drainage in the laboratory test logically corresponds to each case in the field depends on a comparison of loading rate with the rate at which the water can escape or be sucked into the ground. Thus, the permeability of the soil and the drainage boundary conditions in the field together with the loading rate play key roles in determination of the type of analysis and the type of test, drained or undrained, that are appropriate for each case. Field cases with partial drainage can be correctly duplicated in laboratory tests if the effective stress path is determined for the design condition. However, the idea of the CD-, CU- and UU-tests is to make it relatively simple for the design engineer to analyze a condition that will render a sufficient factor of safety under the actual drainage condition, without trying to estimate and experimentally replicate the actual stress path.

It has been determined through experience and common sense that the extreme conditions are drained and undrained with and without consolidation. As a practical matter, in a commercial laboratory it is easier to run an undrained test than a drained test because it is easier and faster to measure pore pressures than volume change. Therefore, even drained parameters are more likely to be estimated from a CU-test than from a CD-test.

1.6.1 Simulation of field conditions

Presented below is a brief review of the three types of tests together with examples of field cases for which the tests are appropriate and with typical strength results shown on Mohr diagrams.

Drained tests

Isotropic consolidation is most often used in the first stage of the triaxial test. However, anisotropic consolidation with any stress ratio is also possible.

The shearing stage of a drained test is performed so slowly, the soil is so permeable and the drainage facilities are such that no excess pore pressure (positive or negative) can exist in the specimen at any stage of the test, that is

$$\Delta u = 0 \tag{1.5}$$

It follows then from the effective stress principle

$$\sigma' = \sigma - u \tag{1.6}$$

that the effective stress changes are always the same as the total stress changes.

A soil specimen always changes volume during shearing in a drained test. If it contracts in volume, it expels pore fluid (usually water or air), and if it expands in volume (dilates), then it sucks water or air into the pores. If a non-zero pore pressure is generated during the test (e.g., by performing the shearing too fast so the water does not have sufficient time to escape), then the specimen will expel or suck water such that the pore pressure goes towards zero to try to achieve equilibrium between externally applied stresses and internal effective stresses. Thus, there will always be volume changes in a drained test. Consequently, the water content, the void ratio, and the dry density of the specimen at the end of the test are most often not the same as at the beginning.

The following field conditions can be simulated with acceptable accuracy in the drained test:

1. Almost all cases involving coarse sands and gravel, whether saturated or not (except if confined in e.g., a lens and/or exposed to rapid loading as in e.g., an earthquake).
2. Many cases involving fine sand and sometimes silt if the field loads are applied reasonable slowly.
3. Long term loading of any soil, as for example:
   a) Cut slopes several years after excavation
   b) Embankment constructed very slowly in layers over a soft clay deposit
   c) Earth dam with steady seepage
   d) Foundation on clay a long time after construction.

These cases are illustrated in Fig. 1.5.

The strength results obtained from drained tests are illustrated schematically on the Mohr diagram in Fig. 1.6. The shear strength of soils increases with increasing confining pressure.

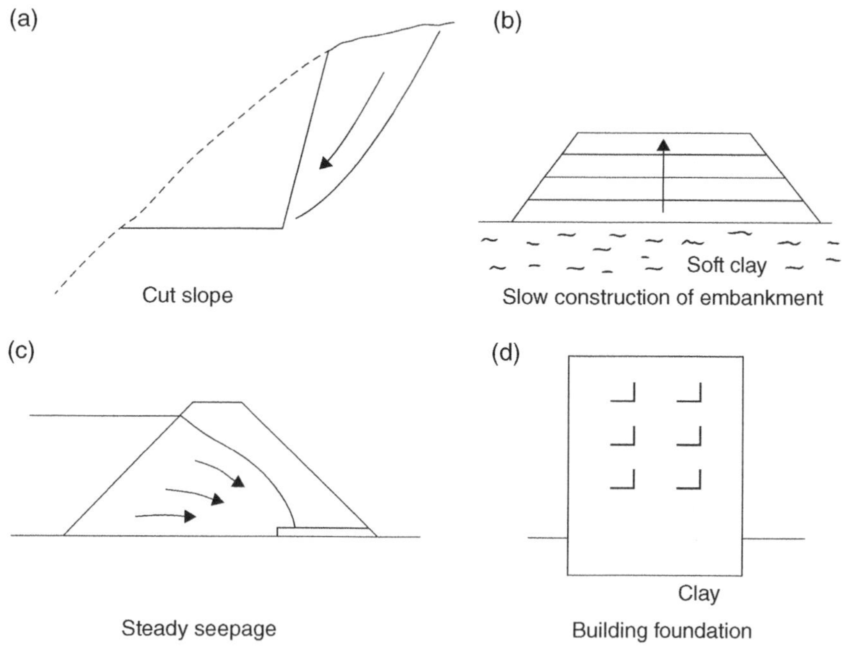

Figure 1.5 Examples of field cases for which long term stability may be determined on the basis of results from drained tests.

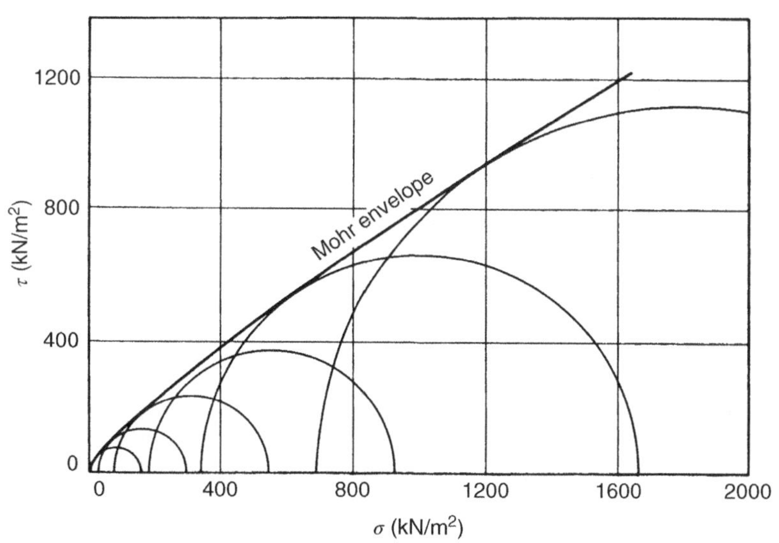

Figure 1.6 Schematic illustration of a Mohr diagram with failure envelope for drained tests on soil.

In the diagram in Fig. 1.6 the total stresses are equal to the effective stresses since there are no changes in pore pressures [Eqs (1.5) and (1.6)].

The effective friction angle, φ', decreases for all soils with increasing confining pressure, and the failure envelope is therefore curved, as indicated in Fig. 1.6. The effective cohesion, c', is zero or very small, even for overconsolidated clays. Effective or true cohesion of any significant magnitude is only present in cemented soils.

The effective stress failure envelope then defines the boundary between states of stress that can be reached in a soil element and states of stress that cannot be reached by the soil at its given dry density and water content.

Consolidated-undrained tests

As in drained tests, isotropic consolidation is most often used in CU-tests. However, anisotropic consolidation can also be applied, and it may have greater influence on the results from CU-tests than those from drained tests. The specimen is allowed to fully consolidate such that equilibrium has been obtained under the applied stresses and no excess pore pressure exists in the specimen.

The undrained shearing stage is begun by closing the drainage valve before shear loading is initiated. Thus, no drainage is permitted, and the tendency for volume change is reflected by a change in pore pressure, which may be measured by the transducer (see Fig. 1.2). Therefore the second stage of the CU-test on a saturated specimen is characterized by:

$$\Delta V = 0 \tag{1.7}$$

and

$$\Delta u \neq 0 \tag{1.8}$$

According to the effective stress principle in Eq. (1.6), the effective stresses are therefore different from the total stresses applied in a CU-test.

The pore pressure response is directly related to the tendency of the soil to change volume. This is illustrated in Fig. 1.7. Thus, there will always be pore pressure changes in an undrained test. However, since there are no volume changes of the fully saturated specimen, the water content, the void ratio and the dry density at the end of the test will be the same as at the end of the consolidation stage.

The following field conditions can be simulated with good accuracy in the CU-test:

1. Most cases involving short term strength, that is strength of relatively impervious soil deposits (clays and clayey soils) that are to be loaded over periods ranging from several days to several weeks (sometimes even years for very fat clays in massive deposits) following initial consolidation under existing stresses before loading. Examples of field cases in which short term stability considerations are appropriate:
   a) Building foundations
   b) Highway embankments, dams, highway foundations
   c) Earth dams during rapid drawdown (special considerations are required here, see Duncan and Wright 2005)These cases are illustrated in Fig. 1.8.
2. Prediction of strength variation with depth in a uniform soil deposit from which samples can only be retrieved near the ground surface. This is illustrated in Fig. 1.9.

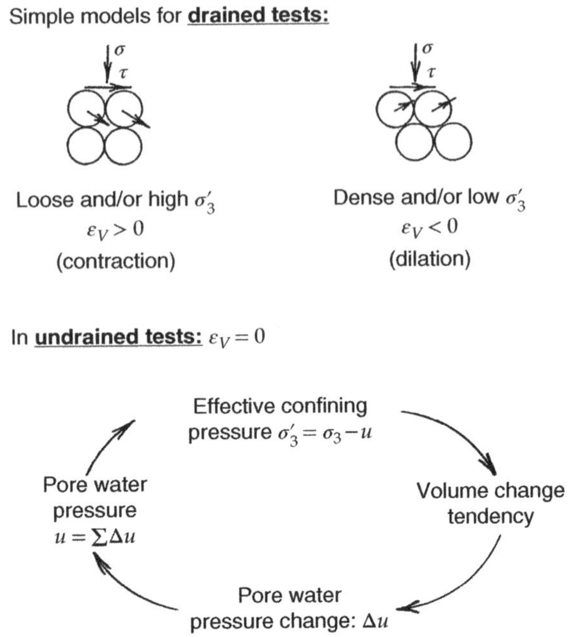

Figure 1.7 Schematic illustration of changes in pore water pressure in undrained tests.

The strength results obtained from CU-tests are illustrated schematically on the Mohr diagram in Fig. 1.10. Since pore pressures develop in CU-tests, two types of strengths can be derived from undrained tests: total strength; and effective strength. The Mohr circles corresponding to

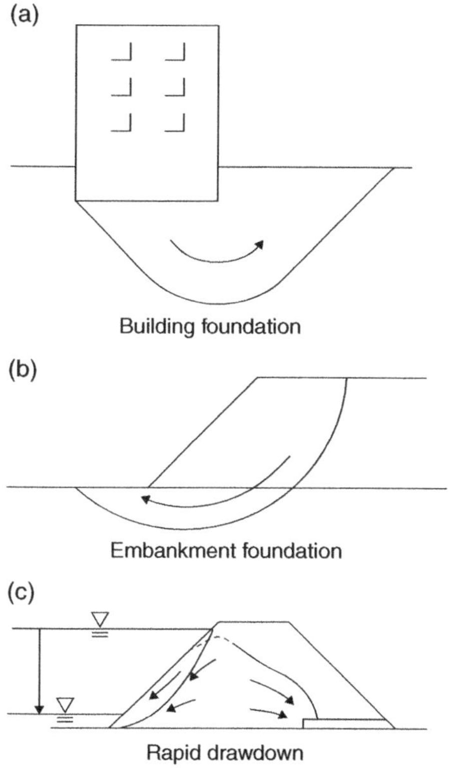

Figure 1.8 Examples of field cases for which short term stability may be determined on the basis of results of CU-tests.

these two strengths will always have the same diameter, but they are displaced by Δu from each other.

Both the total and effective stress envelopes from CU-tests on clays and clayey soils indicate increasing strength with increasing confining pressure. As for the drained tests, the effective friction angle, φ', decreases with increasing confining pressure, and the curvature of the failure envelope is sometimes more pronounced than for sands. In fact, the effective strength envelope obtained from CU-tests is very similar to that obtained from drained tests. Thus, the effective cohesion, c', is zero except for cemented soils. In particular, the effective cohesion is zero for remolded or compacted soils.

The total stress friction angle, φ, is much lower than the effective stress friction angle, φ', whereas the total stress cohesion, c, can have a substantial magnitude. The total stress friction angle is not a friction angle in the same sense as the effective stress friction angle. In the latter case, φ' is a measure of the strength derived from the applied normal stress, while φ is a measure of the strength gained from the *consolidation* stress only. If, for example, the total stress parameters are applied in a slope stability calculation in which a surcharge is suddenly added, then the surcharge will contribute to the shear resistance in the analysis (which is incorrect) as well as to the driving force, because there is no distinction between the normal forces derived from consolidation stresses and those caused by the surcharge. A better approach would be to assign undrained shear strengths (s_u) based on the consolidation stress state by using an approach that involves s_u/σ_v'.

Unconsolidated-undrained tests

In the UU-test a confining pressure is first applied to the specimen and *no drainage* is allowed. In fact, UU-tests are most often performed in triaxial equipment without facilities for drainage. The soil has already been consolidated in the field, and the specimen is therefore considered to "contain" the mechanical properties that are present at the location in the ground where the sample was taken. Alternatively, the soil may consist of compacted fill whose undrained strength is required for stability analysis before any consolidation has occurred in the field.

The undrained shearing stage follows immediately after application of the confining pressure. The shear load is usually increased relatively fast until failure occurs. No drainage is permitted during shear. Thus, the volume change is zero for a saturated specimen and the pore pressure is different from zero, as indicated in Eqs (1.7) and (1.8). The pore pressure is not measured and only the total strength is obtained from this test.

Since there are no volume changes in a saturated specimen, the void ratio, the water content and the dry density at the end of the test will be the same as those in the ground.

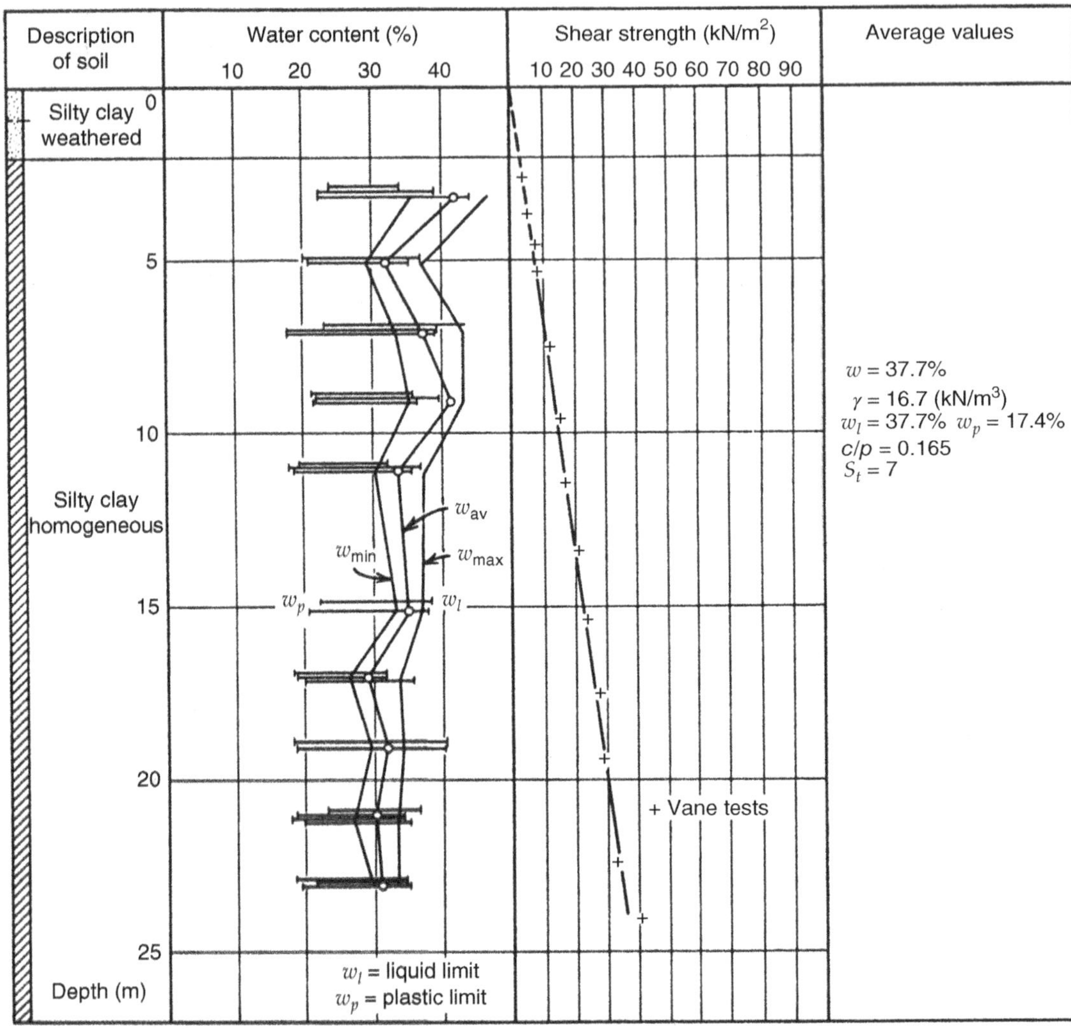

Figure 1.9 Strength variation with depth in uniform soil deposit of Norwegian marine clay. Reproduced from Bjerrum 1954 by permission of Geotechnique.

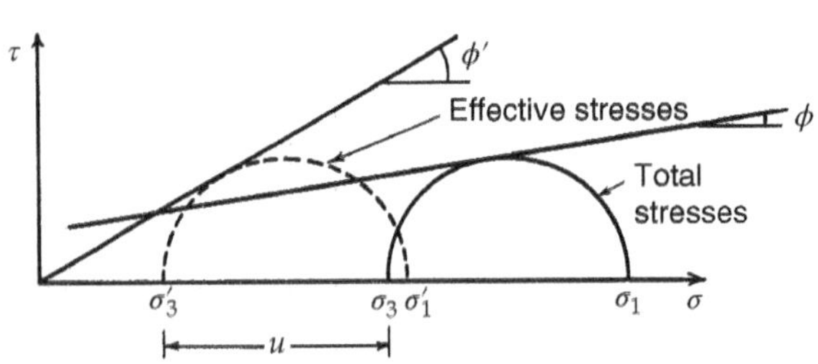

Figure 1.10 Schematic illustration of a Mohr diagram with total stress and effective stress failure envelopes from CU-tests on soil (after Bishop and Henkel 1962).

The following field conditions may be simulated in the UU-test:

1. Most cohesive soils of relatively poor drainage, where the field loads would be applied sufficiently rapidly that drainage does not occur. Examples of field cases for which results of UU-tests may be used:
   a) Compacted fill in an earth dam that is being constructed rapidly
   b) Strength of a foundation soil that will be loaded rapidly

c) Strength of soil in an excavation immediately after the cut is made

These cases are illustrated in Fig. 1.11.

2. Undisturbed, saturated soil, where a sample has been removed from depth, installed in a triaxial cell, and pressurized to simulate the overburden in the field.

The strength results obtained from UU-tests on *saturated* soil are illustrated schematically on the Mohr diagram in Fig. 1.12. The strength obtained from UU-tests on saturated soil is *not* affected by the magnitude of the confining pressure. This is because consolidation is not allowed after application of the confining pressure. Thus, the actual effective confining pressure in the saturated soil does not depend on the applied confining pressure, and the same strength is therefore obtained for all confining pressures. Consequently, the total strength envelope is horizontal corresponding to $\varphi = 0$, and the strength is therefore characterized by the undrained shear strength:

$$s_u = \frac{1}{2}(\sigma_1 - \sigma_3) \tag{1.9}$$

This is indicated in Fig. 1.12.

Since the UU-strength of a *saturated* soil is unaffected by the confining pressure, a UU-test may be performed in the unconfined state. This test is referred to as an unconfined compression test. In order that the unconfined compression test produces the same strength as would be obtained from a conventional UU-test, the soil must be:

1. Saturated
2. Intact
3. Homogeneous

Soils such as partly saturated clay (not saturated), stiff-fissured clays (not intact, fissures may open when unconfined), and varved clays (not homogeneous, cannot hold tension in pore water) do not fulfill these requirements

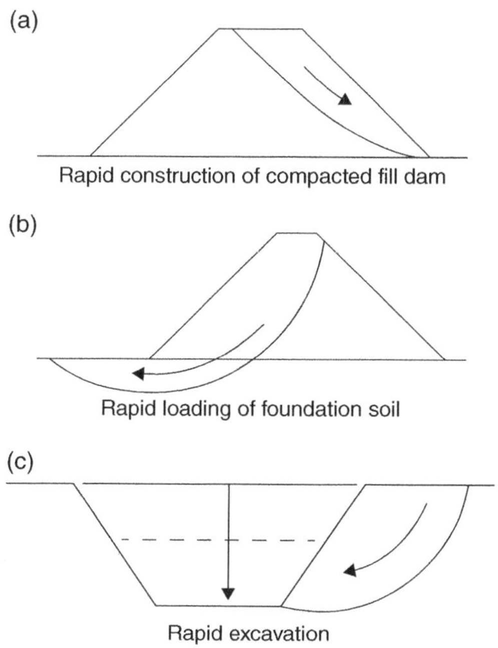

Figure 1.11 Examples of field cases for which short term stability may be determined on the basis of results of UU-tests.

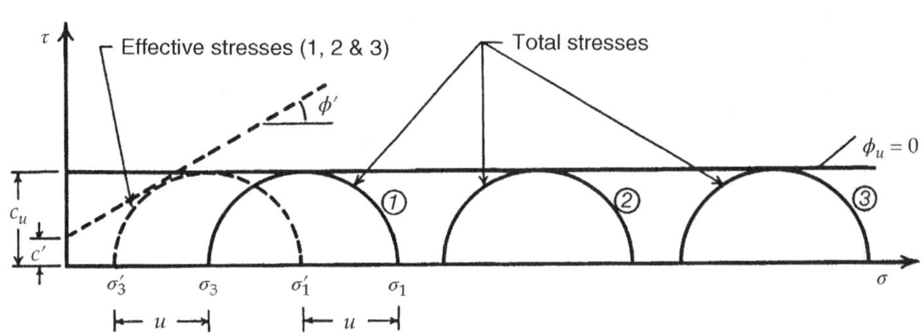

Figure 1.12 Schematic illustration of a Mohr diagram with results of UU-tests on saturated soil (after Bishop and Henkel 1962).

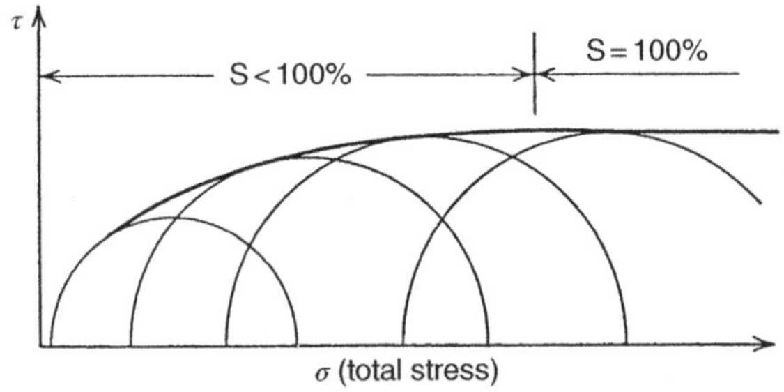

Figure 1.13 Schematic illustration of strength of partly saturated soil obtained from UU-tests.

and should not be tested in the unconfined compression test.

For those soils which qualify for and are tested in the unconfined compression test, the undrained *shear strength* is:

$$s_u = \frac{1}{2} \cdot q_u \quad (1.10)$$

in which q_u is the unconfined *compressive strength*:

$$q_u = (\sigma_1 - \sigma_3)_{max} = \sigma_{1max} \quad (1.11)$$

This is also indicated in Fig. 1.12.

For partly saturated soils the Mohr failure envelope is curved at low confining pressures, as seen in Fig. 1.13. As the air voids compress with increasing confinement, the envelope continues to become flatter. When all air is dissolved in the pore water, the specimen is completely saturated, and the envelope becomes horizontal. The undrained shear strength obtained at full saturation depends on the initial degree of saturation.

1.6.2 Selection of test type

The application of soil properties in analyses of actual geotechnical problems are outside the scope of the present treatment. However, it is important to know in which type of analysis the soil properties are to be used before any testing is initiated. Thus, different types of analyses (total stress or effective stress, short term or long term) may require results from different types of tests or results from different methods of interpretation of the results. In other words, the analysis that is appropriate for each particular field condition dictates the type of triaxial test to be performed.

Generally, soils that tend to contract will develop positive pore pressures during undrained shear resulting in lower shear strength than that obtained from the corresponding drained condition. Short term stability involving undrained conditions would be most critical for such soils. On the other hand, soils that tend to dilate will develop negative pore pressures during undrained shear resulting in higher shear strength than that obtained from the corresponding drained condition. Long term stability involving drained behavior would be most critical for these soils. Field conditions involving partial drainage should be analyzed for the most critical condition(s). For example, an earth dam usually undergoes several different stability analyses corresponding to different phases of construction and operating conditions. Some guidelines may be obtained from the examples given above.

2 Computations and Presentation of Test Results

2.1 Data reduction

Reduction of measured quantities in element tests, such as the triaxial compression test, involves computation of strains, cross-sectional areas, and stresses. Corrections to these quantities may be required to obtain the true behavior of the soil. Corrections to measurements are reviewed in Chapter 9.

2.1.1 Sign rule – 2D

The sign rule employed in soil mechanics has traditionally been opposite to that used in other branches of mechanics in which tensile stress and strains are considered to be positive. This is because most soils exhibit negligible tensile strengths and because deformation and failure most often are produced in response to compressive stresses. To avoid calculations in which the majority of quantities are negative, it is convenient to employ a sign rule in which compressive, normal stresses and strains are positive, as illustrated in Fig. 2.1(a) and (b). This requires a corresponding change in signs for shear stresses and shear strains. Figure 2.1(c) and (d) shows that shear stresses and strains are positive when acting in the counterclockwise direction under two-dimensional (2D) conditions.

As a consequence of this sign rule, the volumetric strains are positive for compression or contraction and negative for expansion or dilation. Thus, the loss of volume in a soil element results in a positive volumetric increment. This may not seem immediately logical, but it is necessary for consistency in the strain computations.

2.1.2 Strains

The strains in a soil element such as a triaxial specimen are calculated from the measured linear and volumetric deformations. Assuming these deformations to be uniformly distributed within the specimen, the strains may be calculated with reference to the original specimen dimensions resulting in "conventional" or "engineering" strains, or they may be calculated with reference to the current dimensions in which case they are referred to as "natural," "logarithmic," or "true" strains.

Engineering strains

The definition of engineering strains is most often employed in soil mechanics. The engineering strains may be converted to natural strains as shown below.

The linear engineering strains of a prismatic volume element with initial side lengths of L_1,

Triaxial Testing of Soils, First Edition. Poul V. Lade.

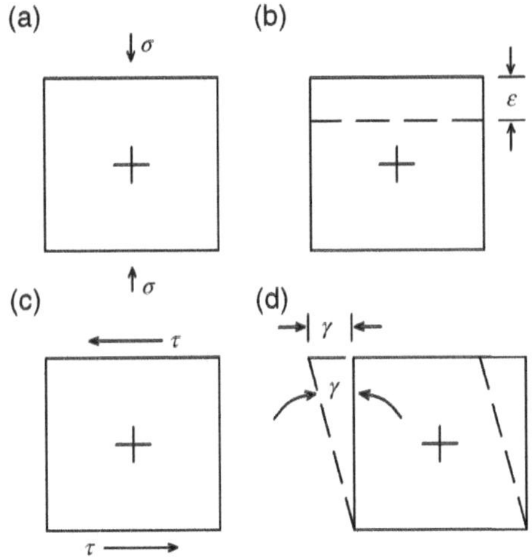

Figure 2.1 Sign rule employed in soil mechanics: compressive normal (a) stresses, σ, and (b) strains, ε, are positive. Shear (c) stresses, τ, and (d) strains, γ, are positive when directed counterclockwise (in two dimensions).

L_2, and L_3 and with incremental changes in these side lengths of ΔL_1, ΔL_2, and ΔL_3 are defined as:

$$\varepsilon_1 = \frac{\Delta L_1}{L_1} \tag{2.1}$$

$$\varepsilon_2 = \frac{\Delta L_2}{L_2} \tag{2.2}$$

$$\varepsilon_3 = \frac{\Delta L_3}{L_3} \tag{2.3}$$

and the volumetric strain of the element, whose initial volume is $V_0 = L_1 \cdot L_2 \cdot L_3$, is calculated from the volume change ΔV as follows:

$$\varepsilon_v = \frac{\Delta V}{V_0} \tag{2.4}$$

The relation between linear and volumetric strains may be derived by expressing the current volume in terms of the current linear dimensions:

$$V_0 - \Delta V = (L_1 - \Delta L_1)(L_2 - \Delta L_2)(L_3 - \Delta L_3) \tag{2.5}$$

Division by $V_0 = L_1 \cdot L_2 \cdot L_3$ and substitution of the expressions for the linear and volumetric strains produces the following relation for a unit volume:

$$1 - \varepsilon_v = (1 - \varepsilon_1)(1 - \varepsilon_2)(1 - \varepsilon_3) \tag{2.6}$$

Further reduction yields a general relation between the strains:

$$\varepsilon_v = \varepsilon_1 + \varepsilon_2 + \varepsilon_3 - \varepsilon_1 \cdot \varepsilon_2 - \varepsilon_2 \cdot \varepsilon_3 - \varepsilon_3 \cdot \varepsilon_1 + \varepsilon_1 \cdot \varepsilon_2 \cdot \varepsilon_3 \tag{2.7}$$

The physical meaning of the terms in Eq. (2.7) is illustrated in Fig. 2.2 for a prismatic element whose initial volume is unity ($V_0 = 1$) and which has undergone contraction in all three perpendicular directions. By adding and subtracting the effects of the linear strains (the three entire slabs), the products of two linear strains (the full lengths of the three bars), and the product of the three linear strains (the small prism), the relation between volumetric and linear strains given in Eq. (2.7) is obtained.

The expression in Eq. (2.7) accounts correctly for the relation between linear and volumetric strains whether these are positive or negative, and it may be used for small as well as large strains. For small strains the second and third order terms become small and may be neglected. Thus, for *small strains* the following expression may be employed:

$$\varepsilon_v = \varepsilon_1 + \varepsilon_2 + \varepsilon_3 \tag{2.8}$$

Using this expression for calculations involving large strains may produce errors whose magnitudes and significance will be considered below.

Natural strains

The definition of "natural" strain was introduced by Ludwik (1909) to obtain a measure of strain with reference to the current dimension of an element undergoing deformations. Thus, the increment in strain referred to the current length is defined as (considering the sign rule in soil mechanics):

$$d\bar{\varepsilon} = -\frac{dL}{L} \tag{2.9}$$

and the total natural strain, $\bar{\varepsilon}$, obtained from the initial length L_0 to the length L is:

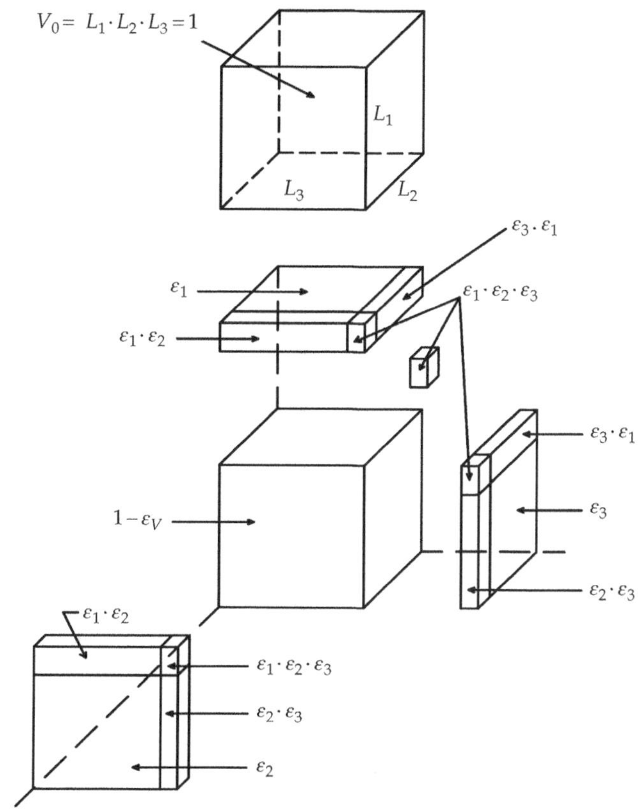

Figure 2.2 Spatial representation of strains in three dimensions.

$$\bar{\varepsilon} = -\int_{L_0}^{L} \frac{dL}{L} = -\ln\left(\frac{L}{L_0}\right) \tag{2.10}$$

This measure of strain represents an average strain obtained during deformation from L_0 to L. Its relation to engineering strain, ε, is readily determined since:

$$\frac{L}{L_0} = \frac{L_0 - \Delta L}{L_0} = 1 - \varepsilon \tag{2.11}$$

and therefore:

$$\bar{\varepsilon} = -\ln(1-\varepsilon) \tag{2.12}$$

Since the engineering strain, ε, is positive for contraction, the natural strain, $\bar{\varepsilon}$, is also positive for contraction, as indicated by Eq. (2.12). For small strains the engineering and the natural strains are practically identical. The natural strains have the advantage of being additive, whereas the engineering strains are not. Taking the natural logarithm on both sides of Eq. (2.6) results in the following simple expression for the natural volumetric strain:

$$\bar{\varepsilon}_v = \bar{\varepsilon}_1 + \bar{\varepsilon}_2 + \bar{\varepsilon}_3 \tag{2.13}$$

This expression is correct for small as well as for large strains. The comparable expression in Eq. (2.8) for engineering strains is correct only for small strains.

Although there are advantages associated with the natural strain definition, engineering strains are most often employed in practice and these will be used in the following.

Strains in a triaxial specimen

The engineering strains in a triaxial specimen are assumed to be uniform and may be calculated assuming the cylindrical specimen deforms as a right cylinder. For isotropic or cross-anisotropic materials with the axis of rotational symmetry in the vertical direction, the two radial, normal strains are equal. For these conditions the linear and volumetric strains are calculated as follows:

$$\text{Axial strain:}\quad \varepsilon_a = \frac{\Delta H}{H_0} \quad (= \varepsilon_1 \text{ for triaxial compression}) \tag{2.14}$$

$$\text{Radial strain:}\quad \varepsilon_r = \frac{\Delta D}{D_0} \quad (= \varepsilon_2 = \varepsilon_3 \text{ for triaxial compression}) \tag{2.15}$$

$$\text{Volumetric strain:}\quad \varepsilon_v = \frac{\Delta V}{V_0} \tag{2.16}$$

in which ΔH, ΔD, and ΔV are the increments and H_0, D_0, and V_0 are the initial height, diameter, and volume, respectively. For this axisymmetric condition, the two perpendicular, radial strains are equal, $\varepsilon_r = \varepsilon_2 = \varepsilon_3$. In a triaxial *compression* test, in which $\sigma_1 > \sigma_2 = \sigma_3$, the axial strain is the major principal strain (positive) and the radial strains are the minor principal strains (negative), as indicated in Eqs (2.14) and (2.15). In a triaxial *extension* test, in which $\sigma_1 < \sigma_2 = \sigma_3$, the axial strain is the minor principal strain (negative) and the radial strains are the major principal strains (positive).

The axial and volumetric strains are most often the basis for calculation of the radial strain as well as the cross-sectional area of the specimen. Setting $\varepsilon_2 = \varepsilon_3 = \varepsilon_r$ in the expression for volumetric strains in Eq. (2.7) produces an expression for ε_r which is valid for small as well as for *large strains*:

$$\varepsilon_r = 1 - \sqrt{\frac{1-\varepsilon_v}{1-\varepsilon_a}} \quad (= \varepsilon_3 \text{ for triaxial compression}) \tag{2.17}$$

The volumetric strain expression in Eq. (2.8) yields a simpler equation for the radial strain which is only valid for *small strains*:

$$\varepsilon_r = \frac{1}{2}(\varepsilon_v - \varepsilon_a) \quad (= \varepsilon_3 \text{ for triaxial compression}) \tag{2.18}$$

These expressions are valid for both compression and extension tests.

Evaluation of small strain calculations

It is convenient to use the small strain expressions in Eqs (2.8) and (2.18) for data reduction, and these expressions are most often employed in practice. The accuracy these expressions provide may be evaluated for various types of axisymmetric test conditions encountered in triaxial testing. To illustrate the difference between the two expressions for the radial strains, the following conditions, often experienced in soil testing, are considered: (1) isotropic compression and expansion of an isotropic material in which the three linear strains are equal; and (2) undrained compression and extension of triaxial specimens in which the volumetric strains are zero.

The diagram in Fig. 2.3 shows the difference between calculated radial strains from Eqs (2.17) and (2.18). The correct volumetric strains

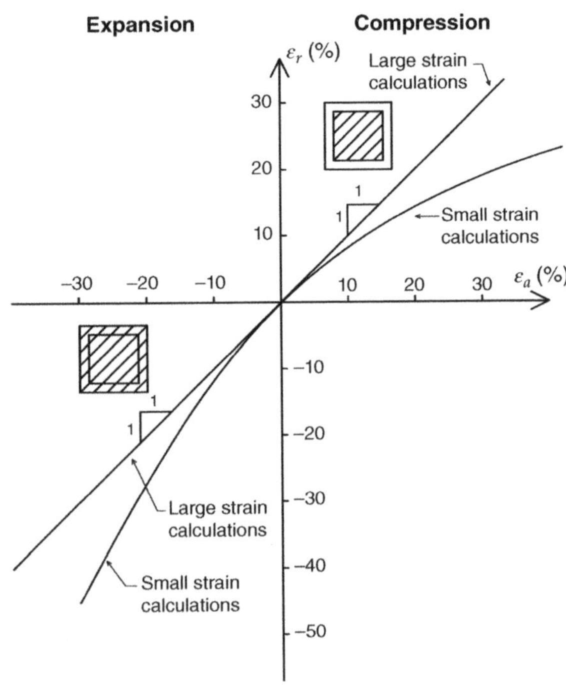

Figure 2.3 Comparison of radial strains calculated from axial and volumetric strains for isotropic compression and expansion of isotropic material.

are obtained from Eq. (2.7) and used in the expressions. The large strain calculations produce the correct radial strains for the isotropic material. The small strain calculations produce radial strains that are too small, whether contraction or expansion. The error is about 1.5% at ±10% axial strain, and it increases to 12% for contraction and 15% for expansion at axial strains of ±30%. In most cases of isotropic contraction and expansion of soil specimens, the linear strains are limited to much smaller values, and the small strain calculations may be sufficiently accurate for practical purposes.

Figure 2.4 shows the radial strains calculated for undrained compression and extension tests on specimens with zero volumetric strains. The large strain calculations produce the correct radial strains. The small strain calculations produce radial strains which, for the compression test indicate too little expansion, and for the extension test show too much contraction. The error is about 0.4% at 10% contraction, and it increases to 4.5% at 30% contraction. For extension, the error is about 0.35% at −10% axial strain, and it increases to 2.7% at −30% axial strain. The axial strain-to-failure is often much smaller in extension than in compression, and the small strain calculations may be sufficiently accurate for extension tests. The axial strain-to-failure is largest in triaxial compression tests (as compared with any other test condition in which the principal stresses are fixed in direction, see also Chapter 11), and it may therefore involve too large inaccuracies to use the small strain calculations for such tests.

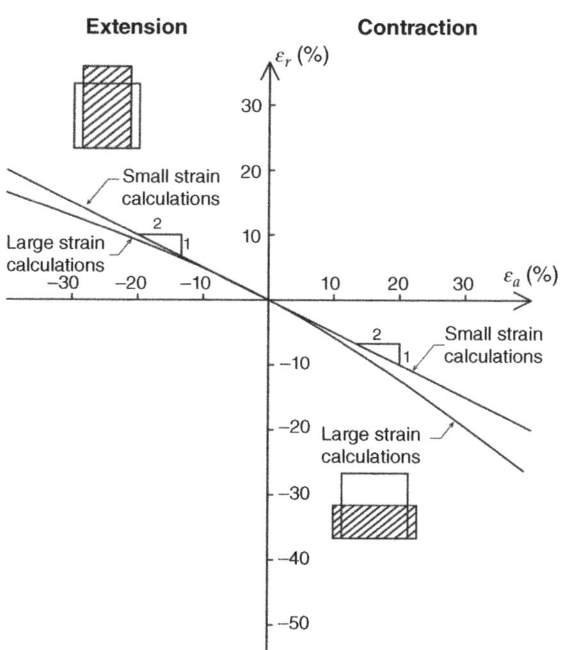

Figure 2.4 Comparison of radial strains calculated from axial and volumetric strains for undrained compression and extension specimens with zero volume change.

Note that if shear banding occurs the strain calculations are no longer valid, because all of the deformation occurs in the shear band.

An overall evaluation of the errors in radial strains produced by small strain calculations for various axisymmetric test conditions is illustrated in Fig. 2.5. In this diagram the initial shape and volume is indicated by a square for each test condition. The deformed shapes are shown by shaded squares or rectangles. Small strain calculations lead to correct radial strains for uniaxial strain or K_0-conditions only. The relative magnitude of errors in all other cases may be evaluated by comparing individual test conditions with those in Fig. 2.5. An indication of the absolute magnitude of errors may be obtained by reference to Figs 2.3 and 2.4.

Because computers or programmable calculators are often employed, the large strain expressions for the volumetric and radial strains in Eqs (2.7) and (2.17) may as well be used for data reduction with resulting greater accuracy in the calculated strains.

Soils with anisotropic behavior

The triaxial test may be used to determine anisotropic soil behavior only for cases in which one of the three axes of material symmetry is aligned with the vertical axis of the triaxial apparatus. For a cross-anisotropic material this includes two possible orientations, and for a material with general anisotropy, three different orientations are possible. These orientations are indicated in Fig. 2.6.

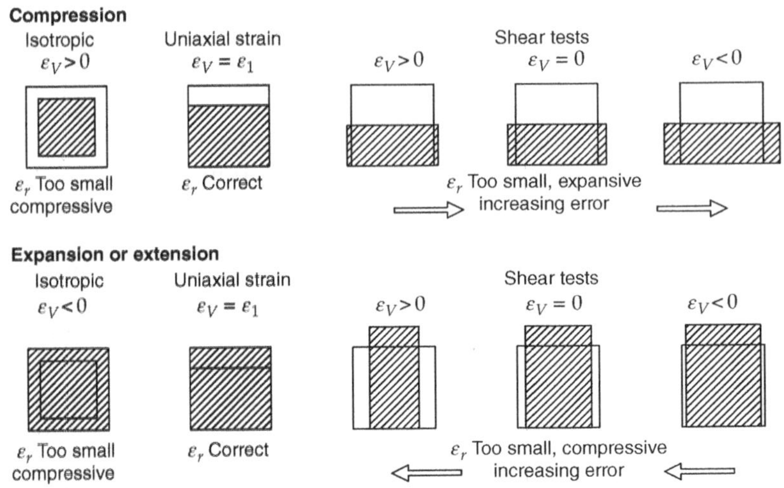

Figure 2.5 Evaluation of errors in radial strains calculated from axial and volumetric strains using small strain calculations.

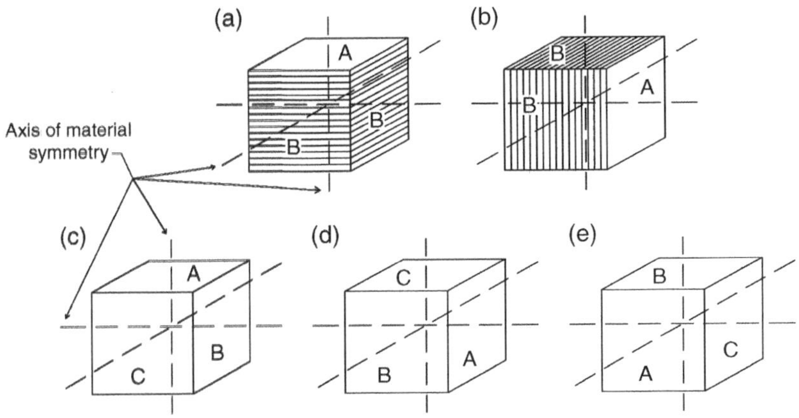

Figure 2.6 Possible orientations in triaxial apparatus of specimens with (a) and (b) cross-anisotropic material and (c), (d) and (e) general anisotropic material.

Except for the specimen in Fig. 2.6(a), the lateral strains in specimens with anisotropic behavior are expected to be different. To determine these lateral strains, it is necessary to measure the deformation in at least one lateral direction. This produces one lateral strain (say ε_2), and the other lateral strain (say ε_3) may be calculated from the expression in Eq. (2.7) as follows:

$$\varepsilon_3 = \frac{\varepsilon_v - \varepsilon_1 - \varepsilon_2 + \varepsilon_1 \cdot \varepsilon_2}{1 - \varepsilon_1 - \varepsilon_2 + \varepsilon_1 \cdot \varepsilon_2} \quad (2.19)$$

in which ε_1 and ε_v are the measured vertical and volumetric strains, respectively. For small strains Eq. (2.8) produces a simple expression for the unknown lateral strain:

$$\varepsilon_3 = \varepsilon_v - \varepsilon_1 - \varepsilon_2 \quad (2.20)$$

Alternatively, both lateral deformations may be measured, and Eq. (2.7) provides a check on the accuracy of the measurements.

Caution

Strains as well as stresses in triaxial tests on specimens prepared with axes of material symmetry inclined relative to the vertical axis of the apparatus are nonuniform and difficult to interpret correctly. Figure 2.7(a) and (b) shows

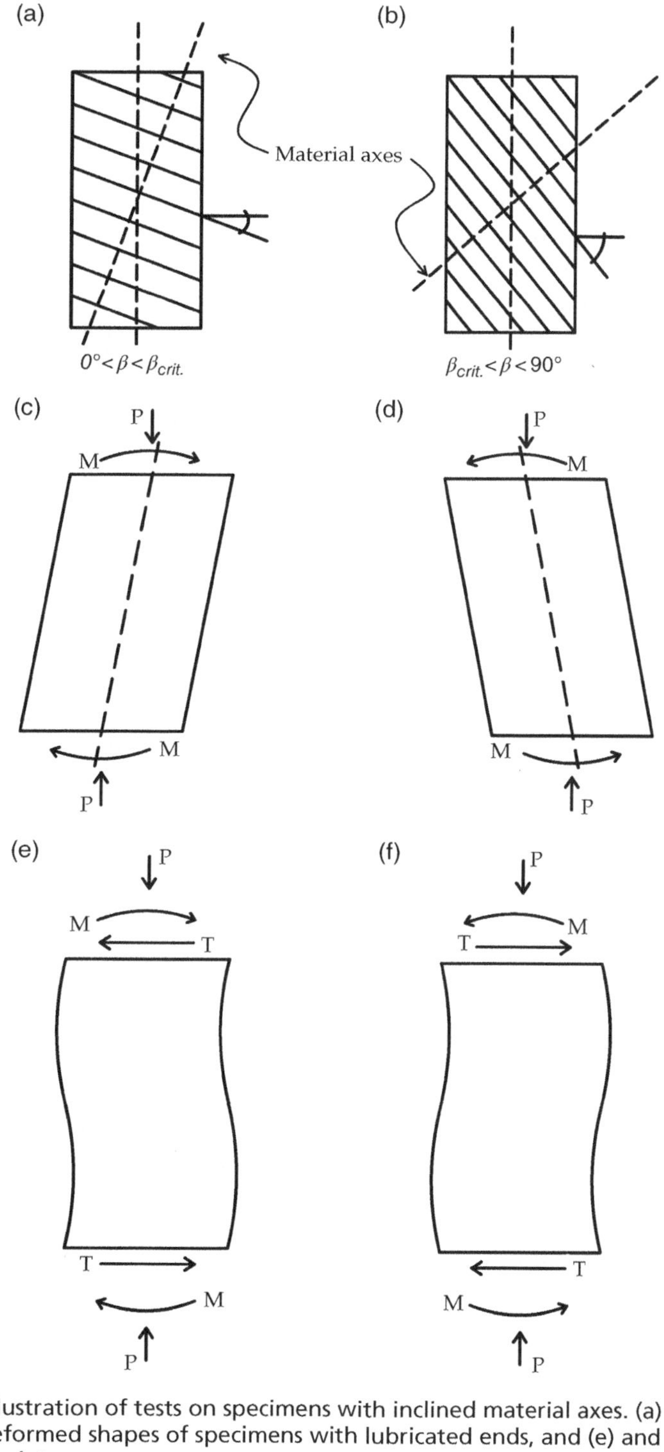

Figure 2.7 Schematic illustration of tests on specimens with inclined material axes. (a) and (b) initial vertical specimens, (c) and (d) deformed shapes of specimens with lubricated ends, and (e) and (f) deformed shapes of specimens with end restraint.

prismatic specimens with inclined bedding planes. For the ideal case in which only normal stresses are applied by horizontal end plates (requiring smooth, lubricated ends), moments are generated at the ends in response to the shear strains developing along the bedding planes. The initially vertical specimen acquires the shape of a parallelogram, and the specimen axis becomes inclined, as shown in Fig. 2.7(c) and (d). The vertical, normal stress distributions at the ends become nonuniform, and the state of stress and the pore water pressures or the volume changes become nonuniform inside the specimen.

If, on the other hand, the deformations of the specimen are restrained at the ends (requiring end plates with full friction), bending moments and shear forces will develop at the ends, causing the specimen to deform nonuniformly, and the states of stress and strain inside the specimen are nonuniform. Typical shapes of the deformed specimens are shown in Fig. 2.7(e) and (f).

Whether the end plates have full friction or are provided with lubrication, triaxial tests on specimens with inclined material axes are at best difficult to interpret, and the results of such tests are questionable. More detailed studies and discussions of these types of tests have been presented by Saada (1970) and Saada and Bianchini (1977). To study the behavior of anisotropic materials it is preferable to incline the principal stress directions rather than incline the specimen. This may be done in equipment in which shear stresses can be applied to the surface of the specimen (see Chapter 11).

Effects of bulging

Triaxial compression specimens with end restraint often exhibit nonuniform deformations during shear. Rather than deforming as a right cylinder, the specimen may bulge at the middle and attain the shape of a barrel. This mode of deformation is particularly pronounced for soils that contract during shear. Vertical, lateral, and volumetric strain distributions as well as the stress distribution inside the specimen become nonuniform, and interpretation of test results are consequently complicated. Although the external shape of the specimen may not indicate the true internal strain distribution, due to conically shaped dead zones near the end plates (for further discussion see Chapter 3), the effects of bulging on the average lateral and vertical strains under various conditions of end restraint may be studied.

Lateral strain distribution

The profile of the barrel-shaped specimen may be described with good approximation as a parabola, as indicated in Fig. 2.8. The following simple analysis indicates the magnitude of nonuniformity in the lateral strain distribution.

For full fixity at the ends, the two shaded areas in Fig. 2.8 are equal. The calculations shown on this figure indicate that the maximum lateral strain at the middle of the bulged specimen is 50% larger than the average lateral strain obtained from the deformed right cylinder. If some lateral deformation occurs at the ends, the nonuniformity decreases, and complete uniformity in lateral strain is obtained when the specimen deforms as a right cylinder.

Average lateral strains

The points at which the parabolic barrel crosses the deformed right cylinder may be obtained by analysis of the parabolic curve in the X–Y coordinate system shown in Fig. 2.8. The results shown in this figure indicate that the crossover points occur at approximately one-fifth of the specimen height from the end plates. The locations of these points are independent of the amount of restraint at the ends. Thus, if some lateral deformation occurs at the ends, the crossover points remain at the same location.

Based on this simple analysis, the average lateral deformation in a specimen that bulges may be measured at the points located one-fifth of the specimen height from the end plates. If measurements of the lateral deformations are performed directly on the cylindrical specimen (see also Chapter 4), and the average lateral deformations are sought, the measurement devices should be attached to the specimen at

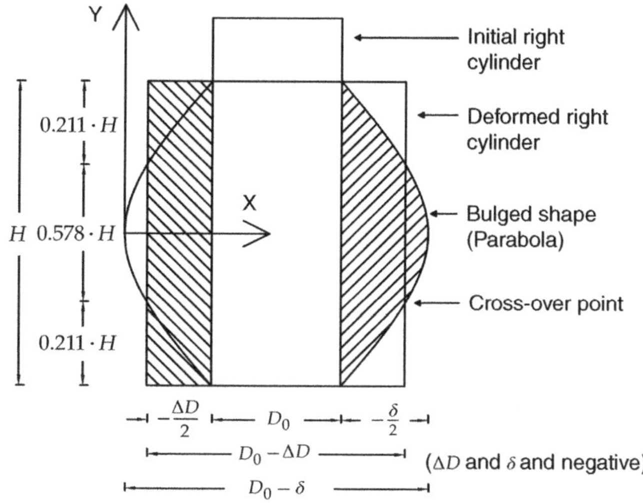

Area of shaded rectangle = area of shaded parabola section

$$H \cdot \left(-\frac{\Delta D}{2}\right) = \frac{2}{3} \cdot H \cdot \left(-\frac{\delta}{2}\right)$$

$$\delta = \frac{3}{2} \cdot \Delta D$$

Expression for parabola: $Y^2 = R \cdot X$

At ends: $Y = \pm\frac{H}{2} \;\; at \;\; X = -\frac{3}{4} \cdot \Delta D$

$$R = \frac{Y^2}{X} = \frac{\left(\frac{H}{2}\right)^2}{\left(-\frac{3}{4} \cdot \Delta D\right)} = \frac{H^2}{-3\Delta D}$$

Cross-over point @

$$X = -\frac{\delta}{2} - \frac{\Delta D}{2} = -\frac{1}{4} \cdot \Delta D$$

$$Y^2 = \frac{H^2}{(-3\Delta D)} \cdot \left(-\frac{1}{4} \cdot \Delta D\right) \;\; \frac{H^2}{12}$$

$$Y = \pm\frac{\sqrt{3}}{6} \cdot H = \pm 0.289 \cdot H$$

Figure 2.8 Analysis of deformations in a barrel-shaped specimen.

these points. Even if lubricated ends are employed and the specimen is believed to deform uniformly, it may be good practice to use these points for measurements.

In extension tests with end restraint, the deformed shape of the specimen resembles a paraboloid whose profile may also be approximated by a parabolic curve. The analysis of the deformed shape then proceeds as indicated above and similar results are obtained. Thus, full friction at the ends results in contractive lateral strains at the middle of the specimen which are 50% larger than the average lateral strains. The points located one-fifth of the specimen height from the ends may be used to obtain the average lateral deformations of triaxial specimens. Note however that for comparable compression and extension tests the strain to peak failure is usually much lower in extension. The parabolic shape of the profile is therefore not likely to be nearly as pronounced in extension as in compression. Further, necking and shear planes tend to develop at an early stage in conventional extension tests, thus invalidating the assumption of the parabolic shape. Conventional extension tests have been shown to be highly unstable and almost always result in erroneous stress–strain and strength results (Yamamuro and Lade 1995; Lade *et al.* 1996). See also below, and Chapter 10.

Vertical strain distribution

The vertical strain distribution is not as clearly visible as the lateral strain distribution. However, measurements along the axis of compression specimens indicate that the vertical strain distribution may also be parabolic with the largest strains near the middle. Even for full fixity, the vertical strains at the ends are not zero, since vertical strains occur in uniaxial strain or K_0-tests. Approximate analyses of vertical strain distributions may, however, be performed in a similar manner as indicated above. The average vertical strain is obtained from measurement of the vertical deformation over the total height of the specimen.

Volumetric strain distribution

The distribution of volume changes follows the pattern indicated above for the linear strains. Due to end restraint the specimen is likely to contract least or dilate most near the middle. The overall volume change measured (e.g., by the amount of water expelled from or sucked into the specimen) represents the average volumetric strain.

Detailed measurements and analyses

The brief review of effects of bulging presented above used relatively simple analysis procedures based on parabolic shapes. More detailed measurements of strain distributions in triaxial specimens may be performed. Detailed analyses based on finite element calculations may also be evaluated.

Development of shear planes

Granular soils that tend to dilate as well as clays in which the platy particles tend to align during shear, with both effects resulting in lower strengths, may develop shear planes or bands when tested in triaxial compression or extension. Once one or more shear planes have initiated in specimens of sufficient height to allow their free development, the deformations become localized to the shear plane, and two essentially solid portions of the specimen move past each other along the shear plane. Figure 2.9 shows a triaxial compression specimen with a shear plane. Very large shear strains occur inside the shear plane, which, due to dilation in granular materials and particle alignment in clays, becomes a weak plane in the specimen. The strains in the specimen become highly nonuniform, and the true relation between stresses and strains cannot be determined from external measurements. The two large volumes of the

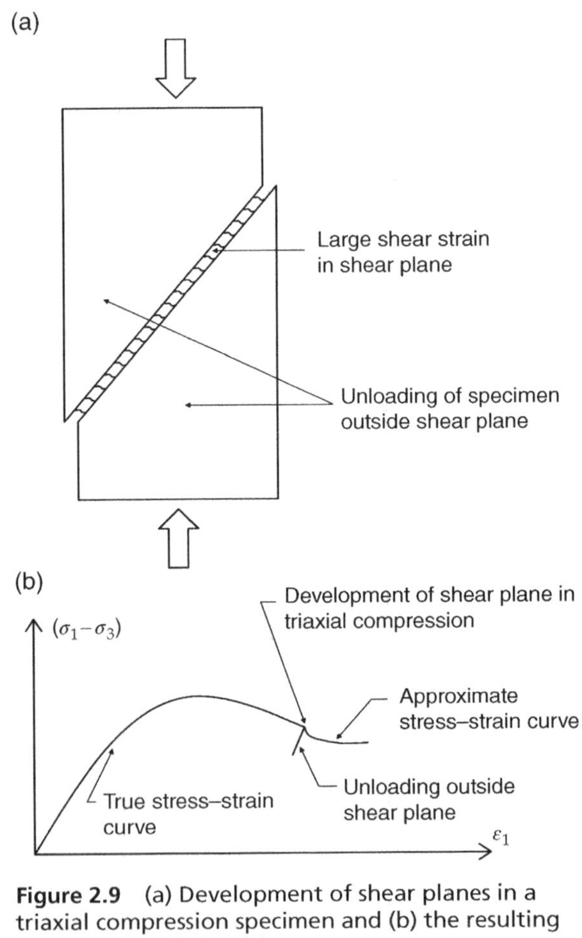

Figure 2.9 (a) Development of shear planes in a triaxial compression specimen and (b) the resulting stress–strain relationship.

specimen outside the shear plane undergo unloading although the specimen is still being compressed and sheared. The development of shear planes occurs after peak failure in triaxial compression and slightly after peak failure in triaxial extension tests on uniform specimens (Lade and Wang 2001; Wang and Lade 2001; Lade 2003), while it occurs before smooth peak failure in plane strain tests. Shear plane development and their effects on the soil behavior are discussed in further detail in Chapters 8 and 11.

It is clear that an attempt to determine the true stress–strain relation after gross development and progression of shear planes is fruitless. Highly nonuniform distributions of stresses and strains prevail inside the specimen.

2.1.3 Cross-sectional area

Once the strains in the triaxial specimen have been determined, the horizontal cross-sectional area may be determined by dividing the current volume by the current height of the specimen:

$$A = \frac{V_0 - \Delta V}{H_0 - \Delta H} = \frac{V_0}{H_0} \cdot \frac{1 - \frac{\Delta V}{V_0}}{1 - \frac{\Delta H}{H_0}} \tag{2.21}$$

Substituting the volumetric and axial strains from Eqs (2.14) and (2.16) and setting $V_0/H_0 = A_0$ (= initial cross-sectional area) produces:

$$A = A_0 \cdot \frac{1 - \varepsilon_v}{1 - \varepsilon_a} \tag{2.22}$$

This expression for the horizontal cross-sectional area is valid for any cross-sectional shape (e.g., circular, square), for small and large strains, for compression and extension tests, for drained and undrained tests ($\varepsilon_v = 0$ for saturated specimens), for consolidation and shearing, and for isotropic and anisotropic soils.

The cross-sectional area may also be obtained from measured lateral strains. Equation (2.6) gives:

$$A = A_0 \cdot \frac{1 - \varepsilon_v}{1 - \varepsilon_a} = A_0 \cdot (1 - \varepsilon_2)(1 - \varepsilon_3) \tag{2.23}$$

in which ε_2 and ε_3 represent the two perpendicular lateral strains. This expression may be used for anisotropic specimens with unequal lateral strains. For isotropic materials or cross-anisotropic materials with a vertical axis of material symmetry, the expression reduces to:

$$A = A_0 \cdot (1 - \varepsilon_r)^2 \tag{2.24}$$

in which ε_r is the isotropic radial or lateral strains.

Caution

If for some reason the radial strains are first calculated from the small strain expression in Eq. (2.18) before substitution in Eq. (2.24), then the cross-sectional area becomes incorrect at large strains. Note that substitution of ε_r for large strains from Eq. (2.17) in Eq. (2.24) results in the expression in Eq. (2.22).

Effects of bulging

The expressions given above are valid for specimens that deform as right cylinders. It may be of interest to calculate the cross-sectional area in the middle of a specimen that bulges during shear. For complete fixity at the ends and parabolic lateral strain distribution, the area in the middle of the specimen may be expressed as the sum of the area of the deformed right cylinder and the area of the ring surrounding the right cylinder (see Fig. 2.8):

$$A_{barrel} = \frac{\pi}{4} \cdot D_0^2 \cdot \frac{1 - \varepsilon_v}{1 - \varepsilon_a} + \pi \cdot (D_0 - \Delta D)\left(-\frac{1}{4}\Delta D\right) \tag{2.25}$$

Reduction and substitution of the large strain expression for the lateral strain in Eq. (2.17) produces:

$$A_{barrel} = A_{cylinder} \cdot \left(2 - \sqrt{\frac{1 - \varepsilon_a}{1 - \varepsilon_v}}\right) \tag{2.26}$$

For an undrained test on saturated soil ($\varepsilon_v = 0$), the cross-sectional area in the middle of a bulging specimen with full end restraint is approximately 5% larger than the area of the right cylinder at 10% axial compression, and it is about 11% larger at 20% axial compression. These values become smaller for a specimen that exhibits volumetric contraction, and they become larger for a specimen that dilates. The values decrease if lateral expansion occurs at the ends, and they are zero for a specimen that deforms as a right cylinder.

In extension tests with full end restraint, the specimen undergoes "necking," and the area in the middle of the deformed specimen (shaped as a paraboloid) may also be calculated from Eq. (2.26). At 10% axial extension in an undrained test, the cross-sectional area in the middle is approximately 5% smaller than the area of the right cylinder, and it is about 11% smaller at 20% axial extension. These values become larger for a specimen that exhibits volumetric contraction, and they become smaller for a specimen that dilates. As in triaxial compression tests, the values decrease if lateral contraction occurs at the ends, and they are zero for a specimen that deforms as a right cylinder.

However, another more important effect in the form of shear banding may occur in conventional extension tests. This is described in Chapter 10.

In the following it is assumed that the triaxial specimen deforms uniformly as a right cylinder, and that all strains represents average strains obtained from the deformed right cylinder. Techniques employed to insure that this mode of deformation is in fact obtained are discussed in Chapter 3.

2.1.4 Stresses

Types of stress measures

Stress is defined in continuum mechanics as the limiting value of force per area as the area reduces to zero:

$$\sigma = \lim_{A \to 0} \frac{P}{A} \tag{2.27}$$

Although forces in soils are transmitted between discrete particles, most problems in soil mechanics involve boundary lengths and loaded areas which are large compared with dimensions of individual particles. The soil can therefore be considered as a continuum, and the principles of continuum mechanics are consequently adopted for solution of problems in soil mechanics.

In analysis of the triaxial test, the specimen is assumed to represent an element of the continuum in the field, and the properties of the specimen are therefore assumed to be representative of the behavior to be encountered in the field. To insure that these assumptions are justified, the relation between maximum grain size and minimum specimen dimensions should fulfill certain requirements as discussed in Chapter 3.

Confining pressure

The stress that confines the specimen in the triaxial apparatus is usually applied as a pressure in the air or fluid present in the cell. This pressure is measured directly by a manometer or pressure transducer and requires no further computation or correction.

Deviator stress

The deviator stress, σ_d, is calculated from the applied deviator load, P, and the current cross-sectional area, A, of the specimen according to Eq. (2.27):

$$\sigma_d = \frac{P}{A} \quad \left[= (\sigma_1 - \sigma_3) \text{for triaxial compression} \right] \tag{2.28}$$

in which the cross-sectional area is obtained as discussed in the previous section. The deviator stress is assumed to act uniformly across the area of the specimen. The vertical, axial stress in the specimen is then calculated from:

$$\sigma_a = \sigma_d + \sigma_{cell} \tag{2.29}$$

In triaxial compression the deviator stress is positive and the major principal stress, σ_1, acts in the axial direction. The other two principal stresses, σ_2 and σ_3 ($= \sigma_{cell}$) are equal in magnitude, and they both act in the horizontal direction. Failure occurs in the axial direction in triaxial compression tests.

In triaxial extension the deviator stress is negative and the axial stress in the specimen becomes the minor principal stress, σ_3. The two horizontal stresses, σ_1 and σ_2 ($= \sigma_{cell}$), are equal in magnitude. Failure occurs in the lateral direction in triaxial extension tests. Note that the presence of a negative deviator stress in extension tests does not imply that the axial stress acting on a horizontal plane in the specimen need be negative. The negative deviator stress simply reduces the axial stress imposed by the confining pressure in the vertical direction. In a material without effective cohesion the specimen always fails under positive, compressive stresses. However, it is possible to achieve negative axial stresses in an extension test on a material with effective cohesion. The maximum value of the negative axial stress is related to the magnitudes of the effective cohesion and the effective confining pressure employed in the extension test. Observation of negative axial stresses requires the end platens to be glued to the specimen.

Pore pressure

Porous materials such as soils in which the pores are interconnected allow flow of fluid and/or gas through their structure. The pore fluid can sustain only hydrostatic pressures under static or pseudo-static loading conditions. All common soil mechanics problems, including dynamic loading during earthquakes and blasting, fall within this category. The pore pressures encountered in triaxial tests are measured directly by a manometer or by a pressure transducer and require no further computation or correction.

2.1.5 Corrections

The stresses and strains determined from triaxial tests, as reviewed above, often require corrections due to the shortcomings of the experimental techniques employed in the tests. These corrections are discussed in Chapter 9. Effects of bulging of the specimen have been reviewed previously in the present chapter.

2.1.6 The effective stress principle

Frictional materials deform and fail in response to changes in shear stresses and normal stresses. Since fluids cannot sustain shear stresses, the applied shear stresses are always effective and require no modification due to pore pressures. However, fluids or gases can sustain normal, hydrostatic stresses, and only that portion of applied normal stresses that is transmitted through the grain structure is effective in causing deformation and failure. The effective normal stresses are equal to the applied total normal stresses reduced by the effect of the hydrostatic pressures generated in the fluid and/or gas in the pores. This effective stress principle can be expressed as:

$$\sigma' = \sigma - \eta \cdot u \tag{2.30}$$

in which σ' is the effective normal stress, σ is the total normal stress, u is the pore pressure, and $\eta \leq 1$, is a factor that determines the relative influence of the hydrostatic pore pressure. The value of η depends on the void ratio and the compressibilities of the soil particles, the soil skeleton, and the pore fluid. For soils at conventional to moderately high stresses the value of η can be taken as unity without significant error (Lade and de Boer 1997).

2.1.7 Stress analysis in two dimensions – Mohr's circle

Figure 2.10(a) shows a general state of stress acting on a soil element in the ground. The stresses on any plane oriented at an angle α with the horizontal may be expressed in terms of the stresses acting on the element:

$$\sigma_\alpha = \frac{\sigma_x + \sigma_y}{2} - \frac{\sigma_x - \sigma_y}{2}\cos 2\alpha - \tau_{xy}\sin 2\alpha \tag{2.31}$$

$$\tau_\alpha = \frac{\sigma_x - \sigma_y}{2}\sin 2\alpha - \tau_{xy}\cos 2\alpha \tag{2.32}$$

These expressions describe a circle, Mohr's circle, in a σ–τ diagram with the same scales on the two axes, as shown in Fig. 2.10(b). The principal

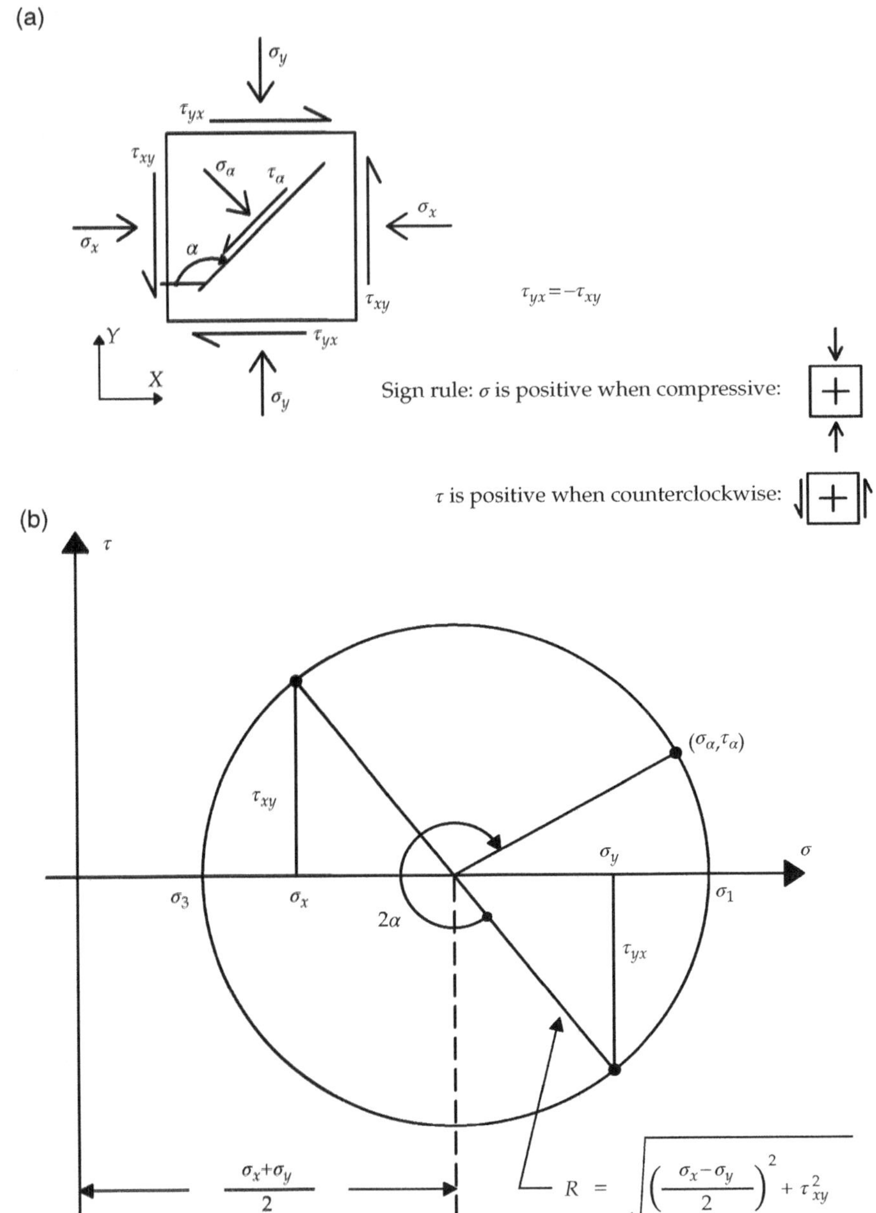

Figure 2.10 (a) Stresses on a material element and (b) construction of Mohr's circle.

stresses, σ_1 and σ_3, are calculated from the following expressions:

$$\sigma_1 = \frac{\sigma_x + \sigma_y}{2} + \sqrt{\left(\frac{\sigma_x - \sigma_y}{2}\right)^2 + \tau_{xy}^2} \qquad (2.33)$$

$$\sigma_3 = \frac{\sigma_x + \sigma_y}{2} - \sqrt{\left(\frac{\sigma_x - \sigma_y}{2}\right)^2 + \tau_{xy}^2} \qquad (2.34)$$

Mohr's circle may also be obtained by construction by setting off the stresses and drawing the circle as shown in Fig. 2.10(b). The principal stresses are located at the intersection between the circle and the σ–axis, and their magnitudes may be scaled directly from the diagram.

Pole method

To find the stresses graphically on any plane through the soil element, the pole method may be used. The pole (origin of planes) is first constructed by the following procedure involving Mohr's circle as shown in Fig. 2.11(a):

Find the pole:

Draw a line through one of the stress points (A or B) parallel to the plane on which the stresses act (vertical for point A, horizontal for point B) to intersection with the circle at point P. This is the pole.

A line [representing a plane, shown in Fig. 2.11(a)] drawn in any direction, β, through the pole crosses the circle at the stresses acting on that plane. The values of normal and shear stresses may be read directly on the axes, as indicated in Fig. 2.11(a).

Conversely, the plane on which a given set of stresses on Mohr's circle (σ_β, τ_β) acts may be obtained by drawing a line through (σ_β, τ_β) to intersect with the circle. In particular, the planes on which the principal stresses act may be determined by drawing lines through the pole and the points where the circle crosses the σ–axis (σ_1, 0) and (σ_3, 0). These lines represent the planes, and the principal stress directions are perpendicular to these planes, as shown in Fig. 2.11(b).

2.1.8 Strain analysis in two dimensions – Mohr's circle

Analyses of strains (or strain increments) in two dimensions follow similar expressions as those given above for stresses. The normal stresses, σ, are replaced by normal strains, ε (or the normal strain increment), and the shear stresses, τ, are replaced by half of the engineering shear strains, $\gamma/2$ (or half of the engineering shear strain increment).

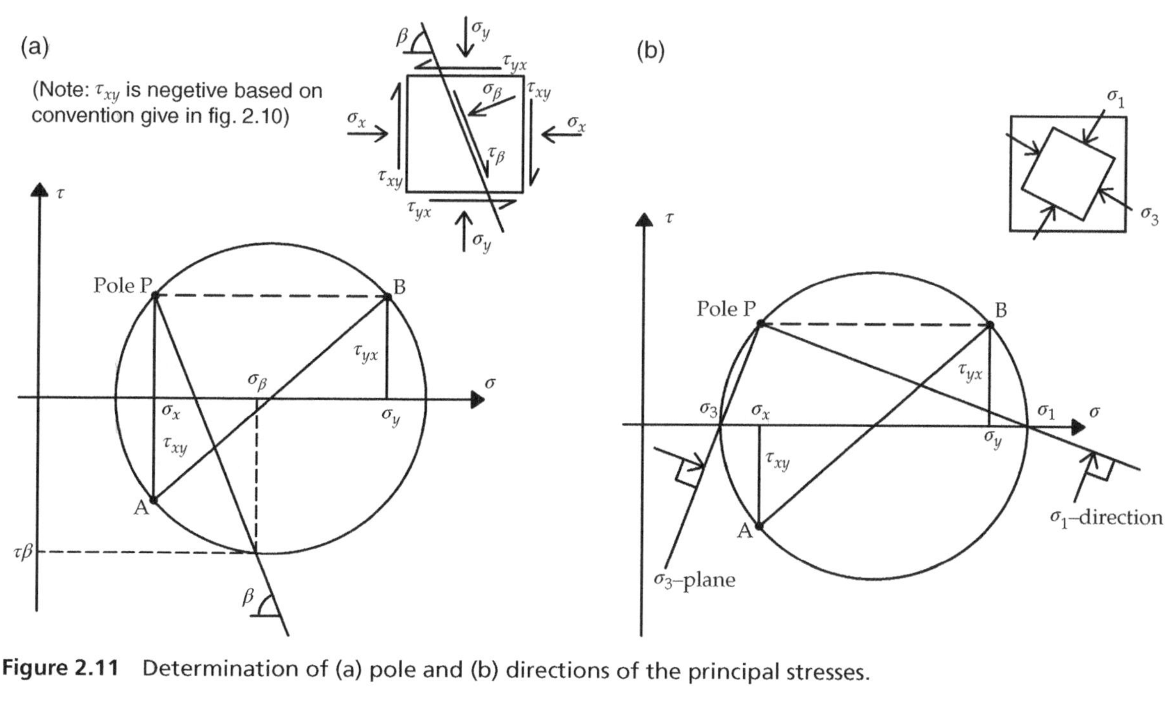

Figure 2.11 Determination of (a) pole and (b) directions of the principal stresses.

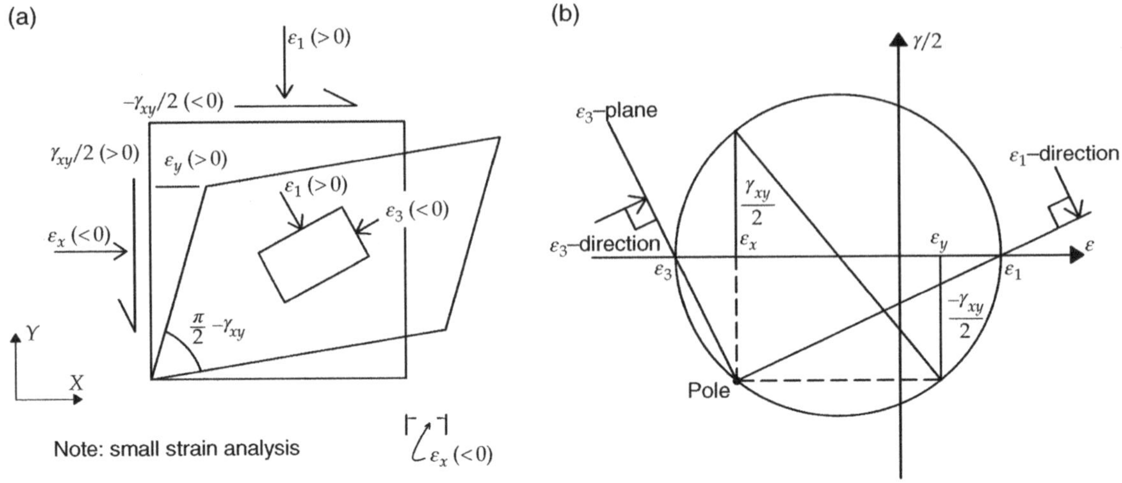

Figure 2.12 (a) Strains in a material element and (b) determination of principal strain directions in a material element.

Similarly, strain analyses may be performed graphically on the Mohr circle for strains (or for strain increments). Figure 2.12(a) shows a soil element compressed in the vertical direction, expanded in the horizontal direction, and sheared as shown. The sign rule for strains follows that for stresses, as also indicated in the diagram. Figure 2.12(b) illustrates the Mohr circle for strains. The location of the pole for strains is determined in a similar manner as for stresses, and the planes and directions of principal strains may be determined as indicated.

2.2 Stress–strain diagrams

2.2.1 Basic diagrams

The results of triaxial tests are presented on stress–strain, volume change and pore pressure diagrams of the types discussed in the following. The initial evaluation of results is best performed on diagrams of directly measured or calculated quantities such as effective confining pressure, σ_3', deviator stress, $(\sigma_1 - \sigma_3)$, axial strain, ε_a (= ε_1 in triaxial compression), volumetric strain, ε_v, and pore pressure, u. These diagrams allow an initial appreciation of the type of behavior exhibited by the soil (plastic or brittle stress–strain relation, etc.), the strain required to reach the maximum deviator stress can readily be determined, and different types of moduli can be evaluated.

The most common, basic diagrams employed for isotropic compression, K_0-compression, drained triaxial compression, and undrained triaxial compression tests are presented below. Typical results of these types of tests are shown to illustrate the use of the diagrams.

Isotropic compression

During isotropic compression of a soil specimen the effective confining pressure, σ_3', the axial strain, $\varepsilon_a = \varepsilon_1$, and the volumetric strain, ε_v, may be determined. The results of this test may be plotted as σ_3' versus ε_v and ε_1 versus ε_v. Examples of such diagrams are shown in Fig. 2.13. The axial and volumetric strains may be expressed as percentages (%) or as pure numbers (mm/mm) (e.g., $\varepsilon_1 = 2.0\% = 0.020$ mm/mm).

The diagram in Fig. 2.13(a) may be used to evaluate the bulk modulus and its variation. The bulk modulus is determined as:

$$K = \frac{\Delta\sigma_3'}{\Delta\varepsilon_v} \tag{2.35}$$

in which $\Delta\sigma_3'$ is the change in isotropic effective confining pressure ($\Delta\sigma_1 = \Delta\sigma_2 = \Delta\sigma_3$) and $\Delta\varepsilon_v$ is the resulting change in volumetric strain expressed as a pure number. The bulk modulus has the

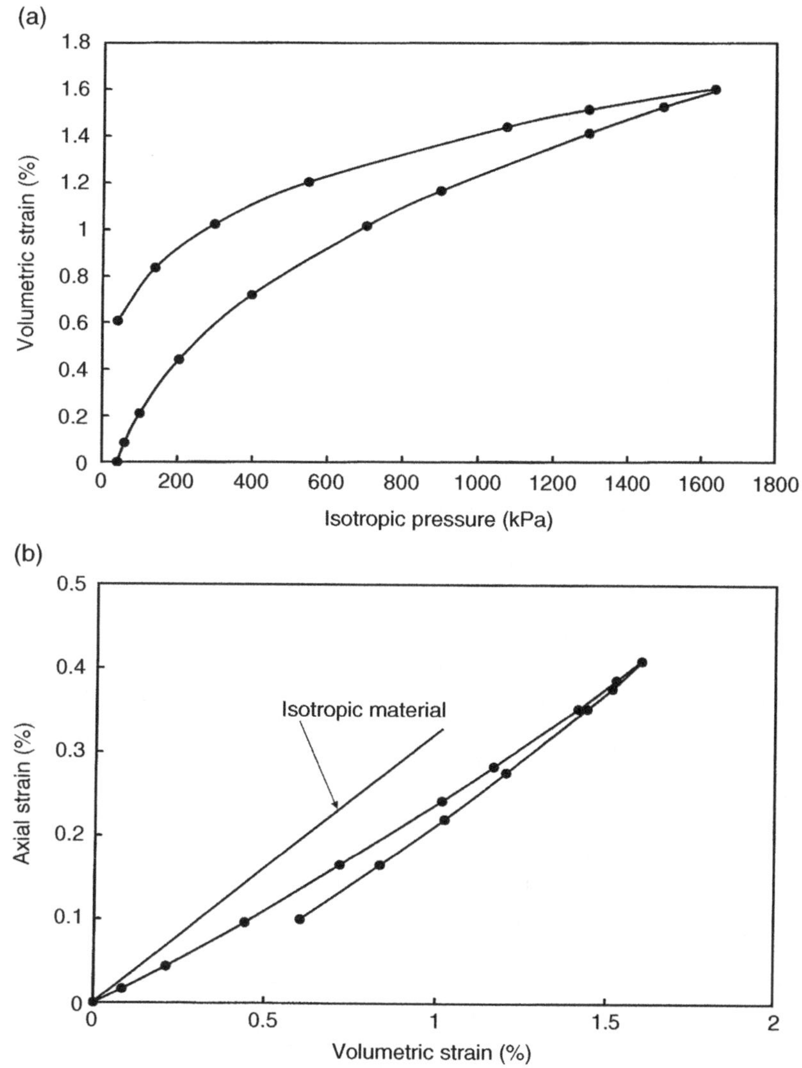

Figure 2.13 Results of isotropic compression and expansion of specimen of air-pluviated dense Nevada sand (D_r = 90%): (a) relation between effective isotropic pressure and volumetric strain; and (b) relation between volumetric strain and axial strain.

same units as those used for $\Delta\sigma_3'$. The bulk modulus for isotropic compression is lower than that for isotropic expansion. The latter expresses essentially the elastic behavior of the soil, whereas the former includes effects of irrecoverable, inelastic strains. Both relations show increasing bulk modulus with increasing isotropic pressure.

Figure 2.13(b) may be used for initial evaluation of the degree of cross-anisotropy of the soil specimen. It is recommended to use the same scales on the two strain axes to obtain the best possible visual perception of the relation between the strains. Isotropic behavior is obtained if the strains follow a relation in which $\varepsilon_v = 3 \cdot \varepsilon_1$ for small strains and $\varepsilon_v < 3 \cdot \varepsilon_1$ for large strains. Equation (2.7) for large strains with $\varepsilon_1 = \varepsilon_2 = \varepsilon_3$ yields:

$$\varepsilon_v = 3 \cdot \varepsilon_1 - 3 \cdot \varepsilon_1^2 + \varepsilon_1^3 \qquad (2.36)$$

The solid line in Fig. 2.13(b) indicates the relation between ε_v and ε_1 for isotropic behavior of the soil. The experimental results shown in Fig. 2.13(b) indicate that the specimen of Fine Silica sand is more compressible in the lateral directions during loading, but it exhibits essentially isotropic behavior during unloading.

K_0-compression

To simulate field loading conditions most correctly, it may be desirable to compress the soil specimen under anisotropic stresses before shearing. One anisotropic compression condition often encountered in the field is the K_0-condition. During K_0-compression of a soil specimen, the lateral strain is maintained at zero, and the vertical strain equals the volumetric strain, $\varepsilon_1 = \varepsilon_v$. The quantities measured directly in a K_0-test performed in the triaxial apparatus are the vertical stress, σ_1', the lateral stress, σ_3', the vertical strain, ε_1, and the volumetric strain, ε_v. The results of this test may be plotted as σ_3' versus σ_1', ε_1 versus ε_v (for general anisotropic compression), and $\varepsilon_1 = \varepsilon_v$ versus σ_1' (for K_0-compression).

It should be mentioned here that in view of the complexities in performing the K_0-consolidation, strength tests may not be required to follow the K_0-path prior to shearing if all that is desired is the effect of the initial stress ratio on strength. If the K_0 response is the focus, then the K_0-path must be followed.

Figure 2.14 shows results of a K_0-test on Fine Silica sand. The diagram in Fig. 2.14(a) may be used to determine the value of K_0, the coefficient of earth pressure at rest, and its variation with stress magnitudes (if any). K_0 is defined as:

$$K_0 = \left(\frac{\Delta\sigma_3'}{\Delta\sigma_1'}\right)_{\varepsilon_r = 0} \tag{2.37}$$

and it is therefore the slope of the curve in Fig. 2.14(a).

The diagram in Fig. 2.14(b) may be used for evaluation of the coefficient of volume compressibility, m_v, and its variation. The definition of m_v is:

$$m_v = \left(\frac{\Delta\varepsilon_v}{\Delta\sigma_1}\right)_{\varepsilon_r = 0} \tag{2.38}$$

The value of m_v is determined as the slope of the $\varepsilon_v - \sigma_1$ curve at any specific value of σ_1. It has dimensions of the reciprocal of stress (e.g., 1/kPa).

Figure 2.14(c) shows a diagram of ε_1 versus ε_v. For K_0-compression $\varepsilon_1 = \varepsilon_v$ while other relations are obtained as indicated for other compression conditions.

It may be noted that oedometer tests in which the soil is compressed vertically inside a metal ring are simpler to perform and they give similar information regarding overconsolidation ratio and modulus. Equipment can be used from which the lateral stress is also obtained.

Drained compression

Figure 2.15 shows the results of isotropically consolidated-drained (ICD) triaxial compression tests on Antelope Valley sand performed at three different, constant confining pressures selected to cover the range of stresses anticipated in the field problem to be analyzed. The stress difference or deviator stress, $(\sigma_1 - \sigma_3)$, and the volumetric strain, ε_v, are plotted versus the axial strain, ε_1. The strains may be expressed as percentages (%) or as pure numbers (mm/mm) (e.g., ε_1 = 3.7% = 0.037). Unloading–reloading stress–strain relations are shown for two tests in Fig. 2.15.

A close association exists between stress–strain and volumetric relations and this association is displayed by plotting the results versus a common variable (ε_1). This provides an opportunity to evaluate the type and magnitude of volume changes in relation to the stress–strain behavior. Usually initial volumetric contraction is followed by dilation (for sand at low confining pressure) or further contraction (for sand at high confining pressure). The best visual perception of the strain relations is obtained by using the same scales on the two strain axes.

Figure 2.16 shows the results of drained triaxial compression tests on laboratory prepared

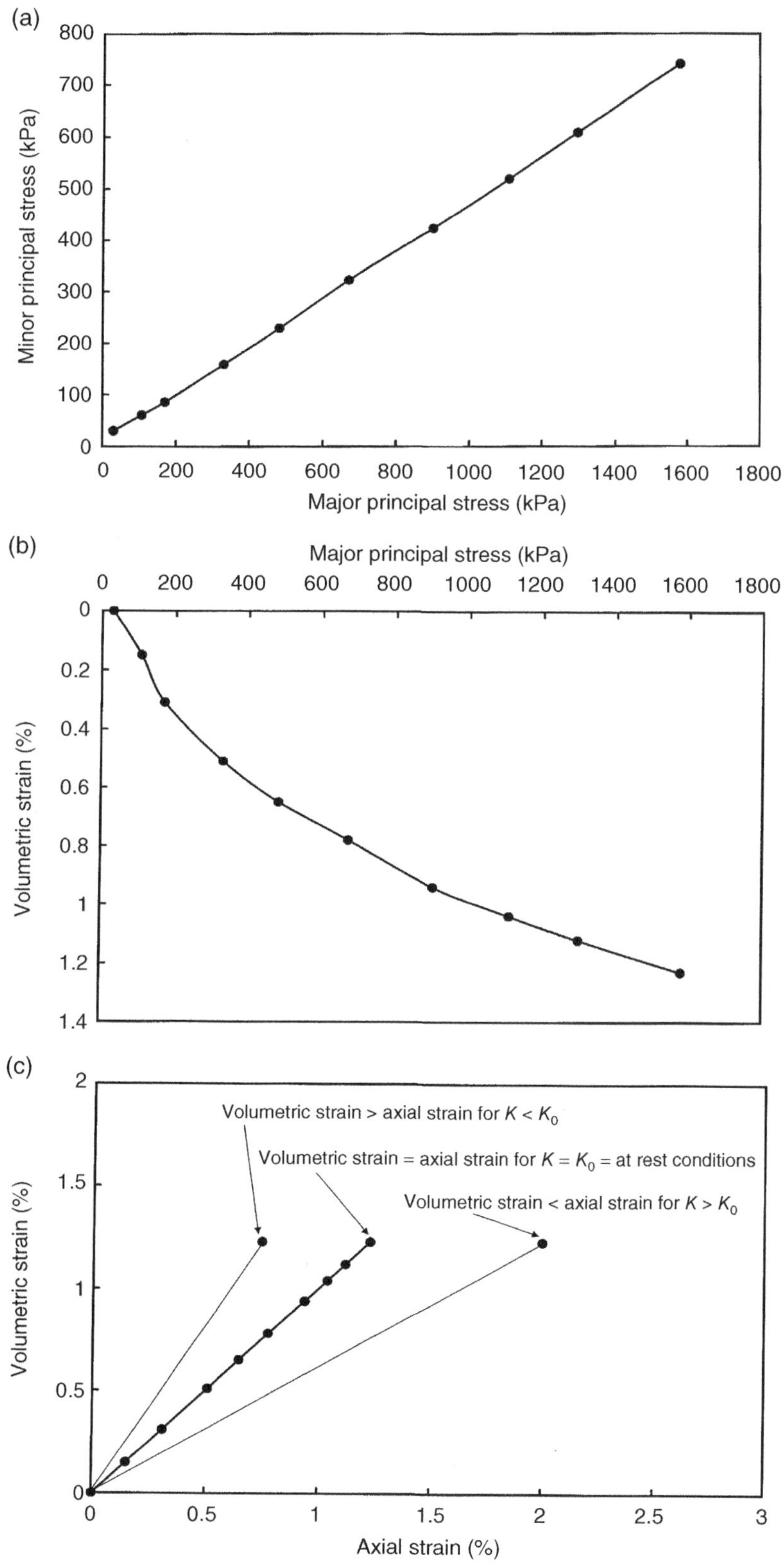

Figure 2.14 Results of K_0-compression of Fine Silica sand: (a) relation between σ_3' and σ_1'; (b) relation between σ_1' and ε_v; and (c) relation between ε_1 and ε_v for K_0 and other constant stress ratio conditions.

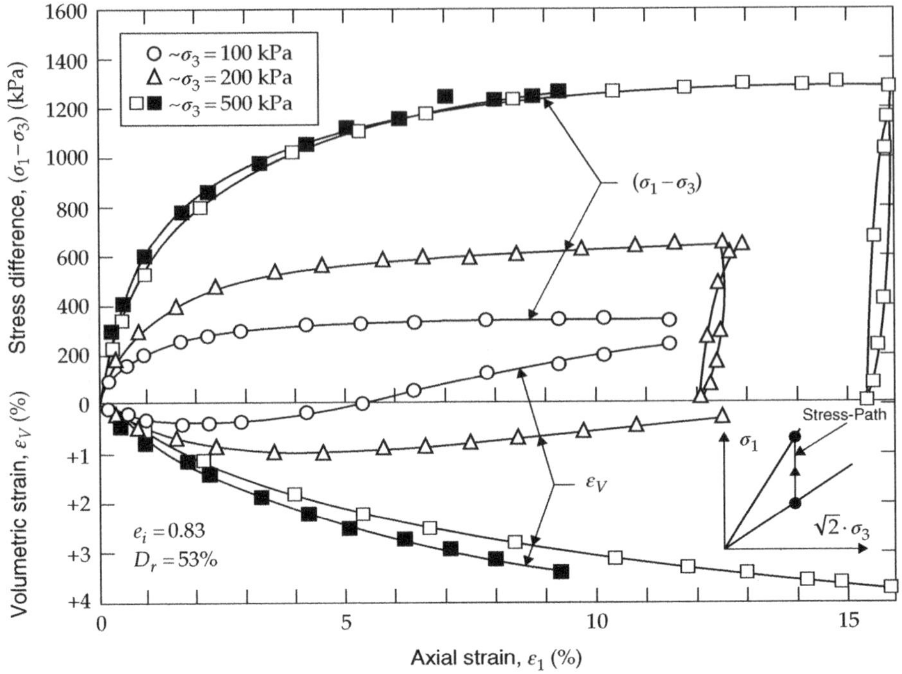

Figure 2.15 Results of drained triaxial compression tests on dense Antelope Valley sand #10–#20.

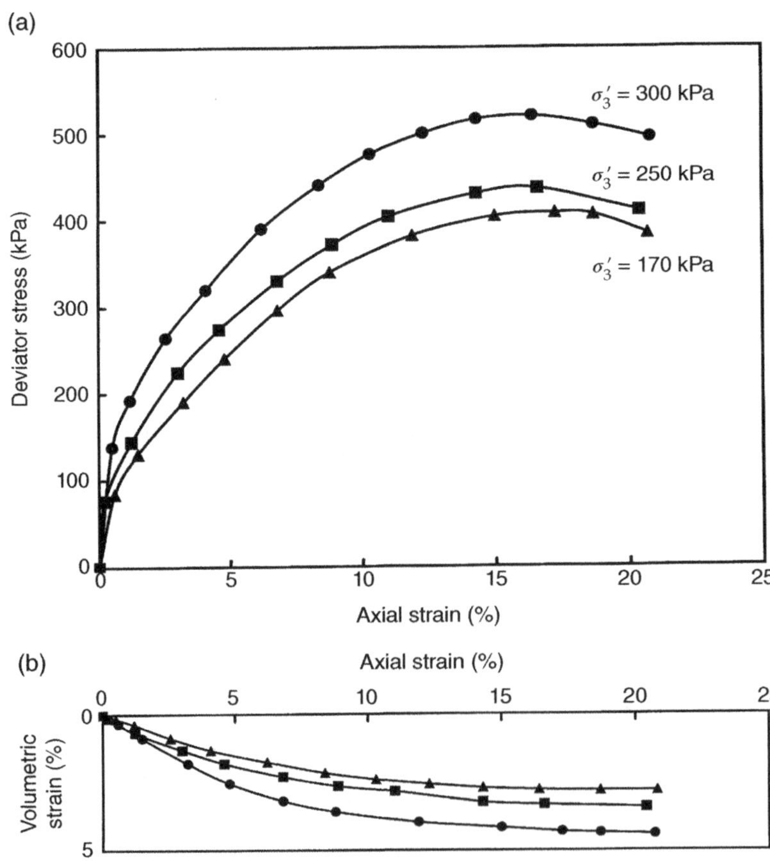

Figure 2.16 (a) Stress–strain and (b) volumetric strain from three drained triaxial compression tests on laboratory prepared intact specimens of normally consolidated Edgar Plastic Kaolinite.

specimens of normally consolidated Edgar Plastic Kaolinite. The shear strength increases with increasing confining pressure, but the volumetric strains indicate contraction throughout the tests. The amount of volumetric strain is essentially the same or it increases a small amount for the three tests with the effective confining pressure.

The results of drained triaxial compression tests on overconsolidated Edgar Plastic Kaolinite are shown in Fig. 2.17. The pattern of volume change behavior appears to be similar to that of sand. Thus, initial volumetric contraction is followed by dilation (for overconsolidated clay at low confining pressure, but at high overconsolidation ratios) or further contraction (for overconsolidated clay at high confining pressure, but at low overconsolidation ratios). The pattern of volumetric strains is different from that of normally consolidated clay and it is produced by and varies with the amount of overconsolidation expressed in the overconsolidation ratio (OCR).

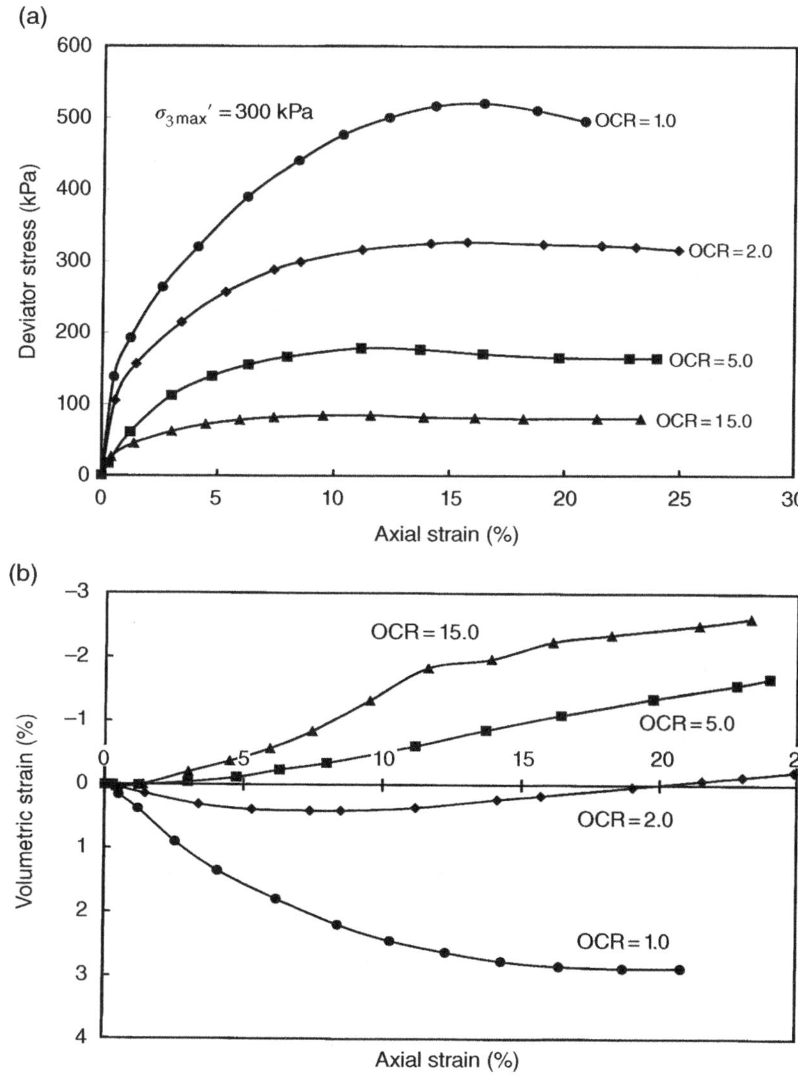

Figure 2.17 (a) Stress–strain and (b) volumetric strain from four drained triaxial compression tests on laboratory prepared intact specimens of overconsolidated Edgar Plastic Kaolinite.

Undrained compression

Figure 2.18 shows the results of isotropically consolidated-undrained (ICU) triaxial compression tests on saturated specimens of Sacramento River sand performed at four different consolidation pressures. The stress difference or deviator stress, $(\sigma_1 - \sigma_3)$, and the pore pressure change, Δu, are plotted versus the axial strain, ε_1. The close association between stress–strain behavior and pore pressure response in undrained tests is displayed by plotting the results versus a common variable (ε_1). The shape of the stress–strain curve may then be evaluated in view of the magnitude and variation of the pore pressure. At low consolidation pressures the pore pressure first increases and then decreases in response to a tendency for dilation of the soil structure. Actual dilation does not occur ($\Delta V = 0$ in undrained tests on saturated specimens) as long as the specimen remains saturated. At high consolidation pressures the pore pressure

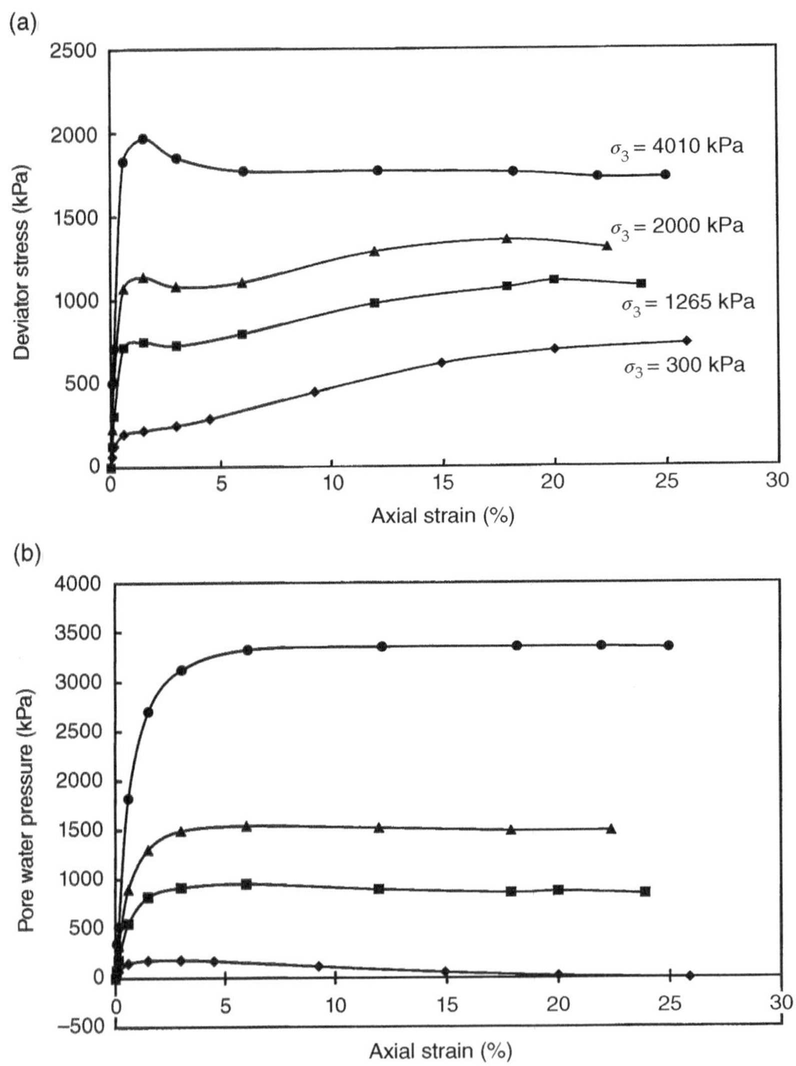

Figure 2.18 Results of isotropically consolidated-undrained triaxial compression tests on loose Sacramento River sand ($e = 0.87$, $D_r = 38\%$): (a) $(\sigma_1 - \sigma_3)$ versus ε_1 and (b) Δu versus ε_1.

increases throughout the undrained test. Because strains are produced in response to effective stresses, the shape of the stress–strain curve depends on the current effective confining pressure, σ_3' ($= \sigma_{3cell} - u$). The shapes of the stress–strain curves are therefore directly related to the magnitudes and changes in pore pressure (for constant cell pressure).

Figure 2.19 shows the results of isotropically consolidated-undrained triaxial compression tests on laboratory prepared, saturated specimens of normally consolidated Edgar Plastic Kaolinite. Both the undrained strength and the pore pressure increase with increasing consolidation pressure, and the stress–strain curves as well as the pore pressure curves form consistent patterns that are related to the magnitude of the consolidation pressure.

The results of isotropically consolidated-undrained triaxial compression tests on laboratory prepared, remolded specimens of overconsolidated Edgar Plastic Kaolinite are shown in Fig. 2.20.

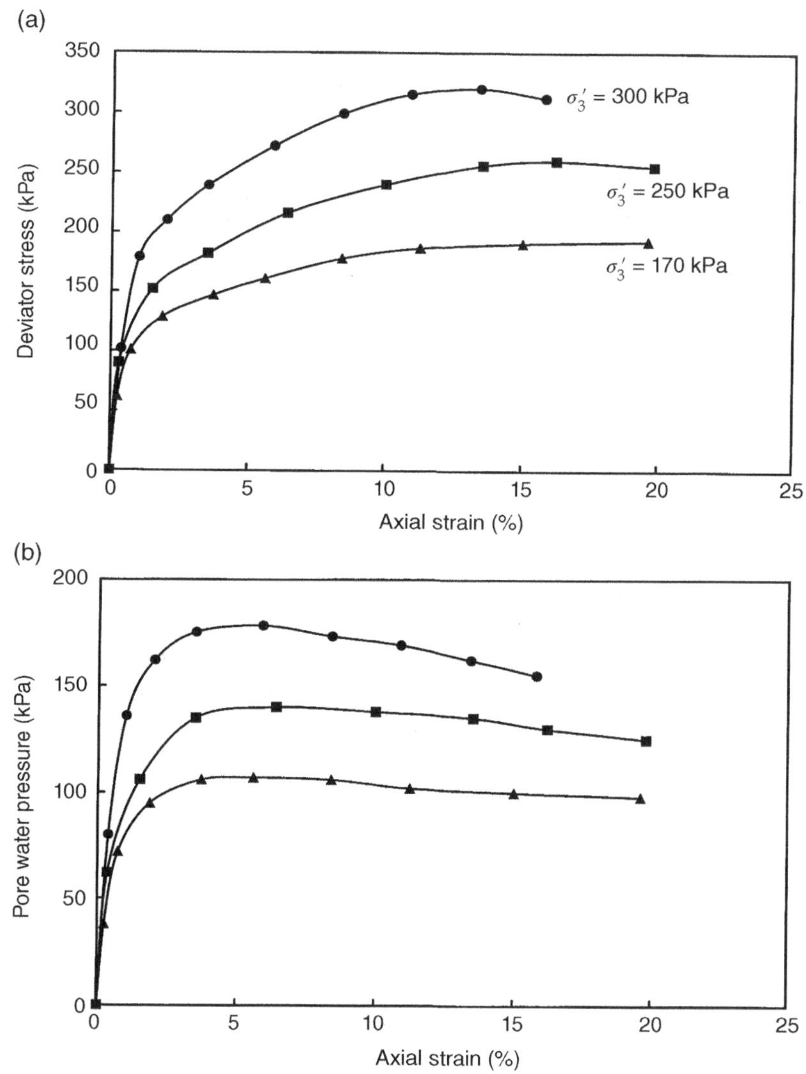

Figure 2.19 (a) Stress–strain and (b) pore water pressure from isotropically consolidated-undrained triaxial compression tests on normally consolidated Edgar Plastic Kaolinite.

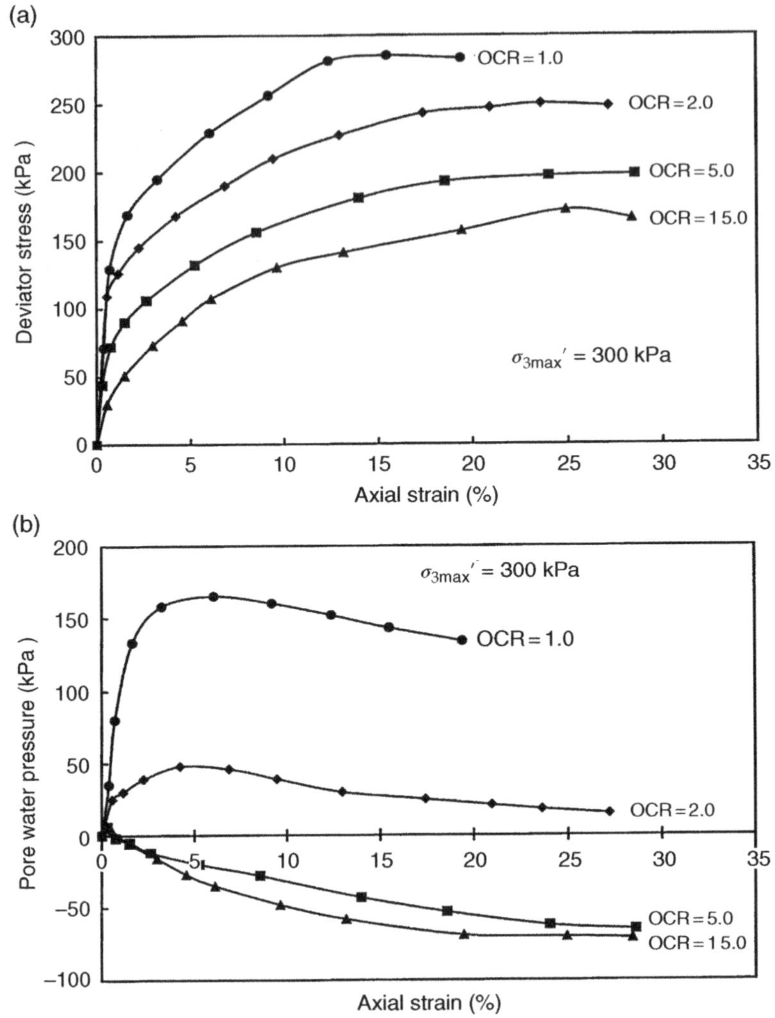

Figure 2.20 (a) Stress–strain and (b) pore water pressure from isotropically consolidated-undrained triaxial compression tests on laboratory prepared, remolded specimens of overconsolidated Edgar Plastic Kaolinite.

The pattern of pore pressure changes appears to be similar to that for sand. Initial pore pressure increase is followed by pore pressure decrease in tests at low consolidation pressures (but high OCRs). At high consolidation pressures (but low OCRs) the pore pressure increases throughout the undrained test.

Tests with initial anisotropic compression

In the case that the initial compression is anisotropic (e.g., K_0-compression), the soil specimen is exposed to an initial deviator stress. The stress–strain relation produced during subsequent shearing of the specimen therefore begins at a location on the $(\sigma_1 - \sigma_3)$-axis corresponding to this initial deviator stress. Figure 2.21 shows the results of undrained shearing of three K_0-consolidated specimens of laboratory prepared, normally consolidated Edgar Plastic Kaolinite. The stress–strain curves begin at the respective deviator stresses applied during K_0-consolidation. Since excess

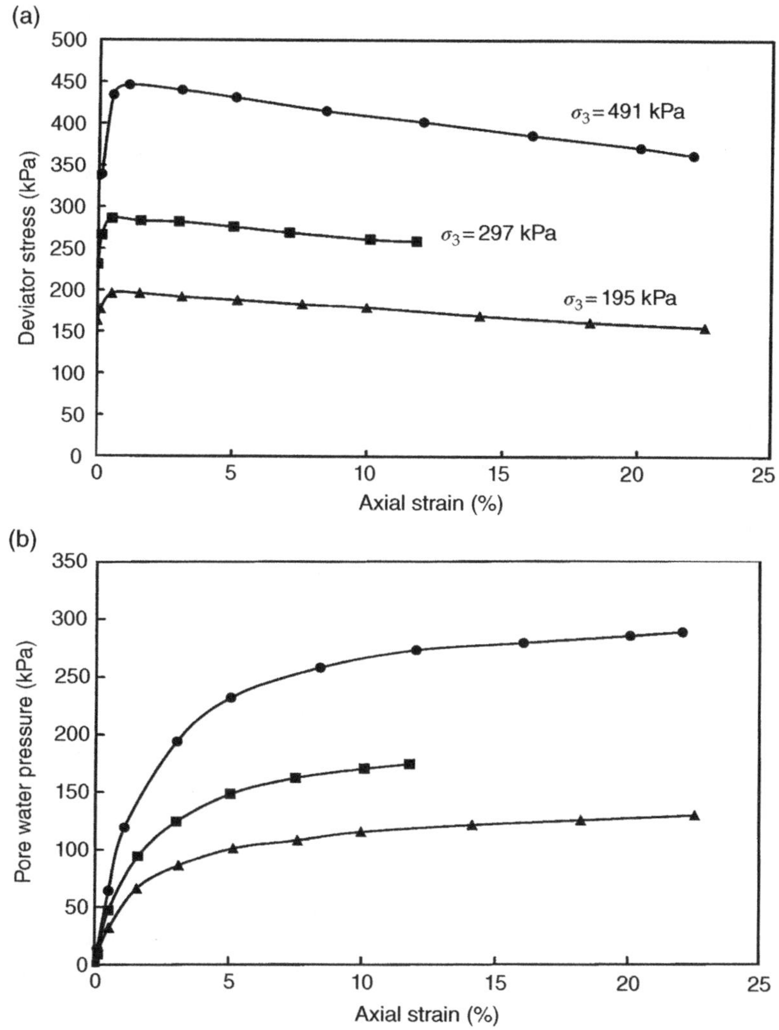

Figure 2.21 (a) Stress–strain and (b) pore water pressure from three normally K_0-consolidated, undrained triaxial compression tests on Edgar Plastic Kaolinite.

pore pressures are not present in the specimens after K_0-consolidation, the changes in pore pressures are initiated at the origin indicating that they are produced entirely due to shearing.

2.2.2 Modulus evaluation

Simple evaluation of deformations of geotechnical structures are most often based on concepts and formulas from the theory of elasticity, because it is the simplest and some close-formed expressions are available. Nevertheless, it is well-established that soils to a large extent behave plastically, that is a large portion of the deformations are irrecoverable and some form of plasticity theory is required to describe this behavior.

To use the elasticity approach some measures of Young's modulus, E, and Poisson's ratio, ν, are required. These quantities are not constant for a given soil, but they may be used to approximate the soil behavior for a given state of stress.

Young's modulus

The modulus may be determined from stress–strain diagrams of the type shown in Figs 2.22, 2.23, and 2.24. The modulus is defined from axisymmetric conditions:

$$E = \frac{\Delta(\sigma_1 - \sigma_3)}{\Delta\varepsilon_1} \tag{2.39}$$

in which $\Delta(\sigma_1 - \sigma_3)$ is the change in deviator stress and $\Delta\varepsilon_1$ is the change in axial strain expressed as a pure number. The modulus has the same dimension as that used for the deviator stress.

Two types of moduli may be determined: the tangent modulus E_t; and the secant modulus E_s. The tangent modulus is defined as the slope of the stress–strain curve at a particular point on the curve, as indicated in Fig. 2.22(a). The secant modulus is defined as the slope of a straight line connecting two separate points on the curve. A secant modulus connecting the origin with a point on the stress–strain curve is indicated on Fig. 2.22(a). The tangent modulus at a point (A) is best evaluated as the secant modulus connecting two points (B and C) at equal (small) distances from the point in question (A), as shown on the inset in Fig. 2.22(a). The variation of the tangent modulus and the secant modulus (initiated at the origin) with axial strain are illustrated in Fig. 2.22(b). Both the tangent and the secant moduli equal the initial modulus, E_i, at the origin of the stress–strain diagram. The tangent modulus decreases with increasing axial strain and becomes negative beyond the peak deviator stress and finally increases again to zero as the residual strength is approached. The secant modulus also decreases with increasing axial strain, but it remains positive as it asymptotically approaches zero at large strains.

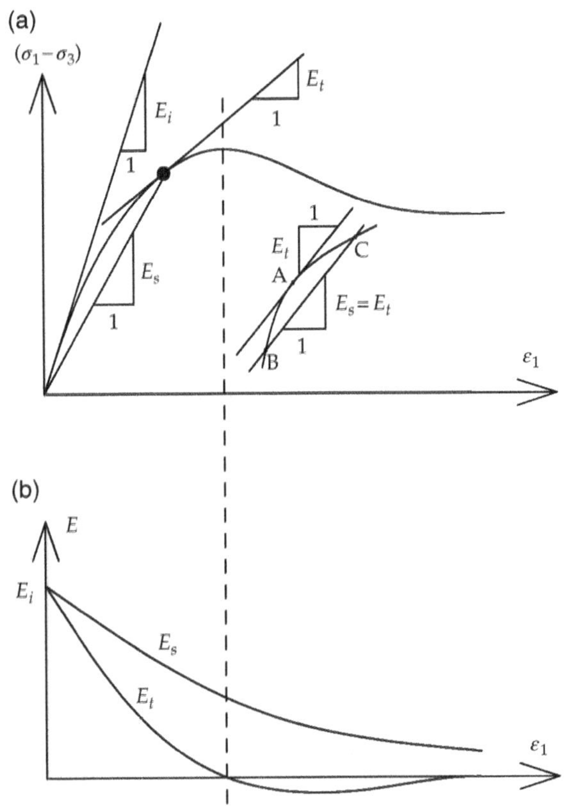

Figure 2.22 Definitions of (a) initial tangent and secant moduli and (b) their variations with axial strain.

It is evident that the tangent and secant moduli do not have unique values, but they vary with the state of stress and the stress increment. It should therefore be cautioned that determination and application of these moduli for field cases can be very ambiguous and at times incorrect. It must be ascertained that the state of stress and the stress increments used in evaluation of moduli are applicable to the particular field case under consideration. It is often more correct to perform tests in which the stress paths correspond better with those in the field than relying on the results of conventional triaxial tests for determination of appropriate moduli.

Neither of the two moduli discussed above may represent the true elastic behavior of the soil, because irrecoverable, inelastic strains occur in the soil from the beginning of loading. A closer approximation to the elastic modulus may be obtained from an unloading–reloading cycle, as illustrated schematically in Fig. 2.23. The slope of the straight line between point A (point where unloading begins) and point B (point where reloading begins) is taken as the unloading–reloading modulus, E_{ur}. This modulus captures the average soil behavior during unloading and reloading of geotechnical structures and it may be sufficiently accurate for many purposes

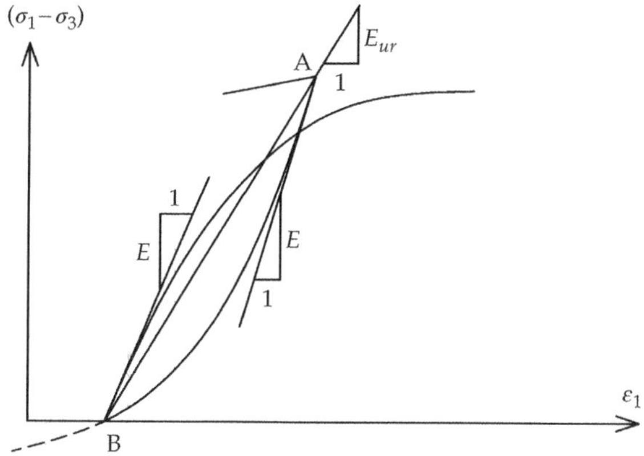

Figure 2.23 Schematic illustration of determination of the unloading–reloading modulus, E_{ur}, and the true elastic moduli, E.

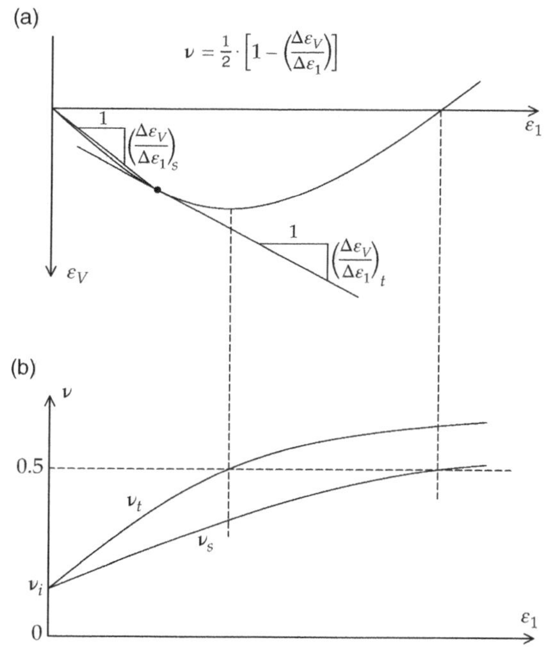

Figure 2.24 Determination of (a) tangent and secant Poisson's ratios and (b) their variations with axial strain.

involving elastic deformations. However, the hysteresis loop obtained from the unloading–reloading cycle reveals that some inelastic behavior is included in E_{ur}. Partial unloading followed by reloading or unloading followed by reloading in the extension regime [negative $(\sigma_1 - \sigma_3)$] indicates the inconsistency in the definition of the elastic modulus.

The closest approximation to the truly elastic behavior is probably obtained from the slopes of the stress–strain curve immediately following unloading and immediately following reloading, respectively. These slopes are also indicated in Fig. 2.23. Note that the modulus values obtained from these portions of the unloading–reloading relations are not equal in magnitude, but depend on the state of stress at which they are obtained. It should be cautioned that these slopes may contain effects of time-dependent behavior such as creep. Soils that exhibit pronounced creep behavior may even produce negative modulus values whose magnitudes depend on the rate of unloading and reloading. Such modulus values clearly do not represent the elastic behavior.

Figures 2.22 and 2.23 show that the various types of moduli discussed above vary with such factors as strain magnitude and state of stress. Analytical modeling of this variation is beyond the scope of this discussion.

Poisson's ratio

The other elastic parameter often sought from basic test results is Poisson's ratio. This nondimensional parameter is defined as:

$$\nu = -\frac{\Delta\varepsilon_3}{\Delta\varepsilon_1} \tag{2.40}$$

in which the strain increments are obtained under axisymmetric conditions as prevail in triaxial compression tests. Poisson's ratio may be determined from the volumetric strain curve. Since the elastic parameters relate to small strains, $\Delta\varepsilon_3$ may be obtained from Eq. (2.18), and Poisson's ratio becomes:

$$\nu = \frac{1}{2}\left(1 - \frac{\Delta\varepsilon_v}{\Delta\varepsilon_1}\right) \tag{2.41}$$

in which $\Delta\varepsilon_v/\Delta\varepsilon_1$ represents the slope of the volumetric strain curve.

Two types of Poisson's ratio may be determined: the tangent Poisson's ratio, ν_t; and the secant Poisson's ratio, ν_s. Their determination and variation with axial strain are illustrated in

Fig. 2.24. Both the tangent and secant Poisson's ratios equal the initial Poisson's ratio at the origin of the volumetric strain diagram. The tangent Poisson's ratio increases and becomes equal to 0.5 at the point where the increment in volumetric strain changes from contractive to expansive. As further expansion occurs, ν_t increases beyond 0.5. The secant Poisson's ratio also increases with axial strain, and it reaches 0.5 at the point where the total volumetric strain changes from contractive to expansive. The value of ν_t increases beyond 0.5 with additional axial strain and expansion.

For undrained tests on saturated soil the volumetric strain is zero. This corresponds to a value of Poisson's ratio of 0.5.

None of these measures of Poisson's ratio represent the true elastic behavior of the soil. For an isotropic material Poisson's ratio is limited in the range of $-1 \leq \nu \leq 0.5$. In practice, Poisson's ratio is usually not smaller than zero (possibly with the exception of some foam materials). However, triaxial compression tests performed at very high confining pressures may exhibit initial values of Poisson's ratio smaller than zero. This indicates that the specimen diameter reduces during the initial phase of loading. This phenomenon may be attributed to crushing of soil grains and can hardly be characterized as elastic behavior.

As explained above for the elastic modulus, the closest approximation to the truly elastic behavior may be obtained immediately following unloading and reloading, respectively. Figure 2.25 shows a schematic illustration of the volumetric strain curve during an unloading–reloading cycle. Unloading is initiated at point A and reloading begins at point B. The linear portion of the curve at point A may be very short, and for some cases it may not exist. However, the linear portion obtained immediately upon reloading at point B is usually of sufficient extent to allow determination of its slope and calculation of Poisson's ratio. For a soil at a given density, the variation of the true elastic Poisson's ratio is usually very small (although the data may show a fair amount of scatter), and it may for many practical purposes be assumed to be constant.

Bulk and shear moduli

Two elastic parameters are required to describe the behavior of isotropic materials. Combinations of Young's moduli and Poisson's ratios of the types illustrated in Figs 2.22 and 2.24 can result in inconsistencies such as negative bulk modulus and grossly incorrect magnitudes of calculated deformations. To overcome some of these problems, it may be advantageous to determine and employ another set of moduli: The bulk modulus, K, and the shear modulus, G. According to Hooke's law for elastic behavior, these are related to Young's modulus and Poisson's ratio as follows:

$$K = \frac{E}{3\cdot(1-2\nu)} \tag{2.42}$$

$$G = \frac{E}{2\cdot(1+\nu)} \tag{2.43}$$

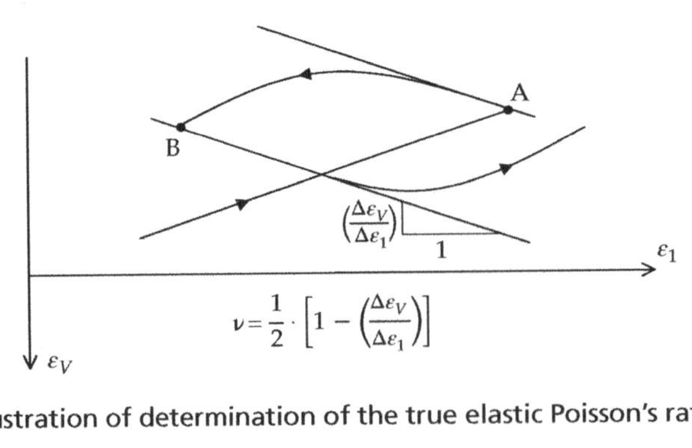

Figure 2.25 Schematic illustration of determination of the true elastic Poisson's ratio.

The bulk modulus may be determined directly from an isotropic compression test, as illustrated in Fig. 2.13(a). Direct determination of the bulk modulus assures that it remains positive as required. The bulk modulus may then be used to determine the elastic strains employing reasonable values of Poisson's ratio ($\nu \le 0.5$) or with the shear modulus. The shear modulus, G, relates the shear strains to the shear stresses, and it may be determined from tests in which shear stresses are applied to the surface of a specimen, as in a simple shear test (see Chapter 11). It may also be determined on the basis of K and ν for the isotropic material:

$$G = K \cdot \frac{3 \cdot (1-2\nu)}{2 \cdot (1+\nu)} \tag{2.44}$$

Another modulus of interest is the constrained modulus, which applies to conditions of K_0-compression. The constrained modulus, D, is defined as the reciprocal of the coefficient of volume compressibility, m_v:

$$D = \left(\frac{\Delta\sigma_1}{\Delta\varepsilon_v}\right)_{\varepsilon_r=0} = \frac{1}{m_v} \tag{2.45}$$

The constrained modulus has the same dimension as that used for the stress increment. In terms of Young's modulus and Poisson's ratio, the constrained modulus may be expressed as:

$$D = E \cdot \frac{(1-\nu)}{(1+\nu)(1-2\nu)} \tag{2.46}$$

2.2.3 Derived diagrams

The results of triaxial tests may also be presented on diagrams in which derived quantities are plotted on the axes. These quantities are derived from the directly measured or calculated quantities discussed above. The derived diagrams are used for convenience or to indicate the variation of certain quantities that are not directly measured. The most common derived diagrams employed for isotropic and K_0-compression and for drained and undrained triaxial compression tests are presented below. Typical results of these types of tests are shown to illustrate the use of derived diagrams.

Isotropic and K_0-compression

The most common, derived diagram employed for results of isotropic and K_0-compression tests is the void ratio e–log(p') diagram. The change in e is determined from the volumetric strain and added to the initial void ratio, e_0, and the current void ratio is plotted versus stress in a semi-log diagram:

$$\Delta e = \varepsilon_v \cdot (1+e_0) \tag{2.47}$$

$$e = e_0 + \Delta e = e_0 - \varepsilon_v \cdot (1+e_0) \tag{2.48}$$

in which the minus in front of ε_v occurs due to the sign rule used in soil mechanics, and ε_v is inserted as a pure number. For isotropic compression the value of $p' = \sigma_1' = \sigma_3' = \sigma_m'$ in which σ_m' is the mean normal stress. For K_0-compression p' is taken as the vertical, effective stress in the field, and this stress is often the major principal stress, σ_1', or it is taken as the mean normal stress, σ_m', defined as:

$$\sigma_m' = \frac{1}{3} \cdot (\sigma_1' + \sigma_2' + \sigma_3') \tag{2.49}$$

For triaxial compression $\sigma_2' = \sigma_3'$ and σ_m' becomes:

$$\sigma_m' = \frac{1}{3} \cdot (\sigma_1' + 2\sigma_3') \tag{2.50}$$

Figure 2.26 shows the results of four isotropic compression tests on Sacramento River sand with four different initial void ratios.

The e–log σ_1' diagram, shown in Fig. 2.26, is often used for a more clear determination of pre-consolidation stress for clays. However, this determination depends on a change in slope of the relation in the e–σ_1 diagram, which will be attenuated in the e–log σ_1' diagram. It should be noted that any straight line with a slope in the straight arithmetic diagram will exhibit a "pre-consolidation" pressure in

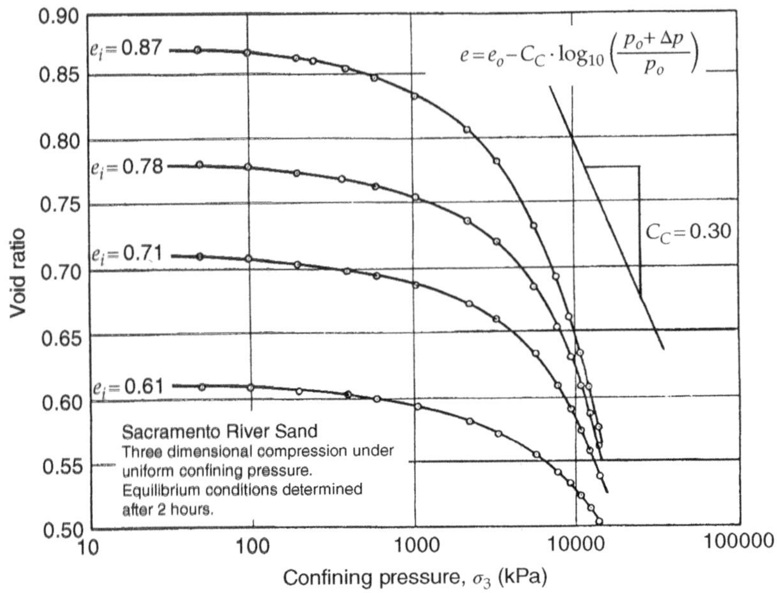

Figure 2.26 Results of four isotropic tests on Sacramento River sand shown on an e–log(p′) diagram. Reproduced from Lee and Seed 1967 by permission of ASCE.

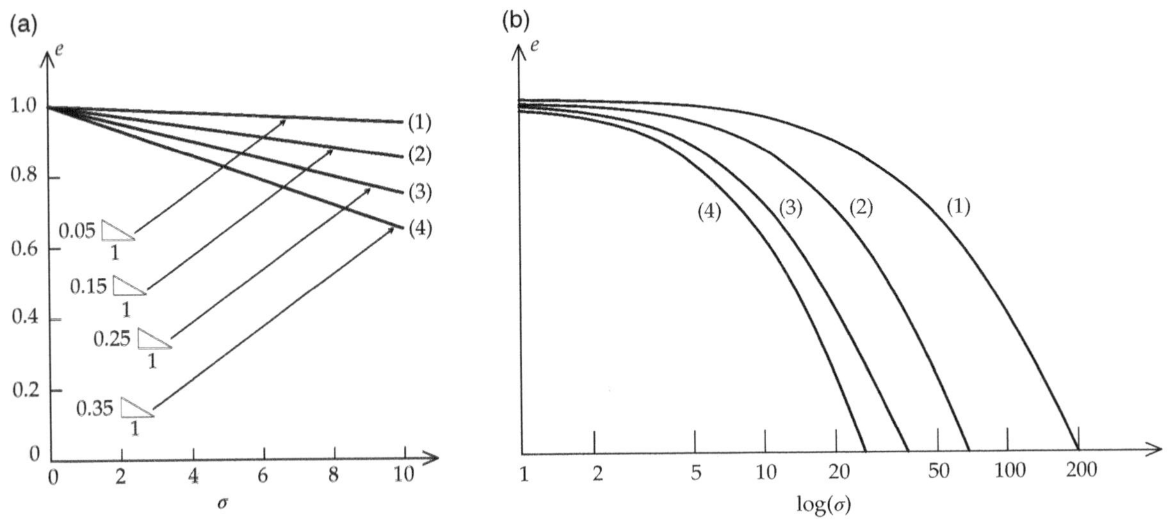

Figure 2.27 Schematic results of four isotropic tests on sand with linear e–p′ relations plotted on (a) an e–p′ diagram and (b) an e–log(p′) diagram.

the semi-log diagram. This is demonstrated in Fig. 2.27. Thus, attention should be paid to the presence or absence of a real pre-consolidation pressure by inspecting the shape of the e–σ_1' relation in the straight arithmetic diagram.

Drained compression

Displaying and evaluating the results of drained tests on a $(\sigma_1 - \sigma_3)$–ε_1 diagram (as in Fig. 2.15) can become inconvenient, especially if the tests are performed with a wide range of constant confining pressures. In such a diagram it becomes difficult to evaluate the results from the tests with low confining pressures, because the scale on the ordinate is made relatively large to accommodate the results from the tests with high confining pressures. Further, it is not easy to compare the stress–strain behavior pattern because the curves are spread out over the diagram and therefore not easy to evaluate.

Evaluation of test results and their pattern is accomplished with greater ease when they are plotted on diagrams with the stress ratio σ_1/σ_3 on the vertical axis. Since the strength increases almost proportionally with the confining pressure, σ_3, for most soils in drained compression, the stress ratios have similar magnitudes in all tests on soils. The stress ratio is calculated from the deviator stress and the confining pressure as follows:

$$\frac{\sigma_1}{\sigma_3} = \frac{(\sigma_1 - \sigma_3) + \sigma_3}{\sigma_3} \tag{2.51}$$

Figure 2.28(a) shows the results of drained triaxial compression tests on Sacramento River sand plotted on a σ_1/σ_3–ε_1 diagram. Confining pressures from 100 to 14 000 kPa were used in these tests. In this diagram the stress–strain curves are much closer together than they would have been in a $(\sigma_1 - \sigma_3)$–ε_1 diagram, and this allows easy evaluation of the internal consistency in the test data. Since pore pressures are zero in drained tests, the peak stress ratios correspond to failure just as the peak deviator stress constitutes failure, that is the occurrences of failure as well as the strain-to-failure are identical whether evaluated in terms of stress ratio or deviator stress. The variation in peak strength with confining pressure indicates the nature of the curved failure envelope discussed in Section 2.3. The variation in shapes of the stress–strain curves may be viewed in terms of the associated volumetric strain pattern shown in Fig. 2.28(b).

To study the variation in void ratios in drained tests, the volumetric strain may be converted to changes in void ratio and added to the initial value according to Eq. (2.48). The void ratio is plotted separately or along with the volumetric strain itself. Figure 2.29 shows an example of variations in void ratios for two tests on dense Santa Monica Beach sand and two tests on loose Santa Monica Beach sand.

Undrained compression

The results of undrained tests may also conveniently be plotted on diagrams similar to those used for drained tests. However, the effective stress ratio σ_1'/σ_3' is used for undrained tests. The effective stress ratio is calculated from measured quantities as follows:

$$\frac{\sigma_1'}{\sigma_3'} = \frac{(\sigma_1 - \sigma_3) + (\sigma_3 - u)}{(\sigma_3 - u)} \tag{2.52}$$

in which σ_3 is the applied cell pressure and u is the total pore pressure consisting of the pore pressure increment due to shearing, Δu, and the back pressure u_{back}, if any (see Chapter 6).

The data from undrained tests on Edgar Plastic Kaolinite previously plotted in Figs 2.19 and 2.20 have been re-plotted on diagrams of σ_1'/σ_3' versus ε_1 in Figs 2.30 and 2.31, respectively. The advantages of this type of diagram, discussed in connection with the drained test, are also present for the undrained test: test data can all be displayed within one diagram since the effective stress ratios have similar magnitudes, the internal consistency of the data can be better evaluated, and the variation in shapes of the curves may be viewed in terms of the associated pattern of pore pressure changes. In addition, the maximum effective stress ratio, which is often taken to represent failure in triaxial tests, may be evaluated directly from these diagrams. Note that the maximum effective stress ratio will most often not occur at the same time as the maximum deviator stress in undrained tests. The axial strains required to reach the maximum effective stress ratios may also be determined from these diagrams.

The development of pore pressures may be seen in view of the magnitude of the deviator stress that produced the pore pressure. Skempton (1954) proposed the widely employed expression for the change in pore pressure Δu as

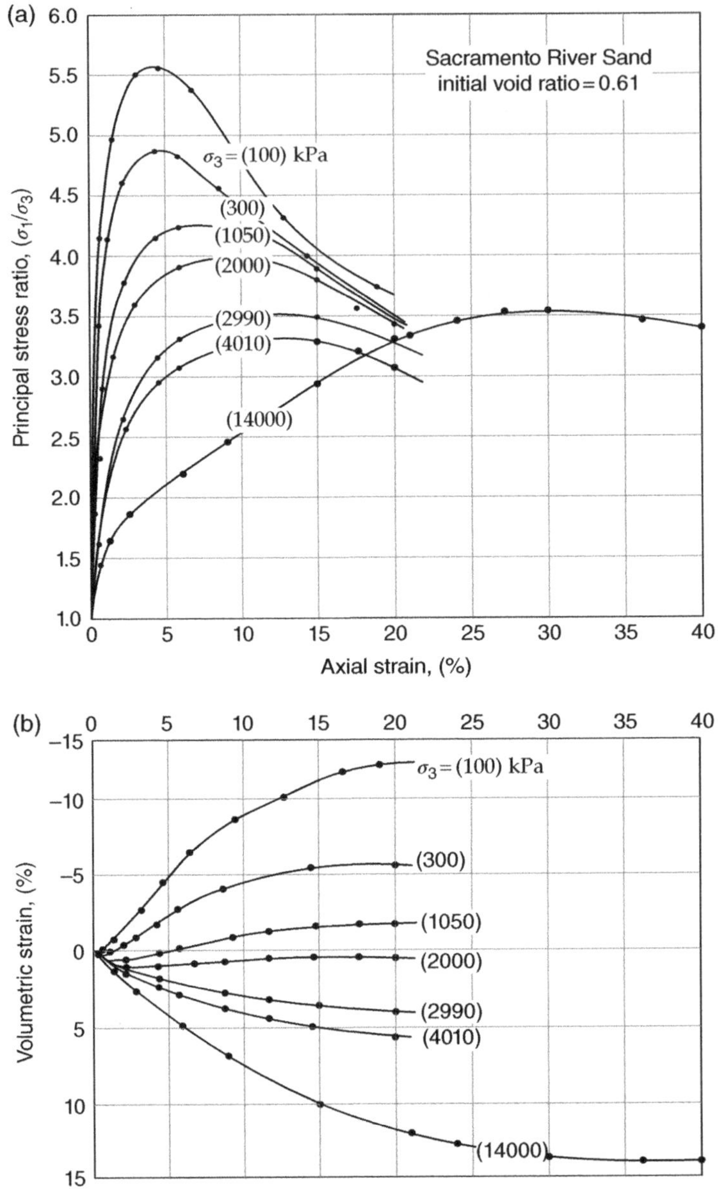

Figure 2.28 (a) Normalized stress–strain behavior and (b) volume change relations for a series of drained tests on dense Sacramento River sand. Reproduced from Lee and Seed 1967 by permission of ASCE.

related to the change in total confining pressure, $\Delta\sigma_3$, and the change in total axial stress, $\Delta\sigma_1$:

$$\Delta u = B \cdot \left[\Delta\sigma_3 + A \cdot (\Delta\sigma_1 - \Delta\sigma_3)\right] \quad (2.53)$$

which, when multiplied out, gives:

$$\Delta u = B \cdot \Delta\sigma_3 + \bar{A} \cdot (\Delta\sigma_1 - \Delta\sigma_3) \quad (2.54)$$

in which B and $\bar{A}(= A \cdot B)$ are empirically obtained pore pressure parameters. The usefulness of the parameter B is discussed in Chapter 6. The variation of parameter $\bar{A}$ is calculated from the experimental results according to:

$$\bar{A} = \frac{\Delta u}{\Delta\sigma_1 - \Delta\sigma_3} \quad (2.55)$$

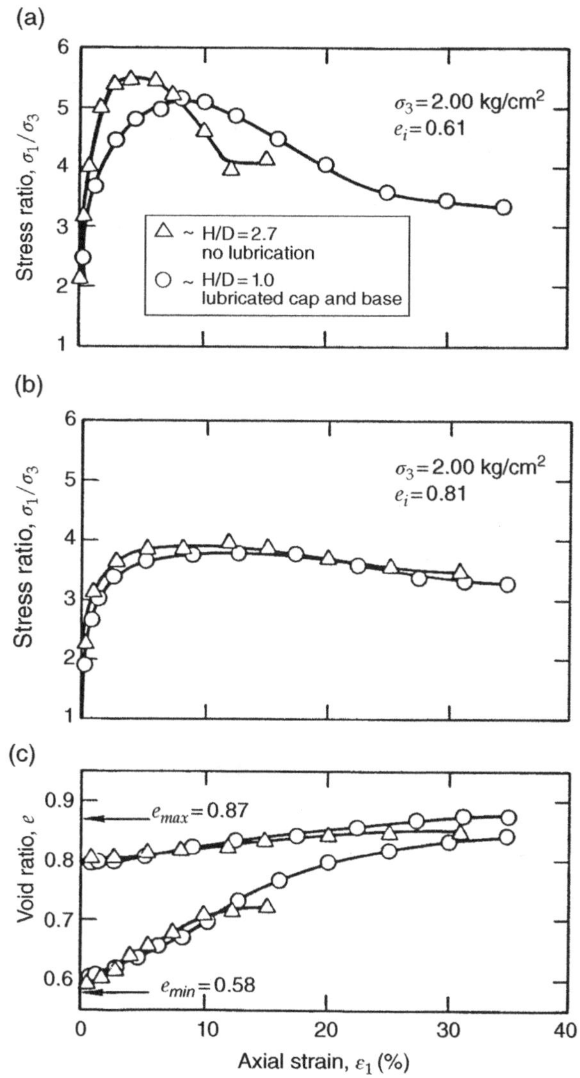

Figure 2.29 Comparison of (a) stress–strain relations, (b) stress ratio–strain, and (c) void ratio changes in triaxial compression tests on specimens with *H/D* = 1.0 and 2.7 for dense and loose Santa Monica Beach sand (after Lade 1982a).

Since the confining pressure is rarely changed in undrained tests on fully saturated specimens, the denominator can be written as:

$$\Delta\sigma_1 - \Delta\sigma_3 = \Delta(\sigma_1 - \sigma_3) \tag{2.56}$$

The variations of $\bar{A}$ in isotropically consolidated undrained tests on normally consolidated clay and in overconsolidated clay are exemplified in Figs 2.30(b) and 2.31(b), respectively. Typically, $\bar{A}$ will first increase and then decrease for a soil that tends to dilate (dense sand, heavily overconsolidated clay) as shown in Fig. 2.31(b), or $\bar{A}$ may decrease slightly or further increase, but at a lower rate for soils that tend to contract during shear (loose sand, normally consolidated clay), as shown in Fig. 2.30(b).

The magnitudes of $\bar{A}$ at failure ($\bar{A}_f$) for fully saturated specimens depend mainly on the soil

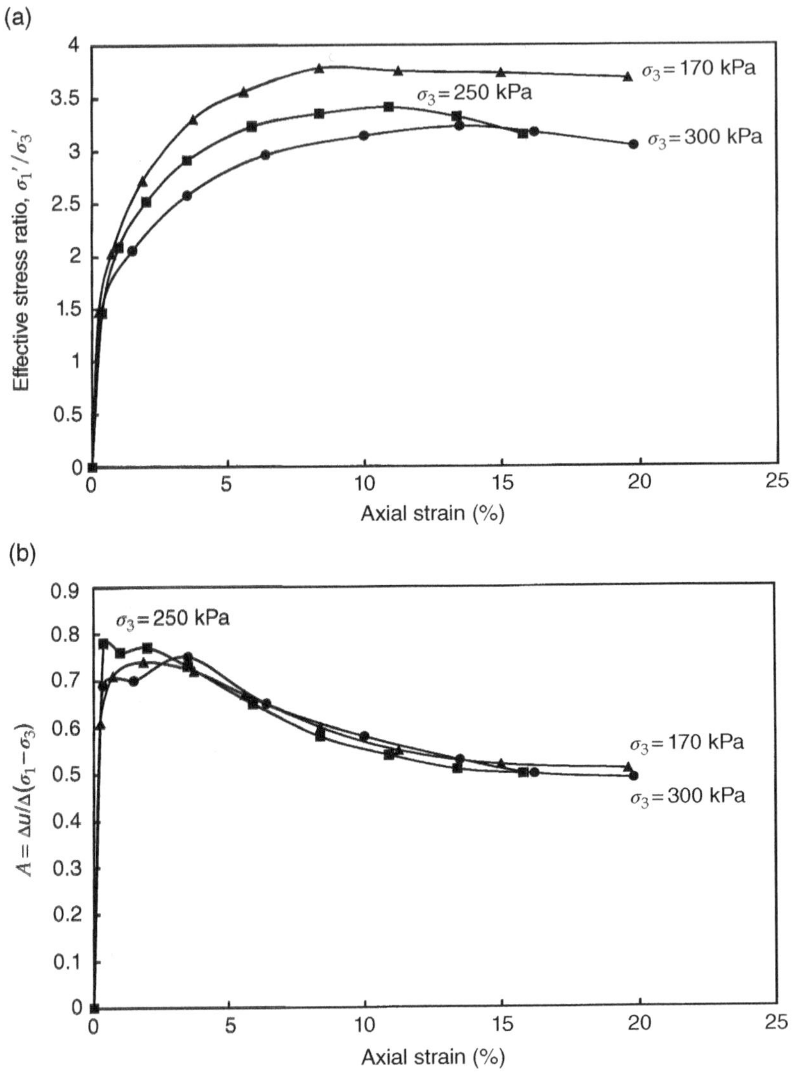

Figure 2.30 Normalized (a) stress–strain behavior and (b) pore pressure behavior of isotropically consolidated-undrained triaxial compression tests on normally consolidated Edgar Plastic Kaolinite. Same results as shown in Fig. 2.19. Note that the sequence of curves is out of order.

type and the overconsolidation ratio. The value of $\bar{A}_f$ may be greater than 1.0 for normally consolidated, sensitive clays, and loose sands tested at high confining pressures, and it may be as low as −0.5 for heavily overconsolidated clays and dense sands tested at low confining pressures. For normally consolidated, insensitive clays and lightly overconsolidated clays the value of $\bar{A}_f$ is typically between 0.5 and 1.0.

Other strain axes

In all diagrams discussed above the axial strain in triaxial compression, ε_1, has been employed on the abscissa. It may sometimes be useful to evaluate test data in terms of octahedral shear stresses and shear strains. The octahedral shear stress for triaxial compression reduces to a value proportional to the measured deviator stress

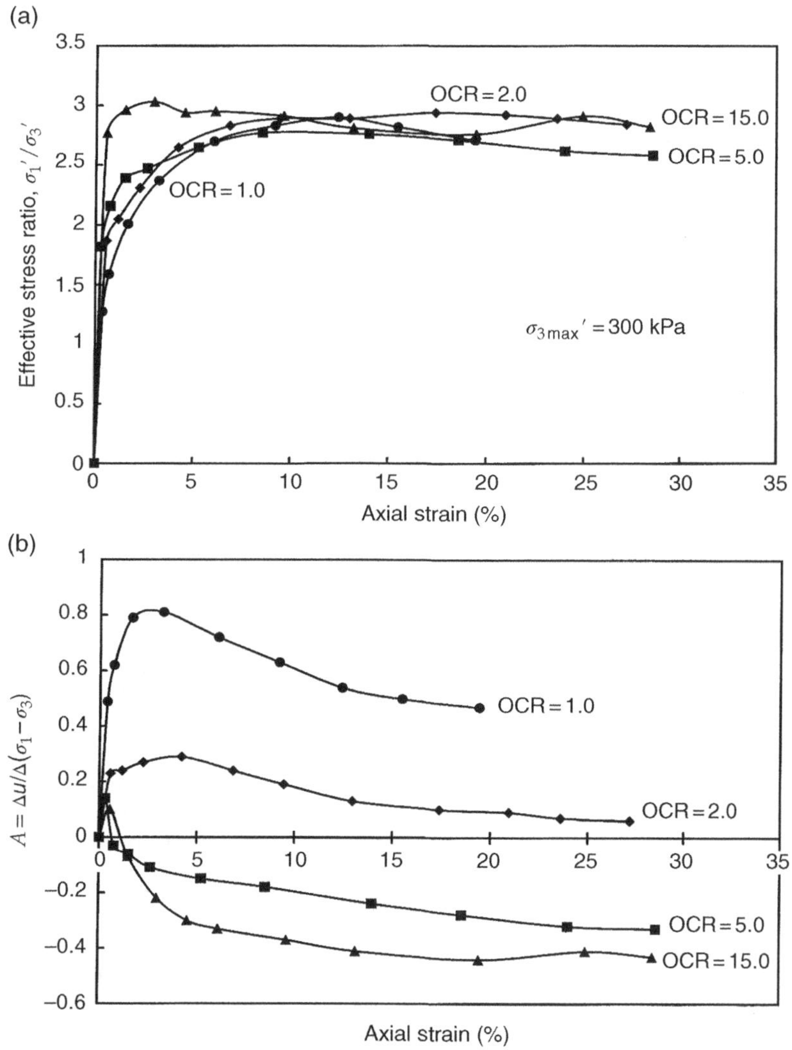

Figure 2.31 Normalized (a) stress–strain behavior and (b) pore pressure behavior of isotropically consolidated-undrained triaxial compression tests on overconsolidated Edgar Plastic Kaolinite. Same results as shown in Fig. 2.20.

(see Section 2.7). The octahedral shear strain is defined as:

$$\gamma_{oct} = \frac{2}{3} \cdot \sqrt{(\varepsilon_1 - \varepsilon_2)^2 + (\varepsilon_2 - \varepsilon_3)^2 + (\varepsilon_3 - \varepsilon_1)^2} \tag{2.57}$$

For triaxial tests in which $\varepsilon_2 = \varepsilon_3$ (this is correct for isotropic or cross-anisotropic specimens with vertical symmetry axis):

$$\gamma_{oct} = \frac{2\sqrt{2}}{3} \cdot (\varepsilon_1 - \varepsilon_3) \tag{2.58}$$

in which ε_3 may be determined from Eq. (2.17) for large strains and from Eq. (2.18) for small strains.

The maximum shear strain occurring in the triaxial specimen may be determined from Mohr's circle for strains shown in Fig. 2.32:

$$\gamma_{max} = (\varepsilon_1 - \varepsilon_3) \tag{2.59}$$

This quantity may also be used on the abscissa of the stress–strain diagram.

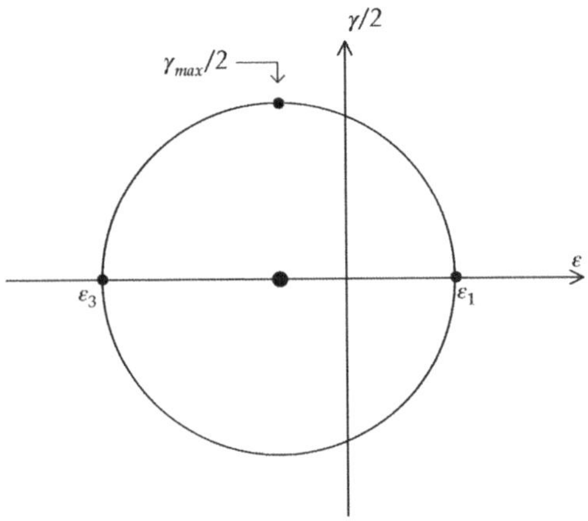

Figure 2.32 Mohr's circle for strains in a triaxial specimen with volumetric expansion.

Another expression for the shear strain emerges from calculation of the amount of work required to shear a triaxial specimen. The increment in work input per unit volume of the specimen is expressed as:

$$dW = \sigma_1 \cdot d\varepsilon_1 + \sigma_2 \cdot d\varepsilon_2 + \sigma_3 \cdot d\varepsilon_3 \quad (2.60)$$

For $\sigma_2 = \sigma_3$ and $d\varepsilon_2 = d\varepsilon_3$ this equation may be reduced and expressed in terms of quantities determined in the triaxial test:

$$dW = \sigma_m \cdot d\varepsilon_v + (\sigma_1 - \sigma_3) \cdot \frac{2}{3} \cdot (d\varepsilon_1 - d\varepsilon_3) \quad (2.61)$$

The second term expresses the distortional work and may be conveniently obtained from a diagram of deviator stress versus shear strain defined as:

$$\gamma = \frac{2}{3} \cdot (\varepsilon_1 - \varepsilon_3) \quad (2.62)$$

All three measures of shear strain are proportional to $(\varepsilon_1 - \varepsilon_3)$. For undrained tests on saturated soil the volumetric strain is zero ($\varepsilon_v = 0$). Using the small strain expression for ε_3 in Eq. (2.18), the three expressions for shear strain become:

$$\gamma_{oct} = \sqrt{2} \cdot \varepsilon_1 \quad (2.63)$$

$$\gamma_{max} = \frac{3}{2} \cdot \varepsilon_1 \quad (2.64)$$

$$\gamma = \varepsilon_1 \quad (2.65)$$

Since the shear strain in all cases are proportional to the axial strain, ε_1, the principal stress–principal strain diagrams for undrained tests with $\varepsilon_v = 0$ can also be considered to be shear stress–shear strain diagrams, which may be used for determination of shear moduli and distortional work.

2.2.4 Normalized stress–strain behavior

The derived effective stress ratio–strain and $\bar{A}$–strain diagrams discussed above are in a sense normalized diagrams in which the deviator stress and the pore pressure have been normalized on the bases of the *current* effective confining pressure σ_3' and the *current* deviator stress increment, respectively. However, the term "normalized stress–strain behavior" has gained particular acceptance in connection with tests on clay (Ladd and Foott 1974). For consolidated-undrained triaxial tests on clay with constant OCR, the deviator stress and the pore pressure at a given strain are proportional to the *initial* isotropic consolidation pressure, σ_c'. Normalization of such test data on the basis of σ_c' therefore produces almost identical stress–strain and pore pressure curves, as shown in Fig. 2.33 for normally consolidated (OCR = 1) and in Fig. 2.34 for overconsolidated (OCR > 1) Edgar Plastic Kaolinite.

Similar but different normalized stress–strain behavior may be obtained from drained tests on clay.

In fact, the normalized stress–strain curves are not completely identical. For a given value of OCR they form a pattern of decreasing maximum normalized deviator stress and decreasing maximum normalized pore pressure with increasing consolidation pressure. This consistent pattern of behavior relates to the curved failure envelope obtained for most soils.

Normalized stress–strain diagrams, in which the *initial* consolidation pressure, σ_c', are employed for normalization of the deviator stress, are particularly useful for evaluation of the undrained behavior of clay. Diagrams in which the effective stress ratio has been obtained from the *current* effective confining pressure are

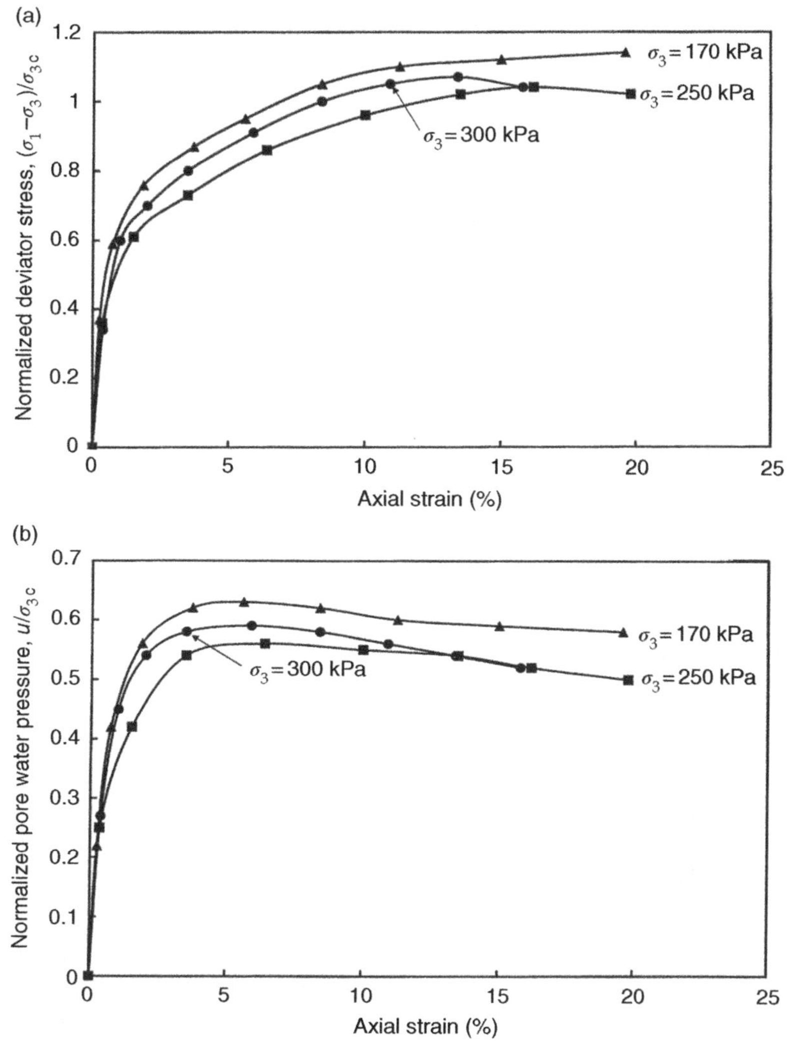

Figure 2.33 Normalized (a) stress–strain behavior and (b) pore pressure behavior of isotropically consolidated-undrained triaxial compression tests on normally consolidated Edgar Plastic Kaolinite. Same results as shown in Fig. 2.19. Note that one test is out of sequence.

more useful for evaluation of soil behavior in terms of effective stresses.

Additional information regarding normalized behavior is given in Chapter 8 and may be obtained from Henkel (1960), Parry (1960), Ladd and Foott (1974), Poulos (1978), Mayne (1985, 1988), and Mayne and Stewart (1988). These articles all deal with the normalized behavior of clay. The concepts of normalization for pore pressures and volumetric strains in tests on sands are not useful, because they do not produce single sets of curves. However, pore pressures and volumetric strains in such tests do follow patterns of their own which can be recognized from the basic diagrams.

2.2.5 Patterns of soil behavior – error recognition

The patterns of stress–strain and pore pressure exhibited in derived or normalized diagrams are particularly useful for recognition of errors

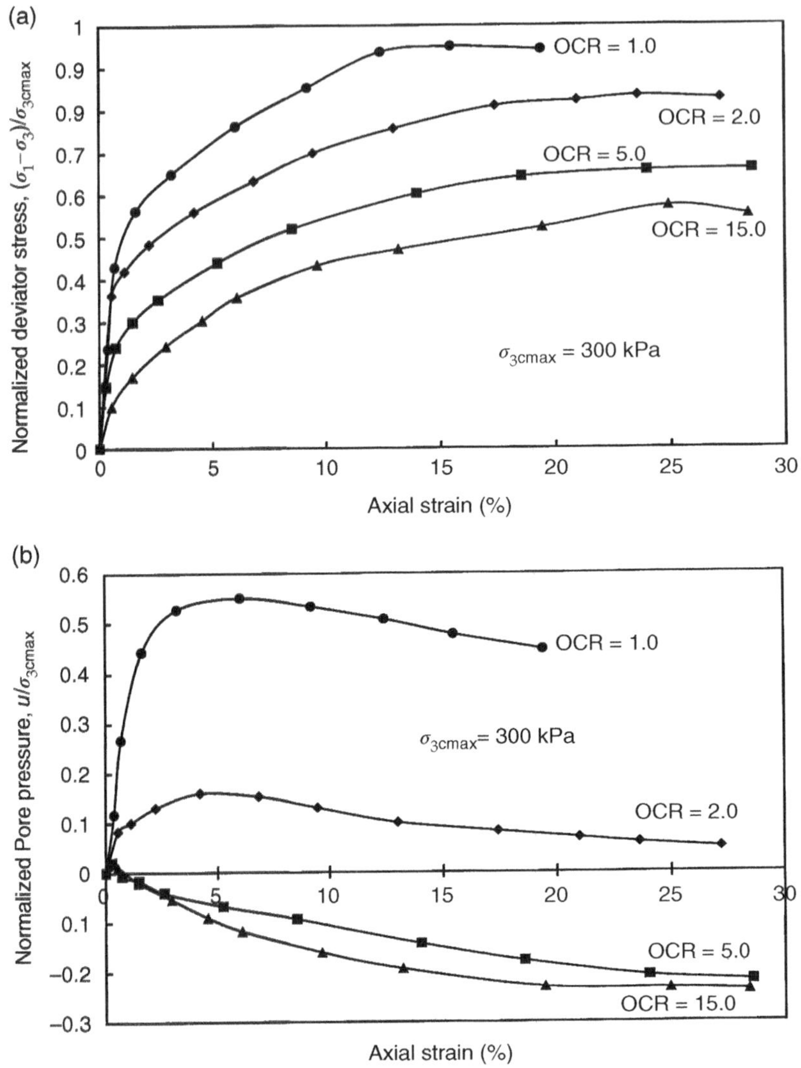

Figure 2.34 Normalized (a) stress–strain behavior and (b) pore pressure behavior of isotropically consolidated-undrained triaxial compression tests on overconsolidated Edgar Plastic Kaolinite. Same results as shown in Fig. 2.20.

or anomalies in test results. In these diagrams the curves are brought into close proximity with each other. Any deviation in the pattern can therefore easily be recognized, and closer inspection of test data that do not conform to the general pattern may be warranted. Pore pressures and volumetric strains in triaxial tests on sand do not conform to patterns that are useful to normalization, but their patterns may be observed on the basic diagrams.

Errors may also be recognized from stress paths plotted on p'-q diagrams (see Section 2.4). Stress paths are particularly sensitive at small strains and may indicate otherwise undetected experimental problems that may require attention.

There are numerous reasons why experiments may not produce satisfactory results. Natural scatter in results from tests on specimens from the field can never be discounted. Gross deviations from expected soil behavior patterns may be anticipated on the basis of specimen inspection during and after preparation as well as after the test has been completed (see Chapters 5 and 8). But errors caused by testing techniques and procedures may be eliminated if they are recognized. It is therefore desirable to identify the reason(s) why a particular triaxial test apparently produced anomalous results. The possibilities for errors are too numerous to list, but an example is shown in Fig. 2.33 to demonstrate the usefulness of derived and normalized diagrams in error detection. The patterns of stress–strain curves clearly indicate anomalous behavior for one test in the diagram.

Note that effects of systematic errors, such as end restraint due to friction on the cap and base, cannot be discovered from normalization, because such errors will tend to superimpose their own systematic and variable effects on the test data.

2.3 Strength diagrams

2.3.1 Definition of effective and total strengths

The strengths obtained from triaxial tests may be expressed as the maximum deviator stress or it may be evaluated in terms of the maximum effective stress ratio. These two quantities are most often employed. They are synonymous and occur simultaneously in drained tests. In undrained tests they usually do not occur at the same time, as shown in Fig. 2.35, and they have different meanings and play different roles in relation to analyses of geotechnical structures. Typically, the maximum effective stress ratio is employed for definition of the effective strength envelope, and the maximum deviator stress is used for determination of the undrained strength. However, these definitions are not entirely consistent.

A failure envelope or surface is most consistently defined as "the locus of stress points defining a line, curve, or surface which separates the stresses that can be reached from the stresses that cannot be reached for a given material." Thus, the outermost surface, as measured away from the hydrostatic state of stress, is implied by this definition. This requires that the peak point on the stress–strain curve is used for location of the failure surface.

Some soils may not exhibit a peak on the stress–strain curve. For such cases the strength may be defined arbitrarily according to some rule which may or may not relate to the particular geotechnical structure to be analyzed. Figure 2.36 shows examples of possible definitions of strength which may be employed for soils whose stress–strain curves do not exhibit peak stress values.

Having determined the strengths from individual tests, these strengths are combined to form a failure envelope for the soil. The failure envelope is then described by the mathematical expression which best approximates the envelope or which is easiest to use in analyses of the geotechnical structure, or both.

2.3.2 Mohr–Coulomb failure concept

Coulomb (1776) proposed in 1776 that the shear strength of a soil can be expressed in the form:

$$\tau = c + \sigma \cdot \tan\varphi \qquad (2.66)$$

in which τ is the maximum shear stress and σ is the normal stress on the failure plane. The parameter c is the cohesion and φ is the friction angle of the soil, both empirical constants to be determined from results of experiments. This empirical relation forms a straight line on a τ–σ diagram as shown in Fig. 2.37.

In 1882, Otto Mohr (1882) presented the concept of Mohr's circle according to which the state of stress on any 2D or three-dimensional (3D) body can be expressed graphically as reviewed for 2D conditions in Section 2.1.7 and indicated for 3D conditions in Fig. 2.38.

The two concepts of stress at failure, by Coulomb and by Mohr, may be combined by plotting the Mohr circles corresponding to stresses at failure and drawing a best-fit straight line tangential to the Mohr circles. This straight line, which forms an envelope to the circles as shown in Fig. 2.39, is then expressible by Coulomb's criterion.

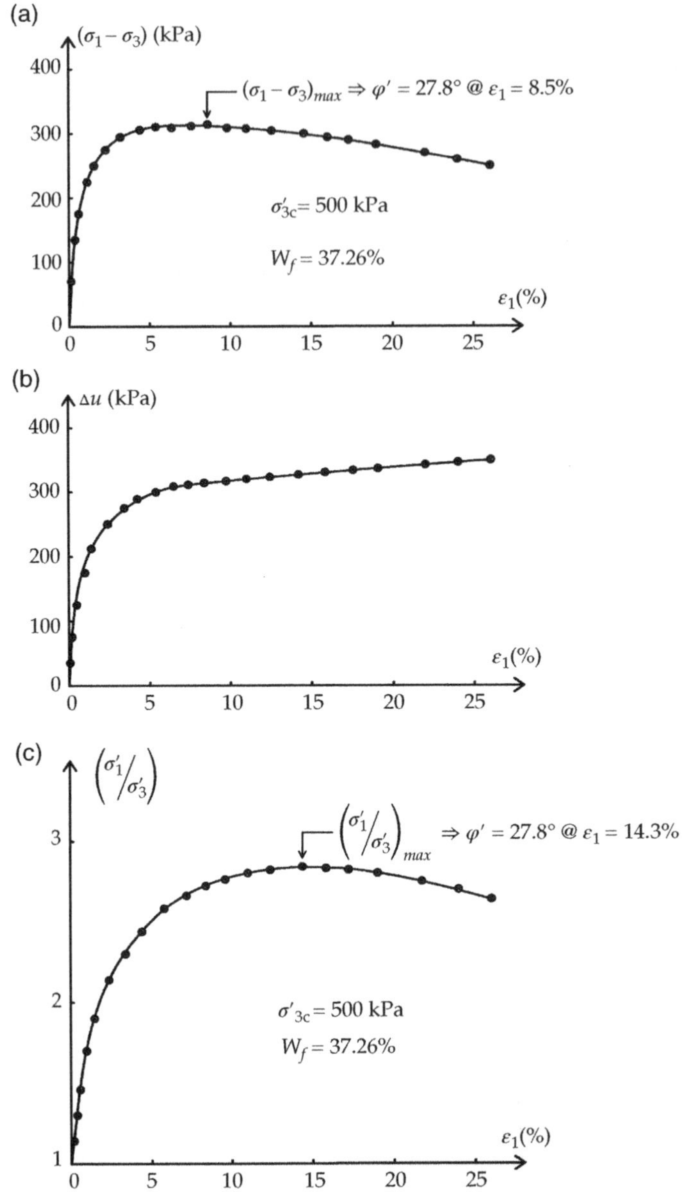

Figure 2.35 (a) Stress–strain, (b) pore pressure and (c) stress ratio–strain relations from undrained triaxial compression tests on Edgar Plastic Kaolinite. Note that the values of $(\sigma_1 - \sigma_3)_{max}$ and $(\sigma_1/\sigma_3)_{max}$ occur at different axial strains.

This constitutes the origin of the Mohr–Coulomb failure criterion. It is derived entirely from observations, it is empirical, it often fits the experimental data well, and it provides a simple representation of failure in soils which is employed in numerous stability analysis procedures.

The intermediate principal stress σ_2 is located between σ_1 and σ_3, and the two smaller circles shown in Fig. 2.38 are below the large circle spanning over the diameter $(\sigma_1-\sigma_3)$. The smaller circles do not touch or push the failure envelope further out, and this suggests, in accordance with the Mohr–Coulomb failure concept,

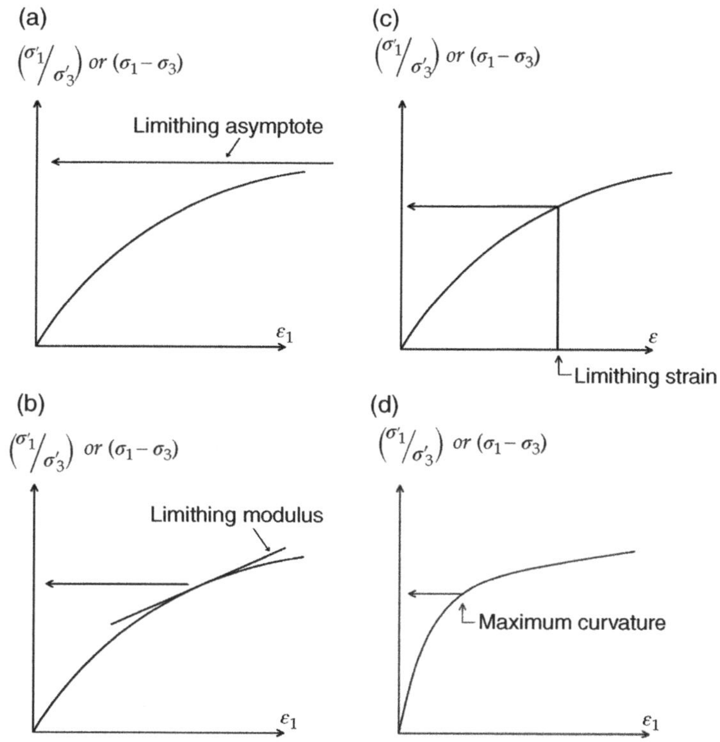

Figure 2.36 Possible definitions of strengths of soils without peak stress: (a) limiting asymptote; (b) limiting modulus; (c) limiting strain; and (d) at maximum curvature.

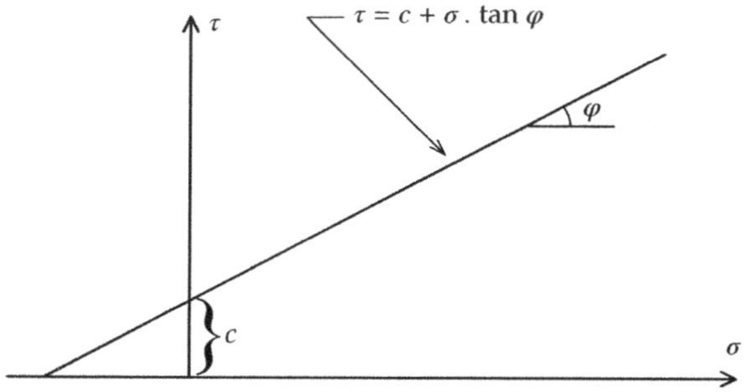

Figure 2.37 Coulomb's failure criterion for soils shown on a σ–τ diagram.

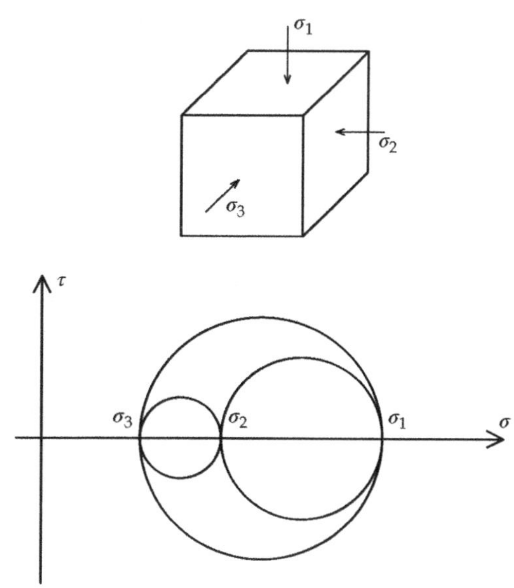

Figure 2.38 Mohr's circle for 3D state of stress (not necessarily at failure).

that the intermediate principal stress has no influence on the strength of soils. Therefore, the intermediate principal stress is often ignored and only the deviator stress, $(\sigma_1 - \sigma_3)$, obtained from triaxial compression tests in which $\sigma_2 = \sigma_3$, is considered. Experimental results, however, show a marked influence of the intermediate principal stress on both stress–strain behavior and strength. This will be discussed in Chapter 11.

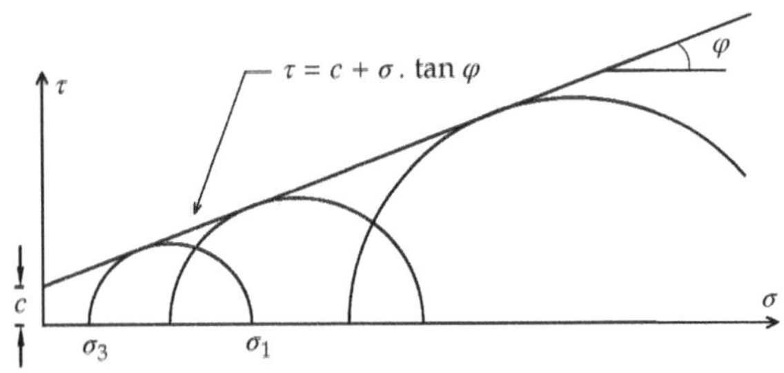

Figure 2.39 Mohr–Coulomb failure criterion for soils.

2.3.3 Mohr–Coulomb for triaxial compression

Failure is associated with a particular plane inclined at the angle α with the σ_1–plane, as indicated in Fig. 2.40(a). Only on this plane do the stresses represent a point on the failure envelope, as shown in Fig. 2.40(b). This may be confirmed by consideration of the origin of planes (see Section 2.1.7). The value of α may then be determined as seen in Fig. 2.40(b). Actually, an infinite number of planes forming angles of $45°+\varphi/2$ with the σ_1–plane are equally critical, and more than one failure plane may form in a specimen.

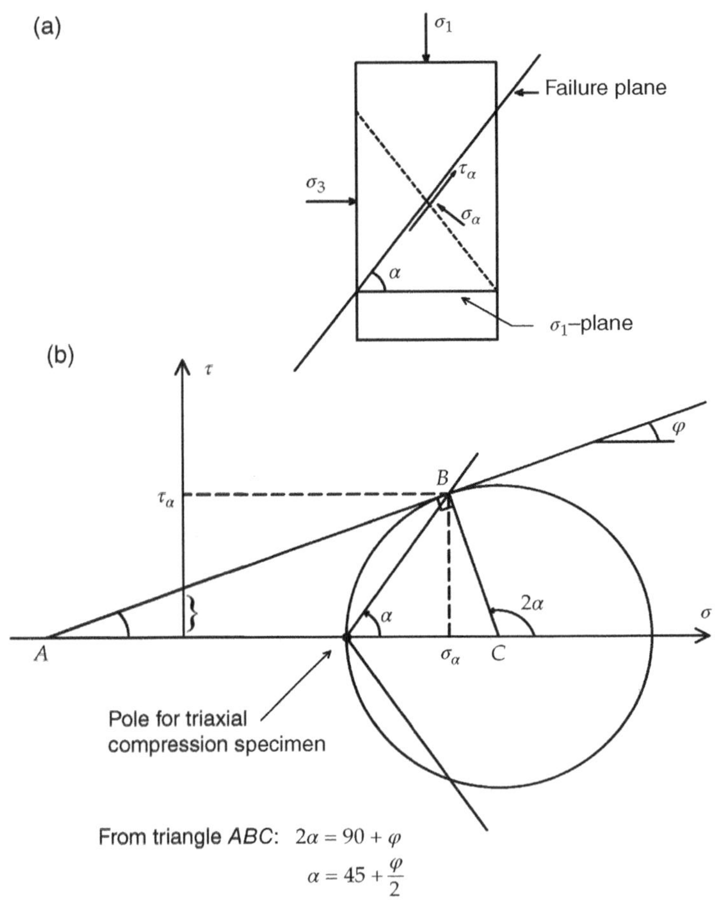

Figure 2.40 (a) Triaxial specimen with failure plane and (b) stresses on failure plane and orientation of failure plane in a triaxial specimen.

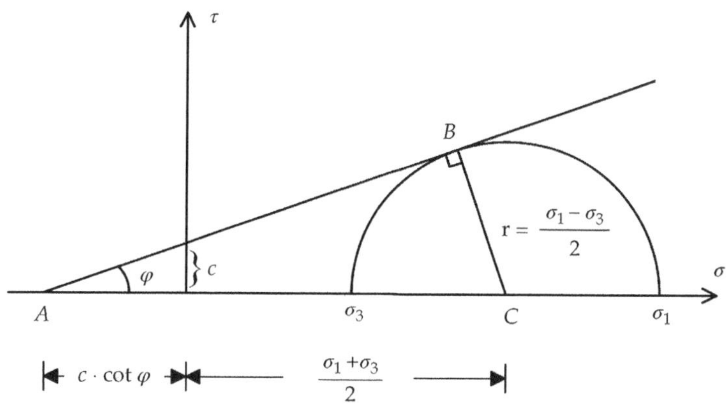

From triangle ABC :

$$r = \frac{\sigma_1 - \sigma_3}{2} = \left(\frac{\sigma_1 + \sigma_3}{2} + c \cdot \cot\varphi\right) \cdot \sin\varphi$$

$$\sigma_1 - \sigma_3 = (\sigma_1 + \sigma_3) \cdot \sin\varphi + 2c \cdot \cos\varphi$$

$$\sigma_1 = \sigma_3 \cdot \frac{1 + \sin\varphi}{1 - \sin\varphi} + 2c \cdot \frac{\cos\varphi}{1 - \sin\varphi}$$

Figure 2.41 Derivation of the Mohr–Coulomb failure criterion.

From geometry of the Mohr circle and the failure surface, the Mohr–Coulomb failure criterion may be derived as shown in Fig. 2.41:

$$\sigma_1 = \sigma_3 \cdot \frac{1+\sin\varphi}{1-\sin\varphi} + 2c \cdot \frac{\cos\varphi}{1-\sin\varphi} \qquad (2.67)$$

or by trigonometric substitution:

$$\sigma_1 = \sigma_3 \cdot \tan^2(45° + \varphi/2) + 2c \cdot \tan(45° + \varphi/2) \qquad (2.68)$$

To determine the material parameters c and φ, three or more triaxial compression tests are performed. The results are plotted in a Mohr diagram, and the failure envelope is drawn to find c and φ, as shown in Fig. 2.42.

The values of c and φ are not soil properties, but simply convenient parameters used for expressing the variation of shear strength with normal stress according to Eqs (2.66), (2.67), and (2.68). The straight line which provides the best representation and therefore the best choices of c and φ depends on:

1. The soil, the fabric of its intact grain structure, its initial dry density and water content

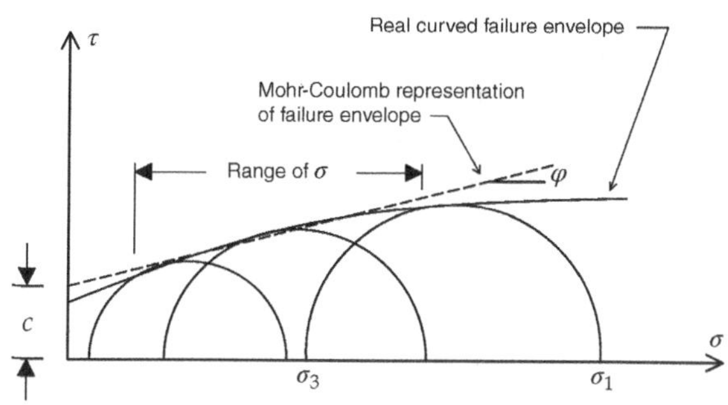

Figure 2.42 Determination of c and φ from results of three triaxial compression tests.

2. The test type
3. The range of pressures in the field (because the real failure envelop is curved as indicated in Fig. 2.42)

Each of these items should therefore be reproduced correctly in the triaxial tests.

2.3.4 Curved failure envelope

There is much evidence to show that real failure envelopes for soils are curved over a large range of normal stresses. Figure 2.43 shows a schematic

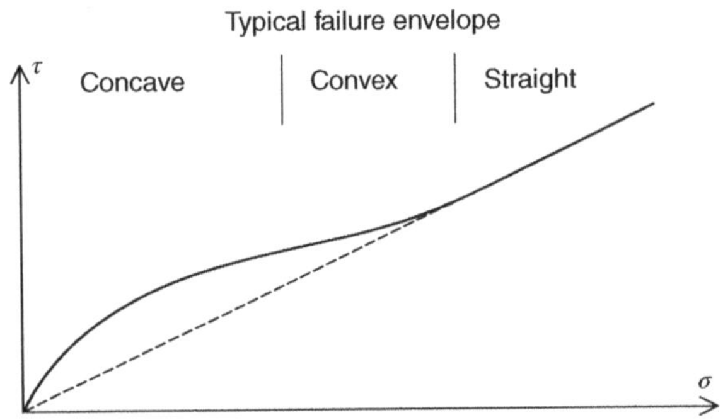

Figure 2.43 Typical failure envelope with concave, convex, and straight sections.

diagram of a typical failure envelope over a wide range of normal stresses. The failure envelope is concave towards the σ-axis in the range from low to medium high normal stresses. At higher stresses it becomes convex and then it straightens out to become linear with a constant slope through the origin. The normal stress values at the transition between these sections of the failure envelope depend mainly on the relative density of the soil and the strength of the grains. Large amounts of particle crushing occur at high stresses where the failure envelope is linear. The normal stresses of engineering significance are most often limited to the range in which the envelope is concave towards the σ-axis.

Whereas the Mohr–Coulomb criterion may describe the failure envelope with sufficient accuracy in a limited range of normal stresses, a better description may be obtained by fitting a power function of the following type:

$$\tau = a \cdot p_a \cdot \left(\frac{\sigma + d \cdot p_a}{p_a} \right)^b \qquad (2.69)$$

in which a (>0), b ($0 \le b \le 1$), and d (≥0) are dimensionless constants, and p_a is atmospheric pressure in the same units as τ and σ. The value of $d \cdot p_a$ expresses the tensile strength of the soil. For $b = 0$ the envelope is horizontal, and $\tau = a \cdot p_a$ indicates the constant shear strength. For b=1, Eq. (2.69) reduces to the Mohr–Coulomb failure criterion with $\tan\varphi = a$ and $c = d \cdot p_a \cdot \tan\varphi$. For $0 < b < 1$ the failure envelope is curved with a vertical tangent at the τ-axis. Thus, the empirical expression in Eq. (2.69) may be suitable for description of the curved failure envelope on the Mohr diagram, and it is convenient to use for stability analyses in which the shear stress is described as a function of the normal stress on the failure plane.

Since τ and σ are stresses on the failure plane in the triaxial specimen, they correspond to the stresses at which the Mohr circle touches the failure envelope. These stresses are not measured in the triaxial test. However, parameter determination does not appear to be very sensitive to accurately estimated values of τ and σ. This is indicated by the following example.

The material parameters for Eq. (2.69) are determined by taking the logarithm on both sides:

$$\log\left(\frac{\tau}{p_a}\right) = \log(a) + b \cdot \log\left(\frac{\sigma + d \cdot p_a}{p_a}\right) \qquad (2.70)$$

This expression describes a straight line in a log–log diagram. The values of τ/p_a and $(\sigma + d \cdot p_a)/p_a$ are therefore plotted on a log–log diagram and the best fitting straight line is located. The intercept of this line with $(\sigma + d \cdot p_a)/p_a = 1$ is the value of a, and b is the geometric slope of the line.

The values of τ and σ to be used for plotting may be estimated from the Mohr circles on a τ–σ diagram. Figure 2.44 shows an example for a soil without tensile strength (d = 0). The best estimates of τ and σ are obtained by sketching a curved envelope to the circles and reading the values of τ and σ at or nearest the tangent points, as indicated at points B in Fig. 2.44. To evaluate the sensitivity of the parameter determination, two additional points on each circle (A and C) have been indicated and used in the following procedure.

The values of τ/p_a and σ/p_a listed in Fig. 2.44 are then plotted on the log–log diagram in Fig. 2.45. Although points B for the three circles represent the best estimate of the tangent points, Fig. 2.45 indicates that points A and C are also located close to the straight line on this diagram. The procedure for parameter determination is

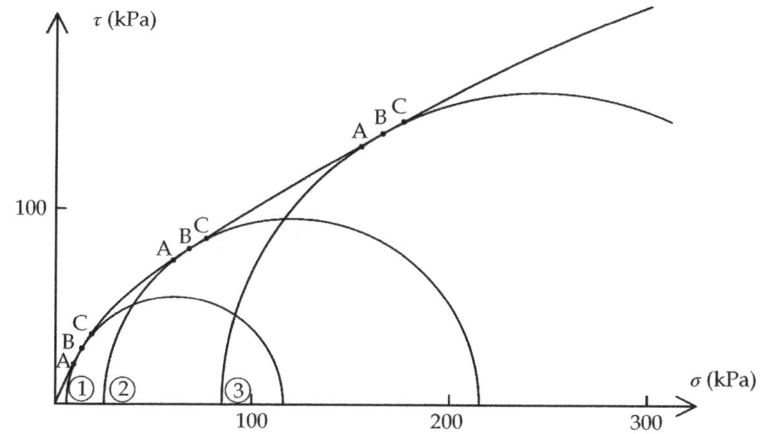

Circle	Point	τ/p_a	σ/p_a
1	A	0.22	0.08
	B	0.32	0.14
	C	0.42	0.20
2	A	0.68	0.50
	B	0.76	0.60
	C	0.84	0.70
3	A	1.38	1.58
	B	1.44	1.70
	C	1.50	1.82

Figure 2.44 Locations of optimal points on Mohr's circles for fitting curved failure envelope.

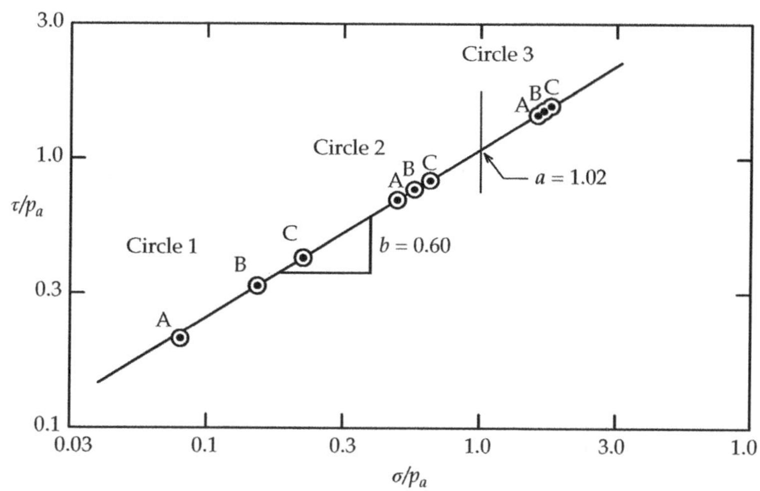

Figure 2.45 Log–log diagram of τ/p_a versus σ/p_a for determining the best fitting curved failure envelope.

therefore not very sensitive to accurate estimates of τ and σ.

For soils with tensile strengths the procedure for parameter determination involves a regression analysis in which $d \cdot p_a$ is added to σ before plotting on the log–log diagram. The value of $d \cdot p_a$ may be determined as that value which produces the best fit straight line corresponding to the highest coefficient of determination, r^2. This procedure is presented in more detail by Lade (1982b) and by Kim and Lade (1984).

Whereas the expression in Eq. (2.69) has been used successfully (see e.g., Dusseault and Morgenstern 1978; Lade 2010), other expressions for curved failure envelopes have been proposed (e.g., Baligh 1976; Lade 1977; Hoek and Brown 1980; Maksimovic 1989). Each of these expressions has its own advantages and limitations.

2.3.5 MIT *p–q* diagram

Often more than three tests are performed on a soil, and it becomes quite difficult to pass a best-fit failure envelope tangentially to all Mohr circles. Figure 2.46 shows an example of a series of Mohr circles from triaxial tests. Passing a straight line tangential to some and through other circles to obtain a best fit with all test data is not easy.

It is much easier to fit a line through a series of points, even if they are scattered. For this purpose each Mohr circle may be represented by its top point, which has the coordinates (see Fig. 2.47):

$$q = \frac{\sigma_a - \sigma_r}{2} \quad \left(= \frac{\sigma_1 - \sigma_3}{2} \text{ for triaxial compression} \right) \tag{2.71}$$

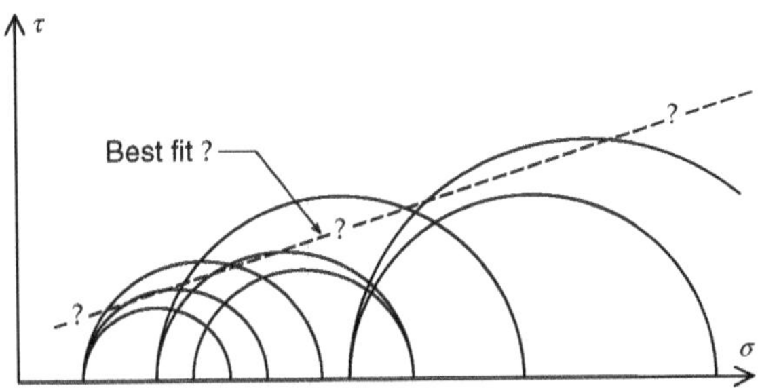

Figure 2.46 Mohr's circles obtained from a series of triaxial compression tests. Location of best fit failure envelope is difficult.

$$p = \frac{\sigma_a + \sigma_r}{2} = \frac{\sigma_a - \sigma_r}{2} + \sigma_r = q + \sigma_r \quad (= q + \sigma_3 \text{ for triaxial compression}) \tag{2.72}$$

These top points are then plotted on a p–q diagram as shown in Fig. 2.48. The expression for the Mohr–Coulomb failure criterion in terms of p and q is derived in Fig. 2.47:

$$q = p \cdot \sin\varphi + c \cdot \cos\varphi \tag{2.73}$$

In this expression $\sin\varphi$ represents the slope and $c \cdot \cos\varphi$ represents the intercept of the straight line with the τ-axis. Determination of c and φ is demonstrated in Fig. 2.48.

The best-fitting straight line on the p–q diagram may be determined by a linear regression analysis. The correct procedure for a regression analysis on a p–q diagram is reviewed in Section 2.5.

The p–q diagram discussed here is often called the modified Mohr diagram. However, the employment of this diagram for displaying stress paths as well as strength data has been popularized largely by the MIT soil mechanics group (Lambe 1964; Lambe and Whitman 1979), and to distinguish it from the p–q diagram discussed below, it will be referred to here as the

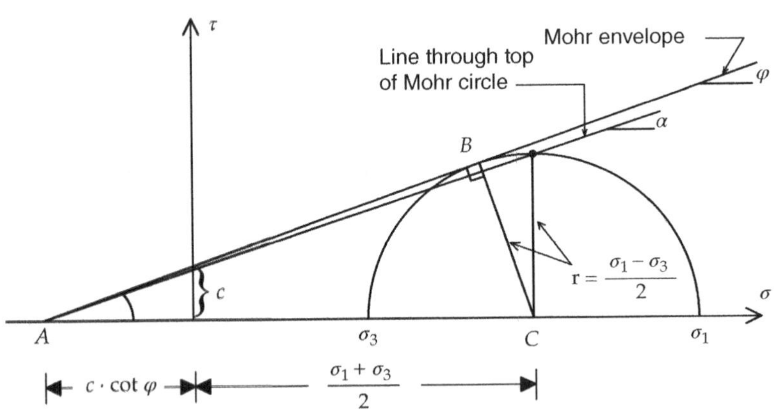

From triangle ABC:

$$r = \frac{\sigma_1 - \sigma_3}{2} = \left(\frac{\sigma_1 + \sigma_3}{2} + c \cdot \cot\varphi \right) \cdot \sin\varphi$$

$$\frac{\sigma_1 - \sigma_3}{2} = \left(\frac{\sigma_1 + \sigma_3}{2} \right) \cdot \sin\varphi + c \cdot \cos\varphi$$

$$q = p \cdot \sin\varphi + c \cdot \cos\varphi$$

Figure 2.47 Derivation of the expression for the Mohr–Coulomb failure criterion in terms of p and q.

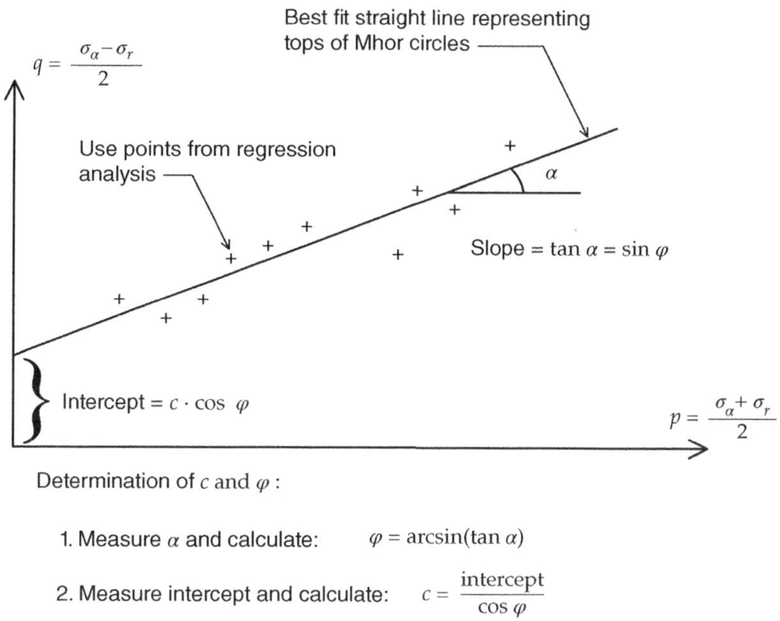

Figure 2.48 MIT p–q diagram with determination of c and ϕ for the results of a series of triaxial compression tests.

MIT p–q diagram. This diagram has also been labeled the t–s diagram in which $t = p$ and $s = q$ (e.g., Wood 1990).

2.3.6 Cambridge p–q diagram

Another type of p–q diagram has been used extensively and popularized by the Cambridge soil mechanics group (see e.g., Schofield and Wroth 1968). Here the definitions of q and p are:

$$q = (\sigma_a - \sigma_r) \quad [= (\sigma_1 - \sigma_3) \text{for triaxial compression}] \tag{2.74}$$

$$p = \frac{1}{3}(\sigma_a + 2\sigma_r) = \frac{1}{3}[(\sigma_a - \sigma_r) + 3\sigma_r] = \frac{1}{3}q + \sigma_r \quad \left(= \frac{1}{3}q + \sigma_3 \text{for triaxial compression}\right) \tag{2.75}$$

Thus, p represents the mean normal stress and q represents the stress difference or the deviator stress in the triaxial test. Figure 2.49 shows a diagram of q versus p for triaxial compression and extension conditions. The straight line failure envelope intersects the p-axis at $p = -c \cdot \cot\varphi$, that is the same intercept value as obtained from the Mohr diagram. The slope of the line is indicated by M (M is taken to be positive for both compression and extension), and the failure criterion in this diagram becomes:

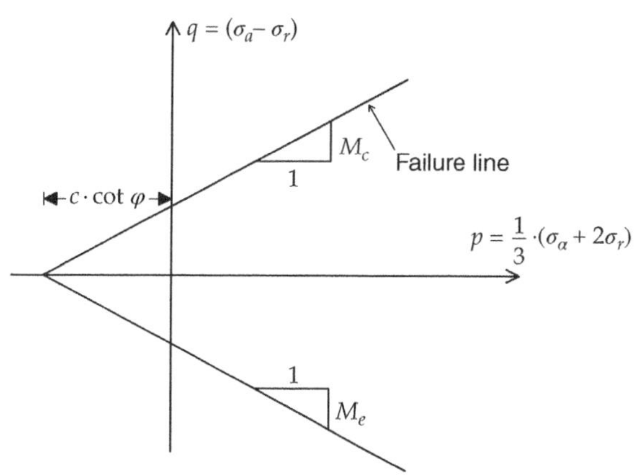

Figure 2.49 Cambridge p–q diagram with determination of c and φ for the results of a series of triaxial compression tests.

$$q = M \cdot (p + c \cdot \cot\varphi) \tag{2.76}$$

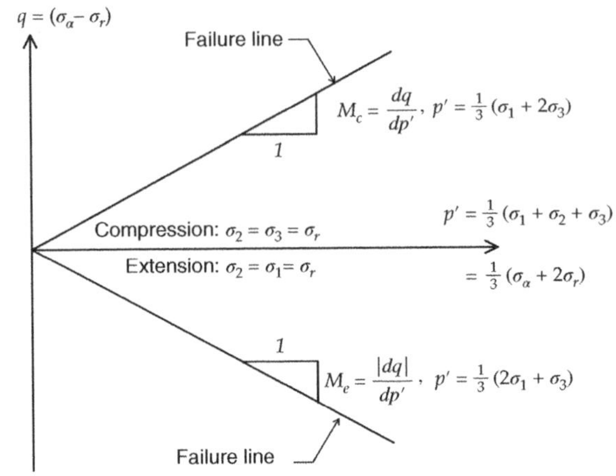

$$M_c = \frac{dq}{dp'} = \frac{(\sigma_1 - \sigma_3)}{\frac{1}{3}\cdot(\sigma_1 + 2\sigma_3)} = \frac{3\cdot(\sigma_1 - \sigma_3)\cdot 2}{(\sigma_1 + 2\sigma_3)\cdot 2} = \frac{6\cdot(\sigma_1 - \sigma_3)}{3\cdot(\sigma_1 + \sigma_3) - (\sigma_1 - \sigma_3)}$$

$$M_c = \frac{6\cdot\left(\frac{\sigma_1 - \sigma_3}{\sigma_1 + \sigma_3}\right)}{3 - \frac{\sigma_1 - \sigma_3}{\sigma_1 + \sigma_3}} = \frac{6\cdot\sin\varphi_c}{3 - \sin\varphi_c} \Rightarrow \sin\varphi_c = \frac{3\cdot M_c}{6 + M_c}$$

$\varphi = 90° : M_{cmax} = 3$

$M_c = 1 \ : \varphi = 25.4°$

$$M_e = \frac{|dq|}{dp'} = \frac{\sigma_1 - \sigma_3}{\frac{1}{3}\cdot(2\sigma_1 + \sigma_3)} = \frac{3\cdot(\sigma_1 - \sigma_3)\cdot 2}{(2\sigma_1 + \sigma_3)\cdot 2} = \frac{6\cdot(\sigma_1 - \sigma_3)}{3\cdot(\sigma_1 + \sigma_3) + (\sigma_1 - \sigma_3)}$$

$$M_e = \frac{6\cdot\left(\frac{\sigma_1 - \sigma_3}{\sigma_1 + \sigma_3}\right)}{3 + \frac{\sigma_1 - \sigma_3}{\sigma_1 + \sigma_3}} = \frac{6\cdot\sin\varphi_e}{3 + \sin\varphi_e} \Rightarrow \sin\varphi_e = \frac{3\cdot M_e}{6 - M_e}$$

$\varphi = 90° : M_{emax} = 1.5$

$M_e = 1 : \ \varphi = 36.9°$

$\varphi = 25.4° : M_e = 0.751$

Figure 2.50 Cambridge *p–q* diagram with failure lines in compression and extension for zero cohesion.

Figure 2.50 shows a Cambridge *p–q* diagram with failure lines in compression and extension for a soil without cohesion. The slopes of the failure lines are indicated by *M* and these are related to the friction angles in compression, φ_c, and extension, φ_e, as follows (see derivations in Fig. 2.50):

For triaxial compression:

$$M_c = \frac{6\cdot\sin\varphi_c}{3 - \sin\varphi_c} \tag{2.77}$$

from which

$$\varphi_c = \arcsin\left(\frac{3\cdot M_c}{6 + M_c}\right) \tag{2.78}$$

For triaxial extension:

$$M_e = \frac{6\cdot\sin\varphi_e}{3 + \sin\varphi_e} \tag{2.79}$$

from which

$$\varphi_e = \arcsin\left(\frac{3\cdot M_e}{6 - M_e}\right) \tag{2.80}$$

The Cambridge *p–q* diagram is not directly related to the Mohr diagram, because different definitions for *p* are used in the two diagrams.

2.3.7 Determination of best-fit soil strength parameters

The correct way to determine the best-fit soil strength parameters depends on the orientation of the stress paths, which is reviewed below. This method is therefore examined at the end of Section 2.5.

2.3.8 Characterization of total strength

The total strength is best characterized by relating the undrained shear strength (top of Mohr's circle) to the consolidation pressure. The reason for this unambiguous characterization is explained in Section 2.4.2. This may be done for both normally consolidated and overconsolidated clays. For normally consolidated clays the ratio between the undrained shear strength, s_u, and the vertical overburden pressure, σ_c', is constant (i.e., $s_u/\sigma_c'=$ constant), while such a simple relationship is not observed for overconsolidated clays.

2.4 Stress paths

Both types of p–q diagrams may be used to show the stress path followed in a triaxial test. For this purpose, values of p and q are calculated for selected data sets from Eqs (2.71) and (2.72) for the MIT p–q diagram and from Eqs (2.74) and (2.75) for the Cambridge p–q diagram. After the points have been plotted in the respective p–q diagrams, they may be connected with smooth curves or straight lines to indicate the stress path followed in the test.

Any stress path that can be followed by variation of cell pressure and deviator stress in triaxial tests may be displayed in either of the two types of p–q diagrams. This includes stress paths produced during isotropic and anisotropic consolidation as well as during drained and undrained shearing.

2.4.1 Drained stress paths

Stress paths for drained triaxial compression and extension tests performed with constant cell pressure or with constant axial stress are shown in Fig. 2.51 for both types of p–q diagrams. The stress paths shown in the MIT p–q diagram represent the variation of the shear stresses and normal stresses on planes inclined at 45° with the principal stress axes, that is on the planes of maximum shear stresses, as indicated in Fig. 2.47. The top points on the drained stress paths correspond to failure, and these points may be combined to form a failure envelope as discussed above.

2.4.2 Total stress paths in undrained tests

Both total and effective stress paths for undrained tests may be shown in the p–q diagrams. Figure 2.52 shows examples of total and effective stress paths plotted for one conventional ICU-triaxial compression test on both types of p–q diagrams. A total stress path basically traces the states of stress measured externally to the specimen. If a back pressure is used in the test (see Chapter 6), two total stress paths may be shown: one indicates the total stresses consisting of the actual, measured values of cell pressure and deviator stress; and the other shows the total stresses reduced for back pressure.

The magnitude of the back pressure may be selected to produce full saturation in the specimen. The cell pressure, consisting of the consolidation pressure and the back pressure, is therefore not related to the in situ stress conditions, and the corresponding total stress path, as well as the point of maximum deviator stress, have no particular meaning relative to the field condition.

On the other hand, the consolidation pressures and the back pressures in each of a series of tests may be chosen to match the pre-shear consolidation pressures and pore pressures at the respective depths in the field. The separation between total and effective stress paths then equals the in situ pore pressures. In this case, the locations of the points of maximum deviator stresses have significance, because they combine to form a total stress failure envelope, which is relevant to a total stress stability analysis.

However, both uncertainty and inflexibility in analysis is associated with matching field conditions in terms of total stresses. Preferable methods of data interpretation and stability analyses involve relating the undrained shear strength to consolidation pressures, both in the tests and in the field. The locations of total stress paths for undrained tests with back pressure are therefore generally of little practical significance.

The location of the total stress path initiating at the point of consolidation (corresponding to zero back pressure) plays a role in displaying the pore pressures produced due to shearing. This is discussed below.

2.4.3 Effective stress paths in undrained tests

Of particular interest are the effective stress paths followed in undrained tests. These are

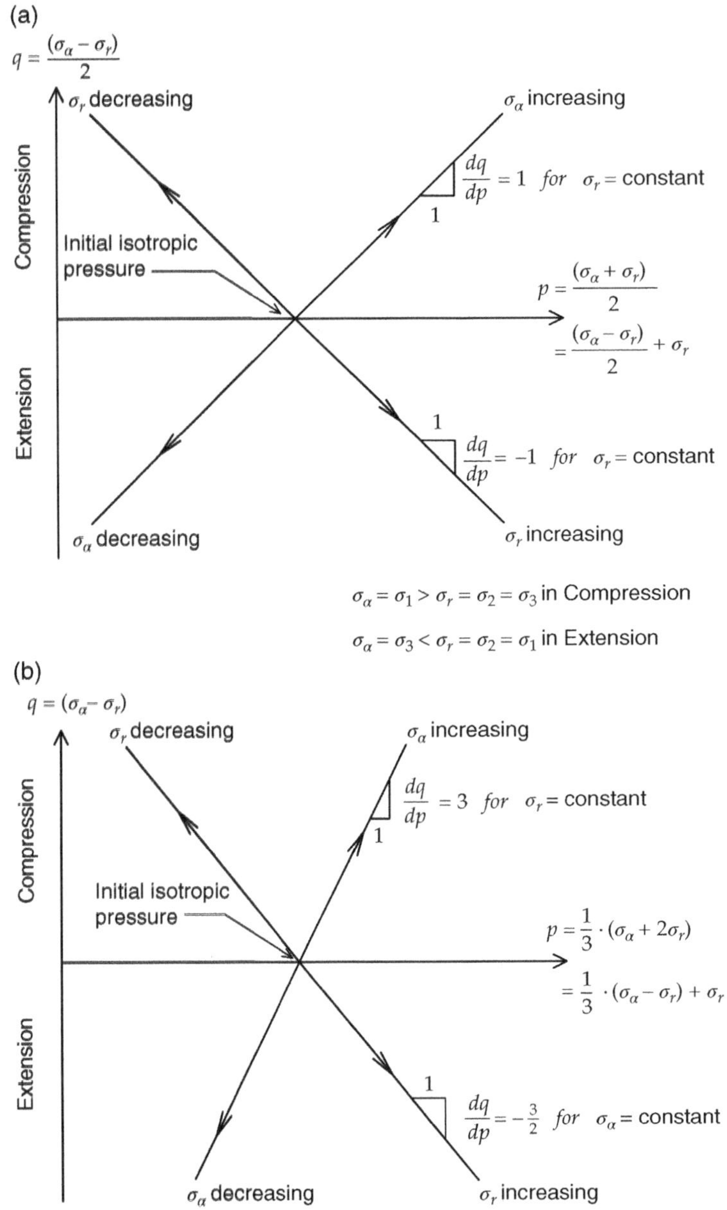

Figure 2.51 Stress paths for drained triaxial compression tests with constant cell pressure (σ_r) or with constant axial stress (σ_a) shown in (a) an MIT *p*–*q* diagram and (b) a Cambridge *p*–*q* diagram.

plotted on the bases of effective stress values p'. For the MIT p'–q diagram:

$$p' = q + \sigma_r' = q + (\sigma_r - \Delta u)$$
$$\left(= \frac{\sigma'_1 + \sigma'_3}{2} \text{ for triaxial compression} \right) \tag{2.81}$$

and for the Cambridge p'–q diagram:

$$p' = \frac{1}{3} \cdot q + \sigma_r' = \frac{1}{3} \cdot q + (\sigma_r - \Delta u)$$
$$\left[= \frac{1}{3}(\sigma'_1 + 2\sigma'_3) \text{ for triaxial compression} \right] \tag{2.82}$$

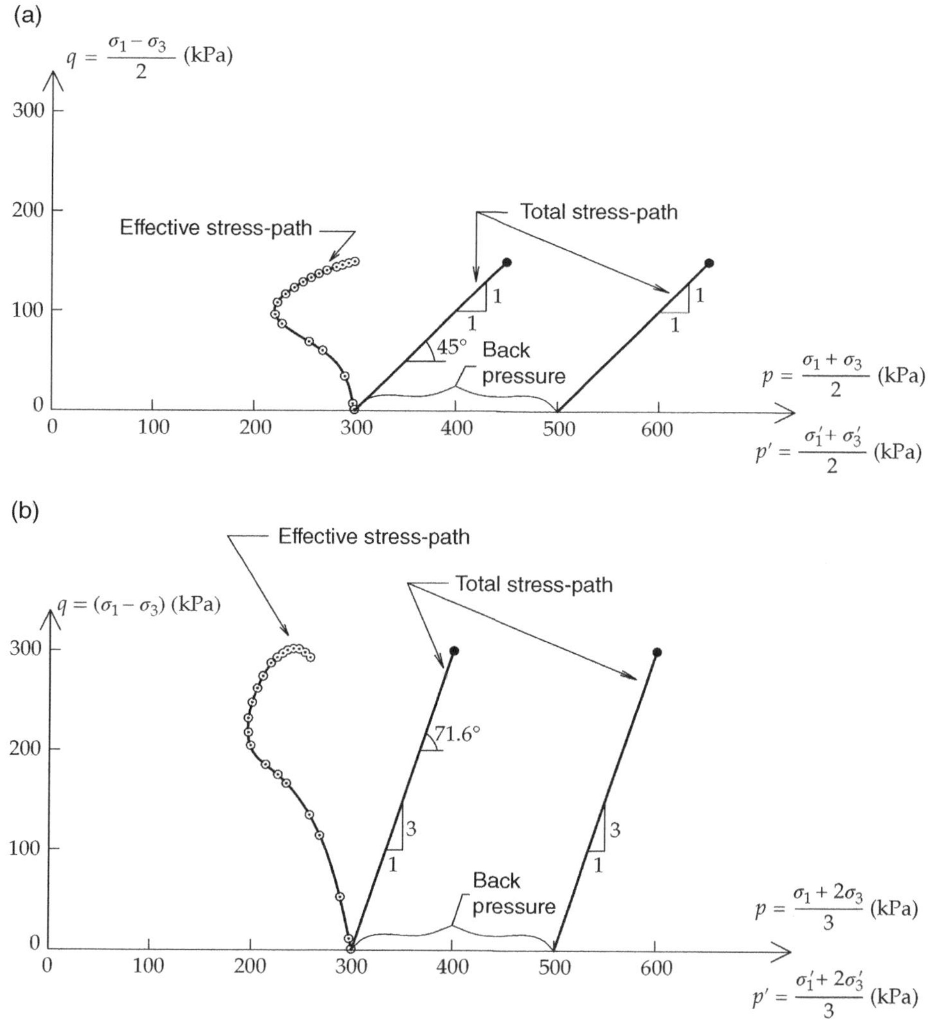

Figure 2.52 Total and effective stress paths for CU-test with back pressure shown in (a) an MIT *p–q* diagram and (b) a Cambridge *p–q* diagram.

The definitions of q indicated in these two equations are those used in the diagrams and given by Eqs (2.71) and (2.74), respectively. The values of q are not affected by the pore pressures. This is indicated in Fig. 2.53(a):

$$q = \sigma_1' - \sigma_3' = (\sigma_1 - \Delta u) - (\sigma_3 - \Delta u) = \sigma_1 - \sigma_3 \tag{2.83}$$

Figure 2.53(a) also shows that the horizontal distance between the top points of the total stress and the effective stress Mohr circles equals the pore pressure due to shearing. Since the total and effective stress paths are indicated by these top points on the MIT p–q diagram, the horizontal distance between the two stress paths in this diagram is equal to the pore pressure in the specimen, as shown in Fig. 2.53(b). This is also indicated by Eq. (2.81) according to which the difference between p' and p equals Δu. Similarly, Eq. (2.82) shows that the difference between total and effective mean normal stress equals the pore pressure Δu. Figure 2.53(c) illustrates this for the Cambridge p–q diagram.

Based on the pattern of pore pressure development indicated in Fig. 2.53, a set of

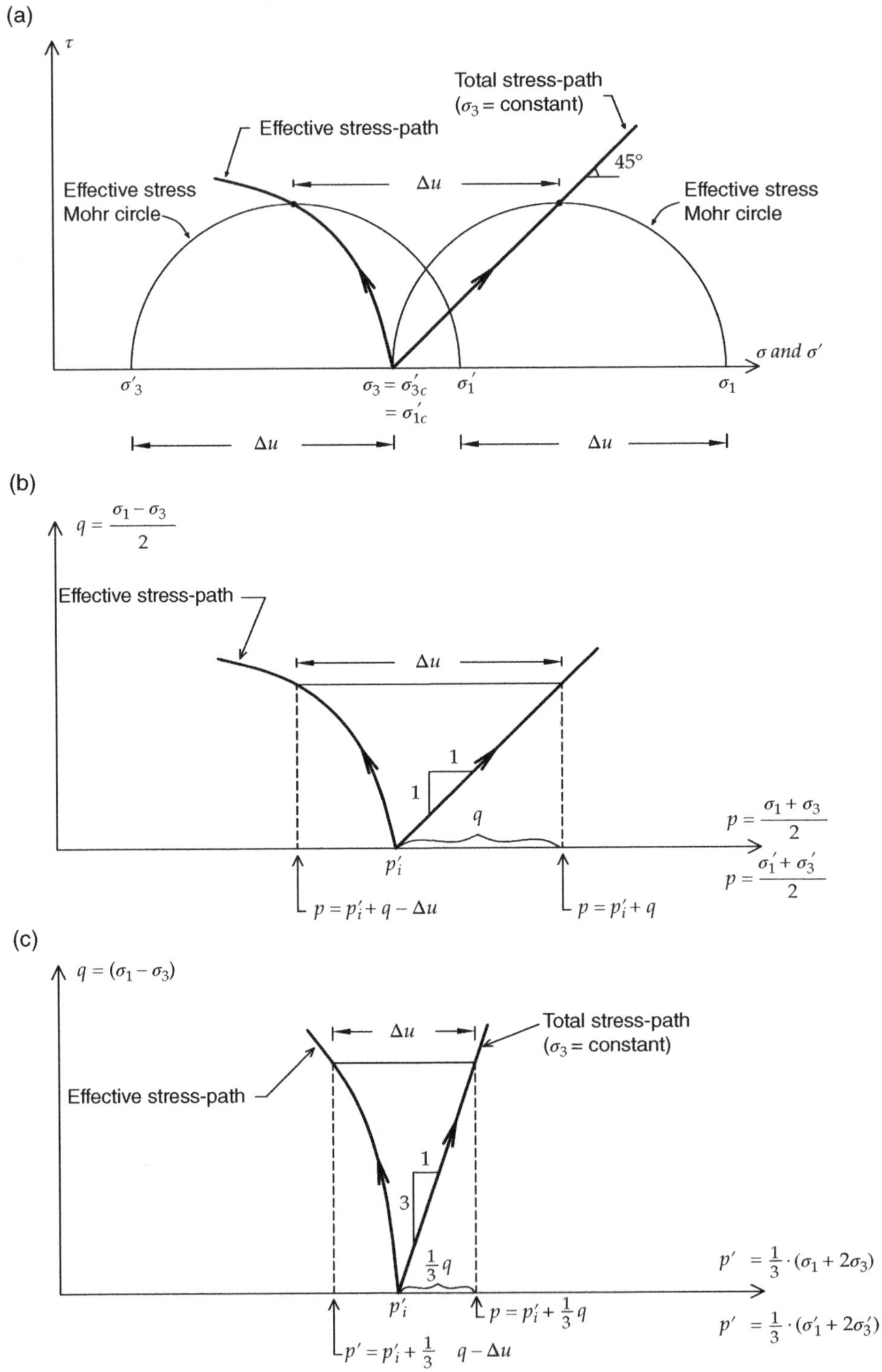

Figure 2.53 Total and effective stress paths for CU-test shown in (a) a σ–τ diagram, (b) an MIT p–q diagram, and (c) a Cambridge p–q diagram.

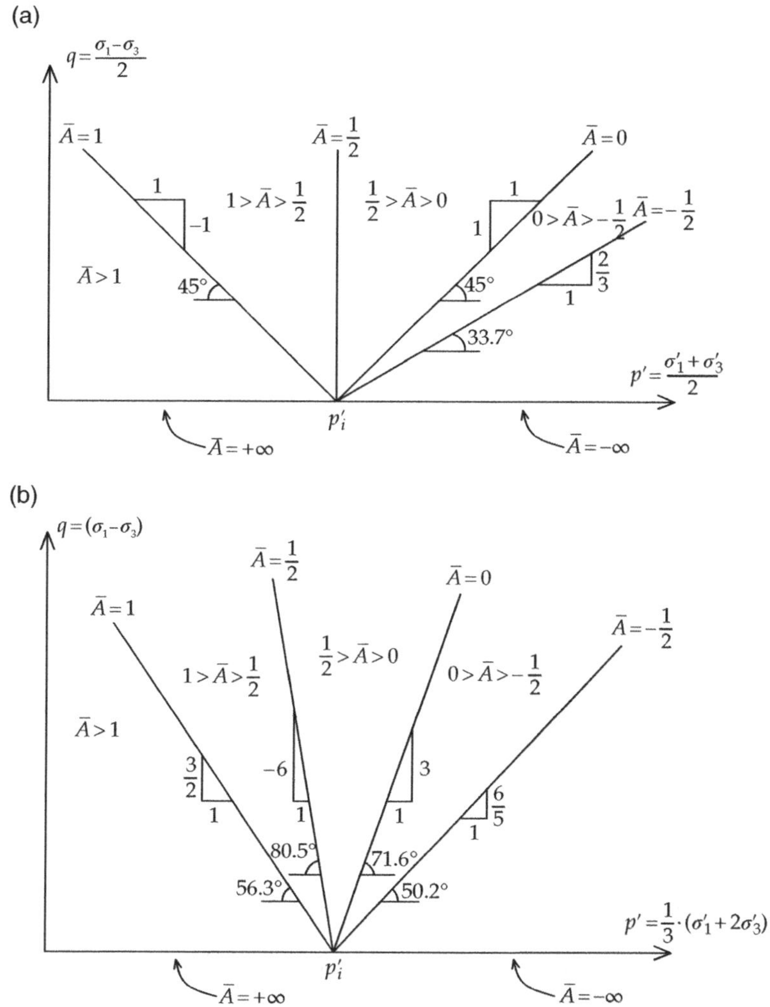

Figure 2.54 Guide lines for effective stresses based on values of $\bar{A} = \Delta u / \Delta(\sigma_1 - \sigma_3)$ shown in (a) an MIT *p*–*q* diagram and (b) a Cambridge *p*–*q* diagram.

guidelines in terms of the pore pressure parameter $\bar{A}$ may be shown on the p'–q diagrams. Figure 2.54 shows lines along which $\bar{A}$, defined in Eq. (2.55), has constant values. As mentioned in Section 2.2.3, the value of $\bar{A}$ at failure may be greater than 1 for normally consolidated, sensitive clays and loose sands tested at high confining pressures, and $\bar{A}$ may be as low as −0.5 for heavily overconsolidated clays and dense sands tested at low confining pressures. This range of values is within the pattern indicated in Fig. 2.54.

All aspects of a triaxial test, except one, are provided by the effective stress path shown in a p'–q diagram. The strains cannot be displayed in this diagram, but discrete strain values may be listed at individual points along the stress path. Figure 2.55 shows axial strains listed at individual points for the undrained tests previously shown in Fig. 2.52. The volumetric strains are zero in this test. Figure 2.55 indicates that the axial strains are small at the beginning of the test (corresponding to an initially steep stress–strain curve) and that very large strain magnitudes are produced for

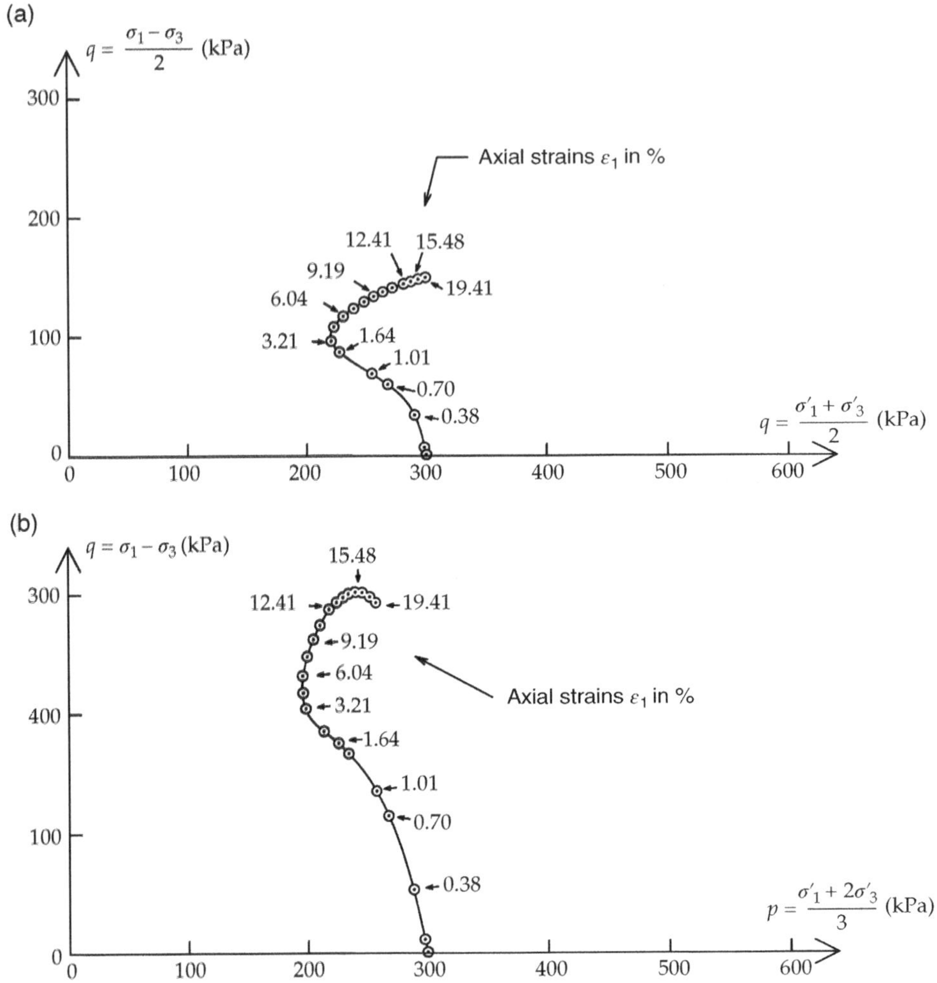

Figure 2.55 Stress paths and axial strains (in %) for CU-test on normally consolidated, remolded Edgar Plastic Kaolinite clay shown in (a) an MIT *p–q* diagram and (b) a Cambridge *p–q* diagram.

smaller changes in stress near the top of the stress path corresponding to the flattening of the stress–strain curve near the maximum deviator stress. Due to this typical strain distribution, the location of the initial portion of the effective stress path is very sensitive to precise measurements at the beginning of the test, that is at the low strain magnitudes. It is therefore important that vertical loads and pore pressures be measured simultaneously or with as short time intervals as possible, especially at the start of the test.

Examples of effective stress paths for undrained triaxial compression tests are shown in Figs 2.56, 2.57, and 2.58 for sand, normally consolidated clay, and overconsolidated clay, respectively. The patterns formed by the effective stress paths are typical for the respective soils. These diagrams may also be useful for detection of errors in the tests, as discussed in Section 2.2.5.

2.4.4 Normalized *p–q* diagrams

The *p–q* diagram may be normalized on the basis of some measure of the consolidation pressure employed for each individual specimen. This allows better comparison of data produced in tests with different consolidation pressures. Normalization of *p–q* diagrams pre-

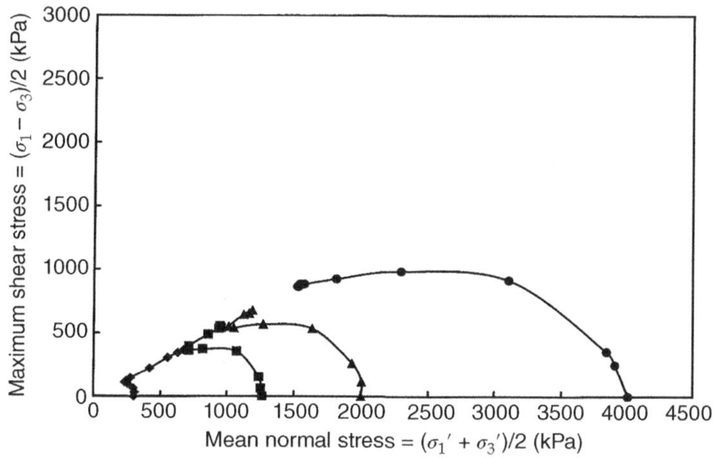

Figure 2.56 Effective stress paths from isotropically consolidated-undrained triaxial compression tests on loose Sacramento River sand ($e = 0.87$, $D_r = 38\%$).

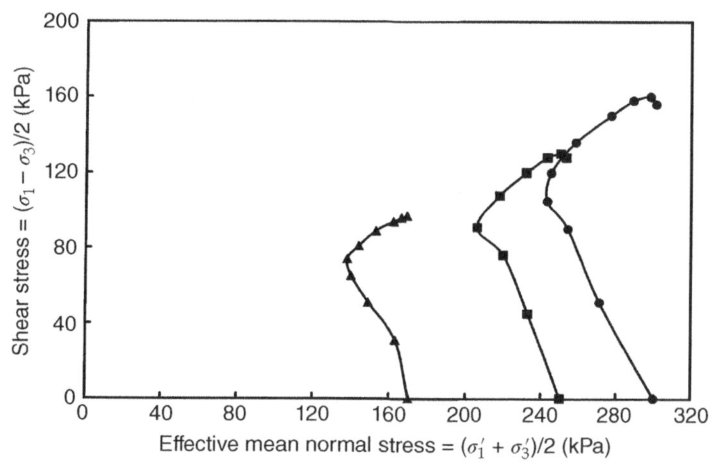

Figure 2.57 Effective stress paths for ICU-test on normally consolidated Edgar Plastic Kaolinite clay shown in an MIT *p–q* diagram.

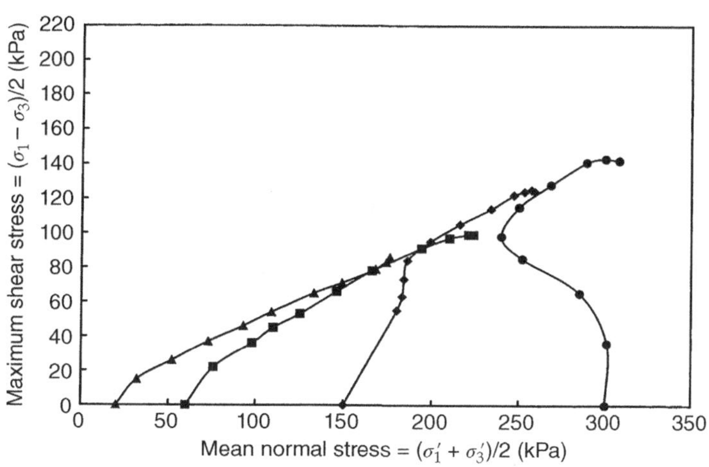

Figure 2.58 Effective stress paths from undrained triaxial compression tests on overconsolidated Edgar Plastic Kaolinite clay shown in an MIT *p–q* diagram.

sent advantages in data interpretation for tests on clay, but does not enhance the interpretation of tests on sand.

Normally consolidated clay

For normally consolidated clay the consolidation pressure is used to normalize the stresses on both axes. Often the equivalent consolidation pressure, $p_{e'}$ is used for the normalization (Hvorslev 1960). The equivalent consolidation pressure is defined as the pressure along the isotropic virgin compression curve that corresponds to the current void ratio, whether the clay is normally consolidated or overconsolidated. The definition of p_e is demonstrated in Fig. 2.59. Figure 2.60 shows normalized p'–q diagrams for the tests results previously presented in Fig. 2.19.

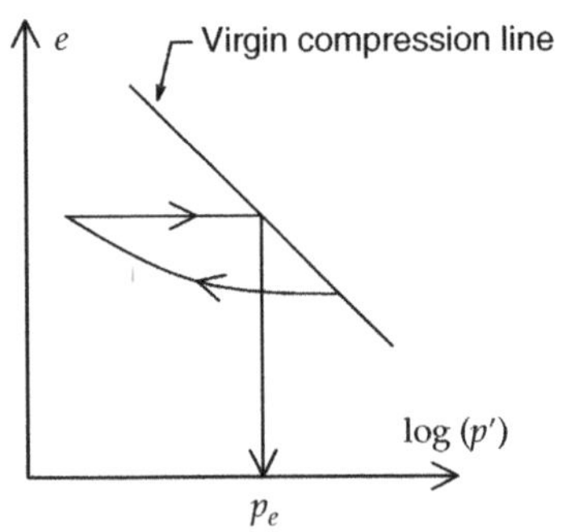

Figure 2.59 Schematic diagram illustrating the definition of the equivalent consolidation pressure, p_e.

Overconsolidated clay

Normalization of test data for overconsolidated clay may be done on the basis of either the maximum consolidation pressure employed in each test or the equivalent consolidation pressure for each specimen. Figure 2.61 shows a normalized p'–q diagram for the test results previously presented in Fig. 2.20.

2.4.5 Vector curves

A vector curve is used to represent the changes in stresses leading to failure. Thus, vector curves are similar to stress paths. However, the stresses in question are the normal and shear stresses on the plane on which failure ultimately occurs. Both total and effective stress vector curves can be displayed.

Figure 2.62 shows the derivation of the normal and shear stresses acting on the plane of failure (not necessarily at failure) inclined at α with the σ_1-plane:

$$\sigma_f{}' = \sigma_3{}' + (\sigma_1 - \sigma_3) \cdot \cos^2 \alpha \tag{2.84}$$

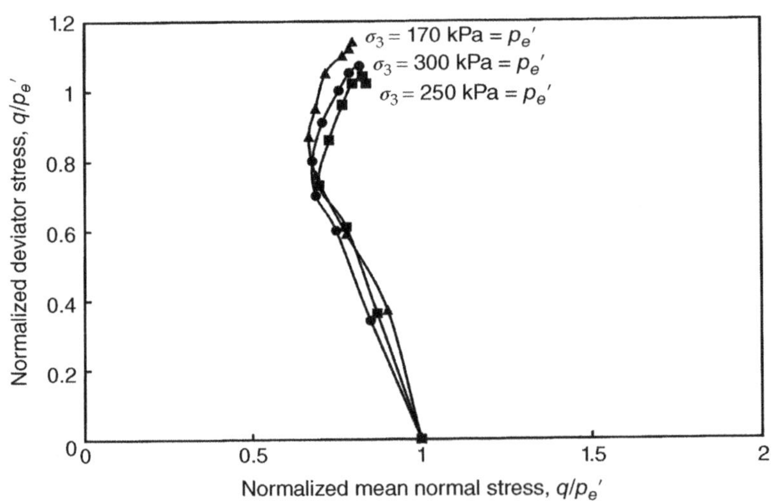

Figure 2.60 Normalized Cambridge p'–q diagram for the test results on normally consolidated clay previously presented in Fig. 2.19.

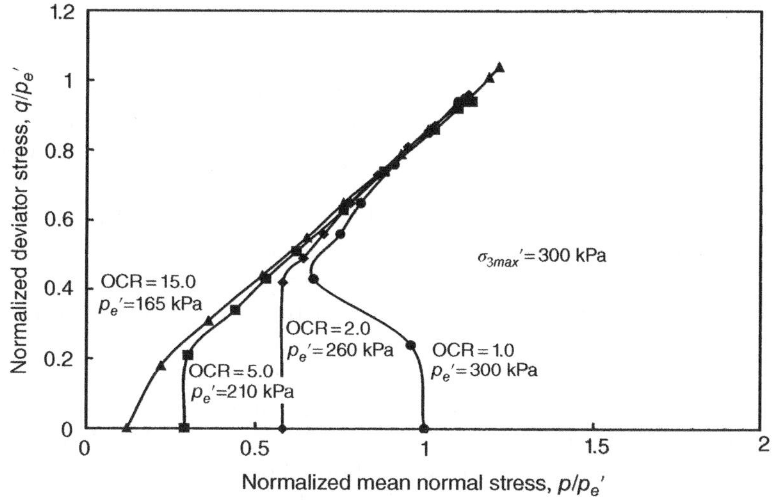

Figure 2.61 Normalized Cambridge p'–q diagram for the test results on overconsolidated clay previously presented in Fig. 2.20.

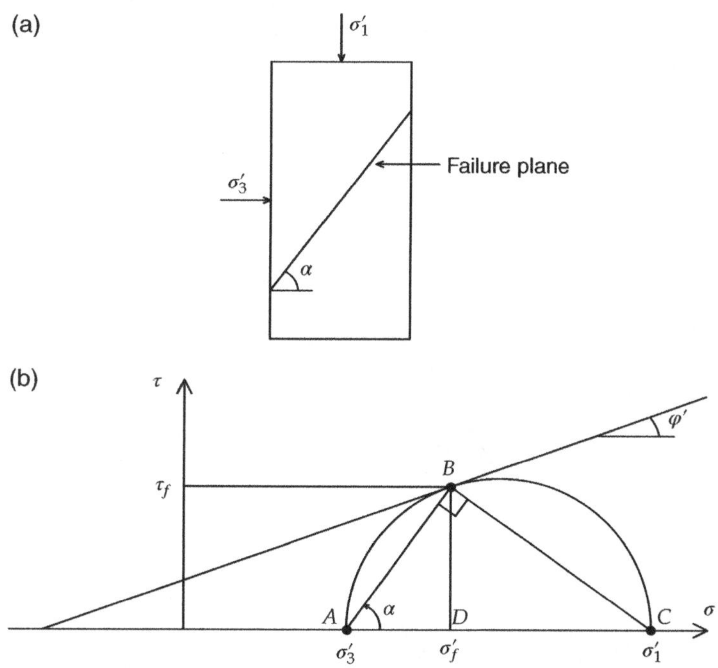

From triangle *ABC:*

$$AB = (\sigma_1 - \sigma_3) \cdot \cos\alpha$$

From triangle *ABD:*

$$\tau_f = AB \cdot \sin\alpha = (\sigma_1 - \sigma_3) \cdot \cos\alpha \cdot \sin\alpha$$

and

$$AD = \sigma_f' - \sigma_3' = AB \cdot \cos\alpha = (\sigma_1 - \sigma_3) \cdot \cos^2\alpha$$

$$\sigma_f' = \sigma_3' + (\sigma_1 - \sigma_3) \cdot \cos^2\alpha$$

Figure 2.62 Derivation of the normal and shear stresses on the failure plane σ_f and τ_f for vector curves. (a) Failure plane in specimen and (b) Mohr circle for stresses in specimen.

$$\tau_f = (\sigma_1 - \sigma_3) \cdot \cos\alpha \cdot \sin\alpha \qquad (2.85)$$

These expressions are valid for triaxial compression in which $\sigma_3' = \sigma_r'$ and $\sigma_1' = \sigma_a'$ as well as for extension in which $\sigma_3' = \sigma_a'$ and $\sigma_1' = \sigma_r'$. The orientation of the failure plane expressed by α may be related to the effective friction angle, φ', such that $\alpha = 45° + \varphi'/2$ (see Section 2.2.3). Deformation and failure in soils are caused by changes in effective stresses, not total stresses, and only the effective friction angle has meaning in this context. Note that this expression for α is determined from force equilibrium considerations without regard to the strains occurring in the soil. Other expressions for the orientation of the failure plane have been presented in the literature (e.g., Hansen 1958; Roscoe 1970; Arthur *et al.* 1977b; Vardoulakis 1980).

Schematic vector curves are shown on the Mohr diagrams for drained and undrained triaxial compression tests with constant cell pressure in Fig. 2.63. The horizontal distance between the total stress and effective stress vector curves equals the pore pressure due to shearing. This is also indicated by Eq. (2.84) according to which the difference between σ_f' and σ_f equals Δu.

As for the p'–q diagrams discussed above, a set of guidelines for the pore pressure parameter $\bar{A}$ may be shown on the Mohr diagram. Figure 2.64 shows lines along which $\bar{A}$ has constant values. Note that these guidelines form a pattern similar to that on the MIT p'–q diagram, but they are rotated counterclockwise by an angle of $\varphi'/2$. Typical values of $\bar{A}$ for various soils are given in Section 2.2.3.

An example of vector curves is shown in Fig. 2.65 for the test data previously exhibited in

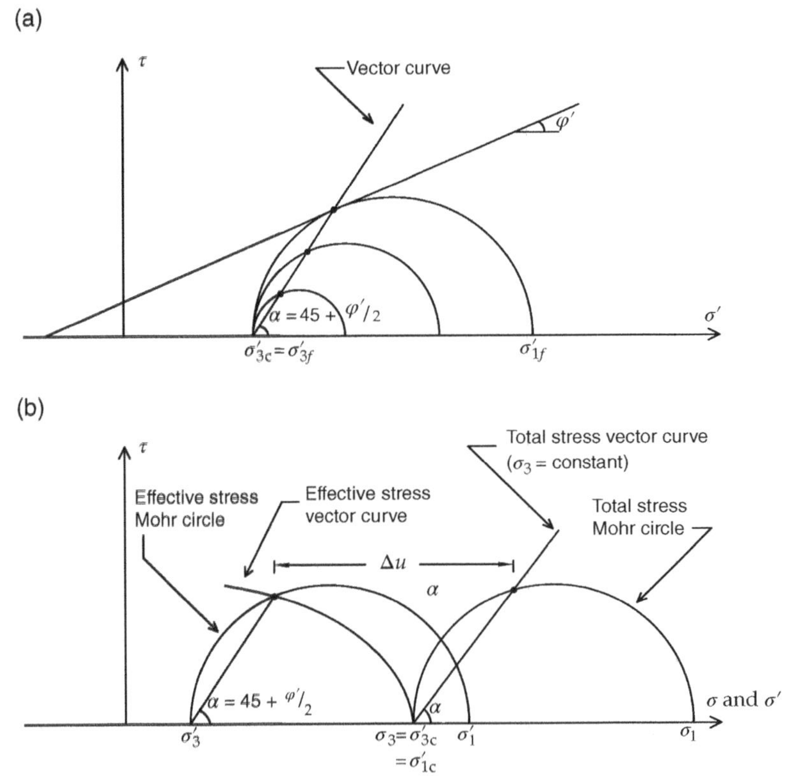

Figure 2.63 Schematic total and effective stress vector curves for drained and undrained triaxial compression test with constant cell pressure. (a) A vector curve goes through points corresponding to stresses on the future failure plane, (b) comparison of total and effective stress vector curves, and (c) calculation of normal stresses along total and effective stress vector curves.

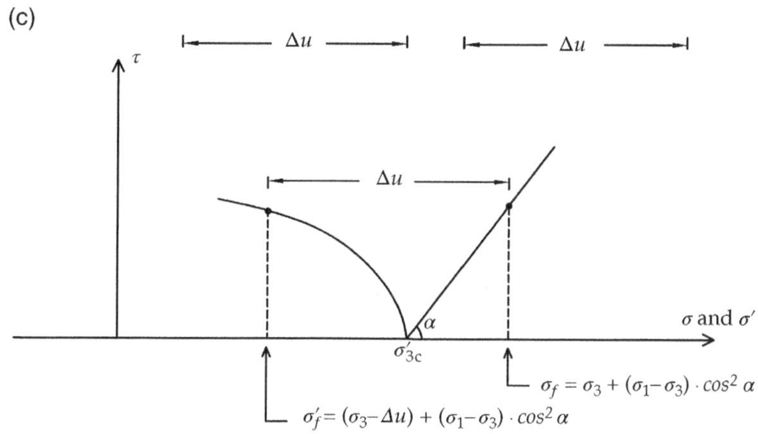

Figure 2.63 (*Continued*)

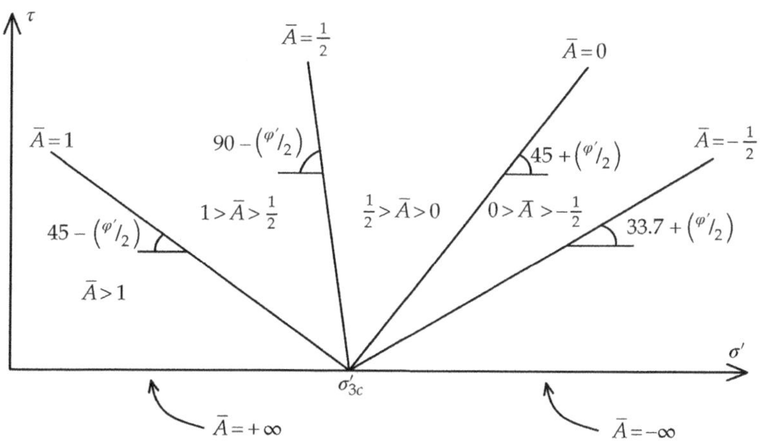

Figure 2.64 Guide lines for effective vector curves based on values of $\bar{A} = \Delta u/\Delta(\sigma_1 - \sigma_3)$ shown in a σ'–τ diagram.

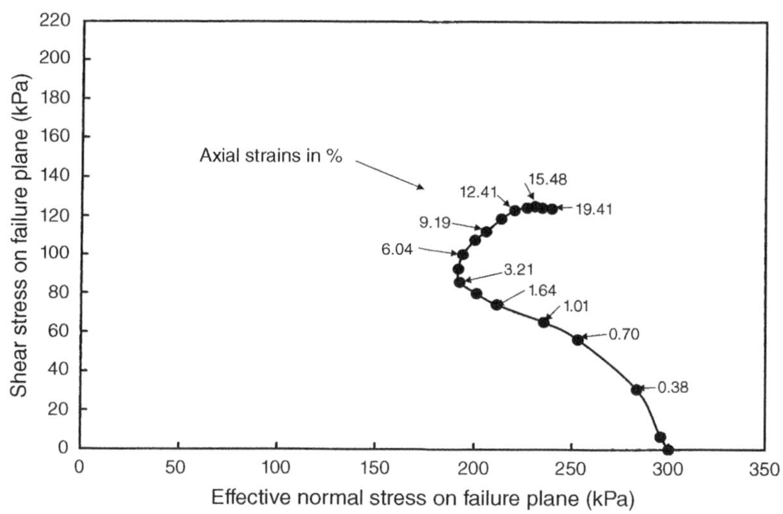

Figure 2.65 Vector curves for ICU-test on normally consolidated, remolded Edgar Plastic Kaolinite clay.

Fig. 2.52. For this test $c' = 0$ and $\varphi' = 28.7°$ and α becomes 59.4°. Both the total stress and the effective stress vector curves are simply rotated counterclockwise by $\varphi'/2$ relative to the stress paths shown on the MIT p'–q diagram. As for the p'–q diagrams, discrete values of strain may be listed at individual points along the vector curves. The discussion in Section 2.4.3 regarding strains associated with the stress paths also pertains to vector curves.

To draw vector curves, the effective friction angle φ' is required. It is therefore not possible to calculate the stresses acting on the plane of failure until the test has been finished. Thus, simultaneous plotting and monitoring of vector curves during progression of a test is not possible. This limitation does not exist for plotting stress paths.

2.5 Linear regression analysis

The statistical treatment of test data follows the patterns covered in standard textbooks (e.g., Taylor 1997). However, it may be of interest to note that it is not straightforward to perform a linear regression analysis for determination of the best fitting straight line on a p–q diagram. This is because p consists of a mixture of dependent and independent variables. The correct analysis for the p–q diagram is reviewed below.

Handy (1981) pointed out that a linear regression analysis requires that the dependent variable is expressed as a function of the independent variable. In a conventional triaxial compression test the confining pressure, σ_3, is controlled by the test operator and is therefore the independent variable. The dependent variable is the resulting strength, $(\sigma_1-\sigma_3)_f$ or just σ_{1f}, which depends on the confining pressure.

To find the best fitting straight line on a p–q diagram requires that the coordinate system be rotated around the origin so the $q_{rotated}$ points in the direction along which σ_3 is constant. In the MIT p–q diagram this corresponds to a clockwise rotation of 45° as seen in Fig. 2.66(a), and for the Cambridge p–q diagram this corresponds to a clockwise rotation of 18.43° as shown in Fig. 2.66(b). The directions of the rotated q-axis corresponds to the directions in which $\bar{A}= 0$ in Fig. 2.54. The p-axes are rotated by similar amounts so the coordinate systems in both cases remain as Cartesian coordinate systems. The regression analysis is then performed in the rotated coordinate system, after which the coordinates are rotated back to the original axes. The actual regression analysis may be done on a handheld calculator. The procedure is given below in example calculations for the MIT p–q diagram (Handy 1981) and for the Cambridge p–q diagram.

2.5.1 MIT p–q diagram

The best fitting straight line is expressed as:

$$y = A \cdot x + B \tag{2.86}$$

which for the MIT p–q diagram corresponds to Eq. (2.73) shown in Fig. 2.47:

$$q = p \cdot \sin\varphi + c \cdot \cos\varphi \tag{2.87}$$

where p is given by Eq. (2.72) and q is expressed in Eq. (2.71). Therefore:

$$A = \sin\varphi = \tan\alpha \tag{2.88}$$

$$B = c \cdot \cos\varphi \tag{2.89}$$

Since the linear regression is performed in the rotated p–q diagram shown in Fig. 2.66(a), the relations between A and A_r and between B and B_r are required. From trigonometric considerations the following relations are obtained [see Fig. 2.66(a)]:

$$A_r = \tan\alpha_r = \tan(\alpha + 45°) \tag{2.90}$$

$$B_r = B \cdot \frac{\cos\alpha}{\sin(45° - \alpha)} \tag{2.91}$$

The results from the linear regression in the rotated coordinate system consist of A_r and B_r, which are then converted to A and B using Eqs (2.90) and (2.91), and these in turn are used to find c and φ from Eqs (2.88) and (2.89).

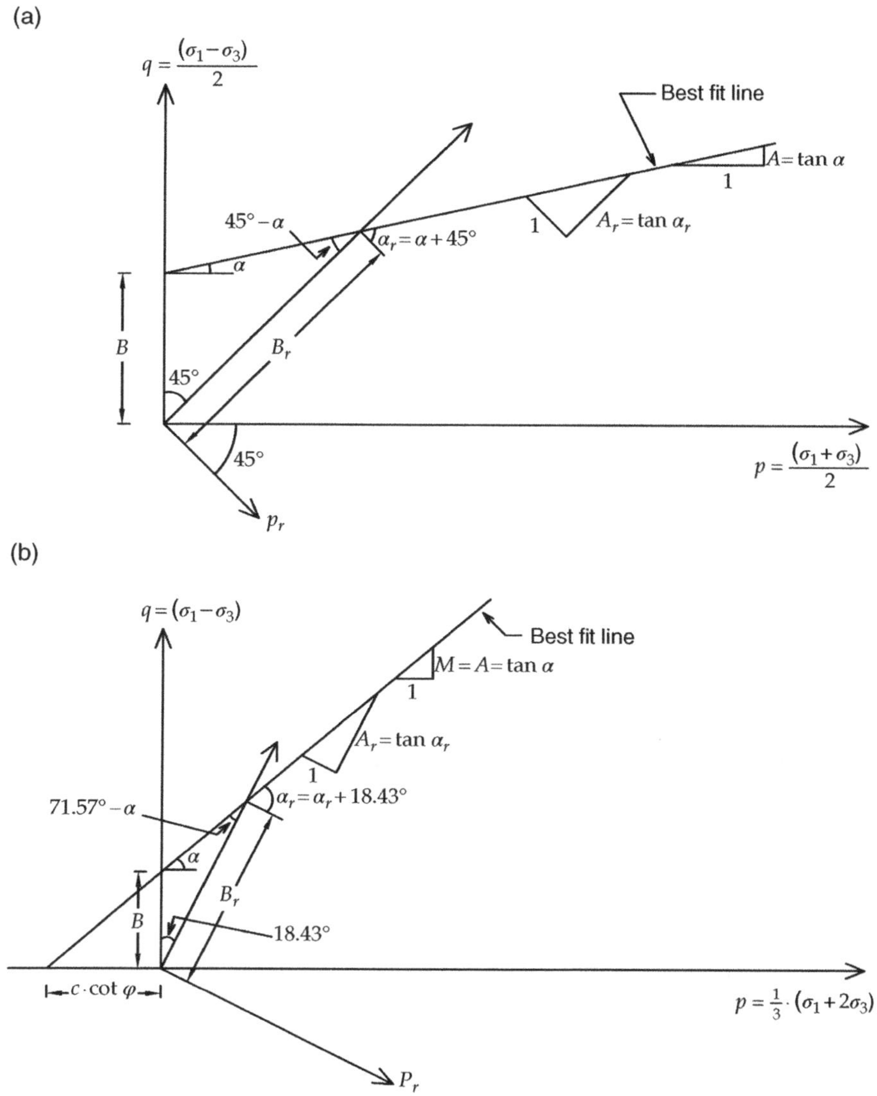

Figure 2.66 Correct determination of best-fit lines in (a) an MIT *p–q* diagram and (b) a Cambridge *p–q* diagram.

Example calculations

A drained triaxial test performed with $\sigma_3 = 100$ kPa produces $\sigma_1 = 300$ kPa.

1. Calculate p and q [Eqs (2.72) and)2.71)]: $p = 200$ kPa; $q = 100$ kPa.
2. Convert p and q to polar coordinates: $r = 223.6$ kPa; $\theta = 26.57°$.
3. Add 45° to θ: $r = 223.6$ kPa; $\theta_r = 71.57°$.
4. Convert back to rectangular coordinates: $p_r = 70.7$ kPa; $q_r = 212.1$ kPa. These are the coordinates in the rotated coordinate system, as shown in Fig. 2.67(a).
5. Enter p_r for the x-value and q_r for the y-value as the first data point in the linear regression.
6. Repeat steps 1–5 for each data point, and then regress q_r on p_r.

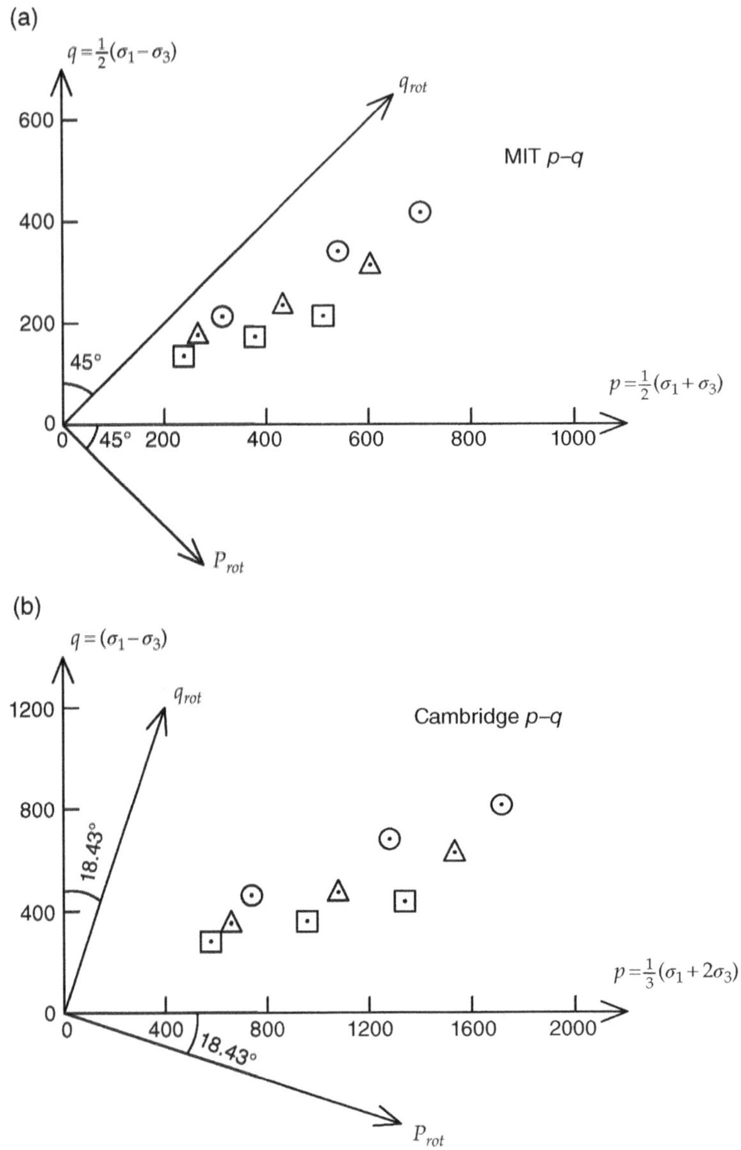

Figure 2.67 Rotated *p*–*q* diagrams for correct determination of best fit values of *c* and φ.

7. The results from the regression with rotated axes are converted back to the original axes by $\alpha = \alpha_r - 45°$ using Eq. (2.90) and B is obtained from Eq. (2.91). These in turn are used to calculate the best fitting values of c and φ from Eqs (2.88) and (2.89).

2.5.2 Cambridge *p*–*q* diagram

The best fitting straight line is expressed as:

$$y = A \cdot x + B \tag{2.92}$$

which for the Cambridge p–q diagram corresponds to Eq. (2.76) shown in Fig. 2.51 (without cohesion):

$$q = M \cdot (p + c \cdot \cot\varphi) \quad (2.93)$$

where p is given by Eq. (2.75) and q is expressed in Eq. (2.74). Therefore:

$$A = M = \tan\alpha \quad (2.94)$$

$$B = M \cdot c \cdot \cot\varphi \quad (2.95)$$

Since the linear regression is performed in the rotated p–q diagram shown in Fig. 2.66(b), the relations between A and A_r and between B and B_r are required. From trigonometric considerations the following relations are obtained [see Fig. 2.66(b)]:

$$A_r = \tan\alpha_r = \tan(\alpha + 18.43°) \quad (2.96)$$

$$B_r = B \cdot \frac{\cos\alpha}{\sin(71.57° - \alpha)} \quad (2.97)$$

The results from the linear regression in the rotated coordinate system consist of A_r and B_r, which are then converted to A and B using Eqs (2.96) and (2.97), and these in turn are used to find c and φ from Eqs (2.78) and (2.95).

Example calculations

A drained triaxial test performed with $\sigma_3 = 100$ kPa produces $\sigma_1 = 300$ kPa.

1. Calculate p and q [Eqs (2.75) and (2.74)]: $p = 166.7$ kPa; $q = 200$ kPa.
2. Convert p and q to polar coordinates: $r = 260.4$ kPa; $\theta = 50.19°$.
3. Add 18.43° to θ: $r = 260.4$ kPa; $\theta_r = 68.62°$.
4. Convert back to rectangular coordinates: $p_r = 94.9$ kPa; $q_r = 242.5$ kPa. These are the coordinates in the rotated coordinate system, as shown in Fig. 2.67(b).
5. Enter p_r for the x-value and q_r for the y-value as the first data point in the linear regression.
6. Repeat steps 1–5 for each data point, and then regress q_r on p_r.
7. The results from the regression with rotated axes are converted back to the original axes by $\alpha = \alpha_r - 18.43°$ [Eqs (2.96) and (2.94)] and B is obtained from Eq. (2.97). These in turn are used to calculate the best fitting values of c and φ from Eqs (2.78) and (2.95).

2.5.3 Correct and incorrect linear regression analyses

An example calculation using synthetic data will demonstrate the difference between correct and incorrect linear regression analyses. Three triaxial compression tests were performed with confining pressures of 100, 200, and 300 kPa on specimens carved from each of three block samples from the same soil, for a total of nine tests. From the three specimens carved from the first block, values of $c = 40$ kPa and $\varphi = 35°$ were obtained. From the second block $c = 60$ kPa and $\varphi = 25°$ were obtained, and from the third block, $c = 80$ kPa and $\varphi = 15°$ were obtained. The values of σ_1 from each of the nine tests may be calculated from Eq. (2.67) or Eq. (2.68), and the corresponding points are shown in the MIT p–q diagram in Fig. 2.68. These stress points are used in the linear regression analyses; the results are given in Table 2.1.

The results of the two p–q regression analyses are shown in Fig. 2.68. It is clear from the location of the failure points in this diagram that the data do not indicate a negative cohesion, but this is predicted by the incorrect regression analysis performed on the non-rotated axes. The friction angle from this incorrect analysis is consequently also in error. A similarly incorrect, but different value of cohesion is also predicted from the incorrect regression analysis in the unrotated Cambridge p–q diagram, and the friction angle is also in error. The correct analyses, in which the confining pressure is treated as the independent variable, all produce the same correct results.

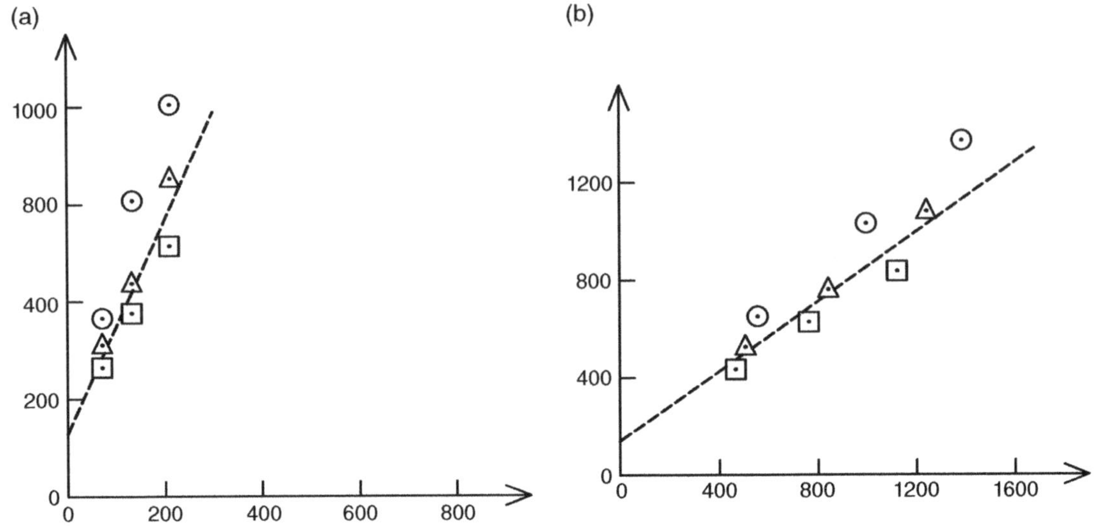

Figure 2.68 Synthetic data points and results of regression analyses shown in (a) a rotated MIT *p–q* diagram and (b) a rotated Cambridge *p–q* diagram.

Table 2.1 Results of linear regression analyses of synthetic data from triaxial compression tests

Regression	Rotation angle, $\Delta\theta$ (°)	Coefficient of correlation, r	Cohesion, c (kPa)	Friction angle, φ (°)	Remarks
σ_1 on σ_3 No rotation	0	0.807	56.7	26.6	Correct
$(\sigma_1 - \sigma_3)$ on σ_3 No rotation	0	0.645	56.7	26.6	Correct
MIT q on p No rotation	0	0.928	−5.1	34.6	Incorrect
MIT q on p Rotated axes	+45	0.807	56.7	26.6	Correct
Cambridge q on p No rotation	0	0.889	6.6	33.1	Incorrect
Cambridge q on p Rotated axes	+18.43	0.707	56.7	26.6	Correct

2.6 Three-dimensional stress states

2.6.1 General 3D stress states

The 3D state of stress at a point is completely defined if six independent quantities are known, for example three mutually perpendicular planes (defined by their normals) and the three stresses (each of which can be decomposed into one normal and two shear components) on these planes. Such a general state of stress is shown in Fig. 2.69. As for 2D conditions, the normal stresses are positive when compressive in soil mechanics. The stress components can be arranged in a symmetric matrix, called a stress tensor, as shown in Fig. 2.69.

2.6.2 Stress invariants

The principal stresses, σ_1, σ_2, and σ_3, are the normal stresses on planes on which no shear stresses act. The principal stresses can be computed from the above six independent quantities. To determine the principal stress magnitudes and their

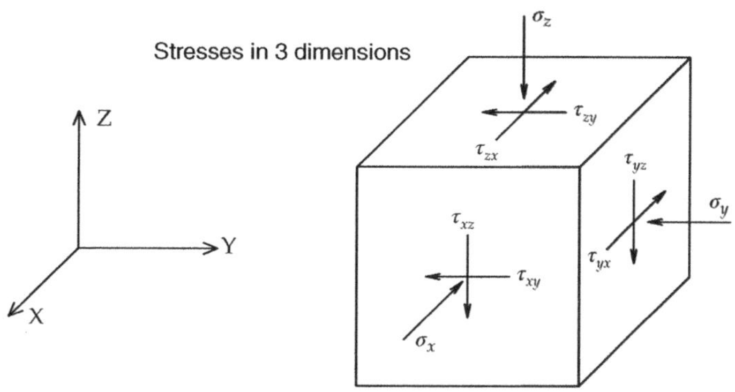

Sign rule: Normal stresses are positive when compressive
$\boldsymbol{\tau_{yx}}$: Shear stress on plane whose normal is x and direction y
3 normal stresses and 6 shear stresses, but only 3 shear stresses are independent since: $\tau_{xy} = \tau_{xy}$, etc.

$$\begin{bmatrix} \sigma_x & \tau_{yx} & \tau_{zx} \\ \tau_{xy} & \sigma_y & \tau_{zy} \\ \tau_{xz} & \tau_{yz} & \sigma_z \end{bmatrix}$$

Figure 2.69 Three-dimensional stress state for material element.

directions, the far corner of the element shown in Fig. 2.69 is cut off by a plane ABC characterized by its normal, as indicated in Fig. 2.70. The normal to plane ABC is characterized by a unit vector $(l, m, n) = (\cos\alpha, \cos\beta, \cos\gamma)$ with the property:

$$l^2 + m^2 + n^2 = \cos^2\alpha + \cos^2\beta + \cos^2\gamma = 1 \tag{2.98}$$

The stresses on the three mutually perpendicular planes are indicated in Fig. 2.70. The stresses acting on plane ABC in the directions of the coordinate axes, p_x, p_y, and p_z, are determined by force equilibrium in the X-, Y-, and Z-directions.

To express equilibrium of forces in the coordinate directions, the areas of triangles OBC, OAC, OAB and ABC are required. They are calculated by setting the area of ABC $= \omega$, as shown in Fig. 2.71, and the areas of the other triangles are expressed by projecting triangle ABC along each of the coordinate axes. Therefore:

$$\text{Area of OBC} = \omega \cdot l \tag{2.99}$$

$$\text{OAC} = \omega \cdot m \tag{2.100}$$

$$\text{OAB} = \omega \cdot n \tag{2.101}$$

Equilibrium of forces in the X-direction therefore gives:

$$p_x \cdot \omega = \sigma_x \cdot \omega \cdot l + \tau_{yx} \cdot \omega \cdot m + \tau_{zx} \cdot \omega \cdot n \tag{2.102}$$

and canceling ω produces:

$$p_x = \sigma_x \cdot l + \tau_{yx} \cdot m + \tau_{zx} \cdot n \tag{2.103}$$

Similar equations are obtained for the Y- and Z-directions:

$$p_y = \tau_{xy} \cdot l + \sigma_y \cdot m + \tau_{zy} \cdot n \tag{2.104}$$

$$p_z = \tau_{xz} \cdot l + \tau_{yz} \cdot m + \sigma_z \cdot n \tag{2.105}$$

These three expressions can be written in matrix form as follows:

$$\begin{Bmatrix} p_x \\ p_y \\ p_z \end{Bmatrix} = \begin{bmatrix} \sigma_x & \tau_{yx} & \tau_{zx} \\ \tau_{xy} & \sigma_y & \tau_{zy} \\ \tau_{xz} & \tau_{yz} & \sigma_z \end{bmatrix} \begin{Bmatrix} l \\ m \\ n \end{Bmatrix} \tag{2.106a–c}$$

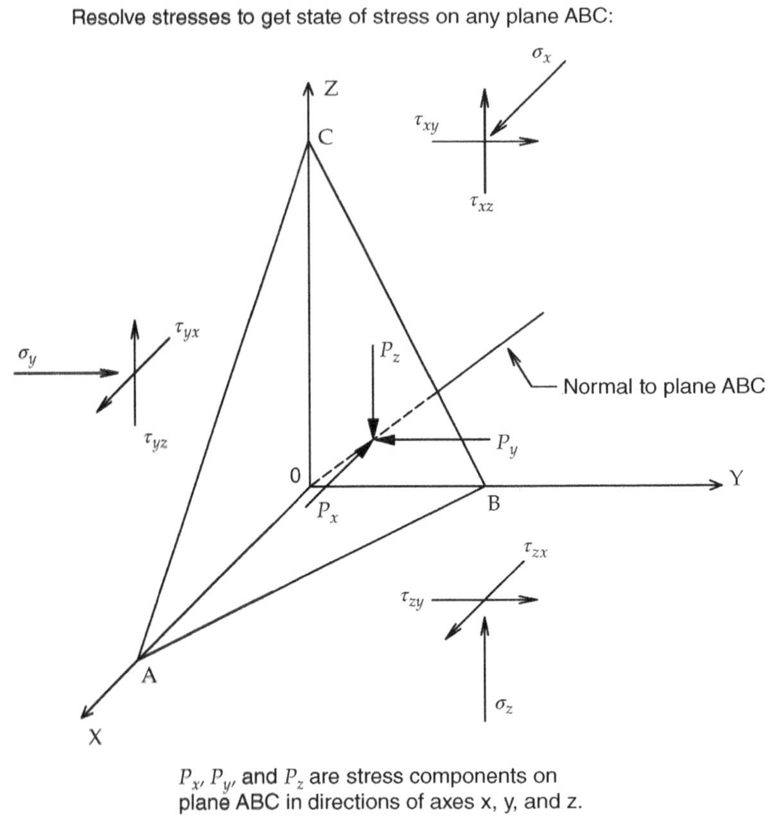

Figure 2.70 Stress state acting on plane ABC indicated by normal oriented relative to the coordinate axes.

The matrix of stresses is referred to as the stress tensor, σ_{ij}. Although the individual components of the stress tensor depend on the coordinate system, the quantity as a whole does not change. Similar to the 2D system represented by the Mohr's circle (in Section 2.1.7 and represented in Figs 2.10 and 2.11), the Mohr's circle represents a tensor in 2D such that the shear and normal components are different on each plane, yet the circle itself and the principal stresses are unchanged. A general tensor has numerical values associated with it called invariants, which for the 2D case corresponds to the values of p and q on the MIT plot. For the 3D case the principal stresses denoted by (σ_1, σ_2, σ_3) are the three solutions to a cubic equation, as shown below.

By rotating the normal to the plane as well as the plane ABC around point O in Fig. 2.71, three locations will be encountered at which no shear stresses act on plane ABC. Such a plane is called a principal plane and the normal stress acting on this plane is a principal stress.

To determine the principal stresses, assume there is one principal plane on which the shear stresses are zero and only a normal (principal) stress = σ_i exists, as shown in Fig. 2.72. Then

$$\begin{Bmatrix} p_x \\ p_y \\ p_z \end{Bmatrix} = \begin{Bmatrix} \sigma_i \cdot \cos\alpha \\ \sigma_i \cdot \cos\beta \\ \sigma_i \cdot \cos\gamma \end{Bmatrix} = \begin{Bmatrix} \sigma_i \cdot l \\ \sigma_i \cdot m \\ \sigma_i \cdot n \end{Bmatrix} \quad (2.107a\text{–}c)$$

Then, since the values of (p_x, p_y, p_z) in Eq. (2.106a–c) and in Eq. (2.107a–c) have to be equal to maintain equilibrium, subtracting the two set of equations from each other will result in zero:

$$\begin{bmatrix} (\sigma_x - \sigma_i) & \tau_{yx} & \tau_{zx} \\ \tau_{xy} & (\sigma_y - \sigma_i) & \tau_{zy} \\ \tau_{xz} & \tau_{yz} & (\sigma_z - \sigma_i) \end{bmatrix} \begin{Bmatrix} l \\ m \\ n \end{Bmatrix} = \begin{Bmatrix} 0 \\ 0 \\ 0 \end{Bmatrix}$$

(2.108a–c)

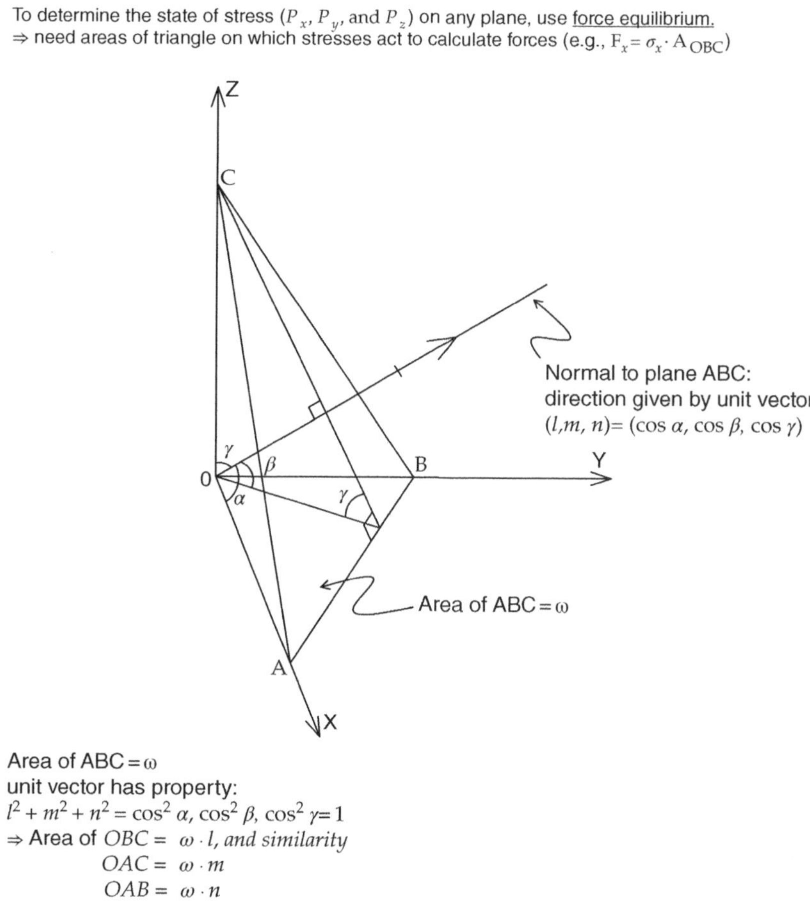

Figure 2.71 Areas of coordinate triangles, OBC, OAC and OAB, on which stresses act.

These are three linear, simultaneous equations for determination of σ_i. To obtain a nontrivial solution, the determinant of the matrix must be zero:

$$\begin{vmatrix} (\sigma_x - \sigma_i) & \tau_{yx} & \tau_{zx} \\ \tau_{xy} & (\sigma_y - \sigma_i) & \tau_{zy} \\ \tau_{xz} & \tau_{yz} & (\sigma_z - \sigma_i) \end{vmatrix} = 0 \qquad (2.109)$$

Calculating the value of the determinant produces the *characteristic equation*:

$$\begin{aligned} &\sigma_i^3 - (\sigma_x + \sigma_y + \sigma_z)\cdot\sigma_i^2 + (\sigma_x\cdot\sigma_y + \sigma_y\cdot\sigma_z + \sigma_z\cdot\sigma_x \\ &-\tau_{xy}\cdot\tau_{yx} - \tau_{yz}\cdot\tau_{zy} - \tau_{zx}\cdot\tau_{xz})\cdot\sigma_i - (\sigma_x\cdot\sigma_y\cdot\sigma_z \\ &+\tau_{xy}\cdot\tau_{yz}\cdot\tau_{zx} + \tau_{yx}\cdot\tau_{zy}\cdot\tau_{xz} - \sigma_x\cdot\tau_{yz}\tau_{zy} \\ &-\sigma_y\cdot\tau_{zx}\cdot\tau_{xz} - \sigma_z\cdot\tau_{xy}\cdot\tau_{yx}) = 0 \end{aligned} \qquad (2.110)$$

or

$$\sigma_i^3 - I_1\cdot\sigma_i^2 + I_2\cdot\sigma_i - I_3 = 0 \qquad (2.111)$$

The magnitudes of the principal stresses, σ_i ($i = 1, 2, 3$), can be found as roots of Eq. (2.111). Since for a given stress state the principal stresses are independent of the choice of coordinate system in which the normal and shear stresses are expressed, the coefficients in the cubical equation, I_1, I_2, and I_3, must also be independent of the coordinate system. These coefficients are therefore invariants with respect to change of coordinate system and they have the same values for all systems. These invariants can consequently also be expressed in terms of principal stresses:

$$I_1 = \sigma_x + \sigma_y + \sigma_z = \sigma_1 + \sigma_2 + \sigma_3 \qquad (2.112)$$

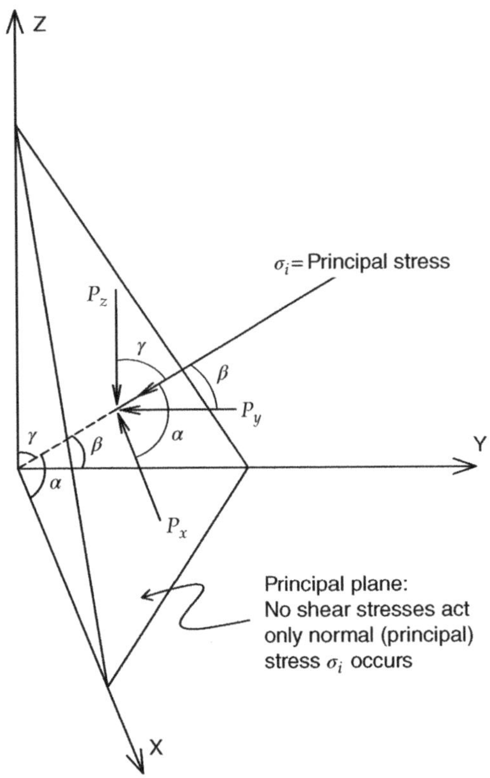

Figure 2.72 Determination of principal stresses acting on material element.

$$I_2 = \sigma_x \cdot \sigma_y + \sigma_y \cdot \sigma_z + \sigma_z \cdot \sigma_x - \tau_{xy} \cdot \tau_{yx} - \tau_{yz} \cdot \tau_{zy} - \tau_{zx} \cdot \tau_{xz} = \sigma_1 \cdot \sigma_2 + \sigma_2 \cdot \sigma_3 + \sigma_3 \cdot \sigma_1 \quad (2.113)$$

$$I_3 = \sigma_x \cdot \sigma_y \cdot \sigma_z + \tau_{xy} \cdot \tau_{yz} \cdot \tau_{zx} + \tau_{yx} \cdot \tau_{zy} \cdot \tau_{xz} - \sigma_x \cdot \tau_{yz} \cdot \tau_{zy} - \sigma_y \cdot \tau_{zx} \cdot \tau_{xz} - \sigma_z \cdot \tau_{xy} \cdot \tau_{yx} = \sigma_1 \cdot \sigma_2 \cdot \sigma_3 \quad (2.114)$$

The three quantities I_1, I_2, and I_3 are called the first, the second, and the third invariants of the stress tensor. Note that the sign used in the cubical equation for the coefficient to σ_i and the consequent sign of I_2 is the user's choice, but it must be used consistently thereafter. Note also that the shear stresses $\tau_{xy} = \tau_{yx}$, and so on in terms of numerical magnitude, but they remain as separate entities in the above expressions for the stress invariants, because they are different from each other with regard to differentiation when used in expressions containing these invariants.

2.6.3 Stress deviator invariants

Decomposition of stress tensor

The symmetric stress tensor, σ_{ij}, can be decomposed into two symmetric tensors, the *hydrostatic stress* (or spherical stress) tensor and the *deviatoric stress* tensor:

$$\sigma_{ij} = \sigma_m \cdot \delta_{ij} + s_{ij} \quad (2.115)$$

where δ_{ij} is Kronecker's symbol ($\delta_{ij} = 1$ for $i = j$ and $\delta_{ij} = 0$ for $i \neq j$). The stress tensor can also be written in matrix form:

$$\begin{bmatrix} \sigma_x & \tau_{yx} & \tau_{zx} \\ \tau_{xy} & \sigma_y & \tau_{zy} \\ \tau_{xz} & \tau_{yz} & \sigma_z \end{bmatrix} = \begin{bmatrix} \sigma_m & 0 & 0 \\ 0 & \sigma_m & 0 \\ 0 & 0 & \sigma_m \end{bmatrix} + \begin{bmatrix} (\sigma_x - \sigma_m) & \tau_{yx} & \tau_{xz} \\ \tau_{xy} & (\sigma_y - \sigma_m) & \tau_{zy} \\ \tau_{xz} & \tau_{yz} & (\sigma_z - \sigma_m) \end{bmatrix} \quad (2.116a–c)$$

The hydrostatic stress or the mean normal stress σ_m is defined as:

$$\sigma_m = \frac{1}{3} \cdot (\sigma_x + \sigma_y + \sigma_z) = \frac{1}{3} \cdot I_1 \quad (2.117)$$

And the deviatoric stress s_{ij} is therefore:

$$s_{ij} = \sigma_{ij} - \sigma_m \cdot \delta_{ij} \quad (2.118)$$

in which the individual components are given in the deviatoric matrix above.

As for the stress tensor, invariant quantities can be determined for the deviatoric stress tensor. The *characteristic equation* is formed as follows:

$$s_i^3 - J_1 \cdot s_i^2 - J_2 \cdot s_i - J_3 = 0 \quad (2.119)$$

in which the invariants of the deviatoric stress tensor are expressed as follows:

$$J_1 = s_x + s_y + s_z = (\sigma_x - \sigma_m) + (\sigma_y - \sigma_m) + (\sigma_z - \sigma_m) \quad (2.120)$$

And with σ_m from Eq. (2.117):

$$J_1 = 0 \tag{2.121}$$

$$\begin{aligned} J_2 &= \frac{1}{2} \cdot s_{ij} \cdot s_{ij} \\ &= \frac{1}{6} \cdot [(\sigma_x - \sigma_y)^2 + (\sigma_y - \sigma_z)^2 + (\sigma_z - \sigma_x)^2] \\ &\quad + \tau_{xy} \cdot \tau_{yx} + \tau_{yz} \cdot \tau_{zy} + \tau_{zx} \cdot \tau_{xz} \\ &= \frac{1}{6} \cdot [(\sigma_1 - \sigma_2)^2 + (\sigma_2 - \sigma_3)^2 + (\sigma_3 - \sigma_1)^2] \end{aligned} \tag{2.122}$$

$$\begin{aligned} J_3 &= \frac{1}{3} \cdot s_{ij} \cdot s_{jk} \cdot s_{ki} \\ &= (\sigma_x - \sigma_m)(\sigma_y - \sigma_m)(\sigma_z - \sigma_m) + \tau_{xy} \cdot \tau_{yz} \cdot \tau_{zx} \\ &\quad + \tau_{yx} \cdot \tau_{zy} \cdot \tau_{xz} - (\sigma_x - \sigma_m) \cdot \tau_{yz} \cdot \tau_{zy} \\ &\quad - (\sigma_y - \sigma_m) \cdot \tau_{xz} \cdot \tau_{zx} \\ &\quad - (\sigma_z - \sigma_m) \cdot \tau_{xy} \cdot \tau_{yx} \end{aligned} \tag{2.123}$$

$$\begin{aligned} J_3 &= (\sigma_1 - \sigma_m)(\sigma_2 - \sigma_m)(\sigma_3 - \sigma_m) \\ &= \frac{1}{27} \cdot (2\sigma_1 - \sigma_2 - \sigma_3)(2\sigma_2 - \sigma_3 - \sigma_1)(2\sigma_3 - \sigma_1 - \sigma_2) \end{aligned} \tag{2.124}$$

The solution to the characteristic equation yields the principal stress deviators s_1, s_2, and s_3. Like the stress invariants, the coefficients J_1, J_2, and J_3 are independent of the coordinate system, and they are called the first, second and third stress deviator invariants, respectively.

The stress deviator invariants may be related to the stress invariants as follows:

$$J_1 = 0 \tag{2.125}$$

$$J_2 = \frac{1}{3} \cdot I_1^2 - I_2 \tag{2.126}$$

$$J_3 = \frac{2}{27} \cdot I_1^3 - \frac{1}{3} \cdot I_1 \cdot I_2 + I_3 \tag{2.127}$$

The principal deviatoric stresses, s_1, s_2, and s_3, coincide in directions with the principal stresses, and the solutions to the characteristic equations are really equivalent, except the principal deviatoric stresses are smaller than the principal stresses by the amount of σ_m.

2.6.4 Magnitudes and directions of principal stresses

Magnitudes of principal stresses

The cubical equation for the principal stresses in Eq. (2.111) [and that for the deviatoric stresses in Eq. (2.119)] has three real roots for a proper 3D stress state. These roots may be determined by graphing the characteristic equation, as indicated in Fig. 2.73, or they may be determined by solving Eq. (2.111):

For a cubical equation in the form:

$$x^3 + A \cdot x^2 + B \cdot x + C = 0 \tag{2.128}$$

the solution is provided by Korn and Korn (1961):

$$p = -\frac{A^2}{3} + B \tag{2.129}$$

$$q = 2 \cdot \left(\frac{A}{3}\right)^3 - \frac{A \cdot B}{3} + C \tag{2.130}$$

$$x_1 = 2 \cdot \sqrt{\frac{-p}{3}} \cdot \cos\left(\frac{\alpha}{3}\right) - \frac{A}{3} \tag{2.131}$$

$$x_{2,3} = -2 \cdot \sqrt{\frac{-p}{3}} \cdot \cos\left(\frac{\alpha}{3} \pm 60°\right) - \frac{A}{3} \tag{2.132}$$

in which

$$\cos\alpha = -\frac{q}{2 \cdot \sqrt{-\left(\frac{p}{3}\right)^3}} \tag{2.133}$$

For Eq. (2.111) the principal stresses are obtained as follows:

$$\sigma_1 = 2 \cdot \sqrt{\frac{I_1^2}{9} + \frac{I_2}{3}} \cdot \cos\left(\frac{\alpha}{3}\right) + \frac{I_1}{3} \tag{2.134}$$

$$\sigma_2 = -2 \cdot \sqrt{\frac{I_1^2}{9} + \frac{I_2}{3}} \cdot \cos\left(\frac{\alpha}{3} + 60°\right) + \frac{I_1}{3} \tag{2.135}$$

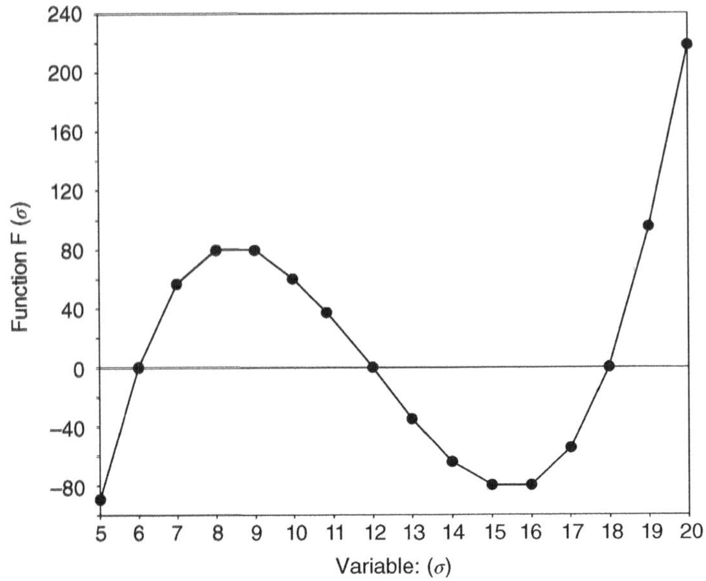

Figure 2.73 The roots of the characteristic equation (= principal stresses) may be determined by graphing the expression for the characteristic equation and finding the values of the normal stress at which this equation is zero (after Nordal 1994).

$$\sigma_3 = -2 \cdot \sqrt{\frac{I_1^2}{9} + \frac{I_2}{3}} \cdot \cos\left(\frac{\alpha}{3} - 60°\right) + \frac{I_1}{3} \tag{2.136}$$

where

$$\cos\alpha = \frac{2 \cdot \frac{I_1^3}{27} + \frac{I_1 \cdot I_2}{3} + I_3}{2 \cdot \sqrt{\left(\frac{I_1^2}{9} + \frac{I_2}{3}\right)^3}} = \frac{3\sqrt{3} \cdot J_3}{2 \cdot J_2^{3/2}} \tag{2.137}$$

For a proper stress state there will always be three real roots to the cubical equation. Note that the roots come out in an ordered fashion such that $\sigma_1 = x_1$, $\sigma_2 = x_2$, and $\sigma_3 = x_3$. Note also that the deviatoric stresses, s_1, s_2, and s_3, equal the first terms in Eqs (2.134), (2.135), and (2.136).

Directions of principal stresses

The directions of the normals to the principal planes, that is the directions of the principal stresses, $(l, m, n) = (\cos\alpha, \cos\beta, \cos\gamma)$, can be found from Eq. (2.108a–c) and the fact that (l, m, n) is a unit vector as expressed in Eq. (2.98):

$$l_i = \cos\alpha = \frac{A_i}{\sqrt{A_i^2 + B_i^2 + C_i^2}} \tag{2.138}$$

$$m_i = \cos\beta = \frac{B_i}{\sqrt{A_i^2 + B_i^2 + C_i^2}} \tag{2.139}$$

$$n_i = \cos\gamma = \frac{C_i}{\sqrt{A_i^2 + B_i^2 + C_i^2}} \tag{2.140}$$

where

$$A_i = (\sigma_y - \sigma_i)(\sigma_z - \sigma_i) - \tau_{zy} \cdot \tau_{yz} \tag{2.141}$$

$$B_i = \tau_{zy} \cdot \tau_{xz} - \tau_{xy} \cdot (\sigma_z - \sigma_i) \tag{2.142}$$

$$C_i = \tau_{xy} \cdot \tau_{yz} - \tau_{xz} \cdot (\sigma_y - \sigma_i) \tag{2.143}$$

The principal deviatoric stresses, s_1, s_2, and s_3, coincide in directions with the principal stresses.

2.7 Principal stress space

To represent a general 3D state of stress, the principal stress space is most often employed. This space consists of a Cartesian coordinate system whose axes represent the three principal stresses σ_1, σ_2, and σ_3. These stresses are positive and compressive in the octant shown in Fig. 2.74(a). The stress condition in a soil element may be represented in principal stress space by a point whose coordinates are given by (σ_1, σ_2, σ_3), as illustrated in Fig. 2.74(a).

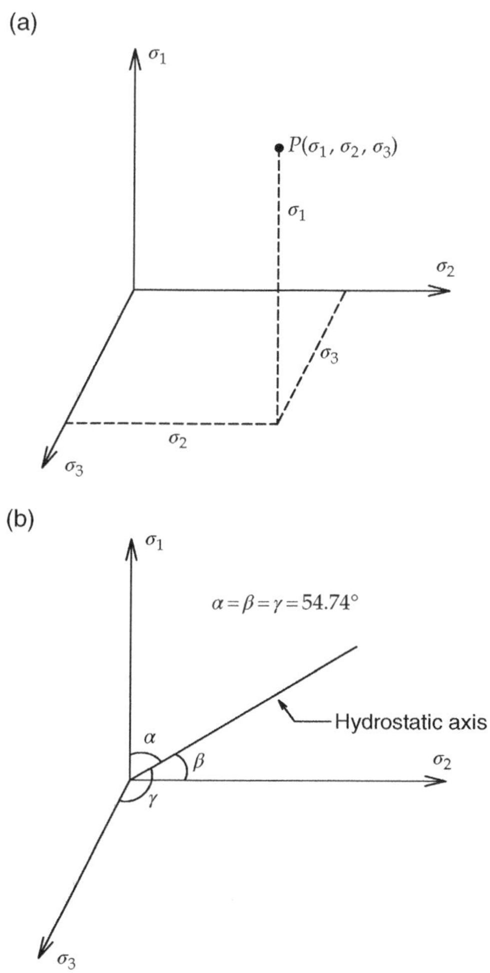

Figure 2.74 The principal stress space with (a) stress point and (b) hydrostatic axis.

The hydrostatic axis or space diagonal is the line in the coordinate system that forms equal angles with the axes, as shown in Fig. 2.74(b). Points on this line represent hydrostatic states of stress corresponding to equal values of the principal stresses ($\sigma_1 = \sigma_2 = \sigma_3$). The hydrostatic axis is characterized by the direction cosines $\bar{n} = (\cos\alpha, \cos\beta, \cos\gamma) = (\frac{1}{\sqrt{3}}, \frac{1}{\sqrt{3}}, \frac{1}{\sqrt{3}})$ and the angles between the hydrostatic axis and the three coordinate axes are therefore $\cos^{-1}(\frac{1}{\sqrt{3}}) = 54.74°$.

It is difficult to work with a 3D stress space on a routine basis. Two planes in the principal stress space are often used for plotting test results. A triaxial plane is a plane that contains the hydrostatic axes and one of the principal stress axes. There are three triaxial planes in the principal stress space. Figure 2.75(b) shows the triaxial plane that contains the σ_1-axis. An octahedral plane is a plane whose normal is the hydrostatic axis. There are eight (octa) planes that together form an octahedron, as shown in Fig. 2.75(a). There are an infinite number of octahedral planes within the octant in which all three principal stresses are positive. One octahedral plane is shown in Fig. 2.75(b).

2.7.1 Octahedral stresses

The normal and shear stresses on any octahedral plane are denoted the octahedral normal stress and the octahedral shear stress and they are expressed in terms of the principal stresses or from a general state of stress or from invariants, as follows:

$$\sigma_{oct} = \frac{1}{3}\cdot(\sigma_x + \sigma_y + \sigma_z) = \frac{1}{3}\cdot(\sigma_1 + \sigma_2 + \sigma_3) = \frac{1}{3}\cdot I_1 \quad (2.144)$$

$$\tau_{oct} = \frac{1}{3}\cdot\sqrt{(\sigma_x - \sigma_y)^2 + (\sigma_y - \sigma_z)^2 + (\sigma_z - \sigma_x)^2 + 6(\tau_{xy}\cdot\tau_{yx} + \tau_{yz}\cdot\tau_{zy} + \tau_{zx}\cdot\tau_{xz})} \quad (2.145a)$$

$$=\frac{1}{3}\cdot\sqrt{(\sigma_1-\sigma_2)^2+(\sigma_2-\sigma_3)^2+(\sigma_3-\sigma_1)^2} \tag{2.145b}$$

$$=\frac{\sqrt{2}}{3}\cdot\sqrt{I_1^2-3\cdot I_2}=\sqrt{\frac{2}{3}\cdot J_2} \tag{2.145c}$$

The octahedral stress components corresponding to the stress point P(σ_1, σ_2, σ_3) can be found as illustrated in Fig. 2.76. The total stress vector $\overline{OP}$ can be decomposed into the components $\overline{OQ}$ on the hydrostatic axis and $\overline{QP}$ in the octahedral plane through P. The length $|\overline{OQ}|$ can be obtained as the projection of $\overline{OP}$ on the hydrostatic axis. The scalar product of $\overline{OP}$ and the unit vector $\bar{n}$ on the hydrostatic axis gives:

$$\begin{aligned}|\overline{OQ}|&=\frac{1}{\sqrt{3}}\cdot\sigma_1+\frac{1}{\sqrt{3}}\cdot\sigma_2+\frac{1}{\sqrt{3}}\cdot\sigma_3\\&=\frac{\sqrt{3}}{3}\cdot(\sigma_1+\sigma_2+\sigma_3)=\sqrt{3}\cdot\sigma_{oct}\end{aligned} \tag{2.146}$$

The deviatoric component $|\overline{QP}|$ of the total stress vector can be obtained as:

$$|\overline{QP}|=|\overline{OP}-\overline{OQ}| \tag{2.147}$$

where

$$\overline{OP}=(\sigma_1,\sigma_2,\sigma_3) \tag{2.148}$$

and

$$\overline{OQ}=|\overline{OQ}|\cdot\bar{n}=\frac{1}{3}\cdot(\sigma_1+\sigma_2+\sigma_3)(1,1,1) \tag{2.149}$$

Performing the calculation indicated in Eq. (2.147) results in:

$$\begin{aligned}|\overline{QP}|&=\frac{\sqrt{3}}{3}\cdot\sqrt{(\sigma_1-\sigma_2)^2+(\sigma_2-\sigma_3)^2+(\sigma_3-\sigma_1)^2}\\&=\sqrt{3}\cdot\tau_{oct}\end{aligned} \tag{2.150}$$

Figure 2.76 shows that the octahedral stress components, σ_{oct} and τ_{oct}, corresponding to the stress point P(σ_1, σ_2, σ_3) are represented in the principal stress space as $\frac{1}{\sqrt{3}}\cdot|\overline{QP}|$ and $\frac{1}{\sqrt{3}}\cdot|\overline{QP}|$,

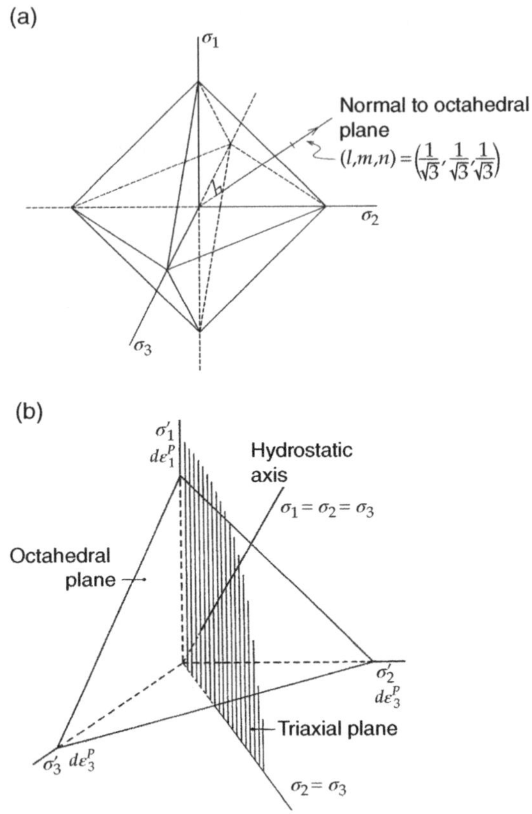

Figure 2.75 Principal stress space with (a) eight octahedral planes forming an octahedron and with (b) triaxial plane and octahedral plane in octant with positive normal stresses.

respectively. The distance from the origin to the octahedral plane that contains τ_{oct} is therefore $\frac{1}{\sqrt{3}}\cdot|\overline{OQ}|$.

Note that all points located on a circle in the octahedral plane have the same value of σ_{oct} and τ_{oct}, as shown in Fig. 2.77.

2.7.2 Triaxial plane

The triaxial plane that contains the σ_1-axis is shown in Fig. 2.78. In this plane the hydrostatic axis forms an angle of $\theta=\arctan(\frac{1}{\sqrt{2}})=35.26°$ with the horizontal axis. Any state of stress that can be produced in a triaxial test in which the

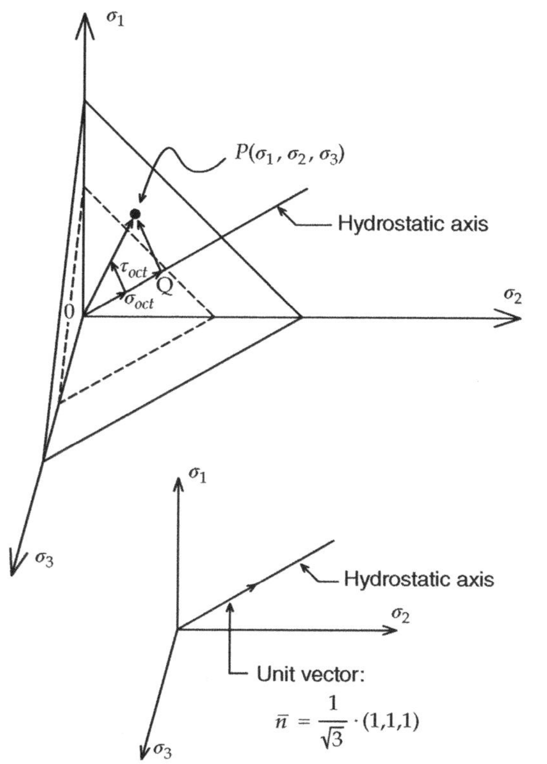

Figure 2.76 Octahedral stress vectors and octahedral planes used in location of stress point P.

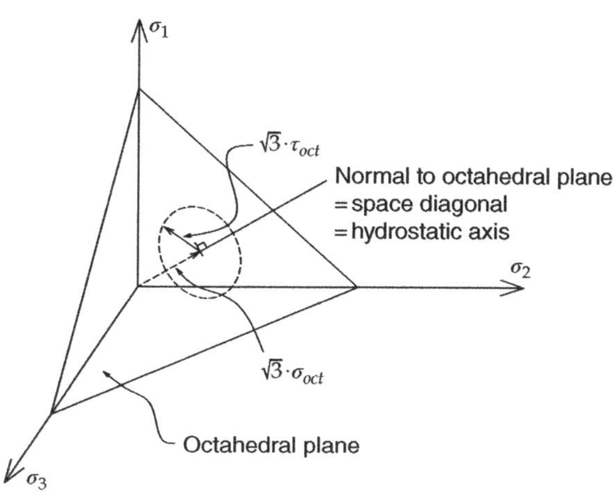

Figure 2.77 Points on a circle in an octahedral plane have the same values of σ_{oct} and τ_{oct}.

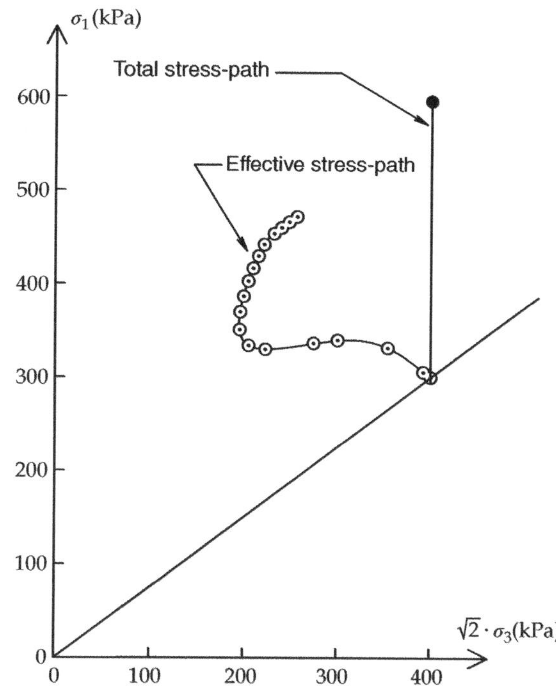

Figure 2.78 Triaxial plane with total and effective stress paths for triaxial compression test on normally consolidated, remolded Edgar Plastic Kaolinite clay.

state of stress is axisymmetric can be shown in a triaxial plane. States of stress in triaxial compression plot above the hydrostatic axis and states of stress in triaxial extension plot below the hydrostatic axis. The total and effective stress paths for an ICU test is shown in Fig. 2.78. These are the stress paths previously shown on the *p*–*q* diagrams in Fig. 2.52.

The stress paths shown in the triaxial plane and the Cambridge *p*–*q* diagram are very similar. In fact, these two diagrams are the same within a linear transformation. Figure 2.79 shows the two diagrams superimposed with the coordinates for a stress point indicated with reference to the triaxial plane. A similar direct comparison cannot be made between the triaxial plane and the modified Mohr diagram, because the abscissa of the latter cannot be changed by a linear transformation to match the distance along the hydrostatic axis.

While the Cambridge *p*–*q* diagram has the advantage of direct transformation into the

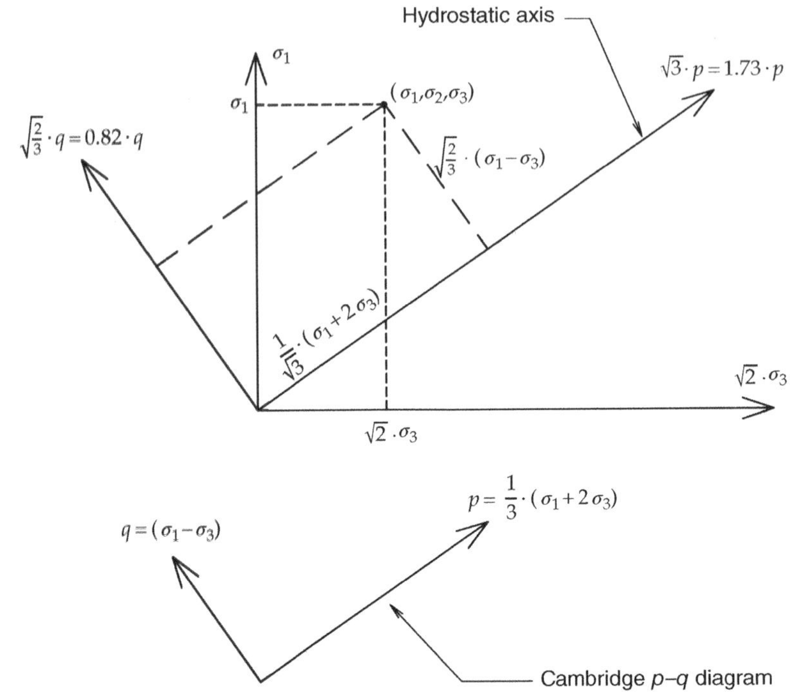

Figure 2.79 Comparison between triaxial plane and Cambridge *p–q* diagram.

triaxial plane and therefore has a direct relation with the principal stress space, the modified Mohr diagram (= MIT *p–q* diagram) is useful for stress analyses in two dimensions, as indicated in Section 2.1.7. In addition, stress paths shown in the Cambridge *p–q* diagram or the triaxial plane are larger, because the slope of the failure surface is steeper in this diagram than in the modified Mohr diagram.

2.7.3 Octahedral plane

The other 2D diagram in the principal stress space that is useful for presentation of test results is the octahedral plane. The results of tests with three unequal stresses are often shown on this diagram.

In the octahedral plane the center point represents the hydrostatic axis, as shown in Fig. 2.80. The three axes are 120° apart and they represent the projections of the principal stress axes on

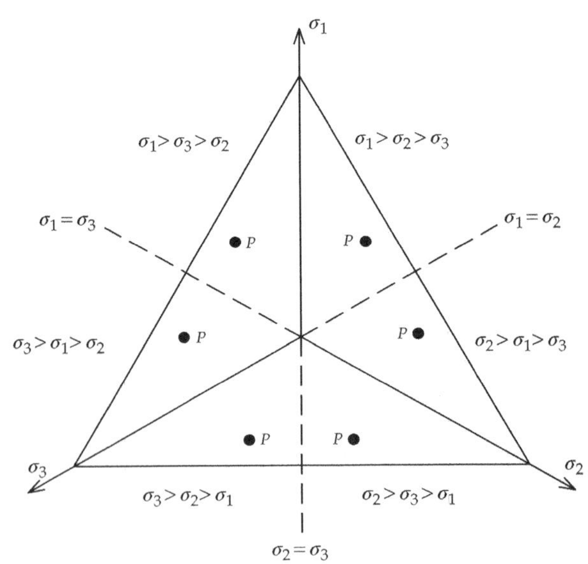

Figure 2.80 Octahedral plane with state of stress represented in each of six sectors.

the octahedral plane. These axes are lines of symmetry such that $\sigma_1 = \sigma_2$, $\sigma_2 = \sigma_3$, and $\sigma_3 = \sigma_1$ represent the traces in the octahedral plane of the three triaxial planes.

If σ_1, σ_2, and σ_3 are taken as the major, intermediate and minor principal stresses, only one-sixth of the octahedral plane is necessary for representing any state of stress. However, any one of three axes could be the major principal stress axis. Thus, interchanging the subscripts (1, 2, 3) one state of stress is represented in each of the six parts of the octahedral plane. The state of stress represented by the point P is shown in each of the six sectors of the octahedral plane in Fig. 2.80.

Figure 2.81(a) shows the Mohr–Coulomb failure criterion in the principal stress space for a material without effective cohesion. The failure surfaces form a cone with the apex at the origin. The cross-sections in the octahedral plane have shapes of irregular hexagons with acute and obtuse angles at the points corresponding to the states of stress in triaxial compression and extension, respectively. The cross-sections change shape as the magnitude of the friction angle changes, as indicated in Fig. 2.81(b). The shape approaches an equilateral triangle for friction angles approaching 90°, and it resembles a regular hexagon at very small friction angles.

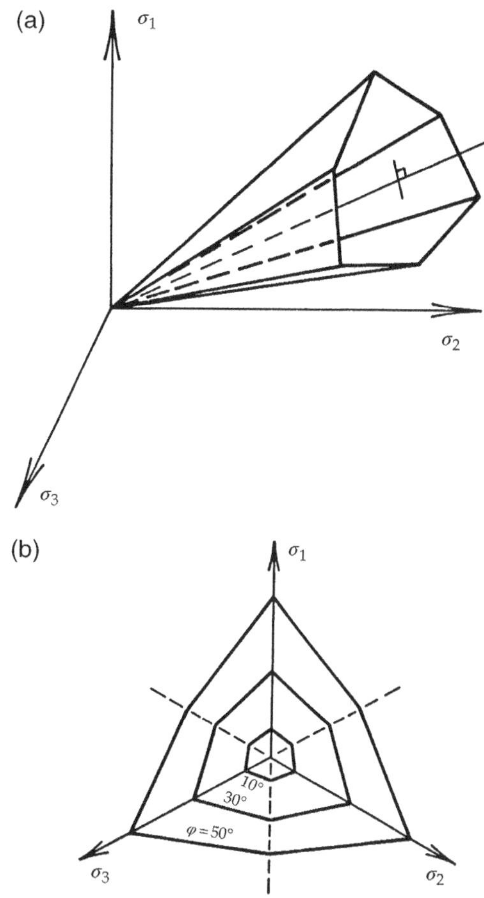

Figure 2.81 (a) Three-dimensional representation of Mohr–Coulomb failure criterion in principal stress space and (b) variation of cross-sectional shape with friction angle in an octahedral plane.

2.7.4 Characterization of 3D stress conditions

b-Value

The magnitude of the intermediate principal stress, σ_2, relative to the major and minor principal stresses, σ_1 and σ_3, may be indicated by the value of b, originally employed by Habib (1953):

$$b = \frac{\sigma_2 - \sigma_3}{\sigma_1 - \sigma_3} \tag{2.151}$$

The value of $b = 0$ when $\sigma_2 = \sigma_3$, as in a triaxial compression test, while $b = 1$ for a triaxial extension test in which $\sigma_2 = \sigma_1$. The value of b increases linearly from 0 to 1 with increasing σ_2-value from σ_3 to σ_1. This parameter may be used for example to show the variation of the friction angle with the intermediate principal stress, as exemplified in the φ–b diagram in Fig. 2.82.

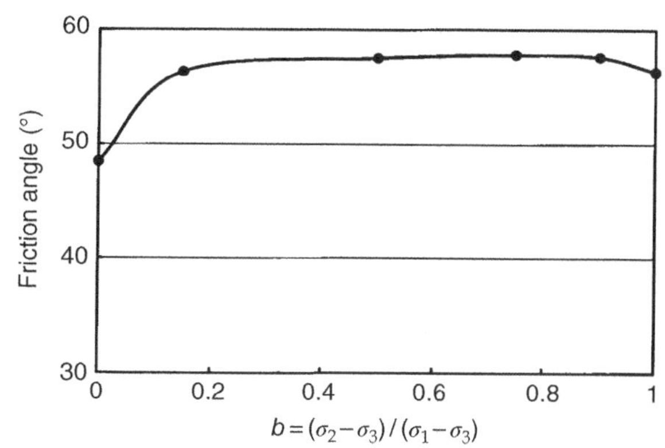

Figure 2.82 Example of φ–b diagram for dense Monterey No. 0 sand.

Lode angle

The Lode angle, θ, relates to the octahedral plane and is defined from Eq. (2.137) (Lode 1926; Nayak and Zienkiewicz 1972) as $\theta = \alpha/3$, and it is given by:

$$\cos(3\theta) = \frac{3\sqrt{3}}{2} \cdot \frac{J_3}{J_2^{3/2}} \tag{2.152}$$

For triaxial compression, $\theta = 0°$ and for triaxial extension, $\theta = 60°$. The value of θ is related to the value of b is indicated in Fig. 2.83, and as given by the following expression:

$$\cos(\alpha) = \cos(3\theta) = \frac{1}{2} \cdot \frac{\left(2b^3 - 3b^2 - 3b + 2\right)}{\left(b^2 - b + 1\right)^{3/2}} \tag{2.153}$$

An example of the use of the Lode angle is indicated by the variation of the friction angle

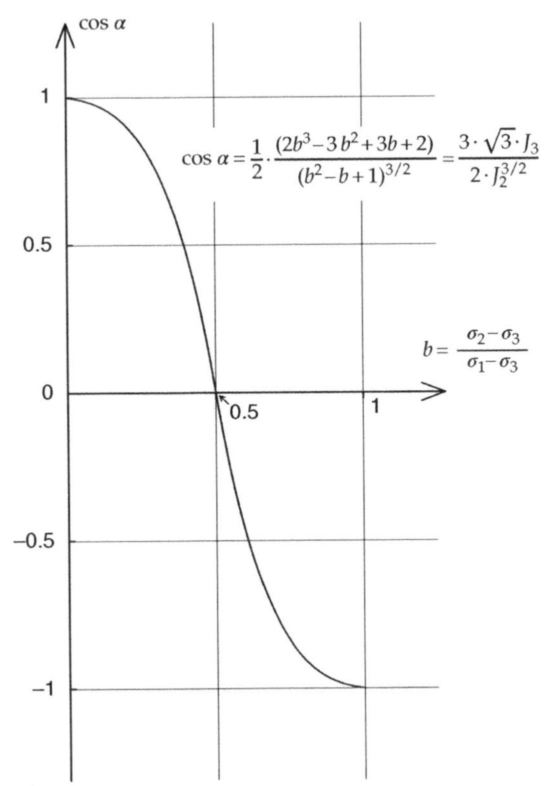

Figure 2.83 Relationship between cosα and $b = (\sigma_2 - \sigma_3)/(\sigma_1 - \sigma_3)$.

in the octahedral plane as shown in Fig. 2.84. This diagram looks very similar to the φ–b diagram, but the Lode angle is rarely employed for this purpose.

b-Values and Lode angles for cross-anisotropic soils

To present results of tests on cross-anisotropic materials, it is important to clearly indicate the directions of stress and strain relative to the principal axes of the material. For this purpose a Cartesian coordinate system is employed as indicated in Fig. 2.85. The X-axis coincides with the axis of rotational symmetry for the cross-anisotropic specimens. The stresses are labeled according to this coordinate system.

The Lode angle, θ, is measured clockwise from the σ_x-axis to the stress point P (σ_x, σ_y, σ_z) as indicated on the octahedral plane in Fig. 2.85(b, c) and is calculated as follows:

$$\tan\theta = \sqrt{3} \cdot \frac{\sigma_y - \sigma_z}{\left(\sigma_x - \sigma_y\right) + \left(\sigma_x - \sigma_z\right)} \tag{2.154}$$

Values of θ are indicated on the stress axes in Fig. 2.85(c). For the ranges of θ between 0° and 60°, 60° and 120°, and 120° and 180°, the b-values are calculated as follows:

$$b = \frac{\sigma_y - \sigma_z}{\sigma_x - \sigma_z} \quad \text{for } 0° \le \theta \le 60° \tag{2.155a}$$

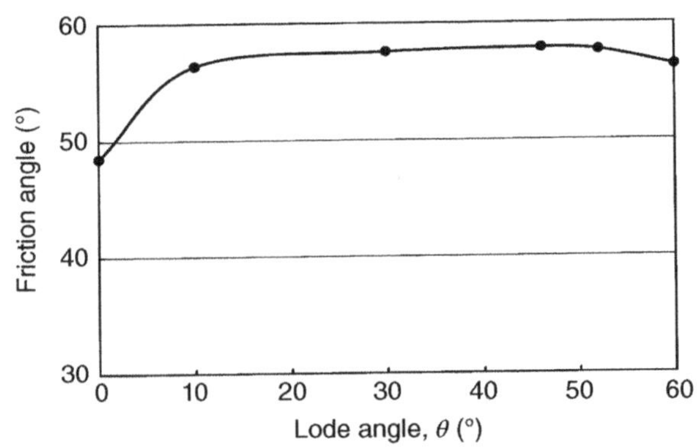

Figure 2.84 Example of ϕ–θ diagram in which θ is the Lode angle in the octahedral plane. Same data for dense Monterey No. 0 sand as shown in Fig. 2.83.

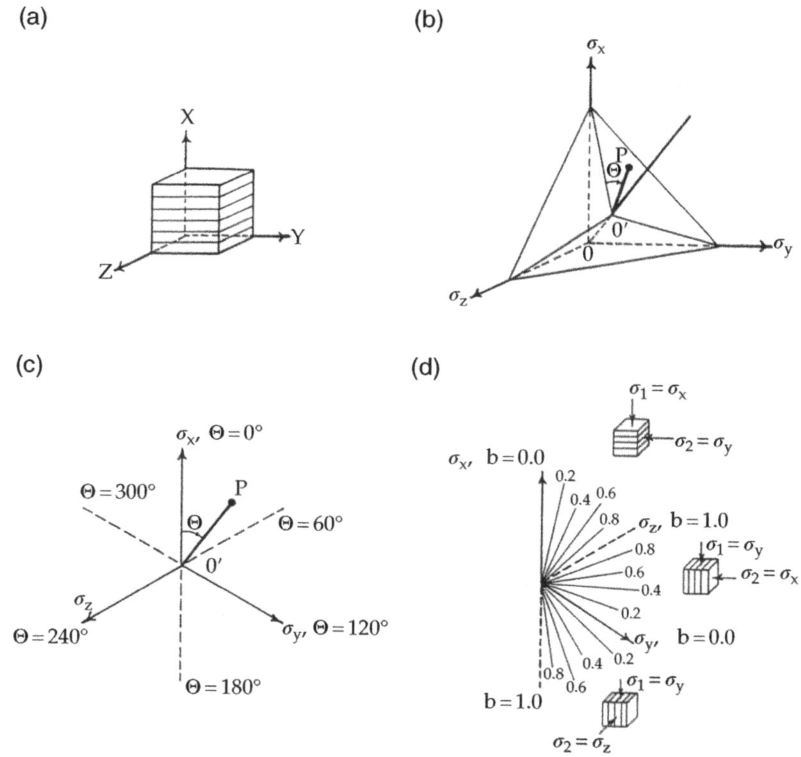

Figure 2.85 Orientation of a cross-anisotropic specimen relative to (a) a Cartesian coordinate system, (b) principal stress space, and (c and d) an octahedral plane. Reproduced from Kirkgard and Lade 1993 by permission of Canadian Science Publishing.

$$b=\frac{\sigma_x-\sigma_z}{\sigma_y-\sigma_z} \quad \text{for } 60^\circ \le \theta \le 120^\circ \qquad (2.155b)$$

$$b=\frac{\sigma_z-\sigma_x}{\sigma_y-\sigma_x} \quad \text{for } 120^\circ \le \theta \le 180^\circ \qquad (2.155c)$$

It is clear that the value of θ is sufficient to indicate the relative magnitudes of the principal stresses, and it also provides information regarding which of the normal stresses are the major, intermediate and minor principal stresses. The value of b varies from 0 to 1 in each of the six sectors of the octahedral plane, and calculation of b requires previous ordering of the principal stresses. However, the parameter b has frequently been used in studies of 3D behavior of soils, and it is convenient to use this parameter together with the parameter θ in discussions of test results.

Figure 2.86 shows experimental results from true triaxial tests on intact, cross-anisotropic San Francisco Bay Mud projected on an octahedral plane.

2.7.5 Shapes of stress invariants in principal stress space

If the stress invariants I_1, I_2, and I_3 are set equal to constant values, then each of these invariants takes very instructive shapes in the part of the principal stress space in which the principal stresses are positive. Thus, I_1 = constant is shown in Fig. 2.87. Figure 2.87(a) shows that I_1 = constant in the octahedral plane is a plane that cuts the principal stress axes at the same values of the principal stresses, while Fig. 2.87(b) indicates the edges in the triaxial plane, and Fig. 2.87(c) shows the plane in the principal stress space.

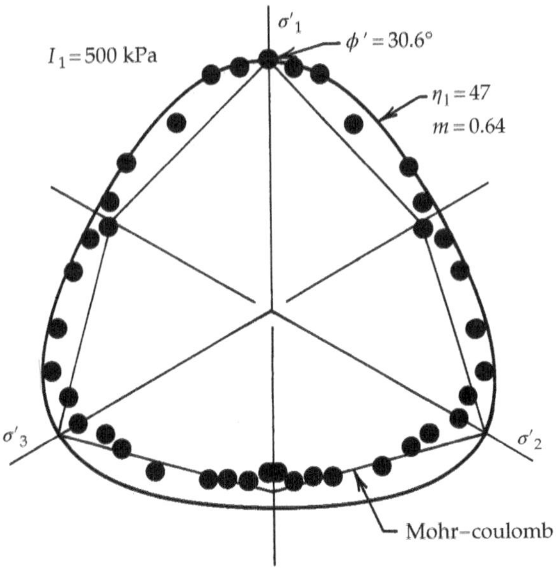

Figure 2.86 Octahedral plane with results from true triaxial tests on intact, cross-anisotropic San Francisco Bay Mud. Reproduced from Kirkgard and Lade 1993 by permission of Canadian Science Publishing.

Figure 2.88 shows the shape of I_2 = constant. Figure 2.88(a) indicates that the traces in the octahedral plane are shaped as concentric circles, whose values are shown on the diagram. Figure 2.88(b) shows the traces in the triaxial plane, which are symmetric around the hydrostatic axis and asymptotic to the major principal stress axis. Figure 2.88(c) shows the combination of these shapes in the principal stress space.

Figure 2.89 indicates the shape of I_3 = constant. The cross-sectional shapes shown in Fig. 2.89(a) are triangular with smoothly curved traces in the octahedral plane. Depending on the value of I_3, the shape is more or less rounded and near the hydrostatic axis the shape becomes circular. For lower values of I_3 the shape becomes more triangular and reaches out in the corners of the octahedral plane. Figure 2.89(b) indicates the traces of I_3 in the triaxial plane. They are asymptotic to the major principal stress axis and to the plane created by the other two principal stresses. Figure 2.89(c) shows the combination of these views in the principal stress space.

2.7.6 Procedures for projecting stress points onto a common octahedral plane

Straight failure envelope

It is often useful to study the shape of failure surfaces in the principal stress space. The cross-sectional shapes of failure surfaces are best shown on an octahedral plane. However, tests conducted with three unequal principal stresses most often do not fail at the same value of the octahedral normal stress, and the results therefore cannot be plotted directly on the same octahedral plane. For soils without cohesion and with straight failure surfaces in meridian planes (i.e., planes that contain the hydrostatic axis), the principal stresses at failure can be modified according to the following expression such that all stress points fall on one octahedral plane:

$$(\sigma_1{}^*,\sigma_2{}^*,\sigma_3{}^*) = (\sigma_1,\sigma_2,\sigma_3)\cdot\frac{\sigma_{oct}}{\frac{1}{3}(\sigma_1+\sigma_2+\sigma_3)} \tag{2.156}$$

where σ_1, σ_2, and σ_3 are the principal stresses measured at failure, and $\sigma_{oct}\left[=\frac{1}{3}(\sigma_1{}^*+\sigma_2{}^*+\sigma_3{}^*)\right]$ is the octahedral normal stress corresponding to the octahedral plane on which the test results are to be plotted. The value of σ_{oct} is of no importance to the cross-sectional shape of the failure surface for soils without cohesion and with straight failure surfaces. Note that only the magnitudes of the principal stresses are modified, whereas the ratios between the principal stresses remain constant. Thus, the friction angle is not changing due to the modification of stresses given in Eq. (2.156).

For soils with cohesion, it is necessary to modify the normal stresses before projecting them along the straight lines going through the stress origin. This is done by translating the principal stress space along the hydrostatic axis to account for the cohesion and tensile strength that can be sustained by such materials. Thus, a constant stress, $a\cdot p_a$, is added to the normal stresses:

$$\bar{\sigma}_1 = \sigma_1 + a\cdot p_a \tag{2.157a}$$

$$\bar{\sigma}_2 = \sigma_2 + a\cdot p_a \tag{2.157b}$$

$$\bar{\sigma}_3 = \sigma_3 + a\cdot p_a \tag{2.157c}$$

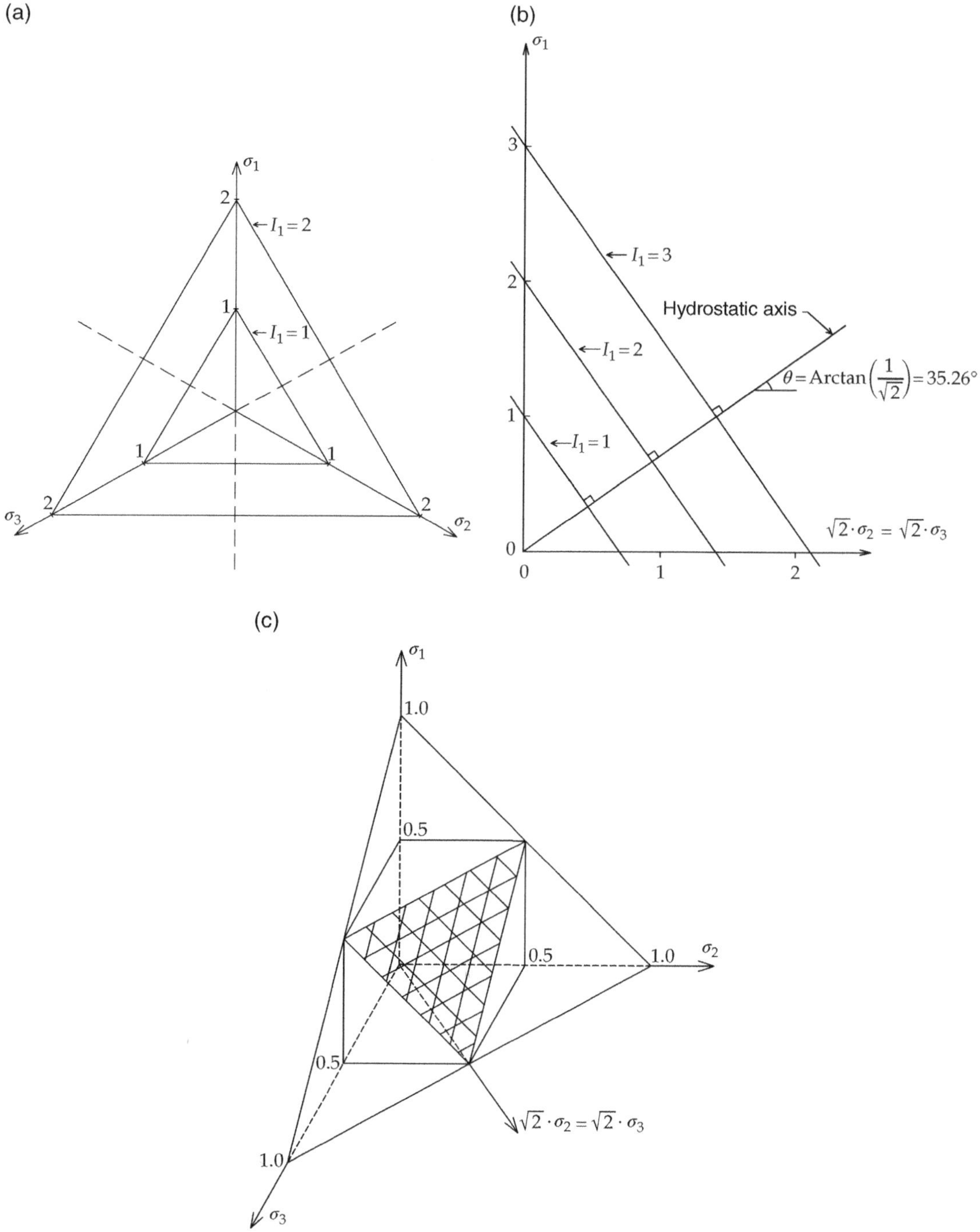

Figure 2.87 I_1 = constant plotted on (a) an octahedral plane, (b) a triaxial plane, and in (c) principal stress space.

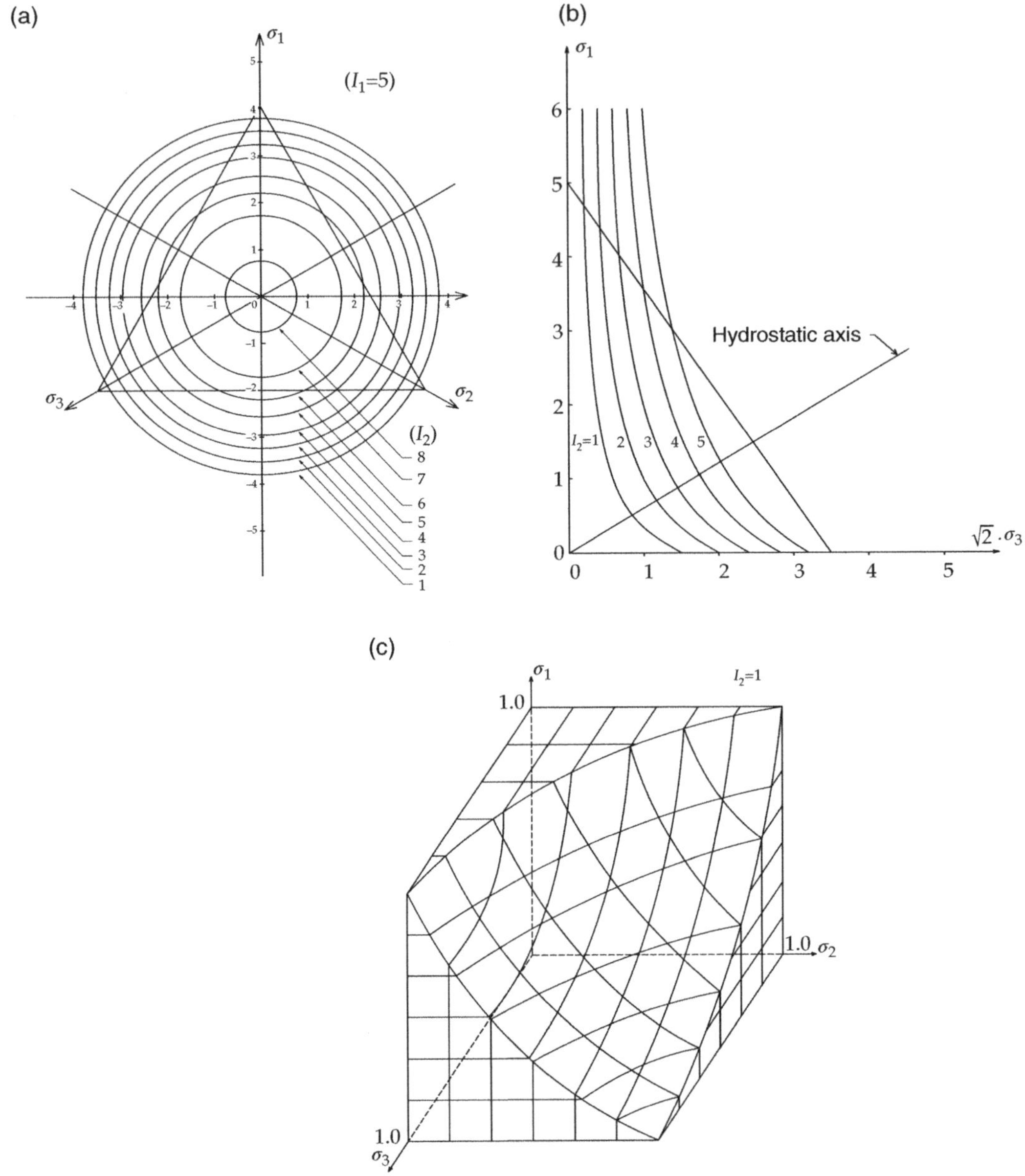

Figure 2.88 I_2 = constant plotted on (a) an octahedral plane, (b) a triaxial plane, and in (c) principal stress space.

This translation along the hydrostatic axis causes the failure surface to go through the translated stress origin, it accounts for the cohesion and tensile strength of the materials as explained by Lade (1982b), and it allows a straight line projection onto the common octahedral plane of the type indicated in Eq. (2.156).

The stress state ($\sigma_1{}^*$, $\sigma_2{}^*$, $\sigma_3{}^*$) for various tests, which has been modified to have a common octahedral normal stress, may now be plotted on the common octahedral plane as shown below.

For soils with curved failure surfaces in meridian planes, it is necessary to project the stress

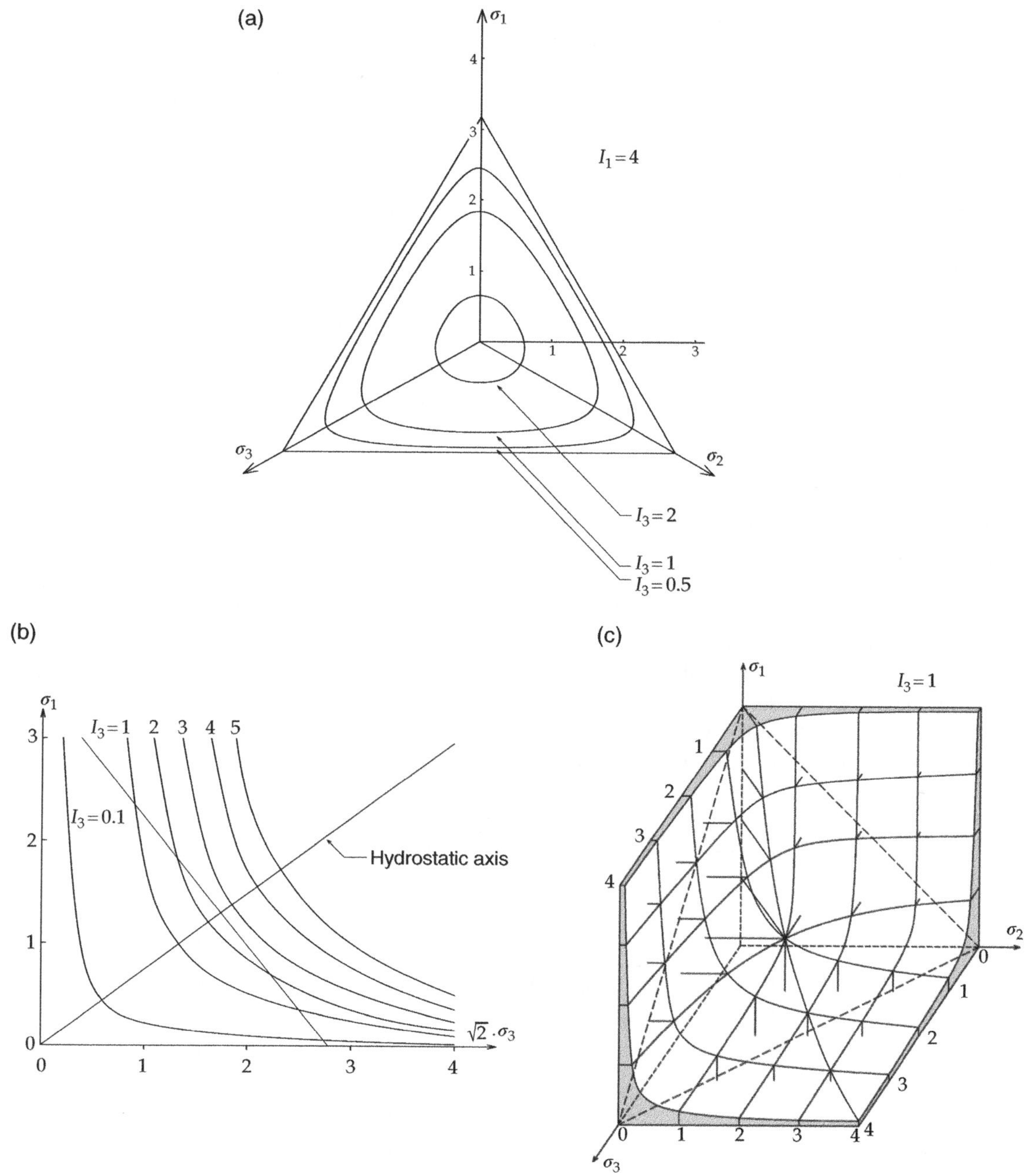

Figure 2.89 I_3 = constant plotted on (a) an octahedral plane, (b) a triaxial plane, and in (c) principal stress space.

points onto a common octahedral plane using a more complex procedure, as indicated below.

Curved failure envelope

The projection of a stress point along a curved failure surface onto a common octahedral plane is more complex, because the expression for the failure surface has to be known a priori, and it may involve solving a cubical equation as in the case presented here. The expression for the curved failure surface employed here is that proposed by Lade (1977):

$$\left(\frac{I_1^3}{I_3}-27\right)\left(\frac{I_1}{p_a}\right)^m=\eta_1 \qquad (2.158)$$

where I_1 and I_3 are the first and the third stress invariants, reviewed in Section 2.6.2, and p_a is atmospheric pressure expressed in the same units as the stresses. The parameters η_1 and m can be determined by plotting $\left(I_1^3/I_3-27\right)$ versus p_a/I_1 at failure in a log–log diagram, as shown in Fig. 2.90, and locate the best fitting straight line. The intercept of this line with $p_a/I_1=1$ is the value of η_1, and m is the slope of the line. For simplicity, it is assumed in the following development that the stresses have already been normalized by p_a and p_a will not be shown.

Figure 2.91 shows the known stress state (σ_1', σ_2', σ_3') at point 1, and the adjusted stress (σ_1^*, σ_2^*, σ_3^*) at point 2 is to be determined by projecting point 1 along a line with slope m to reach the desired common octahedral plane characterized by the first stress invariant I_{12}. The failure state at point 1 is defined by:

$$\left(\frac{I_1^3}{I_3}-27\right)_1\cdot\left(I_{11}\right)^m=\bar{\eta}_1 \qquad (2.159)$$

The failure state at point 2 can also be described by:

$$\left(\frac{I_1^3}{I_3}-27\right)_2\cdot\left(I_{12}\right)^m=\bar{\eta}_1 \qquad (2.160)$$

which may be rearranged to:

$$\left(\frac{I_{12}^3}{I_{32}}\right)=\frac{\bar{\eta}_1}{I_{12}^m}+27 \qquad (2.161)$$

and

$$I_{32}=\frac{1}{\dfrac{\bar{\eta}_1}{I_{12}^m}+27}\cdot I_{12}^3 \qquad (2.162)$$

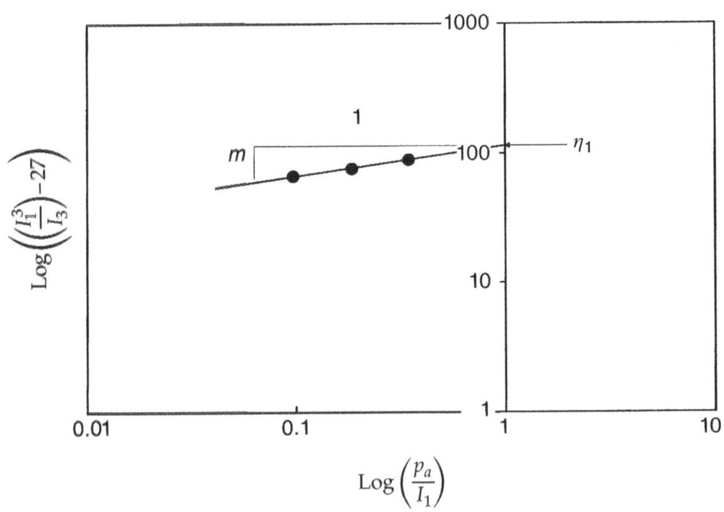

Figure 2.90 Determination of the best fit straight line on a log–log diagram for characterization of the 3D failure criterion.

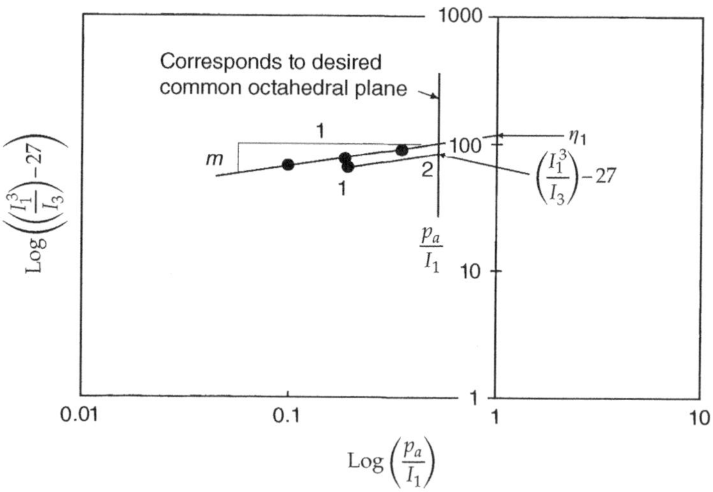

Figure 2.91 Procedure used for projection of stress point along a curved failure surface to a common octahedral plane.

The parameter b is defined as:

$$b = \frac{\sigma_2^* - \sigma_3^*}{\sigma_1^* - \sigma_3^*} \quad (2.163)$$

from which the intermediate principal stress can be expressed as:

$$\sigma_2^* = \sigma_3^* + b\cdot(\sigma_1^* - \sigma_3^*) \quad (2.164)$$

The desired first stress invariant is:

$$I_{12} = \sigma_1^* + \sigma_2^* + \sigma_3^* \quad (2.165)$$

With substitution of σ_2^* from Eq. (2.164) into Eq. (2.165) the following is obtained:

$$\begin{aligned} I_{12} &= \sigma_1^* + \sigma_3^* + b\cdot(\sigma_1^* - \sigma_3^*) + \sigma_3^* \\ &= (1+b)\cdot\sigma_1^* + (2-b)\cdot\sigma_3^* \end{aligned} \quad (2.166)$$

from which the desired major principal stress can be expressed as:

$$\sigma_1^* = \frac{1}{(1+b)}\cdot\left[I_{12} - (2-b)\cdot\sigma_3^*\right] \quad (2.167)$$

The desired third stress invariant is given by:

$$I_{32} = \sigma_1^*\cdot\sigma_2^*\cdot\sigma_3^* \quad (2.168)$$

The expression for σ_2^* in Eq. (2.164) is substituted into Eq. (2.168):

$$\begin{aligned} I_{32} &= \sigma_1^*\cdot\left[\sigma_3^* + b\cdot(\sigma_1^* - \sigma_3^*)\right]\cdot\sigma_3^* \\ &= \sigma_1^*\cdot(\sigma_3^*)^2 + b\cdot(\sigma_1^*)^2\cdot\sigma_3^* - b\cdot\sigma_1^*\cdot(\sigma_3^*)^2 \\ &= (1-b)\cdot\sigma_1^*\cdot(\sigma_3^*)^2 + b\cdot(\sigma_1^*)^2\cdot\sigma_3^* \end{aligned} \quad (2.169)$$

The expression for σ_1^* in Eq. (2.167) is substituted into Eq. (2.169):

$$\begin{aligned} I_{32} &= (1-b)\cdot\frac{1}{1+b}\cdot\left[I_{12} - (2-b)\cdot\sigma_3^*\right]\cdot(\sigma_3^*)^2 \\ &\quad + b\cdot\frac{1}{(1+b)^2}\cdot\left[I_{12} - (2-b)\cdot\sigma_3^*\right]^2\cdot\sigma_3^* \end{aligned} \quad (2.170)$$

Rearrangement of Eq. (2.170) produces:

$$\begin{aligned} I_{32} &= \frac{b\cdot(2-b)^2 - (1+b)(1-b)(2-b)}{(1+b)^2}\cdot(\sigma_3^*)^3 \\ &\quad + \frac{(1+b)(1-b) - 2b(2-b)}{(1+b)^2}\cdot I_{12}\cdot(\sigma_3^*)^2 \\ &\quad + \frac{b\cdot I_{12}^2}{(1+b)^2}\cdot(\sigma_3^*) \end{aligned} \quad (2.171)$$

Further rearrangement produces the following cubical equation:

$$[b\cdot(2-b)^2-(1+b)(1-b)(2-b)](\sigma_3{}^*)^3 \\ +[(1+b)(1-b)-2b(2-b)]\cdot I_{12}\cdot(\sigma_3{}^*)^2 \\ +b\cdot I_{12}^2\cdot(\sigma_3{}^*)-(1+b)^2\cdot I_{32}=0 \qquad (2.172)$$

or

$$(\sigma_3{}^*)^3+\frac{(b^2-4b+1)\cdot I_{12}}{(2-b)(2b-1)}\cdot(\sigma_3{}^*)^2+\frac{b\cdot I_{12}^2}{(2-b)(2b-1)}\cdot(\sigma_3{}^*) \\ +\frac{-(1+b)^2\cdot I_{32}}{(2-b)(2b-1)}=0 \qquad (2.173)$$

This is a cubical equation in (σ_3*). It is in the form of:

$$x^3+Ax^2+Bx+C=0 \qquad (2.174)$$

The solution to the cubical equation is given by Korn and Korn (1961), as reviewed in Section 2.6.4.

For $b \neq 0.5$, the solution to the cubical equation in Eq. (2.173) becomes:

$$b<0.5:\sigma_3{}^*=x_2=-2\cdot\sqrt{-\frac{p}{3}}\cdot\cos\left(\frac{\alpha}{3}+60^\circ\right)-\frac{A}{3} \qquad (2.175)$$

$$b>0.5:\sigma_3{}^*=x_3=-2\cdot\sqrt{-\frac{p}{3}}\cdot\cos\left(\frac{\alpha}{3}-60^\circ\right)-\frac{A}{3} \qquad (2.176)$$

in which A is the coefficient to $(\sigma_3{}^*)^2$ in Eq. (2.173), and p and α are determined from Eqs (2.129) and (2.133), respectively.

For $b = 0.5$, Eq. (2.172) becomes a quadratic equation:

$$\left(b^2-4b+1\right)\cdot I_{12}\cdot\left(\sigma_3{}^*\right)^2+b\cdot I_{12}^2\cdot\left(\sigma_3{}^*\right)-\left(1+b\right)^2\cdot I_{32}=0 \qquad (2.177)$$

The solution to Eq. (2.177) is:

$$\sigma_3{}^*=\frac{1}{3}\cdot I_{12}-\frac{1}{3}\cdot\sqrt{\frac{I_{32}}{I_{12}}}\cdot\sqrt{\frac{I_{12}^3}{I_{32}}-27} \qquad (2.178)$$

Consequently, depending on the value of b, which will be retained from point 1 to point 2, the minor principal stress σ_3* can be calculated from Eq. (2.175), Eq. (2.176), or Eq. (2.178). Then the major principal stress can be calculated from Eq. (2.167), and subsequently, the intermediate principal stress can be obtained, Eq. (2.164).

Because the principal stresses depend on stress invariants, these principal stresses are themselves invariants, and it may be easier to work with them directly.

The stress state (σ_1*, σ_2*, σ_3*), which has been projected to the common octahedral plane, may be plotted on the octahedral plane as shown below.

2.7.7 Procedure for plotting stress points on an octahedral plane

The point P corresponding to the modified principal stresses σ_1*, σ_2*, and σ_3* can be placed on the octahedral plane according to the following procedure. The principal stress space with the octahedral plane that contains the stress point P is shown in Fig. 2.92(a). The procedure for finding the distances between projection of P on the principal stress axes in the octahedral plane and the hydrostatic axis will be demonstrated. The calculations all pertain to the triaxial planes, and the plane containing the σ_1-axis is shown in Fig. 2.92(b). The point in which the hydrostatic axis crosses the octahedral plane is designated O′. The distance between the origin and the projection of O′ on the σ_1-axis is equal to $\sigma_{oct}=\frac{1}{3}\cdot(\sigma_1{}^*+\sigma_2{}^*+\sigma_3{}^*)$, as may be seen from Fig. 2.92(b). The projection of P on the triaxial plane is designated P′. The distance O′P′ = a then becomes [see Fig. 2.92(b)]:

$$a=\left(\sigma_1{}^*-\sigma_{oct}\right)\cdot\frac{1}{\cos\theta} \qquad (2.179)$$

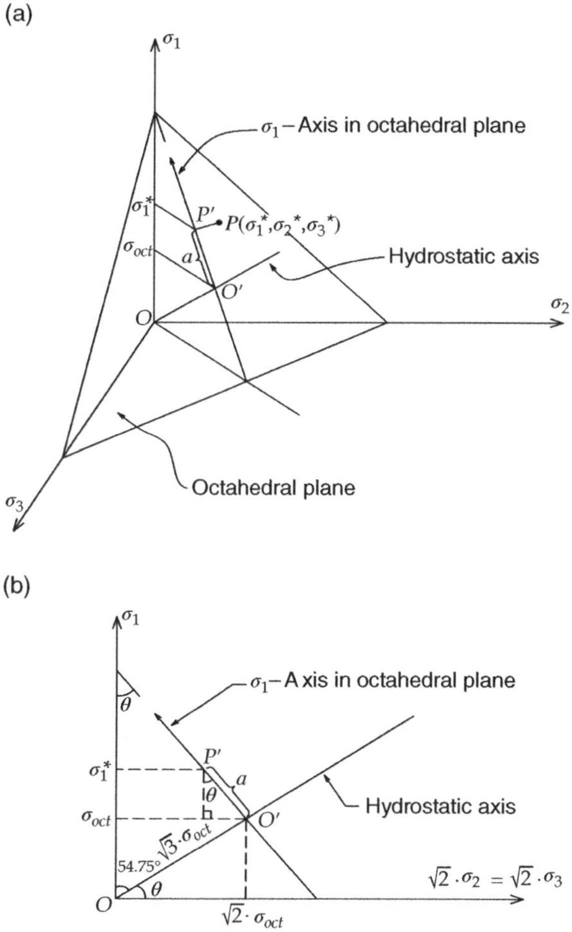

Figure 2.92 Principal stress space with determination of distance *a* on an octahedral plane.

where $\cos\theta = \cos(35.26°) = \sqrt{\frac{3}{2}}$ such that:

$$a = \sqrt{\frac{3}{2}} \cdot (\sigma_1{}^* - \sigma_{oct}) \quad (2.180)$$

Similar expressions can be obtained for the corresponding values of b and c:

$$b = \sqrt{\frac{3}{2}} \cdot (\sigma_2{}^* - \sigma_{oct}) \quad (2.181)$$

$$c = \sqrt{\frac{3}{2}} \cdot (\sigma_3{}^* - \sigma_{oct}) \quad (2.182)$$

With known values of the distances a, b, and c, the point P can be plotted on the octahedral plane, as shown in Fig. 2.93(a). The lengths a, b, and c are marked out on the axes in the octahedral plane and lines perpendicular to the axes are drawn to intersection to give the position of stress point P. It is seen that it is necessary to use only two of the three values a, b, and c.

For the purpose of simplifying the positioning of point P on the octahedral plane, the following trigonometric considerations are made so the coordinates of P are determined in the X–Y diagram shown in Fig. 2.93(b). The y-coordinate is equal to a:

$$y = \sqrt{\frac{3}{2}} \cdot (\sigma_1{}^* - \sigma_{oct}) \quad (2.183)$$

The x-coordinate is determined as shown in Fig. 2.93(b):

$$x = -c \cdot \cos 30° - (a + c \cdot \sin 30°) \cdot \tan 30°$$
$$= -c \cdot \left(\frac{\sqrt{3}}{2} + \frac{1}{2} \cdot \frac{\sqrt{3}}{3} \right) - a \cdot \frac{\sqrt{3}}{3}$$

$$x = \frac{\sqrt{3}}{3} \cdot (-a - 2c) \quad (2.184)$$

Substituting the values of a and c from Eqs (2.180) and (2.181) into Eq. (2.182) and reducing, the x-coordinate becomes:

$$x = \frac{\sqrt{2}}{2} \cdot (\sigma_2{}^* - \sigma_3{}^*) \quad (2.185)$$

Using Eqs (2.183) and (2.185), the point P(σ_1, σ_2, σ_3) can readily be located on the octahedral plane, as shown in Fig. 2.93(b).

2.7.8 Representation of test results with principal stress rotation

For experiments with rotation of principal stresses, shear stresses are applied to the surface of the specimen as is done in torsional, directional shear and direct shear tests. The rotation

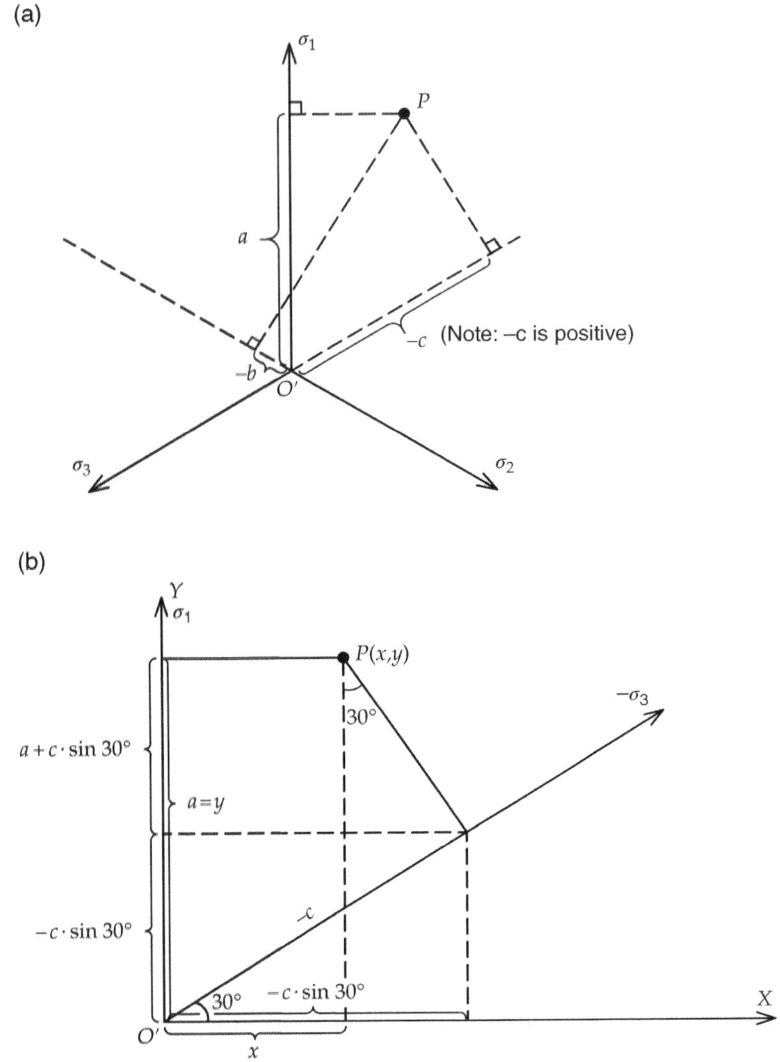

Figure 2.93 Location of stress point P on an octahedral plane from values of (a) *a*, *b*, and *c* or (b) *x* and *y*.

of principal stresses that occur in such tests cannot be indicated in the principal stress space and another diagram may be employed to show the results of such tests. In this diagram the applied shear stress $\tau_{z\theta}$ is plotted versus the stress difference $(\sigma_z - \sigma_\theta)$ expressed in polar coordinates. Such diagrams are shown in Chapter 11.

3 Triaxial Equipment

3.1 Triaxial setup

The equipment necessary for performance of a triaxial test consists of the triaxial specimen setup located inside a triaxial cell filled with fluid, a confining pressure supply and vertical loading equipment, as shown schematically in Fig. 3.1. The principal components employed in the triaxial specimen setup are shown schematically in Fig. 3.2 and they are reviewed below. The instrumentation, measurement systems and control schemes are presented in Chapter 4.

3.1.1 Specimen, cap, and base

Specimen dimensions

Due to the variability of parameters such as unit weights, modulus values, shear strength parameters, and permeabilities, large specimens tend to show less variability than small specimens, and for that reason, large specimens are preferable. In choosing a representative specimen size, the heterogeneities in the field must be considered, as must also the cost of testing larger specimens. Most often the triaxial specimen has a cylindrical shape with diameters varying from 35 mm (1.4 in.) to 150 mm (6.0 in.). However, other cross-sections, such as square and rectangular, may also be employed. Specimens with larger diameters may be required when testing soils with larger grain sizes, such as gravel and rockfill. To avoid inappropriately large grain sizes inside the triaxial specimen, the specimen diameter should be at least equal to six times the largest particle size for uniformly graded material and at least eight times the largest particle size for well-graded material (Marachi *et al.* 1972; Wong *et al.* 1975).

H/D-ratio

The height (H) of the triaxial specimen is usually between 2.0 and 2.5 times the diameter (D). Considerations resulting in this range of H/D-ratios include overcoming effects of end restraint due to friction on the end plates as well as allowing shear bands to develop freely and avoiding interception by the end plates, as shown in Fig. 3.3 (Lade *et al.* 1996; Wang and Lade 2001). Techniques such as using lubricated ends are available to reduce the friction on the end plates, as presented in Section 3.1.8. For such cases it is possible to reduce the specimen height, and an H/D-ratio of unity is often preferred. Experiments to indicate the adequacy of $H/D = 1.0$ are shown in Fig. 3.4.

Triaxial Testing of Soils, First Edition. Poul V. Lade.

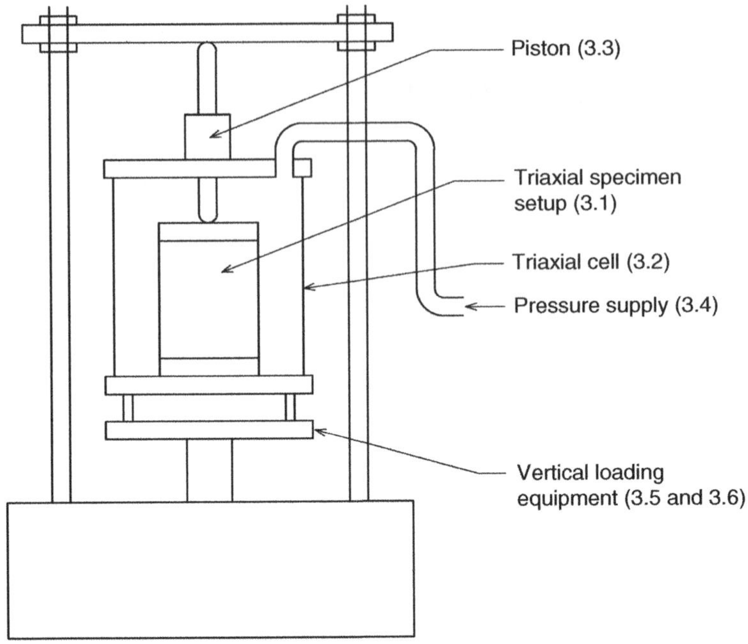

Figure 3.1 Triaxial setup with major components discussed in this chapter.

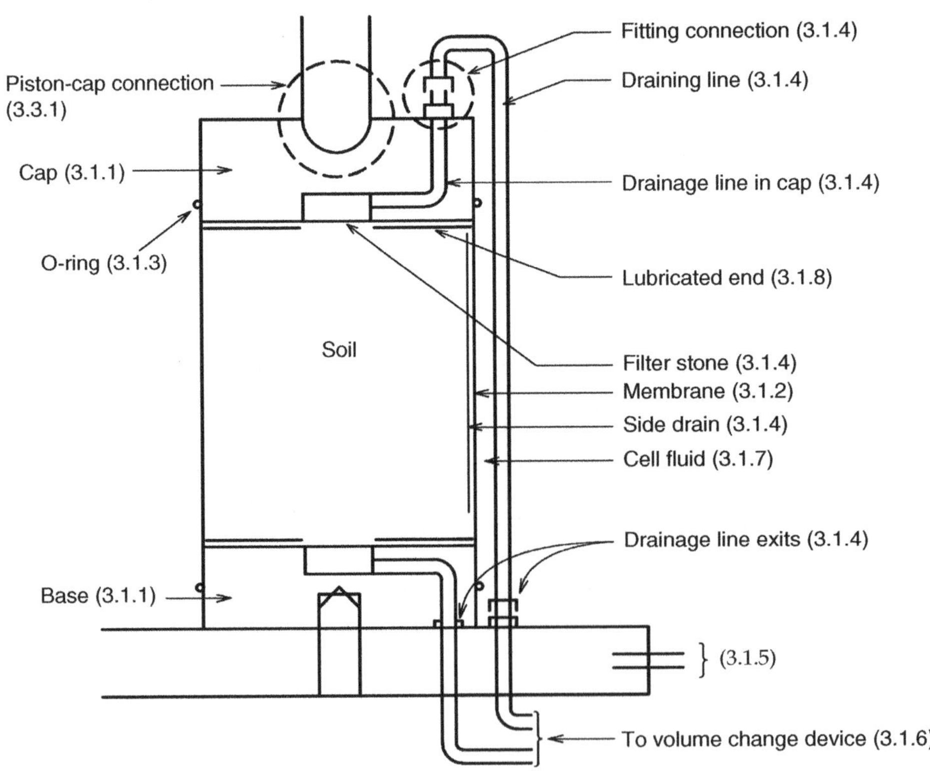

Figure 3.2 Triaxial specimen setup with components discussed in Sections 3.1 and 3.3.1.

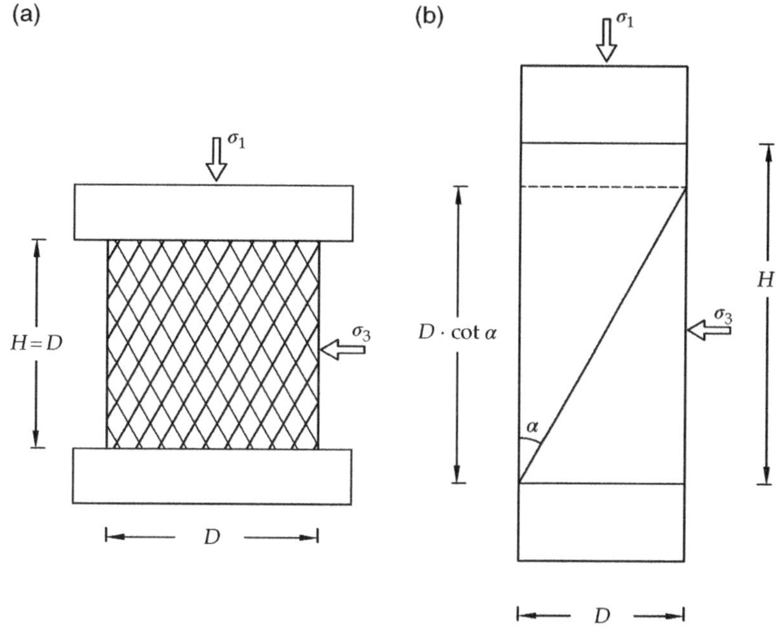

Figure 3.3 Testing techniques employed in triaxial compression tests: (a) short specimen ($H = D$) with lubricated ends and (b) conventional tall specimen. Reproduced from Lade *et al.* 1996 by permission of ASTM International.

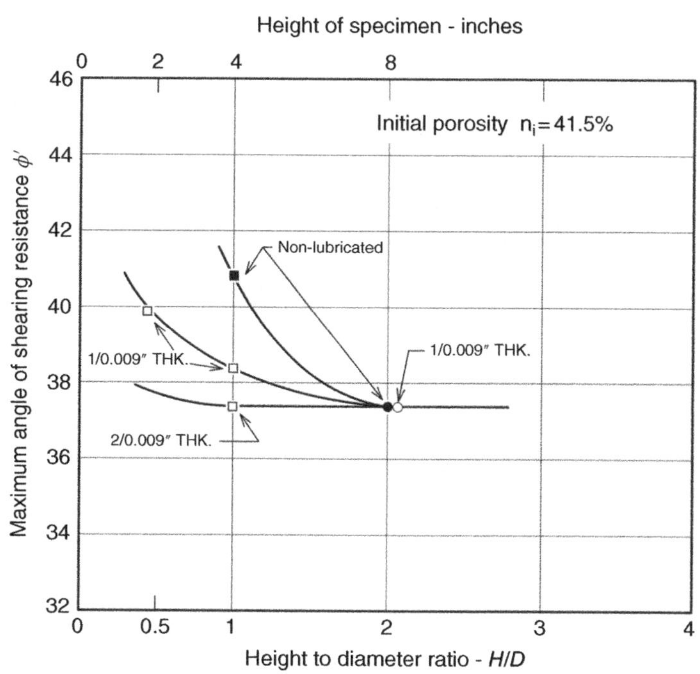

Figure 3.4 H/D-ratio versus φ to indicate adequacy of lubricated ends. Reproduced from Bishop and Green 1965 by permission of Geotechnique.

The choice of $H/D = 1.0$ together with the use of appropriately lubricated ends help promote uniform strains in the specimen and results in an appropriate failure condition that best represents the soil strength at the final, uniform void ratio. Because shear bands do not occur in the hardening regime in triaxial compression tests, this configuration is often preferred.

Successful experiments have been performed on much shorter specimens of intact San Francisco Bay Mud with $H/D = 0.36$ and lubricated ends. These tests resulted in very similar stress–strain relations, including post-peak softening, as obtained from specimens with $H/D = 1.0$ and lubricated ends. Thus, it is feasible to perform triaxial compression tests on very short specimens, as further discussed in Section 8.3.4.

The higher H/D-ratio of 2.5, or possibly slightly higher, is preferred when shear banding is studied (Lade 1982a; Lade *et al.* 1996; Wang and Lade 2001). This avoids interference of shear bands with the end platens. Note that while the shear bands may not attempt to pass through the end plates, their inclinations may be biased by the proximity of the end plates. Thus, for dense specimens to be tested at low confining pressures, the best combination consists of $H/D = 2.5$–2.7 and employment of lubricated ends. Buckling may occur if the H/D-ratio becomes too high.

Cap and base

The specimen is contained between a cap and a base and a surrounding membrane, as seen in Fig. 3.5(a). The cap and base have cross-sectional shapes to fit the specimen. Thus, cylindrical specimens are tested with circular end plates with the same or slightly enlarged diameters relative to the specimen, as shown in Fig. 3.3(a). Because these end plates are used to transfer the axial load to the specimen, they are made of materials that are stiff relative to the soil to be tested. Thus, the end plates may be made of acrylic plastic (Lucite®, Perspex®, Plexiglass®), which is lightweight and easy to machine, for use with soft clays or for soils to be tested at low pressures. Other materials such as aluminum (tends to corrode in water unless it is hard-anodized), brass, or stainless steel (heavy,

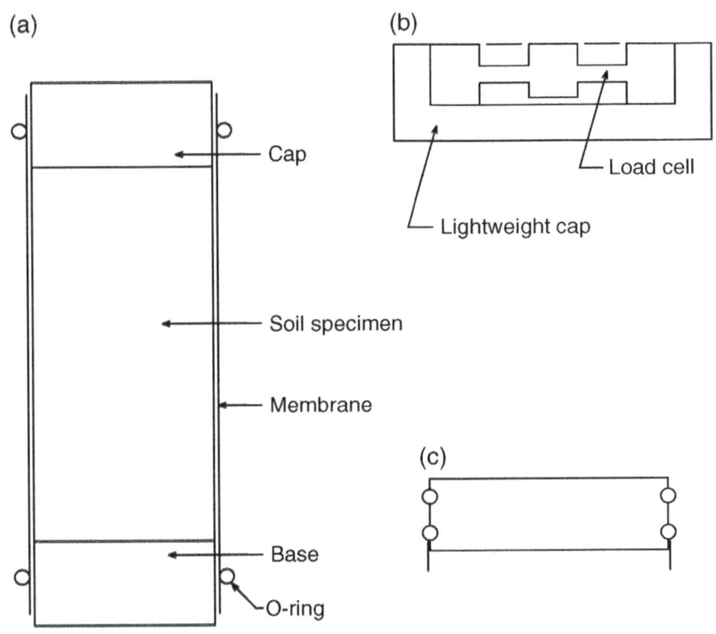

Figure 3.5 (a) Triaxial setup, (b) cap with load cell installed, and (c) grooves for two O-rings.

most difficult to machine) may be employed for testing of increasingly stiff soils at higher confining pressures. The cap may be hollowed out to make it lightweight or to install a load cell, as shown in Fig. 3.5(b). But this may result in reduced stiffness, and it may interfere with possible drainage connections embedded in the cap (see Section 3.1.4).

The cylindrical surfaces of the end plates must be smooth so that they provide good seals with the rubber membrane. Alternatively, grooves may be provided for the O-rings that seal the membrane to the end plates. Two or more O-rings may be used on each end plate to ensure good and safe seals.

If caps and bases with square or rectangular cross-sections are used, as is employed in some true triaxial or plane strain equipment, then the cross-section must transition into a more rounded or a circular shape, as shown in Fig. 3.6, so the O-rings can develop positive seals between the membrane and the end plates.

3.1.2 Membrane

The function of the membrane is to transmit a uniform cell pressure onto the soil specimen and to isolate the specimen and its pore fluid and/or pore air from the surrounding pressurized fluid or air in the triaxial cell. At the same time the membrane should carry a minimum of the applied axial load and not provide any resistance to the deformation of the specimen.

Figure 3.6 Caps and bases with transition of square or rectangular cross-sections into more rounded or circular cross-sections for both true triaxial and plane strain equipment.

Conventional geotechnical pressures

Membranes for soil testing are most often made of latex rubber and typical thicknesses are 0.05 mm (0.002 in.), 0.30 mm (0.012 in.), and 0.64 mm (0.025 in.). The circumference of the cylindrical membrane is made to match the circumference of the specimen. Two membranes of the thin type are often used in conjunction with a thin smear of stopcock silicone grease between them. This reduces the possibility of a leak due to puncture, and it reduces the amount of air that can diffuse through the membrane. Leakage through the membrane may also be reduced by using metal foil between the membranes (J.F. Peters, personal communication, 2015). The foil is cut into pieces an inch or so square. The inner membrane is smeared with silicone grease and the foil pieces are placed, overlapping like fish scales, to provide a barrier to air while providing flexibility. When done on specimens of compacted, partially saturated soil, the need for correction for membrane reinforcement effects is negligible. Leakage through the membrane may also be minimized by choosing an appropriate cell fluid, as discussed in Section 3.1.7.

While clays and fine sands present no problems with regard to membrane perforation, specimens with larger and sharp soil grains may experience puncture when the cell pressure is increased or during the shearing stage. This is best avoided by using two or more medium thick membranes with stopcock silicone grease between them rather than one very thick membrane (Yamamuro and Lade 1996; Bopp and Lade 1997a, b). This is because a puncture of the inner membrane will not propagate to the next membrane, whereas one thick membrane may let such a ripping mode continue through the entire thickness of the membrane.

Most often membranes are made as thin-walled cylindrical tubes and manufactured to fit the cylindrical size of the triaxial specimen by dipping a mandrel into fluid latex, lifting it up and letting it dry. Typically, one dip produces membranes with thickness of 0.30 mm (0.012 in.), and two dips produce a thickness of 0.64 mm (0.025 in.). Latex rubber membranes may be made to fit any shape, and the manufacturing process is detailed in Appendix A.

The much thinner membranes (0.05 mm = 0.002 in.) are made from commercially available, smooth condoms (prophylactics, e.g., Trojan, Ramses). These latex rubber membranes are manufactured to high standards of being leak proof and not easily breakable and they fit specimens with a diameter of 35 mm (1.4 in.). They are particularly useful for clay specimens, but may also be used for fine sand specimens, which are less likely to puncture the thin rubber. Because they are very thin, they carry very little load and experimental results require small to no corrections for loads carried by the membrane (see Chapter 9).

A membrane may also be created directly on the surface of a clay specimen, thus avoiding trapping air or water between the membrane and the specimen. After trimming the specimen to its final dimensions and placing it between the cap and base, the surface of the specimen is first coated with diluted rubber cement using an air brush painting device. Because the diluting agent in the rubber cement is a fluid (solvent and thinner) other than water, the water in the specimen is sealed off by the dried rubber cement layer. This is followed by spraying latex rubber diluted with water on the surface after the rubber cement has dried. Thus, there is no connection between the water in the specimen and the water that evaporates from the latex rubber to create the membrane. Consequently, there is no uncertainty about any intrusion of water into the specimen or any evaporation of water from the specimen during the drying process that creates the membrane.

Because the diluted rubber cement is essentially colorless, a coloring agent such as Oil Red O dye (a powder) may be mixed into the diluted rubber cement so it is possible to see where the specimen has been sprayed. Such a dye is not required in the fluid latex, which is milky white before it dries to a clear slightly yellowish color. Additional layers of fluid latex rubber may be sprayed or painted on the surface to strengthen the membrane.

In a similar fashion a membrane may be created by painting a sturdier and less permeable specimen of overconsolidated clay, such as boulder clay, directly with fluid latex and letting it dry before applying additional layers to thicken the membrane (Jacobsen 1967, 1970). This technique may be used to enclose a less than perfectly smooth specimen, perhaps with a pebble sticking out from the side, thus avoiding trapping air between the membrane and the specimen.

Membranes made of latex rubber are permeable to smaller molecules from the cell fluid or air. This is because the latex rubber consists of polymers, that is long chains of molecules, which when dried, create a network of long strands with holes between them through which smaller molecules can penetrate. Therefore, latex rubber membranes are much more permeable to air than to water. However, using compressed air as the cell fluid can be very dangerous if the cell wall is made of acrylic plastic. This will be further discussed below. Thus, using de-aired water or other cell fluid with larger molecules rather than compressed air has great advantages for long-term experiments, as discussed in Section 3.4. Latex rubber membranes also absorb water, and part of this water may come from the specimen. Berre (1982) recommended saturating the membrane by soaking in de-aired (fresh or salt as appropriate) water for at least 3 days before using it for a triaxial test on clay.

Latex rubber membranes are commercially available with standard diameters of 35 mm (1.4 in.), 51 mm (2.0 in.), 71 mm (2.8 in.), 102 mm (4.0 in.), and 152 mm (6.0 in.). Other sizes can be produced on request by the manufacturers, or by fabricating them in-house, as explained in Appendix A.

Latex rubber membranes weaken in the presence of atmospheric air and ultraviolet light. Membranes are therefore best stored in opaque, closed plastic bags and placed in a refrigerator where the rate of chemical deterioration is reduced.

Membranes have also been manufactured from other materials such as silicone rubber, for example Silastic J-RTV type of silicon rubber (from Dow Corning Co.), which may be more stretchable and have higher tear resistance and lower stiffness than latex rubber membranes. Sture and Desai (1979) explain the procedures for mixing the compounds and forming the silicone rubber membranes. These are more difficult to produce and combining this with the commercial availability of latex rubber membranes results in the use of the latter type for most applications.

High effective confining pressures

For high confining pressure experiments (up to 67 MPa) Yamamuro and Lade (1996) used up to five 0.64 mm thick latex membranes with silicone grease between them to avoid puncture and to avoid leakage at these high effective confining pressures. Up to four O-rings each on the cap and base were required to seal the membranes to the end plates. The O-rings were installed on the outside of all membranes.

Yamamuro and Lade (1996) found that the membranes could be reused after high pressure tests depending on the confining pressure. The inside membrane was usually punctured too much to be of any further use. Generally, the second membrane had some small punctures, but it could be used again as a padding layer in a subsequent test. All other membranes could be reused, either as the inside membrane or as outer membranes.

Vesic and Clough (1968) tested sand at high pressures (2–63 MPa) using 1.3 mm (0.05 in.) thick plasticized polyvinyl chloride membranes with a minimum tensile strength of 12 MPa (1800 psi) and a maximum elongation of 200%. Colliat-Dangus *et al.* (1988) employed a 0.5 mm (0.02 in.) thick high-quality neoprene membrane for tests with confining pressures up to 5 MPa. For higher pressures, two neoprene membranes were used.

Very low effective confining pressures

Finally, experiments may be performed on water saturated clay specimens at low effective confining pressures completely without any membrane by using a cell fluid consisting of paraffin or kerosene. In such experiments, the surface tension between the water inside the specimen and the paraffin in the cell creates a barrier which acts as a membrane that keeps the water inside and the cell fluid outside the specimen. Experiments have been performed on occasion with this paraffin technique (Ramanatha Iyer 1973, 1975; Iversen and Moum 1974; Berre 1982), but it has not been widely used. This topic is further discussed in Chapter 10.

Corrections

Corrections for the load carried by the membrane and to the measured volume change due to membrane penetration are discussed in Chapter 9.

3.1.3 O-rings

O-rings are stretched over the membrane to seal it to the cap and base. Leroueil *et al.* (1988) recommend polishing the sides of the cap and base, and smearing a thin layer of silicone grease on their sides before placing the membrane and the O-rings. Since the O-rings are typically a little smaller than the diameter of the end plates, so that they will press the membrane towards the sides of the end plates, it is necessary to stretch and place them with a minimum of disturbance of the soil specimen. This may be done by using an O-ring stretcher consisting of a small piece of tube with a diameter to fit over the cylindrical end plates with the membranes stretched up and around them. The O-rings are placed on the stretcher, which in turn is placed around the end plate and the O-rings are flipped off the stretcher and onto their position around the cap and base, as shown in Fig. 3.7. This is further discussed in Chapter 5.

Different types of O-ring material are available as indicated in the manufacturers' catalogs.

For conventional triaxial testing where water or gas is used as cell fluid, the bueno-N rubber type is the most stretchable and easiest to work with. However, other types of rubber are resistant to certain chemicals and may be better for special conditions.

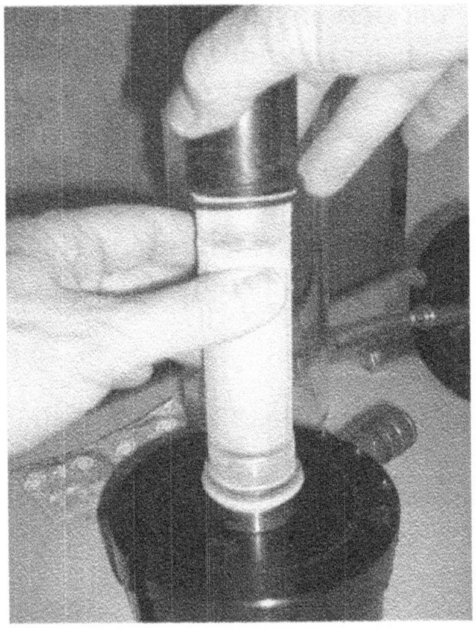

Figure 3.7 Photograph of flipping the O-ring onto the membrane on the cap using the ring finger.

3.1.4 Drainage system

The triaxial specimen setup for tests that are initiated with consolidation, as in CD and CU tests, requires a drainage system, whose function is to connect the water in the specimen with an external volume change device and with a pore pressure transducer, a schematic version of which was previously shown in Fig. 1.2. Basically two types of drainage systems are employed for triaxial tests.

End drains

One system is essentially a nominal system that is used for freely draining soils such as sand and gravel. Such systems may consist of a small hole in the base connected directly to the volume change device. A few strands of steel wool may be placed in the small hole to prevent soil grains from entering and flush out through the hole. This simple drainage system may be augmented with a similar hole in the cap, a small porous stone embedded in each of the end plates, or a stiff filterstone covering each of the end plates. Examples of simple end drainage systems are shown in Fig. 3.8. The simple design

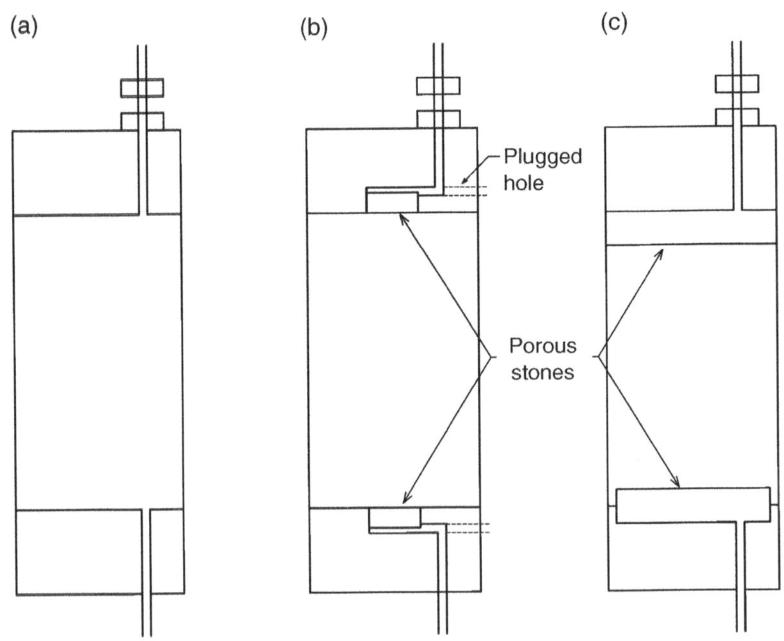

Figure 3.8 Simple drainage systems for freely draining soils.

in Fig. 3.8(a) may be used with free draining sand and gravel and no lubricated ends. The design in Fig. 3.8(b) with the small, centrally located porous stone allows employment of lubricated ends (see Section 3.1.8). To be able to flush the full size porous stones shown in Fig. 3.8(c) as well as the drainage lines, the end plates may be provided with a pattern of groves connected to two drainage ports, as shown in Fig. 3.9.

Porous stones

The porous stones in the form of circular disks may be made of sintered corundum particles, which are extremely hard aluminum oxide particles, or sintered bronze or stainless steel or brass particles, or porous plastic. The latter is the softest material and it comes in sheets with different particle gradations with consequent different hole sizes. The sintered corundum is very hard and therefore difficult to machine, while the porous bronze can be easily machined, and the porous plastic can be cut with a knife or a hole cutting tool. The compressibilities of the sintered corundum and the sintered bronze are so small that their presence inside the triaxial specimen setup may not play any role in the pore pressure development (due to their compression), while the porous plastic is so compressible that it may be necessary to account for its presence. Using full sized porous "stones" made of porous plastic as shown in Fig. 3.8(c) is not advisable, because their compression will generate false pore pressures in the specimen under undrained conditions. However, the porous plastic may be used for the small, centrally located drains in Fig. 3.8(b). They have the advantage that they can easily and inexpensively be replaced if they become clogged.

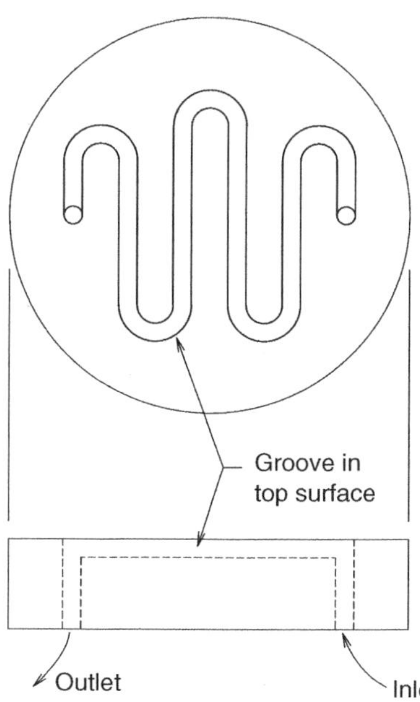

Figure 3.9 Groove pattern in end plates to allow flushing at each end of the specimen.

Unsaturated soils

Testing of unsaturated soils and their requirement for the setup and high air entry filter stones are discussed in Chapter 10.

Side drains

Less permeable soils such as clays often require additional drainage provisions to reduce the time required for testing. As for the testing of granular materials, either small or full size porous stones may be placed in or at the end plates to provide end drainage, as shown in Fig. 3.8(b and c). Additional drainage may be achieved by side drains created by wrapping filter paper or non-woven geotextile around the clay specimen to provide horizontal drainage and decrease the time for consolidation as well as the time for shearing under drained conditions, and to enhance the pore pressure equalization in CU-tests. This may be more effective because the maximum drainage path is equal to the radius, which is typically shorter than the distance to the end drains. It may also be more effective for vertically cored specimens, because the hydraulic conductivity is typically 2–10 times higher in the horizontal direction than in the vertical direction, depending on the clay type (Tavenas *et al.* 1983; Tavenas and Leroueil 1987). Varved clays may have permeabilities that are up to 40 times higher in the horizontal than in the vertical direction (Olson and Daniel 1981).

Because a continuous sheet of filter paper creates a stiff encasement of the specimen due to its shell action, large corrections to the vertical and

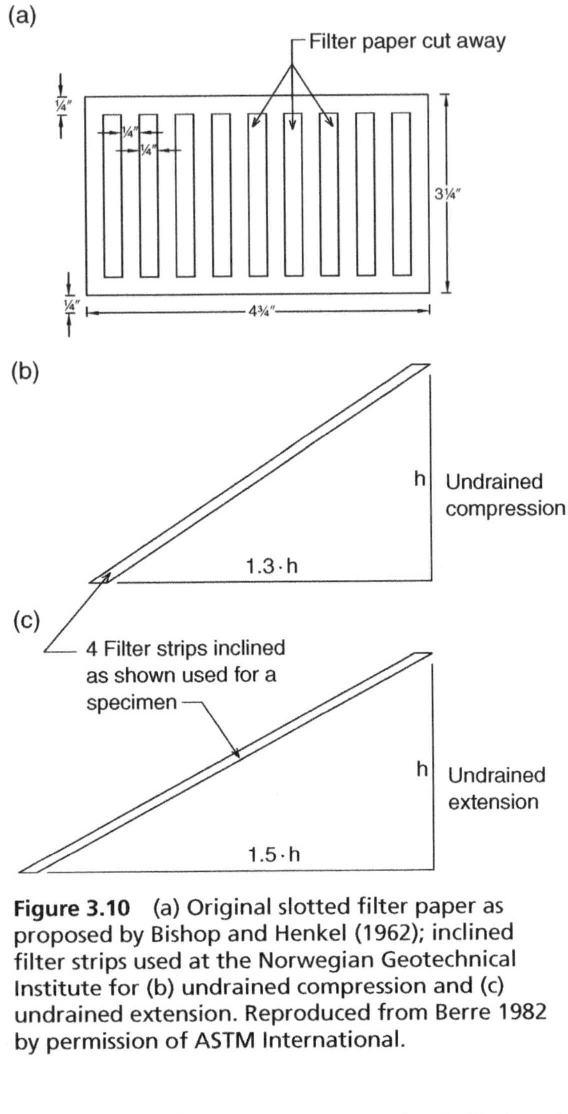

Figure 3.10 (a) Original slotted filter paper as proposed by Bishop and Henkel (1962); inclined filter strips used at the Norwegian Geotechnical Institute for (b) undrained compression and (c) undrained extension. Reproduced from Berre 1982 by permission of ASTM International.

the horizontal stresses are required. To break this shell action, vertical cuts may be made in the filter paper sheet, thus avoiding the correction to the horizontal pressure. Bishop and Henkel (1962) proposed to cut vertical slots in the filter paper, as shown in Fig. 3.10(a). The continuous rims of the filter paper are wrapped around the filterstones at the end plates so as to provide a path out to the external drainage system. The filter paper rim may also be provided with flaps that overlap filter disks located on the sides of the end plates, as shown in Fig. 3.11. A correction for the load carried by the filter paper may be

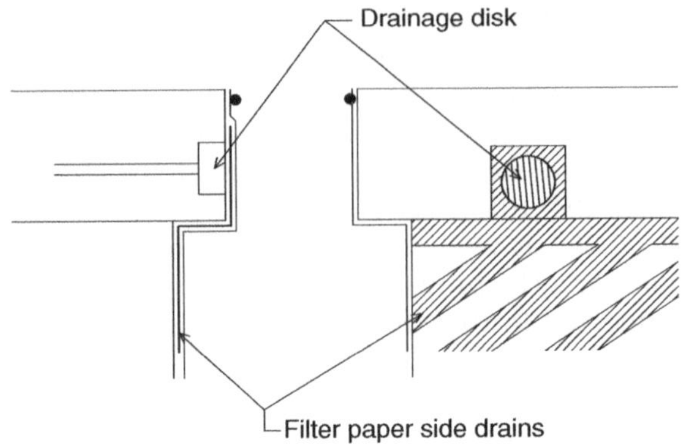

Figure 3.11 Connection between filter paper side drains and drainage disks on the side of the end plate.

required, especially for soft clay specimens. Unfortunately, the strength properties of filter paper are difficult to determine accurately, because it exhibits properties similar to soils, such as stiffness, strength, creep and relaxation, and so on. Besides the effectiveness in draining water become reduced due to clogging and collapse at higher confining pressures.

To reduce the vertical force (and consequent correction to the axial load) that the vertically slotted filter paper carries, Berre (1982) suggested that it is advantageous to incline the strips in the filter paper to a direction that experiences no strain, and since the filter paper is therefore not compressed it will not take any load. For an undrained test on a fully saturated specimen that is sheared in compression or extension, the specimen does not change volume, and for this condition, combined with small strains, the direction of zero strain on the surface of the cylindrical specimen is inclined at $V{:}H = 1{:}\sqrt{2} =$ 1: 1.414. Anticipating that some strains will occur before failure is reached in soft clay specimens, Fig. 3.10(b and c) show examples of inclinations used for undrained compression and undrained extension tests, respectively. For undrained two-way cyclic tests, the strips are inclined in between those for compression and extension. Employment of filter paper strips inclined as shown is considered to result in no correction to the vertical load (Berre 1982). However, this may be

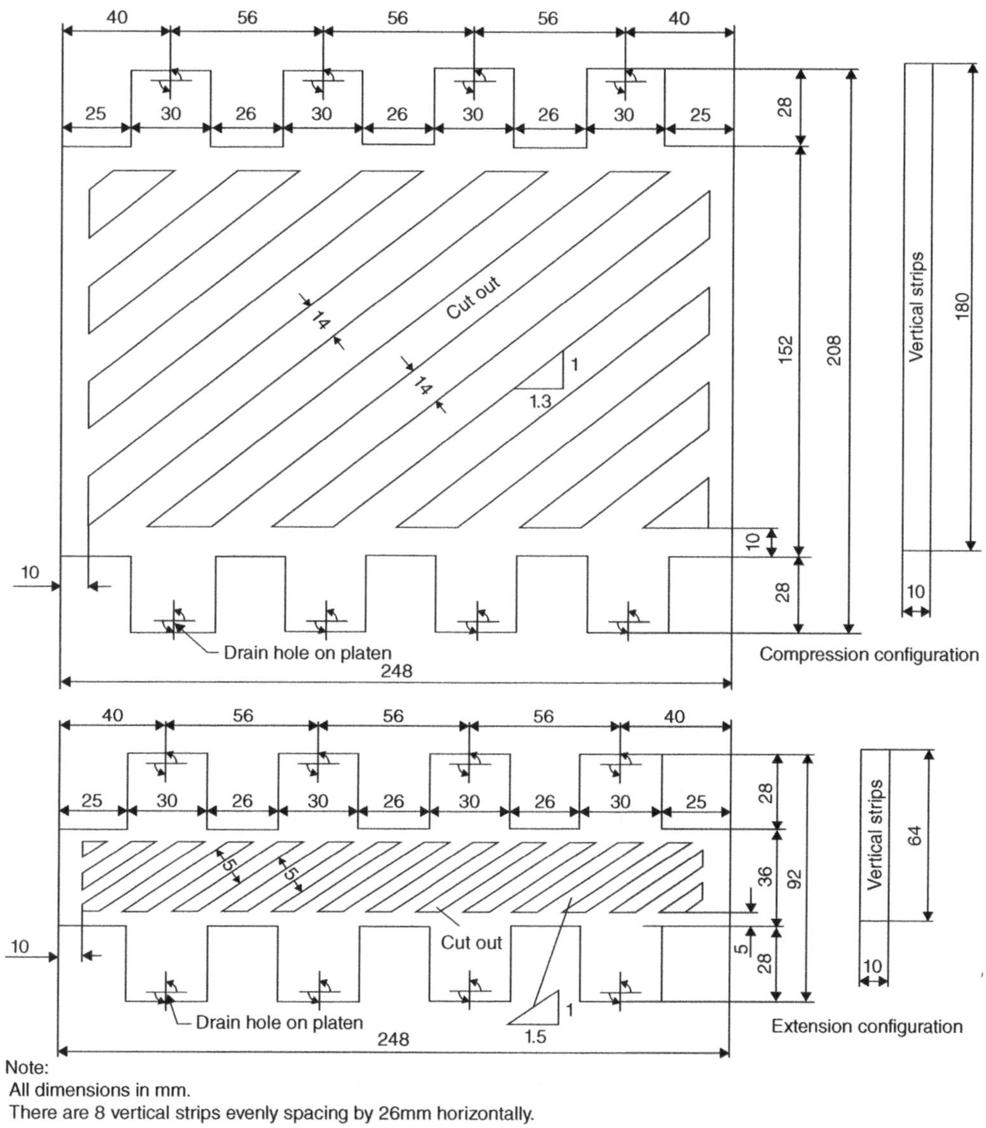

Figure 3.12 Frames with inclined, slotted filter paper (after Yamamuro *et al.* 2012).

reconsidered in view of the results given in Chapter 9. Figure 3.12 shows a frame with inclined filter paper slots for compression tests. For extension tests, Mitachi *et al.* (1988) found the most practical filter paper configuration to consist of a solid sheet with cuts inclined at V:H = 1:1.5.

While Whatman No. 54 is recommended for triaxial testing at conventional geotechnical pressures, it may be advantageous to use lower quality drains (such as Whatman No. 1) for testing weaker, lean clays, because the stiffnesses of these drains are lower than that of Whatman No. 54.

The flow capacity of all types of side drain materials is reduced as the effective confining pressure is increased (Bishop and Gibson 1964; Leroueil *et al.* 1988; Mitachi *et al.* 1988; Oswell *et al.* 1991), while the vertical load correction typically accounts for a reduced percentage of the soil strength. For moderate-to-high confining pressures (up to 1000 kPa), the flow capacity may be increased by using double layers of filter paper such as Whatman No. 54 and 541 (hardened, high-wet-strength, fast filtration rate) or a geotextile drain such as Mirafi 140NS (Oswell *et al.* 1991). Whatman No. 54 filter paper may retain its ability to transmit water for some time, but longer periods of time for consolidation allows the side drains to compress and reduce its flow capacity. A normalized comparison of flow capacities of various filter materials may be achieved by using transmissivity defined as (Giroud 1980; Oswell *et al.* 1991):

$$T = k_p \cdot h = \frac{Q/B}{\Delta p/(\rho_w g L)} \tag{3.1}$$

where

k_p = coefficient of lengthwise permeability in the plane of the drain material (m/s)
h = thickness of the drain (m)
Q = flow rate (m/s)
B = width (combined with of drain strips) (m)
Δp = hydraulic head loss (= inlet pressure minus outlet pressure) (N/m^2)
ρ_w = mass density of liquid (kg/m^3)
g = gravitational constant (= 9.81 m/s^2)
L = length of flow path (m)

The transmissivity reduces with increasing effective confining pressure for all types of filter materials. The Whatman No. 54 filter paper and many other types of filter paper collapse and essentially become impermeable at higher effective confining pressures. For pressures higher than approximately 1000 kPa, it may be necessary to switch from double layers of Whatman No. 54 filter paper to a much stiffer non-woven geotextile filter fabric. For example, the transmissivity of Mirafi 140NS remains relatively high, even at confining pressures near 5000 kPa (Oswell *et al.* 1991). Due to collapse of the filter material with time, the transmissivity reduces with time. Table 3.1 gives initial transmissivity values and values after 100 and 500 h for four filter materials wrapped around a brass specimen (i.e., no clogging from soil particles occurred) investigated by Oswell *et al.* (1991). It is clear that the double layer of Whatman No. 54 performed well and the nonwoven geotextile showed superior performance, but geotextile fabric is also much stiffer than filter paper and therefore requires larger corrections to the vertical load. Corrections to volumes or pore pressures due to compressibility of filter materials may also be necessary. Corrections for filter material side drains are discussed in Chapter 9.

Table 3.1 Comparison of effectivenesses of filter paper materials by their transmissivities. Values obtained by Oswell *et al.* (1991) for side drain materials on a brass specimen covered by a membrane. Confining pressure = 1000 kPa, inlet pressure = 250 kPa, and outlet pressure = 0 kPa

Filter material		Initial thickness h (mm)	Transmissivity (m^2/s · 10^{13})		
			Initial	After 100 h	After 500 h
Filter paper	Single layer Whatman No. 1	0.18	0.30	0.06	0.03
	Single layer Whatman No. 54	0.19	2.9	0.7	0.4
	Double layer Whatman No. 54	0.38	6.2	2.5	1.0
Nonwoven geotextile	Single layer Mirafi 140NS	0.40	490	290	270

Disadvantage of side drains

In addition to the correction to the vertical load due to the strength of the side drains, Carter (1982) pointed out that radial consolidation may result in non-uniform density across the cylindrical clay specimen. This is because the highest degree of consolidation is reached first next to the side drains, thus creating a cylindrical shell whose higher stiffness and strength prevent the inner portions of the clay specimen from consolidating fully corresponding to the applied consolidation pressure. This problem was investigated experimentally by Atkinson *et al.* (1985), and they found that cylindrical specimens of kaolin with side drains showed significant nonuniformity in water content across the specimen diameter. Specimens with diameter of 38 mm showed a variation in water content of 1.5% from the center to the periphery at the end of isotropic consolidation from 55 to 200 kPa. This implies substantial variations in effective stress across specimens with radial drainage. The problem may be avoided by using end drainage only.

Corrections

Corrections for the load taken by the side drains are discussed in Chapter 9.

Drainage lines

While the membrane is installed to isolate the specimen from the cell fluid, drainage lines are provided to connect the fluid in the specimen with the volume change device and the pore pressure transducer. These drainage lines go through the cap and base from the filter stones, connect through fittings to tubings inside the triaxial cell and connect through fittings to tubings attached to the volume change device outside the triaxial cell. The best drainage lines are continuous and connect smoothly and directly to the volume change device through the fewest number of fittings. In particular, the same constant bore tubings and fittings should be employed to avoid cavities where air may be trapped. Therefore, all drainage lines, which are part of the larger specimen fluid volume, should avoid fittings with tapered pipe threads, because they most often cannot be connected without creating a disk-shaped cavity that may be difficult to saturate.

In cap and base

Figure 3.8 shows examples of drainage lines drilled in the cap and base. These may be machined by drilling and plugging holes to direct the drainage line from the central porous stone to fitting near the edge of the cap (to avoid interference with the piston) and the base (to avoid interference with the central bolt used to attach the specimen base to the base plate of the triaxial cell).

Fitting connections to drainage lines

The best quality fittings and valves for the purpose of triaxial testing are made from brass and stainless steel by Swagelok®. Brass fittings and valves are sufficiently strong for most moderate to high pressure applications. However, these fittings come with pressure ratings, and there may be cases of high pressure or corrosive fluid applications where it is necessary to employ stainless steel fittings and valves.

Fittings that provide straight through connections to the drainage tubes are best suited for triaxial testing. As mentioned above, drainage lines should involve as few fittings as possible. Thus, a straight through line under the specimen base, sealed with an O-ring is better than fittings on the side of the base and through the base of the cell. Examples of connections are indicated in Fig. 3.13.

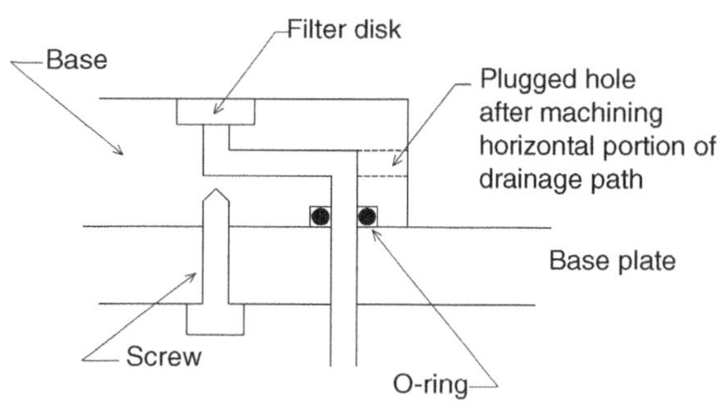

Figure 3.13 Example of connection of a drainage line in a triaxial cell.

Flexibility of tubing

The tubing employed for drainage should have minimal volumetric flexibility and the highest possible bending flexibility. For conventional triaxial testing it may be sufficient to employ tubings with diameters of 3 or 6 mm (1/8 or 1/4 of an inch) made of plastics such as nylon, polyethylene (relatively soft and weak), and polypropylene (relatively stiff and strong). Testing at high pressures may require tubes made of metals such as aluminum, copper and stainless steel. The tubings may be specified in terms of the internal and the external radii and they come with ratings with respect to maximum pressure, temperature range, bending radius, and so on. A bending tool may be required to produce smooth bends on metal tubes while preventing the tubes from buckling.

For conventional triaxial testing, clear, relatively flexible plastic tubes are preferred to be able to see the flow of fluids and possible air bubbles. Such plastic tubing may be bent permanently by heating it, carefully bending the tube, and cooling it down again.

For high pressure testing, thick-walled stainless steel tubes have minimum volumetric flexibility, while their bending flexibility may be enhanced by forming a spiral around the triaxial specimen to minimize the interference with the axial loading of the specimen.

3.1.5 Leakage of triaxial setup

For long-term testing it becomes important to minimize the leakage of the entire triaxial setup to obtain reliable test results. Leroueil *et al.* (1988) pointed out that leakage could occur at many different locations in the triaxial setup. Figure 3.14 shows a schematic diagram with sources of leakage between the pore fluid and the external fluid or air, indicated by numbers on the diagram, as follows: (1) through external fittings; (2) through fittings inside the cell and between the end plates and the membrane; (3) through the membrane due to the pressure difference between the cell and pore pressure and due to osmosis between the pore fluid and the cell fluid; (4) saturation of the membrane; and (5) diffusion inside the back pressure burette. These leakages, both into and out of the specimen, depend on the fittings, the membrane, the cell fluid, the effective confining pressure, and the duration of the test. While these leakages result in erroneous volume changes, they do not

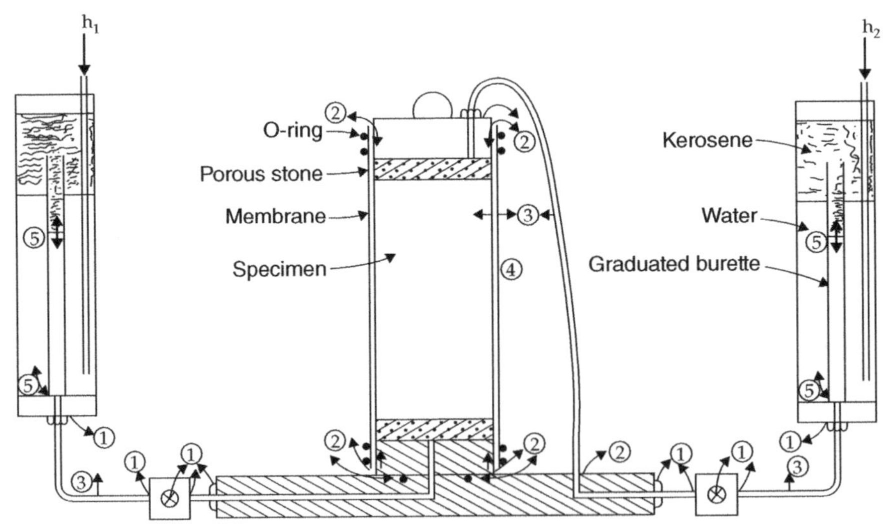

Figure 3.14 Sources of errors in triaxial test installation: (1) leakage in external fittings; (2) leakage in fittings within the cell; (3) osmosis and diffusion through membranes and lines; (4) saturation of membrane; (5) leakage and diffusion within back pressure burette. Reproduced from Leroueil *et al.* 1988 by permission of ASTM International.

affect the pore pressure and therefore do not change the effective stress path in drained tests. In undrained tests the leakage results in changing pore pressure and this in turn changes the effective stress path and therefore the measured soil behavior.

To minimize the problem of leakage from the fittings, Leroueil *et al.* (1988) developed a triaxial cell, shown in Fig. 3.15, in which the volume change and back pressure burette was directly attached to the triaxial cell. Thus, all fittings were enclosed in the back pressure compartment, and this eliminated the leakage through external fittings.

Leroueil *et al.* (1988) recommend polishing the sides of the cap and base, and smearing a thin layer of silicone grease on their sides before placing the membrane and the O-rings. The membrane itself is permeable, but the molecules in various cell fluids have different sizes and some cannot penetrate the latex rubber membrane. This is further discussed in Section 3.1.7.

3.1.6 Volume change devices

The drainage lines are connected to a volume change device and a pressure transducer outside the triaxial cell. The various types and operation principles available for volume change devices are reviewed in Section 4.8.

3.1.7 Cell fluid

The cell pressure is applied uniformly through a flexible rubber membrane by pressurized air or fluid. Conventional latex rubber membranes are permeable to smaller molecules from the cell fluid or air. This is because the latex rubber consists of polymers, that is long chains of molecules, which when dried create a network of long strands with holes between them through which smaller molecules can penetrate. Therefore, latex rubber membranes are much more permeable to air than to water. Using de-aired water or even better, a cell fluid with larger molecules, rather than compressed air, has great advantages for long-term experiments. When the smaller molecules penetrate through the membrane, they result in false volume change measurements in drained tests or in incorrect pore pressures in undrained tests. In addition to compressed air and de-aired water, a number of fluids, including glycerin, castor oil, kerosene, paraffin oil, and silicone oil, which all consist of larger molecules, have been suggested as cell fluids. A brief review of their

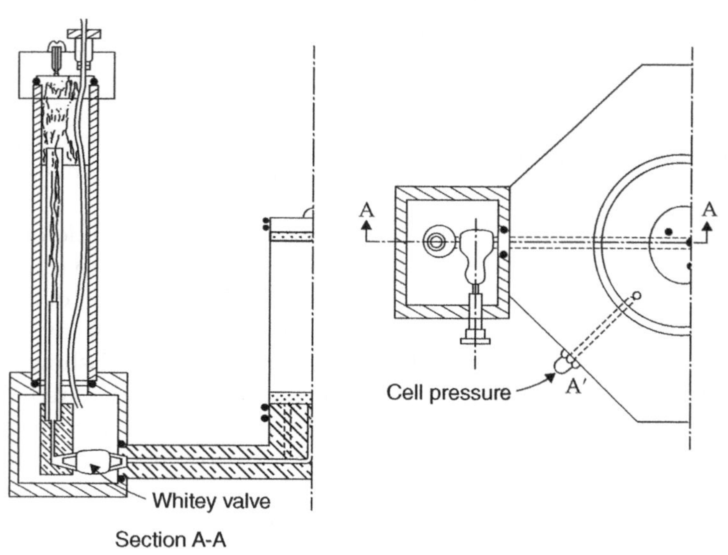

Figure 3.15 Triaxial cell setup for controlling leakage at fittings. Reproduced from Leroueil *et al.* 1988 by permission of ASTM International.

Table 3.2 Properties of some large molecule fluids that may be employed as cell fluid

Fluid	Density, ρ (kg/m^3)	Dynamic viscosity, μ [1 cP = 0.001 kg/(m·s)]	Surface tension, T (mN/m = dyn/cm)
Water	998.2	1.009	72.0
Glycerin	1260	1490	64.0
Castor oil	956	986	36–37
Kerosene	820	2.1–2.2	26–28
Paraffin oil	800	1.9	26
Silicone oil	960–971	48–971	20–21.5
(food grade)	970	340	21.1

All properties are given at room temperature = 20°C = 68°F.
Viscosity is a measure of inability to flow or of resistance to shear deformation. Dynamic viscosity = μ is expressed in centiPoise = 0.001 kg/(m·s) = 1 mPa·s = 0.1 N·s/m^2, while kinematic viscosity = $\nu = \mu/\rho$ is expressed in m^2/s or in Stokes, where 10^4 Stokes = 1 m^2/s.

properties and usefulness as cell fluids are given below, and Table 3.2 summarizes their physical properties such as densities, viscosities, and surface tensions at room temperature.

Air

Compressed air may be employed as the cell medium and air at atmospheric pressure may be used as the external medium in vacuum triaxial tests. Latex rubber membranes are very permeable to atmospheric air, and any testing at elevated effective confining pressures may result in erroneous results. This medium is therefore rarely used, or used only for rather low effective confining pressures and for relatively short-term tests. Compressed air inside a cell with an acrylic cell wall is very dangerous, because breakage of the wall results in an explosive reaction which will send sharp plastic pieces out in the environment with great velocity.

De-aired water

De-aired water is most inexpensive, easiest to clean up, most practical to work with and is therefore most often used as cell fluid. Given the high permeability of latex rubber membranes to air molecules, it is important to remove as much dissolved air as possible from the water used inside the specimen and as cell fluid in the triaxial cell. It is therefore useful to have available a reservoir with de-aired water at any time in the laboratory. Such a reservoir may be located as high up under the laboratory ceiling as possible so the water can conveniently run into the cell by gravity rather than by pumping.

Production of de-aired water

Air and other gases dissolve into water in small amounts. The higher the pressure the more air can be dissolved. On the other hand, increasing the temperature drives the air out, and by boiling the water, first the air and then water vapor bubbles out of the water. Since the boiling temperature of water reduces with reducing pressure, as seen in Fig. 3.16, it is possible to remove the dissolved air by applying a vacuum to the water until the water boils at room temperature thus first removing dissolved air and then water vapor.

Based on the principle of boiling the water at room temperature, two methods have been devised for production of de-aired water. In one method the water is sprayed into a tank under a vacuum. Figure 3.17 shows a laboratory setup for production of de-aired water using a deep vacuum and a tap-water supply. It will take some time to fill the tank with de-aired water by spraying, and the tank should be under continuous vacuum to prevent air from entering the water again during storage. The vacuum is replaced with atmospheric pressure when de-aired water

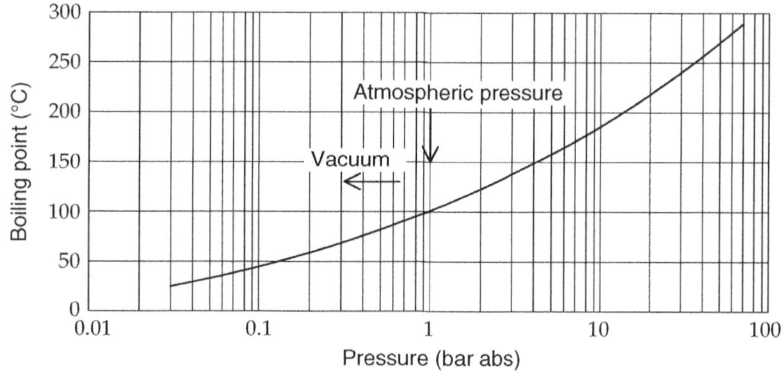

Figure 3.16 Temperature at the boiling point of water as affected by pressure and vacuum.

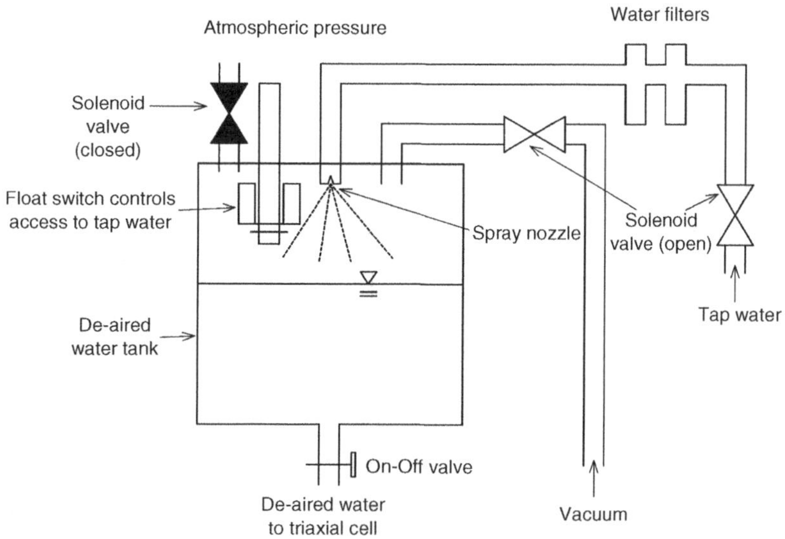

Figure 3.17 Schematic drawing of production of de-aired water by spray-through-vacuum.

is required for the triaxial experiment. Placing the tank high up under the laboratory ceiling allows gravity flow of de-aired water into the triaxial cell.

In the manual method shown in Fig. 3.17 the filling of the tank is initiated by drawing a vacuum on the tank through the solenoid valve for 1–5 min, then opening the water line to allow water to be sprayed through the vacuum in a mist that produces a large surface area which allows the vacuum to remove air from the water. As the de-aired water fills the tank to a predetermined level, the float switch is activated, and this stops the flow of water. The vacuum is allowed to remain in the tank to prevent air from entering the water again. When de-aired water is required, the vacuum is removed, atmospheric air pressure is allowed at the top of the tank and the on–off valve at the bottom is opened to allow de-aired water to run into the triaxial cell. A computer-controlled system was devised by Aydelik and Kutay (2004) by which continuous production of de-aired water was made possible by turning a second system on when the first was in use.

In the other manual method, the Nold principle is used. A quantity of water is placed under vacuum in an appropriate tank and a disk

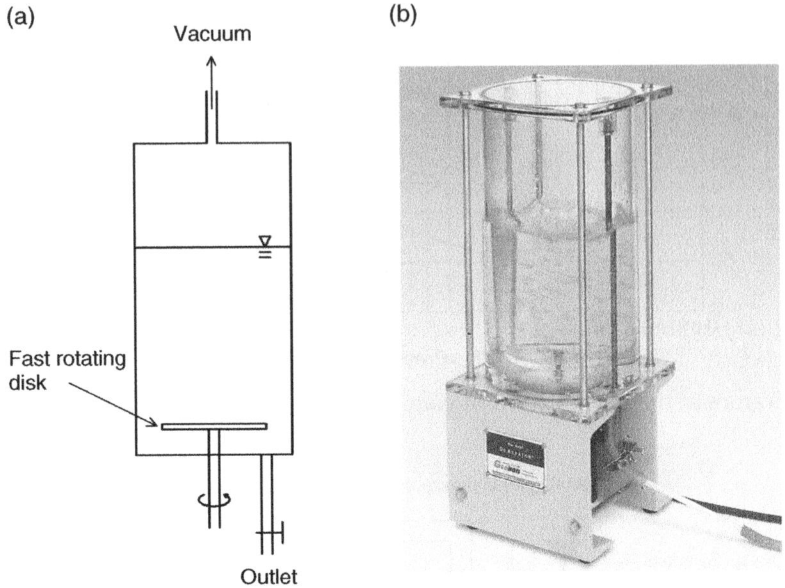

Figure 3.18 Nold principle de-aerator accomplished by cavitation and nucleation. (a) Schematic diagram and (b) photograph of apparatus.

mounted near the base of the tank is spun fast to agitate the water, which causes cavitation and subsequent vaporization into a mist. The dissolved air is thereby released from the water and it bubbles up to the surface, where it is removed by the vacuum. Figure 3.18 shows the operation principle of the Nold de-aerator. This device can produce de-aired water much faster than the spray-through-vacuum method in Fig. 3.17. As with the water reservoir in Fig. 3.17, the vacuum is replaced with atmospheric pressure when de-aired water is required for the triaxial experiment.

To minimize the amount of suspended solids, the tap water may be passed through filters with openings of 5 and 1 μm before it enters into the tank. The effectiveness of the water de-airing system may be checked by measuring the content of dissolved oxygen with a desirable goal of less than 6.0 mg/l.

While the two methods described above are manual and each produce one tank of de-aired water at a time, a continuous method of de-aired water production may be set up by using two tanks. Aydelik and Kutay (2004) explain a method in which the solenoid valves attached to each of the two tanks operate in opposite mode, thus filling one tank while drawing de-aired water from the other. The operation is controlled by a computer program developed in LabView^{R}.

Other methods of producing de-aired water have been developed as indicated by Klementev and Novak (1978).

Glycerin

Glycerin is a byproduct from the production of soap. Soap is made from animal and vegetable fats, which contain 7–13% glycerin. It is a thick, colorless liquid that is sweet-tasting and chemically neutral with a high boiling point. It can be dissolved in water or alcohol, but not in oils. Glycerin is hygroscopic, that is it absorbs water, and it is a good solvent, because it can dissolve many compounds easier than water and alcohol. In addition to being the basis for production of nitroglycerin, glycerin is used for conservation of fruit, lubrication of molds, cake and sweet making, and it is a basis for lotions, and clear soaps that dissolve easily in water.

Castor oil

Castor oil is a vegetable oil made from castor seeds. It is a colorless or very pale yellow liquid with mild to no odor or taste. Its boiling point is 313°C (595°F). It contains approximately 90% fatty acid chains. Castor oil is used as the basis for many products in the transportation, cosmetic, pharmaceutical and manufacturing industries. Its applications include hydraulic and brake fluids, machine oil, and lubricants.

Kerosene

Kerosene is made from the distillation of crude oil and petroleum and it is a combustible hydrocarbon liquid. Kerosene is known as paraffin in the UK, South East Asia, and South Africa. It is a thin, clear liquid that consists of a mixture of carbon chains. The flash point of kerosene is 37–65°C (100–150°F) and its auto-ignition point is at 220°C (428°F). It is widely used as fuel for heating and as jet propulsion.

Paraffin oil

Paraffin consist of alkane hydrocarbons with the general formula C_nH_{2n+2}, the simplest of which is methane for which $n = 1$, a gaseous compound at room temperature. Octane has $n = 8$ and forms a mineral oil at room temperature. Paraffin oils have different names depending on their composition of heavier alkane hydrocarbons. They are used as lubricants in mechanical mixing, as a laxative for chronic constipation, as coating for fruits, as a release agent in the baking industry, and as fuel.

Silicone oil

Silicone oil is the best cell fluid to be used for triaxial testing. It is a man-made chemical that is the silicon analog of carbon based organic compounds. Silicone oil consists of long and complex molecules which form a clear, colorless, odorless, non-flammable and inert fluid. It has excellent thermal stability (i.e., properties do not change much with temperature), it comes with a wide range of viscosities, and it has low surface tension (see Table 3.2). Silicone oil is essentially non-toxic and is therefore safe to use. It is used as lubricant for plastics and elastomeric surfaces and as a hydraulic fluid. It is compatible with rubber and used in medical facilities and in the food industry. Silicone oil is soluble in a wide range of solvents, but due to the large difference in surface tension between water and silicone oil, it is water repellent and does not mix with or imbibe water. These properties make it almost ideal as cell fluid for triaxial testing. The food grade silicone oil listed in Table 3.2 with a dynamic viscosity of 340 cP (kinematic viscosity of 350 cSt) has been used for this purpose (Leroueil *et al.* 1988). However, as with any fluid, it will cover all surfaces inside the triaxial cell and it is not as easy to clean up after the experiment as water. Besides, it is more expensive to purchase than any of the other fluids.

Flow through membrane

Experiments were performed by Leroueil *et al.* (1988) to study flow through latex rubber membranes. They covered a dummy specimen, 5 cm in diameter and 10 cm high, with filter paper and subsequently placed a membrane over this specimen assembly. Commercially available 0.3 mm thick latex rubber membranes and 0.07 mm thick Ramses prophylactics were tested. Fluid flow can occur into and out of the specimen due to hydraulic as well as osmotic pressure differences. An effective pressure difference of 100 kPa was applied to the cell fluid and the amount of fluid moving through the membrane was observed over long periods of time. The different types of cell fluids reviewed above were employed in the experiments, and the results are shown in Table 3.3.

Table 3.3 shows that when de-aired water is used as cell fluid, both inflow and outflow from the specimen were observed. It was speculated that this could be due to differences in chemical composition of the pore and cell water, resulting in an osmotic effect, as well as the effect of the hydraulic pressure difference imposed across the membrane. However, these very small amounts

Table 3.3 Flow through membranes (after Leroueil *et al.* 1988)

Type of membrane	Cell fluid	Range of measured flow through the membrane (cm^3/week)
Latex rubber membrane	De-aired water	−0.23 to +0.25
Thickness = 0.3 mm	Glycerin	−1.45 to −1.76
Diameter = 5 cm	Castor oil	−0.25 to −0.41
Height = 10 cm	Paraffin oil	−0.04
	Silicone oil	−0.04 to −0.05
Ramses membrane	De-aired water	+0.01
Thickness = 0.07 mm		
Diameter = 3.8 cm		
Height = 7.7 cm		

+, Flow from cell to specimen (inflow); −, flow from specimen to cell (outflow).

of flow (−0.23 to +0.25 cm^3/week for the 0.3 mm membrane and +0.01 cm^3/week for the Ramses membrane) indicate that de-aired water is an excellent cell fluid for triaxial testing.

The experiment with glycerin as cell fluid showed relatively large amounts of outflow from the specimen and this is because glycerin is hygroscopic, that is it absorbs water. Thus, water molecules are pulled out from the pore water in the specimen. This would result in apparent dilation effects in drained tests and in decreasing pore pressures in undrained long-term tests. It is therefore not recommended to use glycerin as a cell fluid.

Castor oil show similar but less pronounced properties as glycerin. Thus, due to the osmotic pressure difference, water molecules are attracted into the cell fluid, as indicated in Table 3.3. Castor oil is also not ideal as a cell fluid in triaxial testing.

Paraffin oil appears to cause very small amounts of flow out of the specimen. However, paraffin oil and kerosene interact chemically with the latex rubber membrane to such an extent that they make the membrane useless, even for short-term experiments. Kerosene and paraffin oil are therefore unsuitable as cell fluids when a latex rubber membrane is present. However, they may be used for experiments at low confining pressures in which a membrane is not present. The difference in surface tension for these fluids and water is relatively high and therefore creates a barrier to flow between the pore water and the kerosene or paraffin oil. This is further discussed in Chapter 10.

Silicone oil also causes very small amounts of flow out of the specimen (−0.04 to −0.05 cm^3/week). Combining this with all the other beneficial properties of silicone oil reviewed above makes it the ideal cell fluid for triaxial testing.

Setup for long-term and/or high pressure testing

Several possible setups for long-term and/or high pressure testing in which the permeability of the latex rubber membrane is counteracted by the choice of cell fluid or by other measures are reviewed below.

Silicone oil

To avoid fluid passing through the membrane, the entire cell may be filled with silicone oil and pressurized. This has the advantage that it works well for the experiment. For this purpose a special container for silicone oil may be mounted under the laboratory ceiling so the oil will enter through the base of the triaxial cell by gravity flow. At the end of the experiment, the triaxial cell may be pressurized slightly to make the silicone oil flow back up into the container. The disadvantage of this technique is that all internal surfaces of the triaxial cell and the surface of the specimen will be covered with a thin layer of silicone oil, which makes it more difficult to clean up at the end of the experiment than if using de-aired water.

Silicone oil on top of de-aired water

To minimize the oil coverage of the specimen, de-aired water may be used to surround the specimen in the bottom of the triaxial cell and the top is filled with silicone oil (Jacobsen 1970).

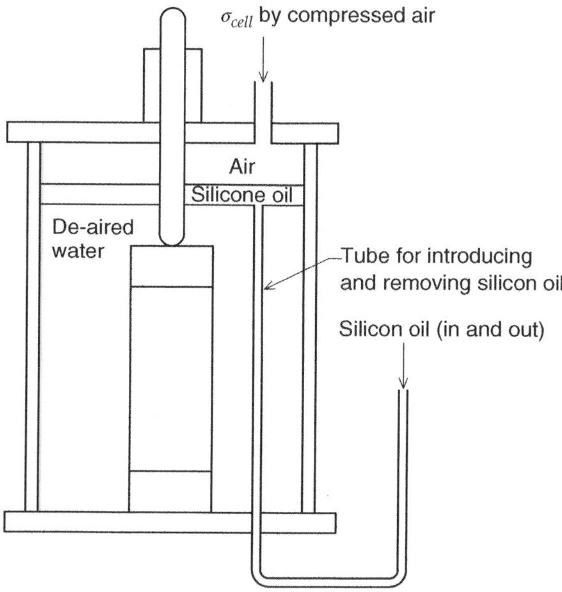

Figure 3.19 Use of silicone oil on top of de-aired water to avoid air penetration of latex rubber membrane.

To employ this technique, the de-aired water is first filled into the cell to near the top of the specimen cap followed by filling silicone oil on top of the water through a tube installed through the base, as shown in Fig. 3.19. Since the silicone oil has a slightly lower density than water, as shown in Table 3.2, it will remain floating on top of the water. Pressurization of the cell occurs through the silicone oil, thus avoiding any air entering into the de-aired water. At the end of the test, the silicone oil is first pushed out through the tube in the base, thus never coming into contact with the specimen setup. Only the upper portion of the cell is covered with a thin layer of silicone oil, but these surfaces are relatively easy to clean up. The silicone oil may contain small amounts of water and may have to be filtered after the experiment.

Several membranes and de-aired water

It is also possible to retard the flow of fluid or air into or out of the specimen by using several membranes with a smear of silicone grease between the membranes. De-aired water may then be used in the cell. The required number of membranes may be determined by trial-and-error. This system works well for high pressures and relatively short-term experiments. Figure 3.20 shows the results of a high pressure triaxial test in which air begins to enter the specimen and clearly changes the measured volume change characteristics.

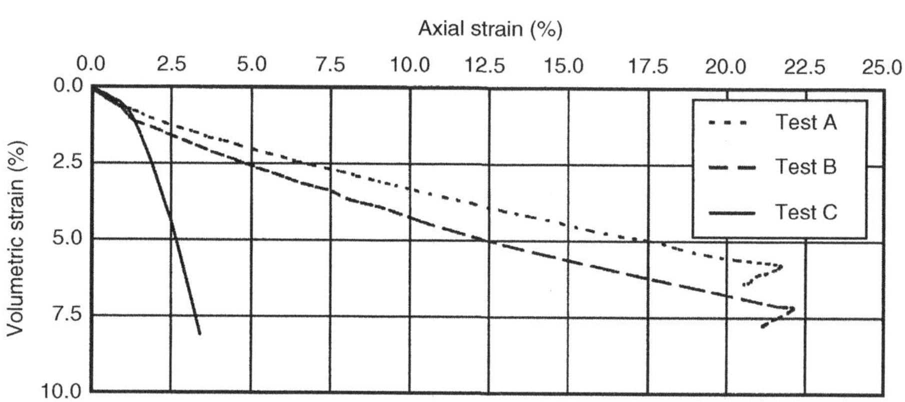

Figure 3.20 Effect of fluid (air, water) penetration into a specimen on measured volume change (after Karimpour 2012).

Long access tube with de-aired water

To minimize the amount of air that arrives at the surface and penetrates the membrane during an experiment, a long access tube (plastic or metal) with de-aired water may be used for pressurization. The triaxial cell and the long access tube is filled with de-aired water and pressurized. It is assumed here that pressurization occurs by compressed air or nitrogen (worst case), but other means of pressurization are possible, as reviewed in Section 3.4. Because the pressurized gas molecules move by diffusion through the initially de-aired water, a longer tube will retard the arrival of the air molecules at the specimen surface more than a short tube. The length of tube to be used may be determined by trial-and-error. The tube may be wound up in a spiral to take as little space as possible. A coiled 1/4 in. or 1/8 in. tube can be made from standard tubing by winding it around a heavy-duty cardboard mailing tube and heating it to set the shape. The heating can be done with a heat gun or by placing it in an oven set to low heat (e.g., a water content oven). Note that the heating weakens the tube material and the lines may not be suitable for pressures over conventional house-line pressures and for long durations. For higher pressures a 1/8 in. stainless steel tube may be wound up in a similar coil and used for the purpose. This is further discussed in Section 3.4.6.

3.1.8 Lubricated ends

To reduce the effects of end restraint to negligible amounts, lubricated ends may be employed (Rowe and Barden 1964; Lee 1978; Norris 1981; Tatsuoka *et al.* 1984; Tatsuoka and Haibara 1985). If used correctly, the specimen will deform as a right cylinder during shear. This results in uniform strains and uniformly distributed pore pressures in undrained tests.

Conventional geotechnical pressures

Enlarged caps and bases are usually employed to support the specimen fully during shear so that no overhang occurs when using lubricated ends. Figure 3.21 shows examples of setups for undrained tests on clay. As indicated, short specimens are often, but not always, used in conjunction with lubricated ends. Note also that to allow for lubricated ends in the setup with filter paper, the drainage connections are located on the sides of the cap and base. Flaps of filter paper cover the filterstones and connect the filter strips with the drainage system.

To produce lubricated ends the surfaces of the end plates should be hard and smooth. These surfaces may consist of polished stainless steel or smooth glass plates glued to the faces of the end plates by epoxy (Jacobsen 1970). They are coated with thin layers of silicone grease (Dow Corning High Vacuum Grease) and covered with rubber sheets cut out of discarded membranes. These may be rolled with a cylindrical rod to push trapped air bubbles out from under the membrane, as shown in Fig. 3.22. Figure 3.23 shows comparisons of stress–strain and volume change relations for triaxial specimens with lubricated and conventional ends and with different H/D-ratios.

Additional lubricating rubber sheets may be added to further improve the lubrication system. Equal amounts of lubrication on the two ends often results in slightly less expansion near the cap than at the base, possibly due to the weight of the specimen itself. It may therefore be necessary to provide a little more lubrication on the cap than on the base to maintain the shape of a right cylinder during the test. This may be achieved by adding one more lubricating sheet or providing a slightly thicker layer of silicone grease on the cap than on the base. On the other hand, because the rubber sheets have a Poisson's ratio of 0.5 and a low elastic modulus, it is possible to add too many lubricating sheets resulting in splitting or flaring at the ends and premature failure of the specimen. The objective is simply to produce a specimen that deforms as a right cylinder.

It appears that the effect of the lubricating sheets is to prevent the soil particles from being pressed through the sheets and develop frictional contacts with the end plates (which should be hard and smooth). Norris (1981) found that an initial choice of the total thickness of

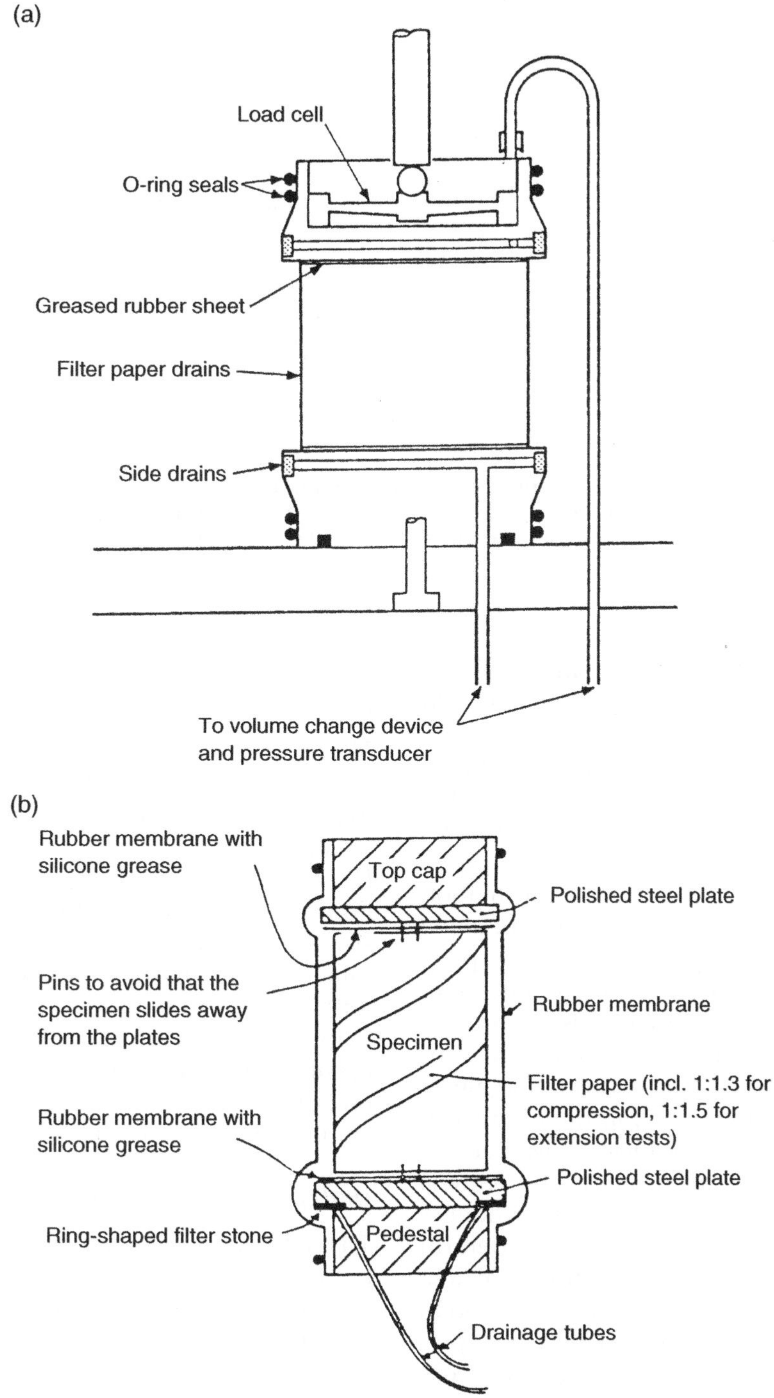

Figure 3.21 (a) Use of lubricated ends and short specimens for undrained tests on clay and (b) arrangement used at the Norwegian Geotechnical Institute for undrained tests on clay. (b) Reproduced from Berre 1982 by permission of ASTM International.

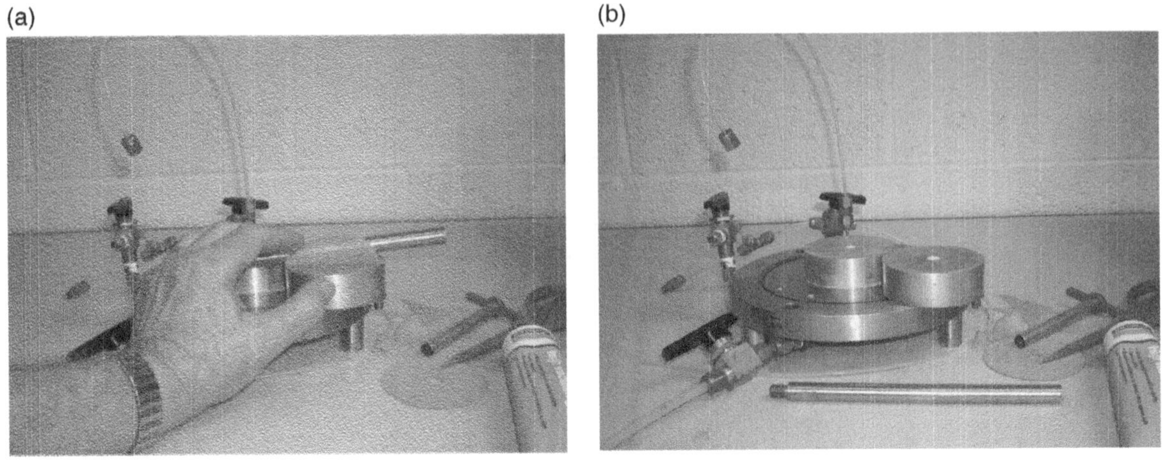

Figure 3.22 (a) Creating lubricated ends involves smearing a thin coat of silicone grease on the end plate, then rolling out a pre-shaped membrane on top and smoothing it by rolling a rod across it. Additional coats of grease and additional membranes may be applied. (b) End plates with lubricated ends.

the lubricating system should be in the order of 1.5 times the average grain diameter to produce good results. Thus, for clay and fine sand, the lubricated ends may be made of one or more thin prophylactics sheets. However, it is often a matter of trial-and-error to find the optimum thickness of the silicone grease coating and the number of lubricated sheets on each of the two end plates.

The correction to the vertical deformation measured outside the triaxial cell (but not to the deformation measured directly on the specimen) increases and becomes more uncertain (due to natural scatter) with increasing number of lubricating sheets. Thus, to maintain reasonable accuracy in the vertical deformation measurements, it is desirable to limit the number of lubricating sheets used in a test. However, the guiding principle for using lubricated ends is that the specimen deforms as a right cylinder.

Since the specimen is expected to deform away from its central axis, the centrally located, small porous disk is not required to be smooth. In fact, the roughness of this disk may help prevent the specimen from sliding out from between the cap and base. It is at times necessary to use taller disks, so they act as dowels that stick up into the specimen to hold on to the specimen. They will have negligible influence on the soil behavior.

Tests on very short specimens

Successful experiments have been performed on specimens of intact San Francisco Bay Mud with H/D = 0.36 and lubricated ends. These tests resulted in very similar stress–strain relations, including post-peak softening, as obtained from specimens with H/D = 1.0 and lubricated ends and specimens with H/D = 2.5 without lubricated ends. Figure 3.24 shows comparisons of stress–strain and pore water pressure from tests on intact specimens of San Francisco Bay Mud with three different H/D-ratios. Thus, if the amount of end lubrication is adequate, it is feasible to perform triaxial compression tests on very short specimens, that is with a shape similar to a hockey puck.

High pressure triaxial tests

Triaxial compression tests performed on sand at high confining pressures require more lubrication due to the correspondingly high deviator stresses, which will cause the sand grains to penetrate through the lubricated ends to create friction on the surfaces of the end plates. Thus, for the experiments performed at confining pressures up to 67 MPa by Yamamuro and Lade (1996), two layers of latex rubber sheets, each with thickness of 0.64 mm (0.025 in.), were

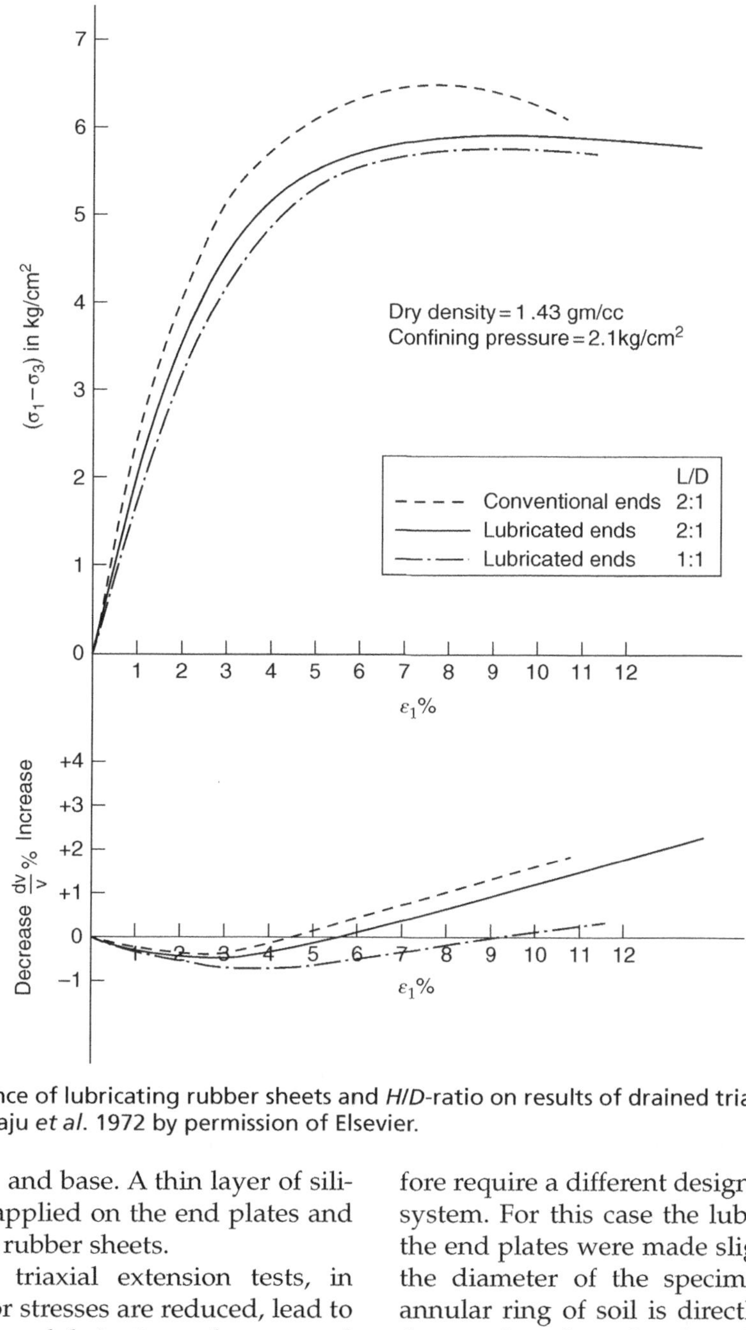

Figure 3.23 Influence of lubricating rubber sheets and *H/D*-ratio on results of drained triaxial tests on sand. Reproduced from Raju *et al.* 1972 by permission of Elsevier.

placed on the cap and base. A thin layer of silicone grease was applied on the end plates and between the latex rubber sheets.

High pressure triaxial extension tests, in which the deviator stresses are reduced, lead to unloading of the lubricating sheets, and Yamamuro and Lade (1995) found that the high confining pressure will cause the surrounding membrane to intrude laterally under the cap. This in turn caused the lubricating sheet to buckle laterally and separate the cap from the specimen. High pressure extension tests therefore require a different design of the lubrication system. For this case the lubricating sheets on the end plates were made slightly smaller than the diameter of the specimen. Therefore, an annular ring of soil is directly in contact with the metal surface of each end plate, thus forming a protective barrier of soil around the lubricated ends. Some friction will therefore develop between the end plates and the soil. Since the vertical stress is decreasing during a conventional extension test the amount of friction is not nearly as large as in a compression test.

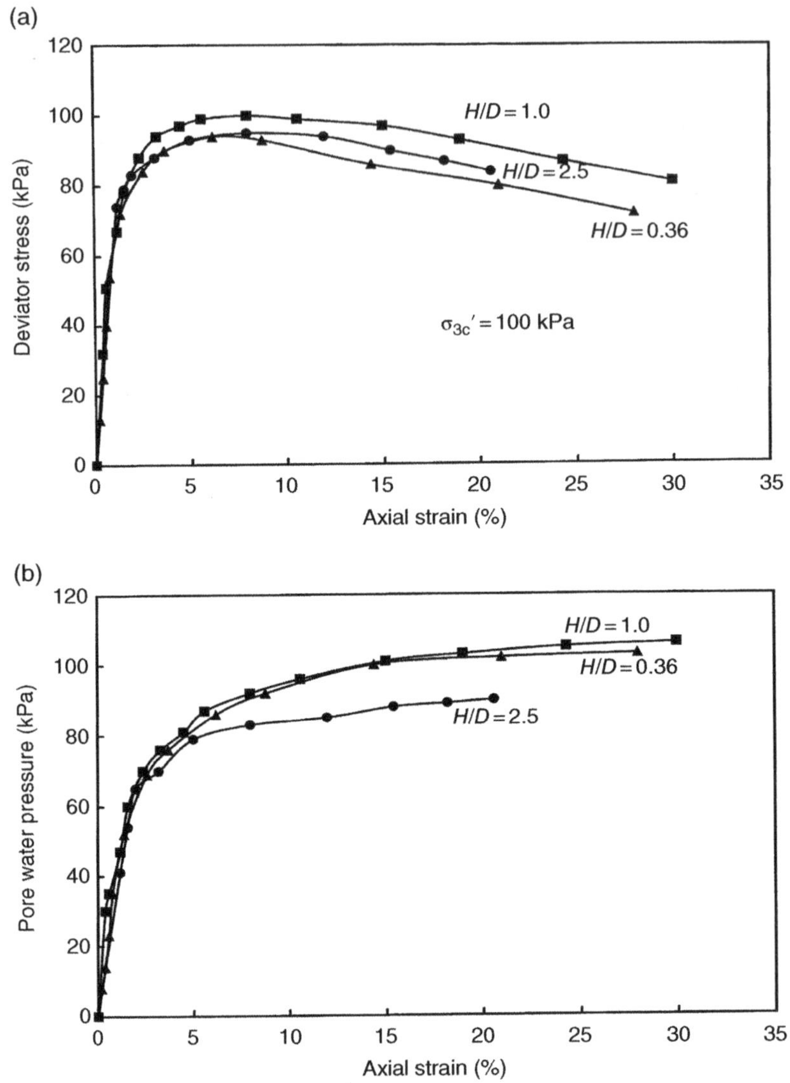

Figure 3.24 Results of CU triaxial compression tests on cylindrical specimens of intact San Francisco Bay Mud with different shapes and end conditions: (a) stress–strain curves; and (b) pore water pressure relations. They indicate that short specimens may be tested with results similar to those from taller specimens when adequate lubrication is supplied at the end plates.

Besides, the goal is to maintain a cylindrical shape thereby ensuring uniform stresses and strains in the specimen. Specifically, for an extension specimen with diameter of 7.1 cm, a single sheet of latex rubber with thickness of 0.64 mm (0.025 in.) and diameter of 4.6 cm (1.8 in.) in diameter was employed on each of the cap and base. A thin smear of silicone grease was employed to provide some lubrication and to ensure that air was not trapped within the lubricated ends.

Long-term and high strain rate tests

The effectiveness of lubricated ends reduces in long-term tests, because the grease tends to be squeezed out from between the end plates and the lubricating sheets. This produces increasing friction as the test progresses in time, and full friction may be expected to develop before failure has been reached.

Lubricated ends have been found to be ineffective in tests performed with very high strain rates, because the high viscosity of the grease prevents the system from working as intended at the high deformation rates (Abrantes and Yamamuro 2002; Yamamuro *et al.* 2011).

Corrections

Corrections to vertical deformations and volume changes due to lubricating sheets are discussed in Chapter 9.

3.2 Triaxial cell

3.2.1 Cell types

The triaxial specimen setup discussed above is located inside a triaxial cell consisting of a base plate, a top plate with piston bushing, a cell wall, and tie-rods. These principal components are shown schematically in Fig. 3.25, and they are reviewed below. Three principles can be distinguished in design and assembly of the triaxial cell. Two of these principles are different by their use of external and internal tie-rods, and the principle for the third triaxial cell involves an integrated loading system. This latter type is explained in Section 3.6.

Cell with external tie-rods

In the simplest triaxial apparatus, shown in Fig. 3.26, the cylindrical cell wall is carefully guided down over and around the previously mounted triaxial specimen. After the cell wall has been placed on the O-ring in the bottom plate, the top plate is positioned with its O-ring on top of the cell wall and the triaxial cell is tied together with three or more tie-rods outside the cell. The tie-rods may be permanently

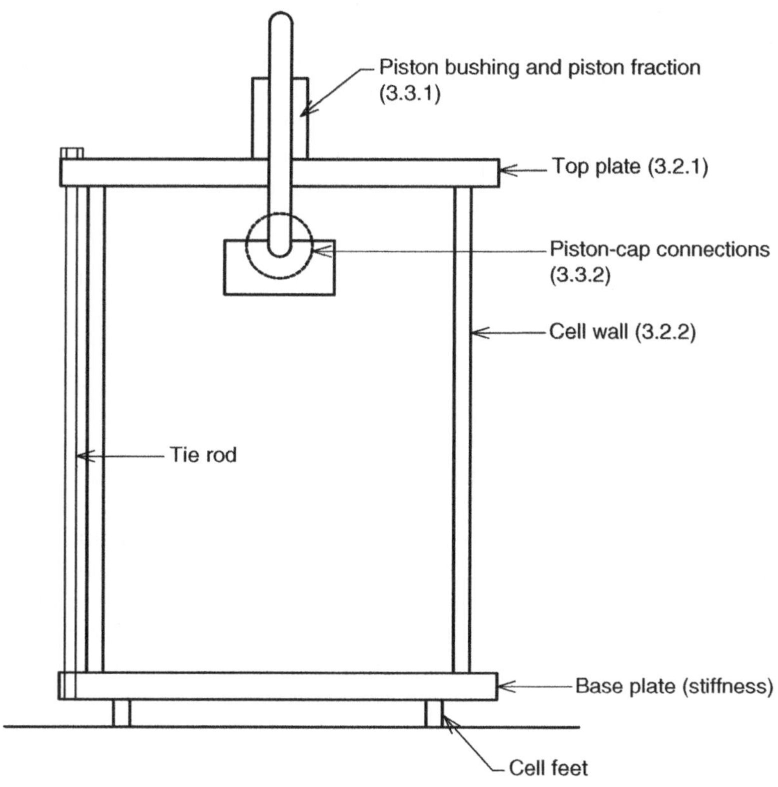

Figure 3.25 Triaxial cell with components discussed in Sections 3.2 and 3.3.

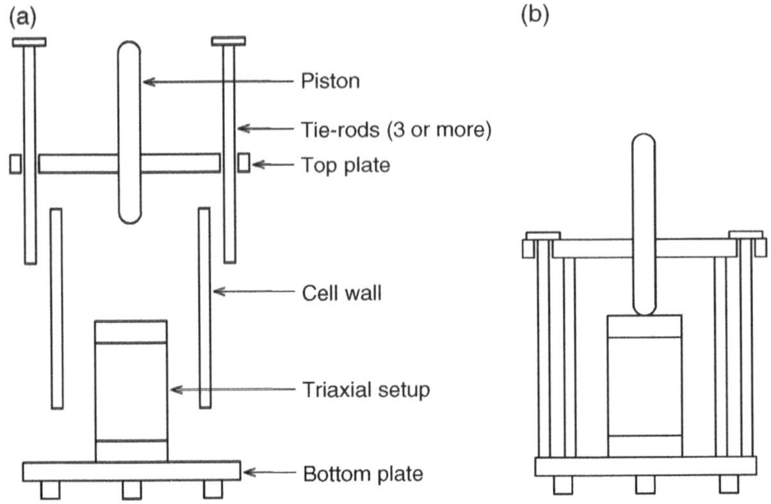

Figure 3.26 Triaxial cell type 1: (a) in disassembled parts; and (b) assembled.

fixed to the bottom plate and nuts are used to tie the cell together, or the tie-rods are simply long bolts, as shown in Fig. 3.26. This latter cell principle with the long bolts allows a completely open working space above the base on which to mount an existing clay specimen or to deposit a sand specimen to be tested. The disadvantage is that all electric and other connections have to go through the base, or be permanently tied to the top plate, which is not practical. It is also more difficult to connect the piston to the cap in case of misalignment of the specimen setup.

Cell with internal tie-rods

In the second design of the triaxial cell, shown in Fig. 3.27, the central portion of the top plate with the piston bushing is fixed to the bottom plate by two or more columns sitting inside the cell. Assembly of this triaxial cell then consists of first mounting the top plate with the piston to the columns and then carefully lowering the cell wall over the top plate and triaxial specimen and bolting the cell wall to the bottom plate. The advantage of this design is that the entire triaxial specimen setup, with connections to the piston and with drainage lines and electric wires from gages mounted on the specimen, can go through the top plate. This allows complete mounting and adjustment of all measurement devices as necessary before assembly of the cell.

Other design considerations

It is practical to design the bottom plate of the triaxial cell with a small recess, because disassembling the cell often leads to spilling of cell fluid (water), and this may be avoided by containing the last little amount in the bottom recess for easy clean-up after complete disassembly.

The base plate may be sitting on short feet, as shown in Fig. 3.26, to allow drainage lines and electric wires to be led out through the base plate. The base plate may be outfitted with a central foot to support the central setup of the specimen and avoid deflection of the base plate. However, the drainage lines may also be led out through a thick base plate as previously discussed in Section 3.1.4.

Introduction of cell fluid, most often de-aired water, is best accomplished through the bottom plate, because entrance through the top plate results in splashing, which in turn leads to air bubbles attaching themselves to inside surfaces. Thus, the cell fluid should enter through the base plate, because reintroduction of air into the water is largely avoided. An outlet in the top plate is required for escape of air as cell fluid is introduced

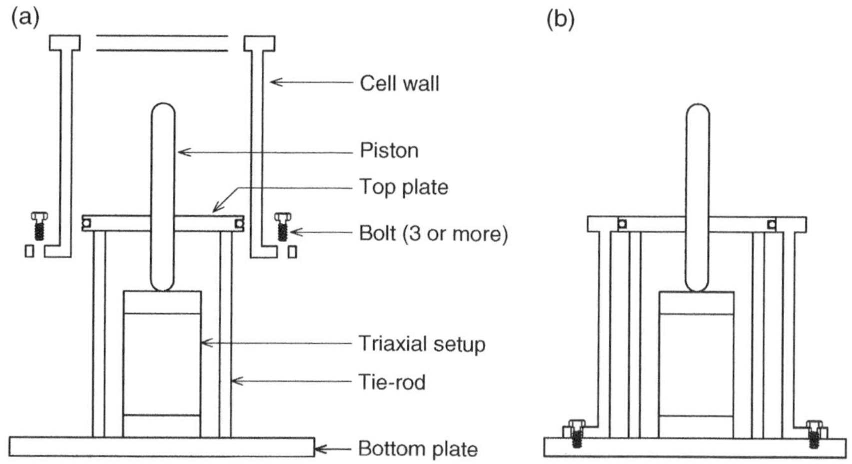

Figure 3.27 Triaxial cell type 2: (a) in disassembled parts; and (b) assembled.

in the cell. The inside ceiling surface of the top plate may be slightly tapered towards the point of the exit fitting to help collect and push the air out of the cell as the fluid enters from the bottom.

3.2.2 Cell wall

It is useful to be able to observe the triaxial specimen during an experiment. Thus, any deviation from the expected performance of the specimen may potentially be corrected or the experiment may be terminated before any unrecoverable damage to the specimen occurs. Besides, the uniformity of deformations (e.g., does the specimen exhibit bulging or barreling? Do the lubricated ends work correctly?) may be monitored and the time of shear banding may be recorded. In addition, some measurements may be made visually through the cell wall.

Conventional cell wall

Cell walls for conventional tests are most often made of clear, acrylic plastic or, more recently, clear polycarbonate.

Extruded acrylic plastic may have a slightly higher tensile strength (near 76 MPa) and is stiffer (E_{tens} = 3800 MPa) than polycarbonate. However, the behavior of acrylic plastic is brittle, and it therefore fractures relatively easily. Consequently only about 10% of the nominal tensile strength should be counted on for design. Scratches on the surface may be points of initiation of brittle fracture of acrylic plastic. It is also advisable to use water (very stiff) as the cell fluid to avoid the explosiveness of compressed air. The acrylic plastic cell wall may be reinforced by stiffer hose bands or glass fiber reinforced bands to be able to take higher cell pressures. Commercially available cell walls of this type can take pressures up to 3.5 MPa for smaller specimens and cell diameters and up to 2.0 MPa for larger specimens and cell diameters.

Cell walls made of extruded polycarbonate (Lexan®) exhibit ductile behavior, and polycarbonate has high impact strength, a tensile strength of 65 MPa, and an elastic modulus in tension of 2300 MPa. This material can be easily machined and a much higher percentage of its strength may be used for design because of its ductile behavior. Extruded pipes with diameters greater than 30.5 cm (12 in.) can now be obtained.

High pressure cell wall

Cell walls for high pressure triaxial testing may be made from the cylindrical walls from compressed gas bottles, or they may be machined from 4340 steel. The maximum pressures to be used inside cell walls from compressed gas bottles may be as high as the rated pressures for the gas bottles. 4340 steel is particularly strong and is suitable for production of high pressure cell walls with relatively small thicknesses.

Window in cell wall

Clear cylindrical walls filled with cell fluid will exhibit serious amounts of parallax, so it will be impossible to see the actual shape of the specimen or to take measurements through the wall. However, this may be remedied by attaching a compartment on the outside of the cell wall and filling it with water, as shown in the drawing in Fig. 3.28. This water is not pressurized and the side compartment may be filled after assembly of the triaxial cell. A very clear, non-distorted view of the specimen is obtained through the window, and accurate measurements of deformations between points or pins stuck into the specimen, say by a telescope, may be performed through the window (Lade and Liu 1998).

If visibility of the specimen in a high pressure cell is desirable, it is necessary to build in a porthole with strong glass to support the high pressure. An additional hole is required to provide for some type of illumination of the specimen.

3.2.3 Hoek cell

The Hoek cell (Hoek and Franklin 1968) is designed for the testing of cylindrical rock cores with particular diameters such as those coming out of rock coring operations. The Hoek cell is shown in Fig. 3.29 and it consists of two stainless steel pistons that have the same diameter as the rock core, and a membrane sitting around the specimen that is made to fit inside the stainless steel cell which is assembled around the specimen. Strain gages may be attached directly on the specimen and the Hoek cell allows the wires to exit under the membrane and out to the outside instrumentation without complications. Thus, strains may be measured directly on the specimen to determine moduli and Poisson's ratios and their variation as the test progresses. The fabrication of synthetic rubber sleeves for the rock specimen is also described by Hoek and Franklin (1968).

3.3 Piston

The axial load is transferred from the loading device to the specimen through an axial piston that protrudes into the triaxial cell through a bushing. The piston is usually made of

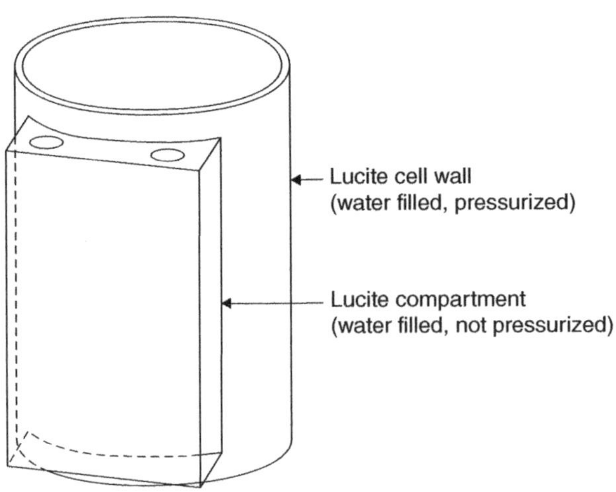

Figure 3.28 Lucite compartment attached to the outside of a Lucite triaxial cell allowing the undistorted view of a triaxial specimen (after Lade 2004).

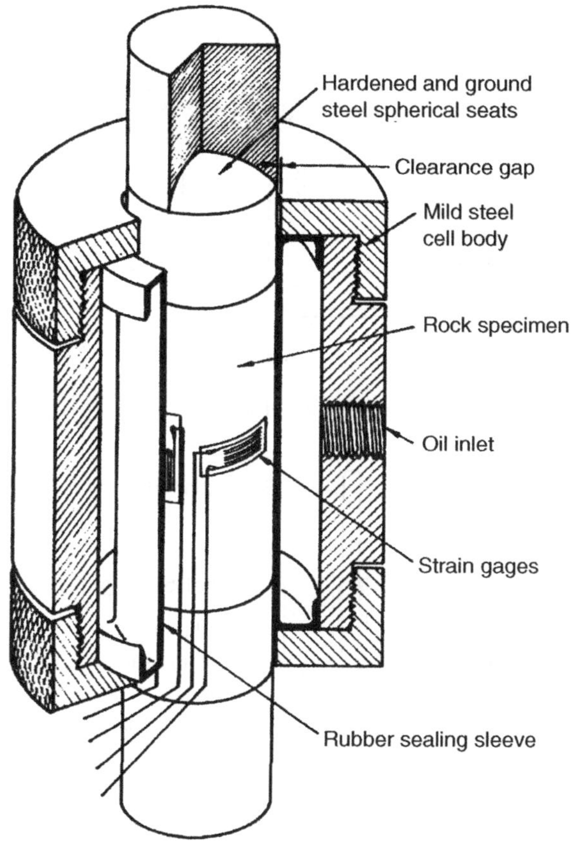

Figure 3.29 Schematic of the high pressure triaxial cell referred to as the Hoek cell (after Hoek and Franklin 1968).

hardened and ground steel or it is hard-chromed to fit the specifications of the ball bushings that support its alignment. The upper end of the piston is usually rounded to fit loosely with a load cell or a loading machine. A type of seal between the piston and the top plate is incorporated to keep the pressurized cell fluid inside the cell.

3.3.1 Piston friction

Friction along the piston where it is introduced into the triaxial cell through a bushing can lead to considerable errors in measurement of the vertical deviator load, especially for soft clays and other weak soils (Duncan and Seed 1967). The problem can be entirely circumvented by mounting the load measuring device inside the triaxial cell. However, this may lead to other types of problems requiring attention (e.g., protection of load cell against corrosion, water proofing of electric wires). Thus, reduction of piston friction is of considerable interest, especially for routine testing.

Piston friction reduction

Several different designs have been introduced to minimize the amount of piston friction for which corrections may be necessary when the axial force is measured outside the triaxial cell. Some of the principles used to reduce the piston friction are indicated in Fig. 3.30 together with the references in which they are presented and discussed.

The system proposed by Seed *et al.* (1960) consists of two ball bushings spaced apart to guide the piston rod centrally through the top plate. A rubber quad ring was recommended for sealing around the piston, but a conventional O-ring works as well for the seal. This system works quite well for routine testing.

Bishop and Henkel (1962) suggested a piston guide consisting of a long bronze bushing, with a clearance to the ground stainless steel piston of approximately 0.0076 mm (0.0003 in.). Without oil in the clearance, the piston will fall through the bushing by its own weight, while it will fall slowly when the clearance is filled with oil. To provide the constantly, but slowly seeping oil along the piston, a gutter and a collecting groove is constructed outside the cell and a reservoir for the oil is created inside the cell around the piston, as seen in Fig. 3.30. The lighter oil will float on the cell water and be pressed up along the piston to lubricate and create a seal with the piston. The small clearance between the piston and the bushing ensures that the cell fluid is kept inside the cell and just a small amount of oil is pushed out of the cell.

A system consisting of a bushing rotating around the piston to break any piston friction in the vertical direction was described by

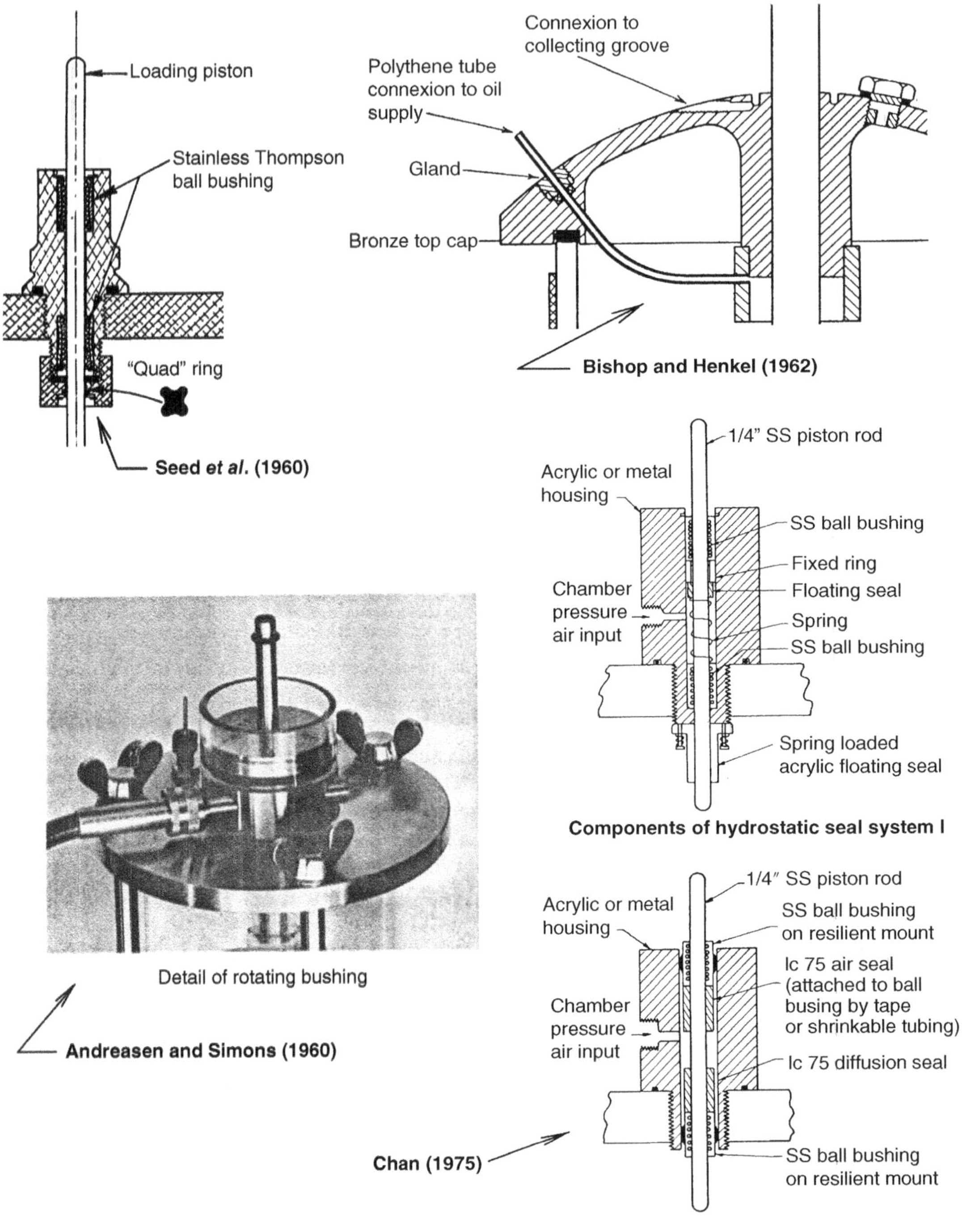

Figure 3.30 Examples of devices used to reduce piston friction in triaxial cells. After Bishop and Henkel 1962, and reproduced after Seed et al. 1960 by permission from ASCE, after Andreasen and Simons 1960 by permission from ASCE, and after Chan 1975 by permission from ASCE.

Andreasen and Simons (1960). This system uses a remote motor, which through a flexible axle turns the rotating bushing during the triaxial experiment, thus avoiding or reducing the friction in the vertical direction. The piston is hard-chromed, the bushing is nitride-hardened, and the clearance between the piston and the bushing is minimized, thus reducing the leakage of oil from the cell (Berre 1982).

Chan (1975) proposed two versions of a hydrostatic seal, which resulted in negligible bushing friction. These represent improvements of the system proposed by Seed *et al.* (1960). In addition to the two guiding ball bushings, seals are provided in both ends, and air pressure to match the cell pressure is applied inside the bushing guide in both types of seal. In the upper end of system I, shown in Fig. 3.30, the air pressure is sealed from the outside by a fixed ring that fits inside the bore for the ball bushings and by a floating ring that fits around the piston. These two rings create a floating seal ring. The matching surfaces of these two rings are ground and supplied with a drip of oil to provide the seal. The two ground surfaces are pressed against each other by a spring supported in the other end by the lower ball bushing, as seen in Fig. 3.30. The seal towards the cell is created by matching the air pressure to the cell pressure and with a spring loaded acrylic floating seal. Because the air pressure and the cell pressure are of similar magnitude, only the force from the small springs causes the floating seal to press up against the underside of the bushing housing. System II, shown in Fig. 3.30, works on a similar basis as the first system, but the seals are created by resilient mounts for the two bushings.

The first system appears to be more reliable, but it has been found practical to exchange the inside seal in system I with the mouth piece of a rubber party balloon. This is sealed in one end with an O-ring around the piston and in the other end with an O-ring around the end of the bushing housing, as seen in Fig. 3.31. Because it is now difficult to rotate the piston in the bushing, the connection with the cap is either fitting straight into the socket in the cap or is screwed into the cap with very few turns of the piston. For the latter purpose, the piston is turned as much as the balloon allows in the counterclockwise direction before it is screwed clockwise into the cap. The bushing housing is now pressurized by air pressure separately from the pressure supply for the cell pressure. The magnitude of the air pressure is slightly different from the cell pressure, and the goal is to make the balloon look completely relaxed with the cell fluid pressure on the outside and the bushing pressure on the inside. This system provides a completely frictionless system for which no correction is required.

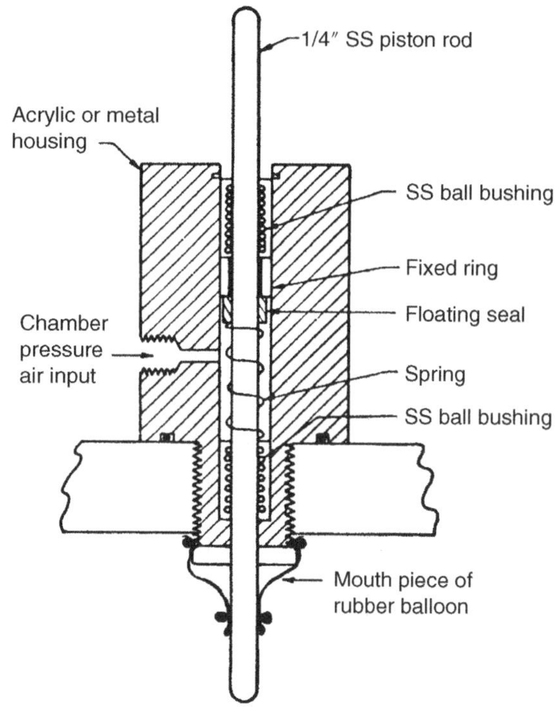

Figure 3.31 Piston system with rubber balloon to seal between air and water (modified after Chan 1975 by permission from ASCE).

Piston friction avoidance

An alternative to reducing the piston friction is to completely avoid the problem by mounting the axial load cell at the end of the piston inside the triaxial cell or under the specimen base. Examples of such internal load cells are shown in Fig. 3.32. However, this requires the load cell and the electric wires to be water proof, and it requires the electric wire to go through the top

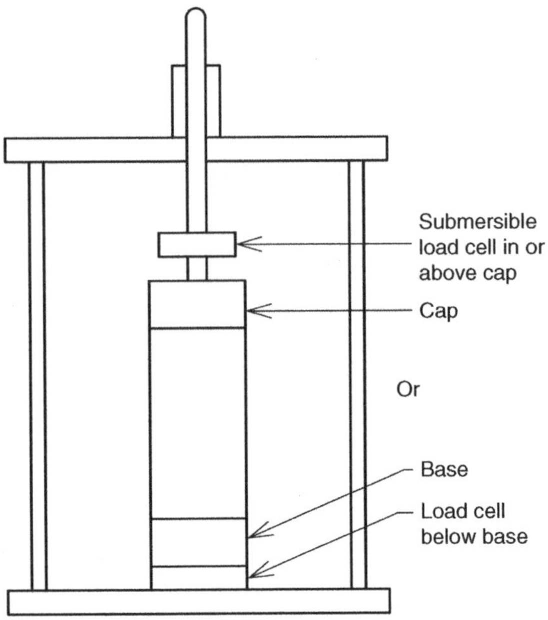

Figure 3.32 Load cell mounted inside a triaxial cell to avoid piston friction effects on axial load measurements. Load cell mounted either on top of the specimen or below the base.

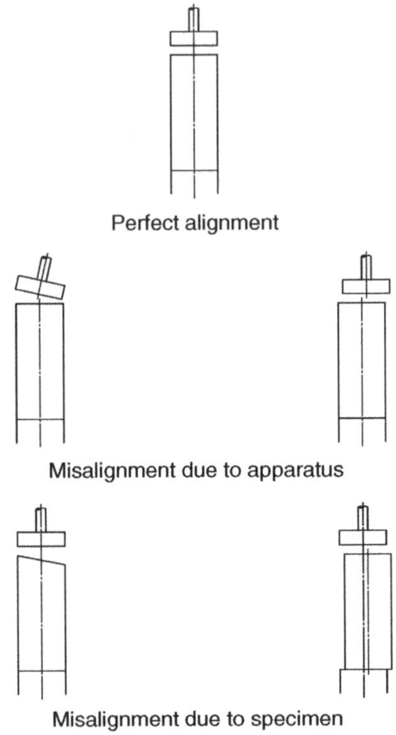

Figure 3.33 Perfect alignment and examples of misalignment of specimen setup with piston. Reproduced from Baldi *et al.* 1988 by permission of ASTM International.

or base plate. The option of placing the load cell under the base was employed for dynamic tests by Yamamuro *et al.* (2011) in which the correct axial load was measured, free of piston friction effects and including effects of inertia from the specimen acceleration. The potential response of an internal load cell to changes in cell pressure should be checked.

3.3.2 Connections between piston, cap, and specimen

The connections between the piston, cap, and specimen should align along the central axis of the triaxial apparatus, and the cap interface with the specimen should be perpendicular to the alignment axis. These alignments and the connections between the piston and cap have been discussed by Bishop and Henkel (1962), Berre (1982), and Baldi *et al.* (1988). This is important for uniform application of the axial deviator stress and for correctly measuring the axial deformation of the specimen.

Alignment

The condition of perfect alignment and examples of initial misalignment due to the apparatus and due to the specimen are shown in Figs 3.33 and 3.34. Misalignment may also develop during the test due to nonuniform response of the specimen. In all such cases the measurements and derived behavior of the soil is questionable. Good alignment may be easier to achieve in the triaxial cell with internal tie-rods (see Section 3.2.1), because all adjustments between the specimen, the piston, and the cap can be made before the cell wall is positioned and fastened. Clearly, it is necessary that the specimen is shaped as a cylinder with perpendicular ends in order that it aligns correctly. Similarly, the individual parts of the triaxial cell have to align to provide proper loading and deformation measurement conditions for the

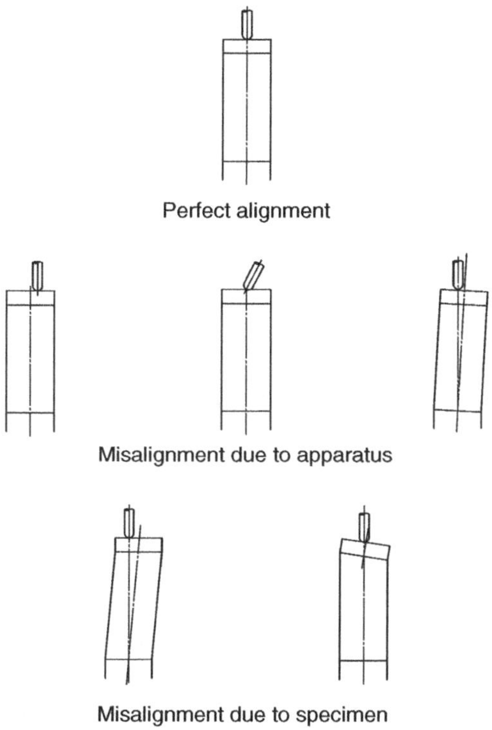

Figure 3.34 Perfect alignment and examples of misalignment of specimen setup with end plates. Reproduced from Baldi *et al.* 1988 by permission of ASTM International.

experiment. Limits on acceptable platen and load rod alignments are given in ASTM Standard D5311 (2014).

Connections

Examples are shown in Fig. 3.35 of the connections between the piston and the cap. These connections will initially produce loading centrally through the specimen, but they allow the cap to pivot and therefore they rely on the specimen being uniform such that the cap remains horizontal during the experiment. If, however, the specimen is not uniform, the cap may begin to tilt, and the stress–strain results will not be reliable.

Stiffer connections between the piston and the cap are obtained with the connections shown in Fig. 3.36, in which either the smooth piston end fits loosely into a guide hole in the cap or the thread at the end of the piston is screwed stiffly into the cap, possibly with a nut or washer to maintain alignment and a definite seating stop. The latter connection and the suction cap connection shown in Fig. 3.36 may be used for compression as well as extension tests. While the latter two types of connections require corrections for piston uplift, the first, loose connection shown in Fig. 3.36 allows truly isotropic compression of the specimen without correction for piston uplift, which may be achieved when the piston is not rigidly attached to the cap. Figure 3.37 shows a setup that allows vertical load transfer to the triaxial specimen while the cap is allowed to slide horizontally. This setup also requires correction for piston uplift.

Corrections

Corrections due to piston friction and piston uplift are discussed in Chapter 9.

3.4 Pressure supply

Several different types of pressure supply systems are being used to generate the confining pressure in the triaxial cell and the back pressure applied to the specimen.

3.4.1 Water column

For very low confining pressures a column of water may be used to apply the confining pressure in the water-filled triaxial cell. Because this method depends entirely on the gravitational potential of the water, the pressure is limited by the height of the water column. Considering also that the triaxial apparatus is usually placed in a loading machine at a height of approximately 1 m above the floor, the height of the water column is usually severely limited in a conventional laboratory. However, the confining pressure provided by a water column can be measured very precisely by the physical length of the water level above the middle of the specimen.

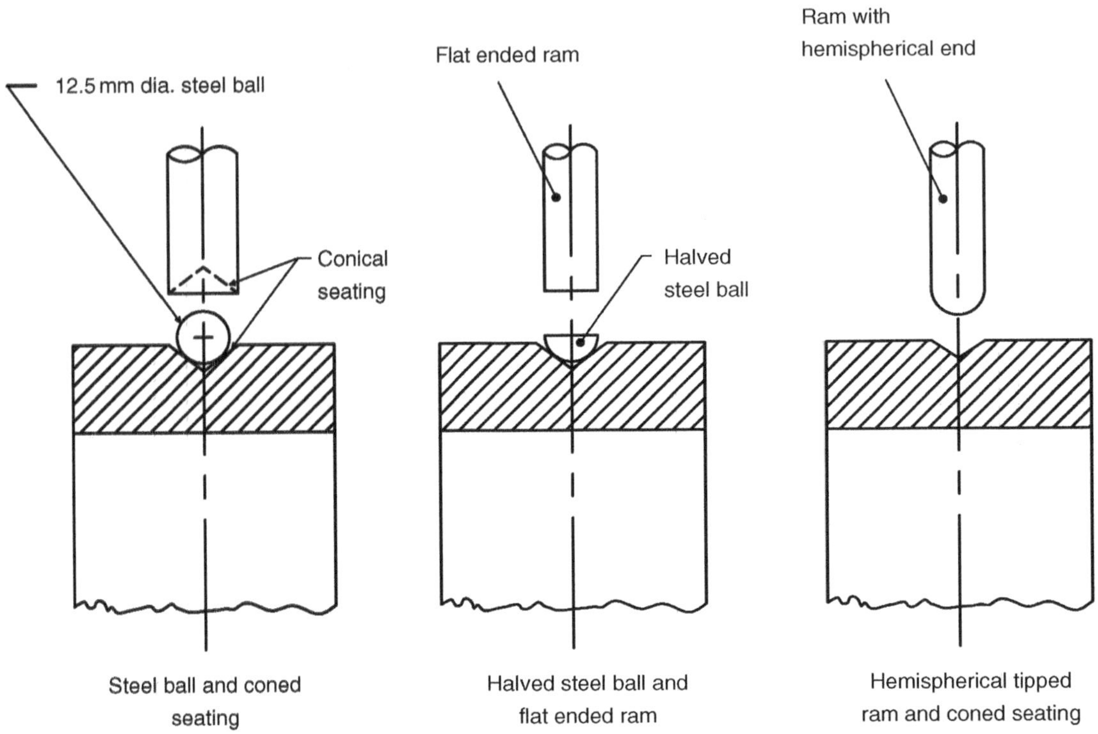

Figure 3.35 Examples of connections typically used with triaxial cells with external tie-bars. Reproduced from Baldi *et al.* 1988 by permission of ASTM International.

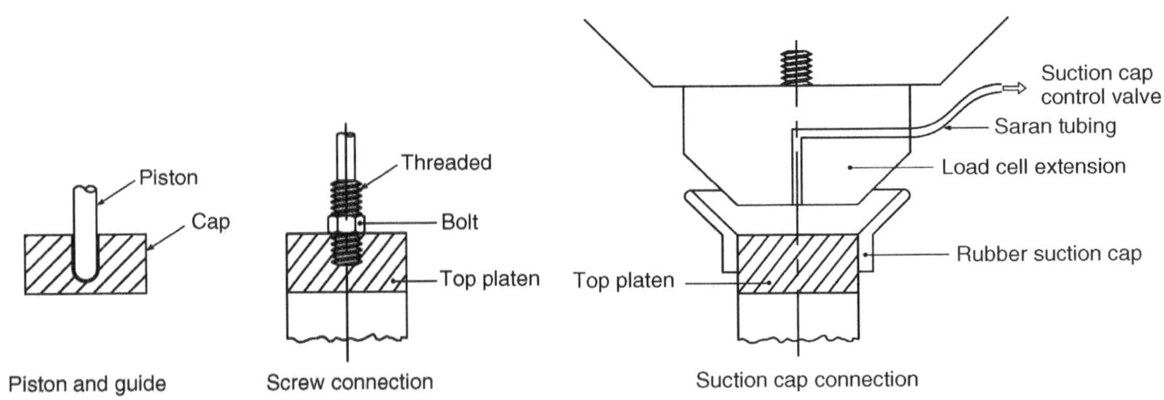

Figure 3.36 Examples of possible connections for compression/extension tests (modified after Baldi *et al.* 1988 by permission from ASTM International).

3.4.2 Mercury pot system

Figure 3.38 shows a diagram of the mercury pot system. This method of pressure generation employs the very high specific gravity of mercury (13.55 versus 1.0 for water) to generate pressure in the cell fluid. A pot of mercury is connected through a lower pot to the triaxial cell and lifted up above the cell to generate a hydrostatic pressure in the cell fluid. The generated pressure relates to the height of the mercury pot above the triaxial cell as well as to the density of the mercury. This system requires a fluid filled triaxial cell to produce a stiff back-up, so the mercury does not run out of the pot and into the cell. By hanging the upper pot in a calibrated spring the system is self-compensating

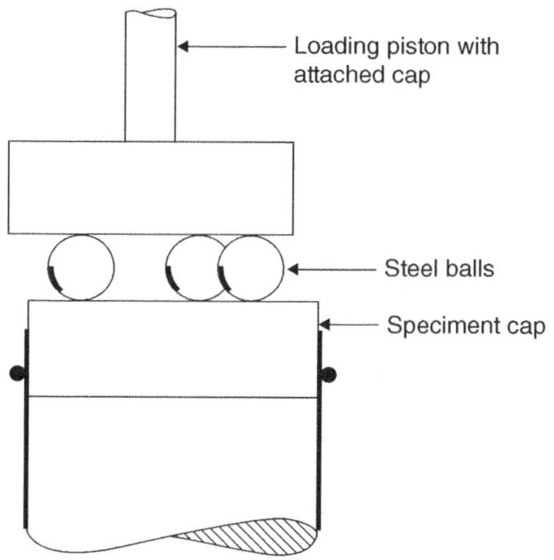

Figure 3.37 Vertical load transfer to a triaxial specimen allowing horizontal sliding of the specimen cap. The three steel balls are initially held in place by a dab of silicone grease (after Lade 2004).

and maintains a very steady pressure. The mercury pot system may be used to supply the confining pressure in the triaxial cell and a separate system may be employed to provide the back pressure to the specimen. The details of this system have been explained by Bishop and Henkel (1962).

Note

Due to the poisonous nature of the fumes from mercury as well as the mercury itself, mercury pot systems are no longer employed to generate pressure in triaxial cells.

3.4.3 Compressed gas

Compressed gas may also be used to supply the confining pressure. Two sources of compressed gas are available. A motor-driven air compressor accumulating and compressing air within a preset range of pressures (e.g., 750–800 kPa) in a tank is often used to supply a houseline in the laboratory with compressed air. Air pressure regulators are able to provide reasonably steady pressures, but careful

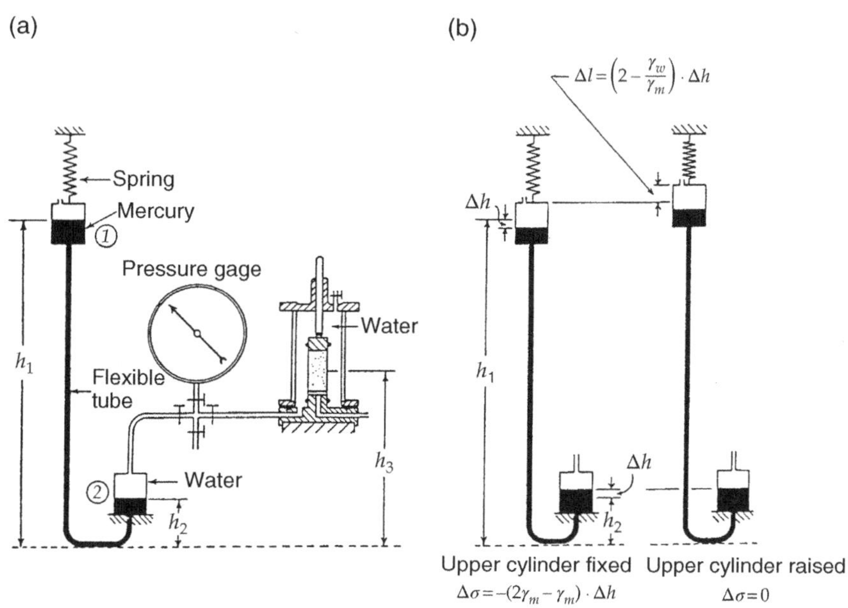

Figure 3.38 The principle of the self-compensating mercury control (after Bishop and Henkel 1962): (a) setup; and (b) movement of the upper cylinder in response to volume change from the triaxial cell.

measurements with a pressure transducer will reveal that the regulated pressure fluctuates with the tank pressure, but with much lower amplitude than that observed in the tank. Regulating the air pressure through several air pressure regulators (producing lower and lower pressures) produces relatively steady pressures, but the frequency of the tank pressure fluctuations can still be discerned on a sensitive pressure transducer.

Note that a filter or air dryer may be necessary at the outlet from the tank to reduce the amount of water that enters the compressed air houseline. This is because compression of humid, atmospheric air causes water to precipitate and this water accumulates in the tank and may potentially enter the houseline, which therefore should be provided with an outlet at its lowest point. Even then, a filter may be required locally and immediately before the compressed air enters the air pressure regulators, where the humid air/water may cause the regulators to malfunction.

The other source of gas pressure consists of bottled gas (nitrogen, air) regulated down to appropriate pressures for use in conventional tests. Bottled gas supplies very steady pressure and may be used to supply very high pressures up to approximately 20 MPa. Note that some gas pressure regulators vent gas to the atmosphere as part of the regulation scheme. Such regulators will exhaust the compressed gas bottle over a rather short period of time. Special non-venting regulators are necessary when working with bottled gas.

3.4.4 Mechanically compressed fluids

It is possible to generate pressure in a fluid-filled triaxial cell by compressing the fluid inside a cylinder by a motor and a gear that converts rotary motion to linear motion, as seen schematically in Fig. 3.39. It is devised to control either pressure or volume through the liquid applied from the device to the components of the triaxial cell, such as (1) the triaxial specimen in which either the back pressure or the specimen volume is controlled while the other is measured, (2) the triaxial cell in which the pressure is controlled, and (3) the axial loading device in which the pressure is controlled.

The working principle of the digital pressure/volume controller is shown in Fig. 3.39 and described by Menzies (1988). The liquid in the hydraulic piston (de-aired water) is pressurized by a piston that is pushed or pulled by a stepper motor through a ball-screw that guides the piston rod linearly a certain amount for each turn. The stepper motor is outfitted with a gear so as to be able to advance or retract the piston at different rates, and a pressure transducer measures the liquid pressure produced by the stepper motor action with feedback to the digital controller. The digital controller responds to the measured pressure so as to increase, decrease or maintain constant

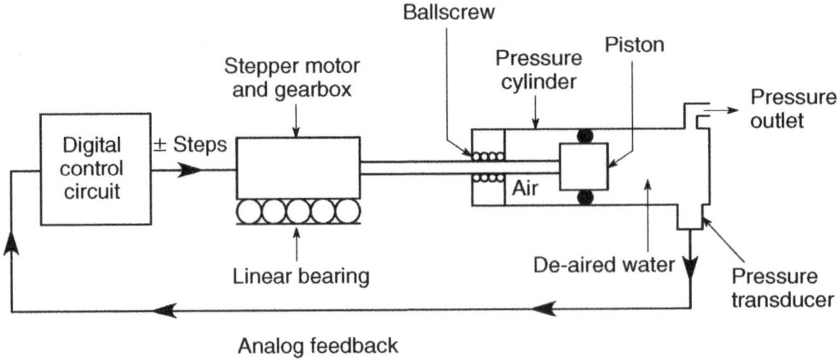

Figure 3.39 Schematic diagram of a digital controller for generation of mechanically compressed fluid. Reproduced from Menzies 1988 by permission of Geotechnique.

output pressure, as desired, and it measures the volume of fluid pushed into or retracted from the specimen by the number of turns by the stepper motor multiplied by a suitable calibration factor. Very precise volume measurements may be obtained by such systems. The volumetric capacity depends on the piston diameter and travel (e.g., 200 cm^3 and 1000 cm^3), and resolution down to 0.001 cm^3/step of the stepper motor (Menzies 1988), and the pressures generated may be resolved to 0.2 kPa and controlled to 0.5 kPa and varied over a wide range up to 64 MPa (Menzies 1988).

Similar pressure controller systems were employed by Yamamuro and Lade (1993b) for high pressure triaxial testing in which both the cell pressure and the vertical load were generated and controlled. Cell pressures up to 70 MPa were generated and supplied to the triaxial cell.

3.4.5 Pressure intensifiers

If limited air pressures or fluid pressures are available in the laboratory, it may be possible to generate higher pressures by using a pressure intensifier, the principle of which is shown in a schematic diagram in Fig. 3.40. The lower, available pressure is acting on the larger cross-sectional area of the piston and generating a higher pressure at the end with the smaller cross-sectional area. The pressure may thus be magnified by a factor equal to the ratio of the two cross-sectional areas.

Several possibilities are available for constructing a pressure intensifier, as indicated in Fig. 3.40. The one limitation of this device is that it may exhaust the stroke available, and then it will have to be regenerated to continue operation. Thus, using compressed air tends to quickly exhaust the stroke, and a way to regenerate the available stroke is required. High pressure intensifiers or boosters that recharge the air on the active pressure side are commercially available.

3.4.6 Pressure transfer to triaxial cell

Air as cell fluid

Regulated compressed air may in principle be used directly in the triaxial cell. However, this is only done for tests conducted at very low pressures and over very short periods of time (e.g., for UU-tests on soft clay). This is because of (1) the danger involved in containing compressed air in an acrylic plastic vessel that may shatter, and (2) the much higher permeability of latex rubber membranes to air than to water (Pollard *et al.* 1977). On the other hand, air may be used in part of the cell employed for cyclic triaxial tests. These tests require a very compressible component in or near the triaxial cell to cushion the volume changes generated by the specimen and the piston moving in and out of the cell. Without the compressible component, a false confining pressure would be generated in the cell. An accumulator connected to the triaxial cell may also be used as a cushion for the cell pressure in cyclic tests.

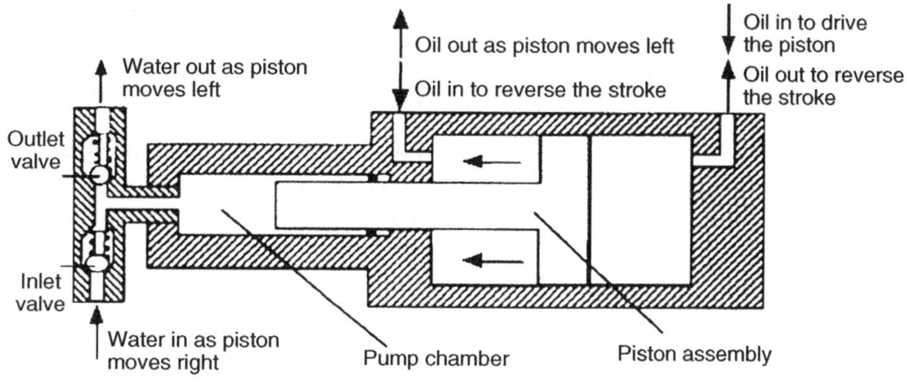

Figure 3.40 Pressure intensifier working with oil, water, or air.

De-aired water as cell fluid

The regulated compressed air may be used directly at the top of the triaxial cell that is filled with de-aired water. This setup (1) avoids the danger associated with large amounts of accumulated compressed air, and (2) air does not surround and begin to enter the specimen right away. Tests at low confining pressure with duration of 1–2 days may be performed this way without any problems resulting from air entering the specimen.

Most often, however, the regulated air pressure is transferred to water pressure in a small reservoir or a tank, as shown in Fig. 3.41. The triaxial cell is filled completely with de-aired water and directly connected to the air/water tank. Note that the pressure control is done in the compressed air by an air pressure regulator.

The most common problem arising from using regulated compressed air over water comes from air entering into the specimen causing it to be less than 100% saturated. This occurs in the following way: air dissolves in the water at the air/water interface, because the water is able to dissolve increasing amounts of air with increasing pressure. The dissolved air then travels by diffusion in the water to the surface of the specimen where it permeates through the membrane. Inside the specimen the pore pressure is lower than the cell pressure and the air comes out of solution, because the pore water is unable to maintain such large amounts of dissolved air at the lower pressure.

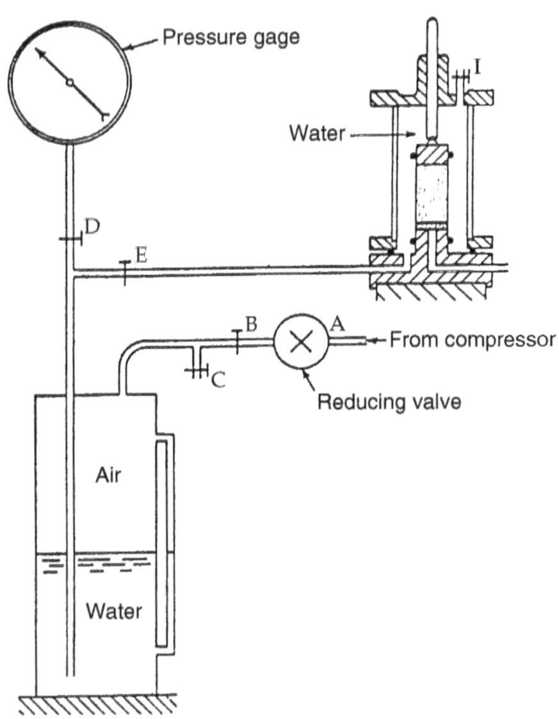

Figure 3.41 Transfer of regulated air pressure to water pressure (after Bishop and Henkel 1962).

This problem may be overcome by one of three methods, as illustrated in Figs 3.19 and 3.42. In the first method the dissolved air is prevented form reaching the specimen within the duration of the test. Movement of air in still water occurs by diffusion and by connecting the air/water reservoir with the triaxial cell through a long, small bore tube, as illustrated in Fig. 3.42(a), the time required for the air to reach the specimen would exceed the time for testing.

In the second method, a layer of silicone oil is floated on top of the de-aired water, which surrounds the specimen, as shown in Fig. 3.19. Since silicone oil does not dissolve air or water, it creates a barrier against the applied air pressure at the top of the triaxial cell.

In the third method, the air pressure acts on the surface of glycerin enclosed in the transfer tank, as shown in Fig. 3.42(b). Glycerin has almost no solubility for air. To further guard against air entering the water, a rubber balloon filled with de-aired water is submerged in the glycerin and connected to the triaxial cell or pore pressure (Winter and Goldscheider 1978).

3.4.7 Vacuum to supply effective confining pressure

The vacuum triaxial test is convenient, because a triaxial cell is not required, and it is possible to have direct access to the specimen during the test. The vacuum triaxial test may be performed by applying a vacuum to the interior pore space of a specimen. The difference between atmospheric pressure and the vacuum can be regulated to supply effective confining pressures from zero to 1 atm. Experiments may be performed on saturated, partly saturated or dry

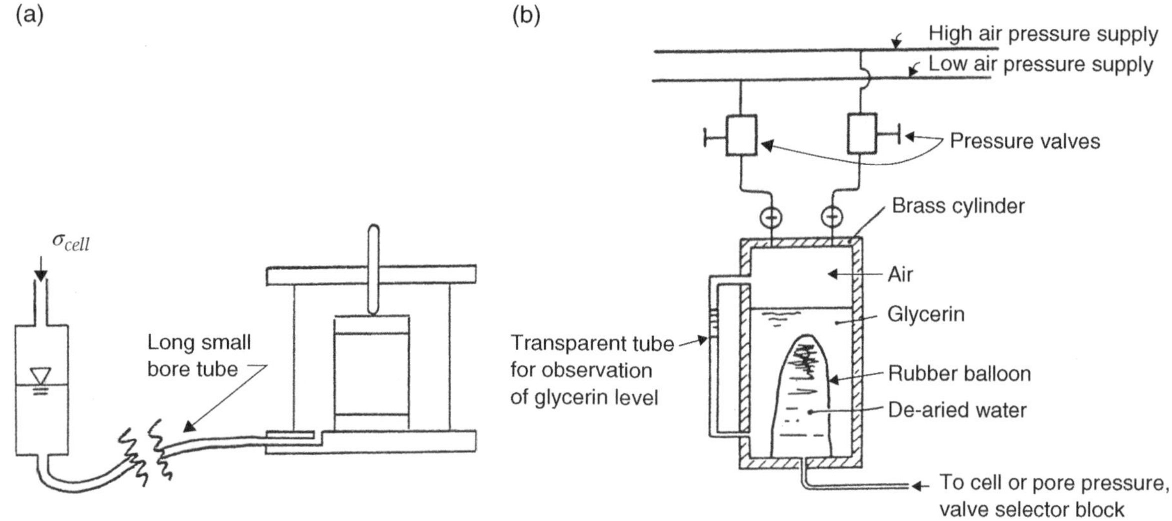

Figure 3.42 Methods of preventing air from reaching the triaxial specimen: (a) use of long, small bore tube between air/water reservoir and triaxial cell; and (b) use of improved air/water transfer tank. Reproduced from Berre 1982 by permission of ASTM International.

specimens of any soil. By including a volume change device in the line from the regulated vacuum to the specimen, it is possible to perform drained tests with volume change measurements. For dry specimens, the volume change device may consist of a horizontal, clear, small bore, calibrated tube with a water bubble that moves in response to changes in air pressure caused by changes in volume of the specimen. This volume change device is further explained in Chapter 4.

Due to their simplicity, vacuum triaxial tests are instructive and handy for students. One can readily convert a setup for unconfined tests and the equipment is safe, because there is no positive pressure involved. Deformations can be measured directly using a PI-tape and the students can get a feel for the test mechanics before being introduced to the complications of real triaxial tests.

3.5 Vertical loading equipment

A triaxial specimen may be loaded vertically under deformation and strain control or under load control or stress control. These methods have advantages and limitations, and they may be required for certain types of tests. Most conventional triaxial tests are performed under deformation or strain control, which is necessary to obtain post-peak behavior.

3.5.1 Deformation or strain control

Vertical loading under deformation control may be accomplished in a motor driven loading machine or in a hydraulic loading machine such as an MTS system. In both cases the machine may be set to load the specimen at a constant deformation rate. Since the vertical deformation is divided by the initial constant height to produce engineering strain, the test is also strain controlled. Most loading machines for soil testing have a rather large range of deformation rates available such that specimens may be loaded to failure and beyond within minutes, or the test may last several months, as illustrated in Fig. 3.43. The vertical load is measured by a load cell in series with the piston in the triaxial cell.

Vertical loading under strain control is useful for shearing the specimen under all but the circumstances mentioned below. Drained and undrained shearing to peak failure and beyond may be accomplished using strain control of the vertical load. Various paths may be followed by regulating the confining pressure to match the

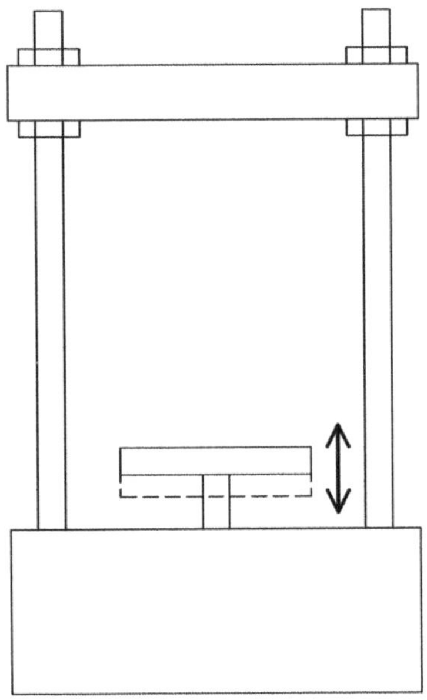

Figure 3.43 Schematic of a deformation control loading machine operated by a stepper motor or gear motor moving the platform for a triaxial cell up or down at a constant deformation rate.

desired stress path, or to follow a given strain path (e.g., strains are controlled in a K_0-test).

3.5.2 Load control

The vertical load may be applied under load control using a hanger system with deadweights or a loading cylinder with compressed air, as shown in Fig. 3.44. The system shown in Fig. 3.44(a) requires space and access below the table or platform supporting the triaxial cell, whereas the hanger system in Fig. 3.44(b) is located entirely above the table. There is a practical upper limit to the magnitude of the vertical force that can be supplied by deadweights. In practice this limit may be reached at 500–1000 N.

A single acting loading cylinder operated by air pressure may be used to supply much higher forces under load control. The loading cylinder may be placed under the table and the force is transferred through a yoke to the specimen, or it may be sitting on the cross-bar of a loading

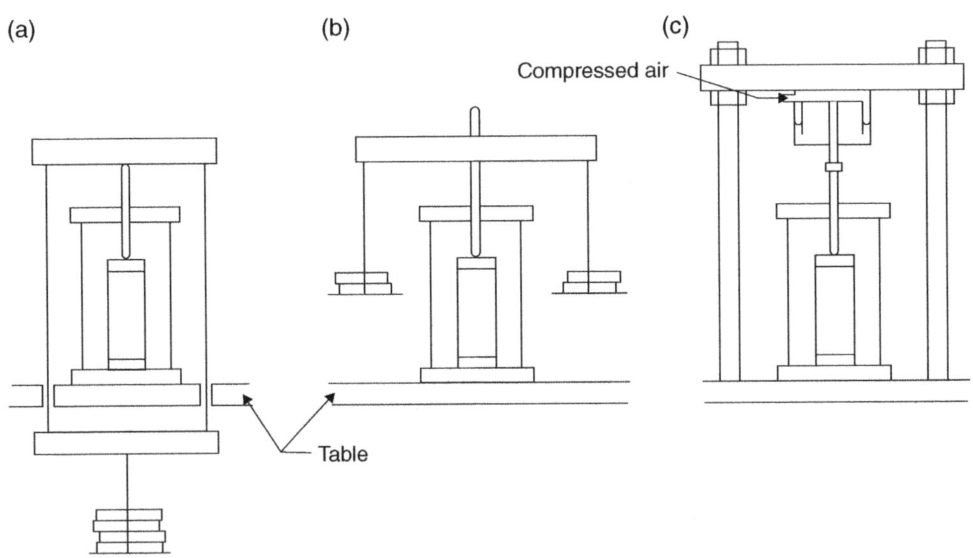

Figure 3.44 Methods of applying vertical loads to a triaxial specimen under stress control (or load control) using a hanger system with dead weights (a) below the table top and (b) above the table top, or (c) loading cylinder with compressed air.

frame, as shown in Fig. 3.44(c). For cyclic loading of the specimen the single acting piston is replaced by a double acting piston.

Vertical loading under load control is useful for anisotropic consolidation, studies of creep, and studies of instability of soils. However, if a specimen is sheared to failure under load control, the portion of the stress–strain relation occurring after peak failure cannot be obtained, unless special provisions are made for recording the dynamic response. This is because the specimen cannot sustain the applied vertical load, and this in turn cannot be reduced fast enough to avoid collapse of the soil specimen.

3.5.3 Stress control

For specimens that change cross-sectional area during testing, it is not possible to control the deviator stress unless a feedback system is used to adjust the vertical load in response to the calculated area so as to follow a desired stress path (e.g., constant vertical stress). Calculating the cross-sectional area requires determination of vertical strain and volumetric strain (see Sections 2.1.2 and 2.1.3). This may require a computer for measuring deformations and volume changes, calculating the cross-sectional area, and adjusting the vertical load. For slower tests, this may be done by hand calculations and manual adjustment of the vertical load, but it requires the presence of an operator for continuous control of the experiment. This may not be practical for, say, creep tests that last longer than 8 h.

3.5.4 Combination of load control and deformation control

There is often a need for being able to use both load control and deformation control within the performance of one experiment. For example, if a specimen is to be K_0-consolidated before shearing, the initial vertical stress should be load controlled so that proper K_0-consolidation can proceed under stress control in both vertical and horizontal (cell pressure) directions. Following K_0-consolidation the shearing phase

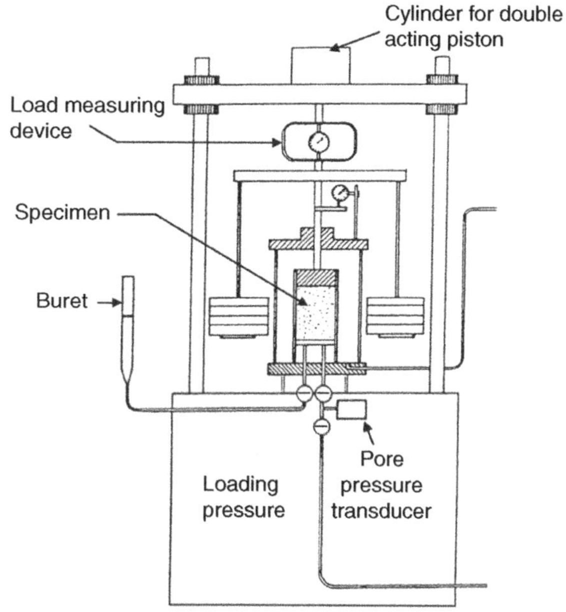

Figure 3.45 Vertical loading facilities used at the Norwegian Geotechnical Institute. Reproduced from Berre 1982 by permission of ASTM International.

is most often performed under deformation control. Thus, a switch from one to the other loading method is necessary, and this requires that both loading options are available in the same loading machine. Figure 3.45 shows an example of a loading setup that allows switching from load control to deformation control. It requires that the load control portion can be locked off so it reacts stiffly when the deformation control loading is initiated.

Another situation where a switch is required is exemplified by a creep test that is initiated after first increasing the deviator load under deformation control. Following the loading up to a desired deviator stress, the creep test requires axial deformation under load control. Stopping the deformation control loading machine usually provides a stiff response as the continued creep proceeds under load control. However, to produce a smooth continuation from deformation control deviator loading into the creep stage requires that the load from the pressure cylinder be brought up to its proper

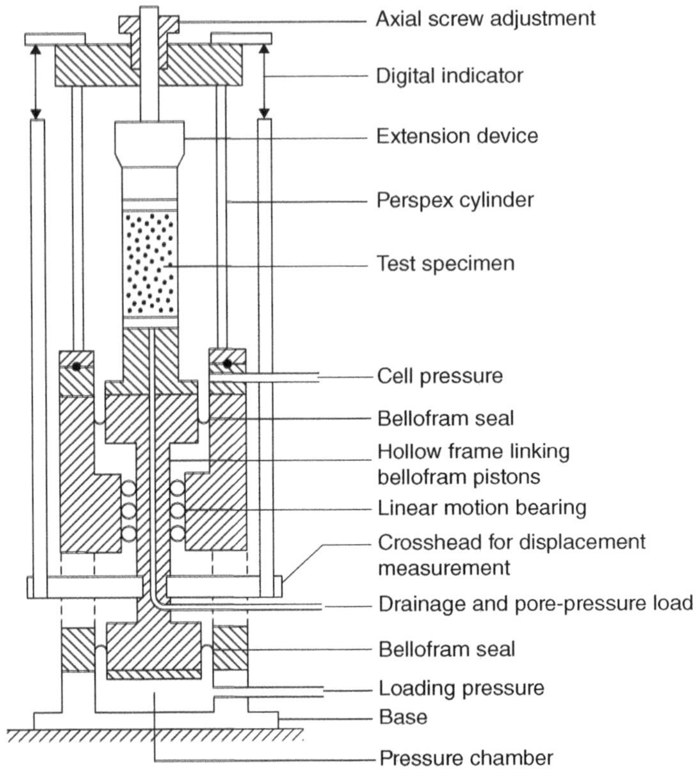

Figure 3.46 Schematic drawing of hydraulic triaxial apparatus (after Bishop and Wesley 1975).

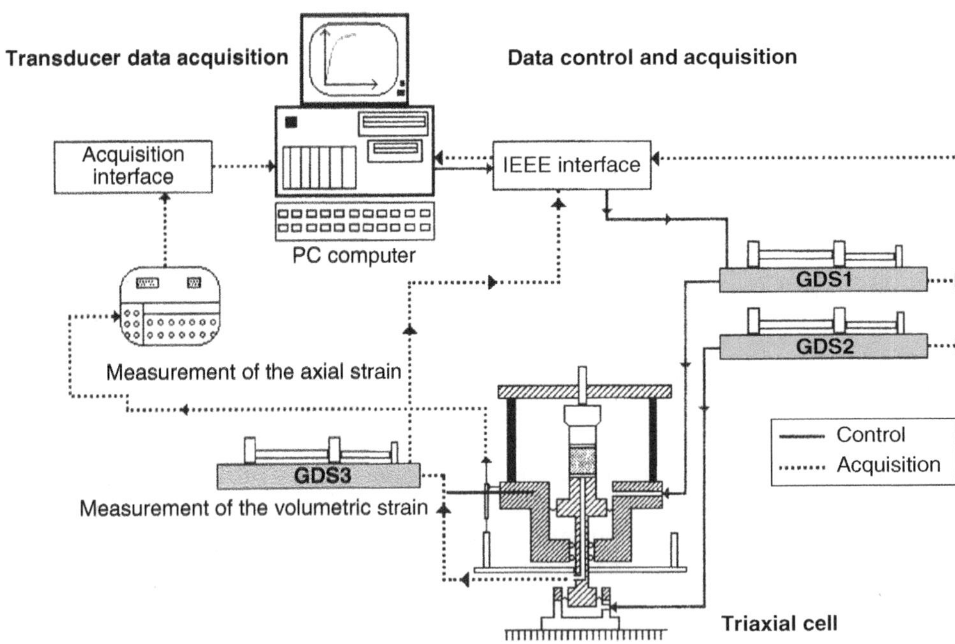

Figure 3.47 Triaxial testing setup with Bishop–Wesley hydraulic loading apparatus and stepper motors used for test control and for data acquisition. Reproduced from Hattab and Hicher 2004 by permission of Elsevier.

magnitude (to match the vertical deviator load generated by the deformation control loading machine) to be able to continue smoothly with the creep stage.

3.5.5 Stiffness requirements

Requirements for the stiffness of the loading machine become important when testing stiff and brittle materials in which the declining part of the stress–strain curve observed past peak failure may be very abrupt. The reason is that during loading the loading frame will deform elastically and store energy. This stored energy will be released if the specimen is brittle and the load–deformation relation of the specimen in the softening regime is steeper than the load–deformation relation for the loading machine. The release of stored energy will occur abruptly and cause the specimen to collapse and possibly to send parts of the material laterally out as projectiles (Hudson *et al.* 1972).

One method of obtaining a very stiff testing machine is to place a strain-gaged metal tube outside the specimen. The declining branch of the stress–strain curve may then be obtained under stable conditions. Effectively, the specimen is loaded in parallel with the tube, which takes the load as the specimen softens. The tube is calibrated so that its load can be subtracted from the total load to obtain the load on the specimen.

3.5.6 Strain control versus load control

Granular soils which may become unstable inside the effective stress failure surface under undrained conditions may exhibit effective stress paths that are different under strain and load control. The reason for this phenomenon is explained by Lade and Karimpour (2010), and it occurs in a transition zone in which the soil behavior under strain control exhibits instability and temporary instability. It does not occur for soils in which loading always entails increasing deviator stresses up to peak failure, and differences in results from strain control and load control tests are not expected.

3.6 Triaxial cell with integrated loading system

Triaxial equipment with integrated cell pressure and axial loading capability was designed and built by Bishop and Wesley (1975). Because this triaxial apparatus comes with a self-contained loading system, an external loading frame is not required. Figure 3.46 shows a schematic diagram of this hydraulic apparatus, which requires hydraulic pressures supplied to the cell and to the axial loading cylinder that forms the pedestal. Friction in the axial direction is minimized by using frictionless rolling diaphragms to contain the cell fluid (water) and to contain the hydraulic oil in the axial loading piston. By varying the cell pressure and the pressure in the axial loading cylinder relative to each other in a predetermined manner, it is possible to produce and control any desirable stress path in triaxial compression.

Supply of pressures may come from any of the sources reviewed above. Often the Bishop–Wesley stress path apparatus is outfitted with devices that create mechanically compressed pressures in fluids such as de-aired water and/or oil. Digital pressure controllers (Menzies 1988) allow control of pressures and measurement of volume changes, thus providing for automatic control as well as datalogging through a computer. The operational principles of the digital controller were reviewed in Section 3.4.4.

In addition to triaxial compression tests, triaxial extension tests may be performed with the apparatus shown in Fig. 3.46 by providing the extension device shown in Fig. 3.36, which stiffly attaches the cap to the top plate, thus preventing any cell pressure effect in the vertical direction. This allows the axial pressure to be reduced below the cell pressure and therefore creating an extension condition in the specimen.

Three digital controllers are required for a typical triaxial setup with the Bishop–Wesley device; one for the axial load or displacement, one for the cell pressure, and one for the back pressure/volume change measurements. Such a setup is shown in Fig. 3.47.

4 Instrumentation, Measurements, and Control

4.1 Purpose of instrumentation

The purpose of instrumentation is to measure the physical processes that occur in a test to be able to describe it. The physical quantities of interest (e.g., stresses, strains, temperature) require transformation into more usable and more easily measured quantities. Thus, the purpose of instrumentation is to transform one physical quantity into another physical quantity that can be measured. This process is called transduction.

An example of a simple transduction is shown in Fig. 4.1. A bucket of water is suspended from a spring that will stretch due to the weight of the bucket of water. Since the amount of stretch is related to the weight, it is possible to determine the weight by measuring the change in length of the spring and multiplying this by the calibration constant for the spring (= spring constant). Thus, the weight has been transformed into a change in length, that is a transduction has taken place to measure the weight.

Figure 4.2 shows another example of a transducer, an electrical strain gage. Loading of the steel plate to which the strain gage is attached causes it to strain, in the beginning very little. The straining of the strain gage causes a change in electrical resistance. When supplied with an input voltage, the change in resistance results in a voltage change that can be measured. Thus, the strain is transformed into a voltage change, that is the strain gage has transformed one type of signal (the uniform deformation of the steel plate) into another type of signal (the change in resistance) which in turn can be measured.

Note that no instrumentation and therefore no transduction was required to determine the vertical load applied to the specimen under load control in Fig. 3.44(a) and (b). In this case the vertical load was determined by simply counting the weights.

4.2 Principle of measurements

Because physical processes may be measured by many different types of instruments, it is necessary to determine which ones are appropriate, their availability, and so on. Whether an instrument is appropriate for measuring a particular quantity may be determined from the principle of *maximum signal for minimum interference* in the physical process. It is necessary to take energy from the process which is to be measured to activate the instrument. This should be done according to the above principle, that is

Triaxial Testing of Soils, First Edition. Poul V. Lade.

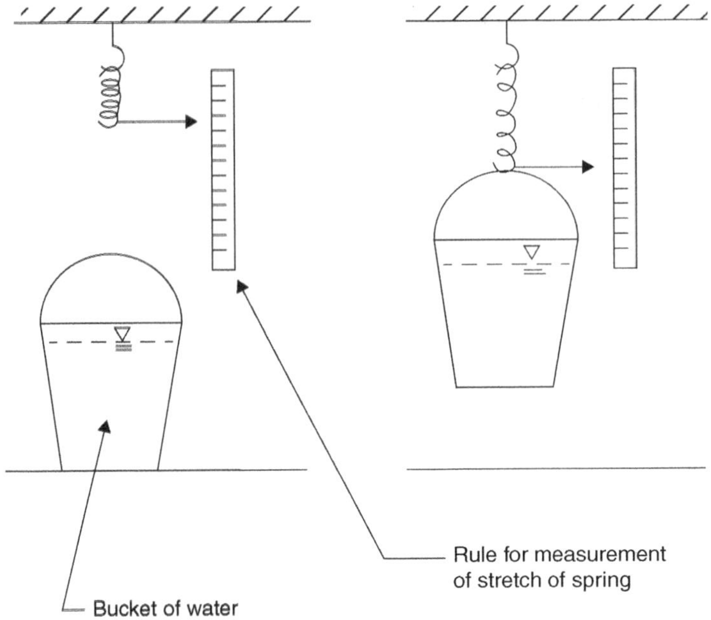

Figure 4.1 Spring scale: example of a simple transducer.

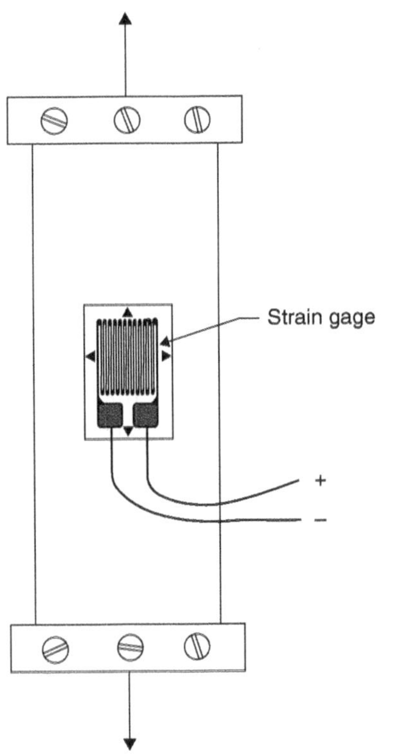

Figure 4.2 Electrical strain gage: example of a simple transducer.

the physical process should be disturbed as little as possible by the measurement method.

Two examples are given in which the instruments used for measurements interfere too much in the physical processes. Figure 4.3 shows a small bottle of boiling water. To measure the boiling temperature of the water (= 100°C = 212°F), a conventional mercury thermometer is inserted in the water. The measured temperature is 80°C (= 176°F). It is clear that too much energy is taken out of the process to obtain a signal that can be measured on the temperature scale.

In the second example, shown in Fig. 4.4, a Bourdon tube pressure gage is used to measure the pore water pressure in a CU-test on a triaxial specimen. The volume change required to make the gage respond correctly is considerable, and in this case it is too large: the specimen has to expel too much water to make the gage deflect, thus producing a drop in pore pressure in the specimen. The system consisting of a triaxial specimen, tubing, and a Bourdon gage is a closed system during undrained shear. The volume of water that has to enter the gage to make it deflect is excessive in relation to the

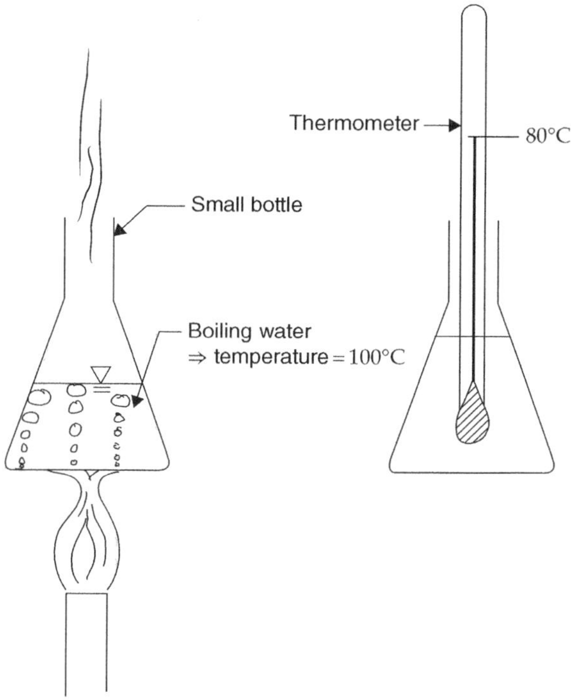

Figure 4.3 Temperature measurement in a small bottle of boiling water: example of too much interference in the physical process.

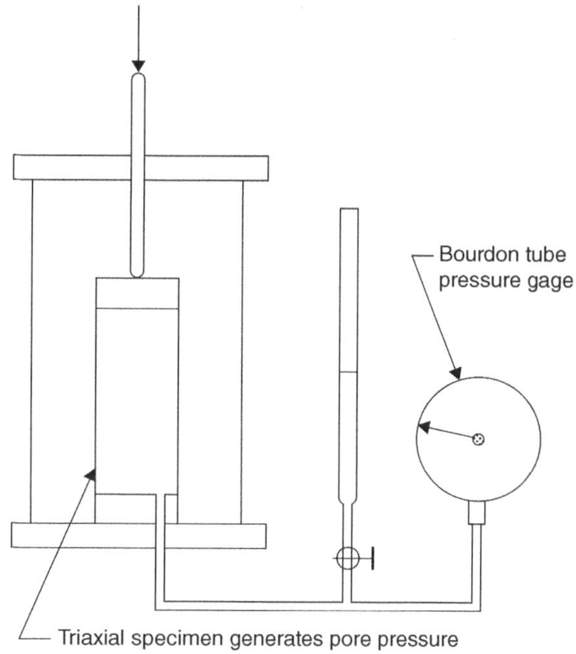

Figure 4.4 Pore water pressure measurement in a CU-test on a triaxial specimen: example of too much interference in the physical process.

total available amount in the specimen. Much too low values of pore water pressure are measured in the test, because too much energy was drained from the process to obtain the signal. Note, however, that the Bourdon gage is suitable for measuring pressure in a system in which a large supply of pressurized medium is available or in an open system (e.g., air in a houseline supplied by an air compressor).

4.3 Instrument characteristics

An instrument performs three functions, as indicated in Fig. 4.5(a). The signal measurement is done by the prime sensor, and the signal is then modified by the signal conditioner before it is read out on the end device.

An example of these three functions in an instrument is illustrated by the Bourdon tube pressure gage in Fig. 4.5(b). The application of an air pressure of 417 kPa at the bottom of the gage causes the Bourdon tube to stretch out an amount in proportion to the applied pressure. Thus, the Bourdon tube measures the pressure; it acts as the prime sensor. The measured signal (stretch of the tube) requires modification by the signal conditioner as performed by the rack-and-pinion. The end device consists of an arm that has been rotated to point at 417 kPa on the pressure scale so the pressure can be read.

Two other examples are shown in Fig. 4.6. The functions of the parts of the dial gage and the proving ring are indicated on the diagram. These devices are used to measure deformation and force, respectively.

The instruments discussed above are suitable for measuring relatively slow processes that allow readings to be taken and manually written down in columns on a piece of paper. These instruments may be used in all types of static triaxial tests. However, they are not suitable for

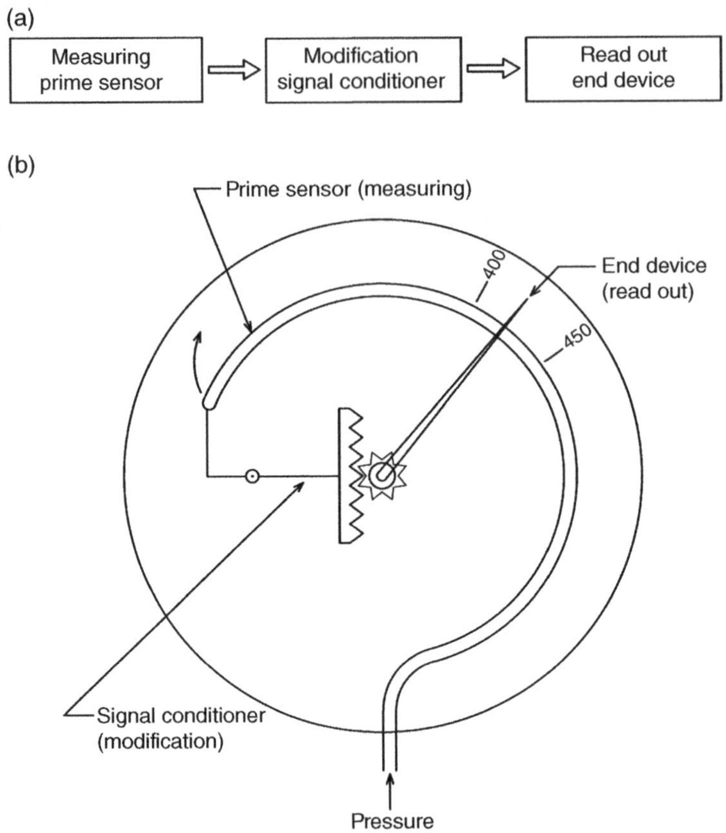

Figure 4.5 (a) Functions of an instrument for measuring physical processes and (b) example of a measurement instrument: Bourdon tube pressure gage.

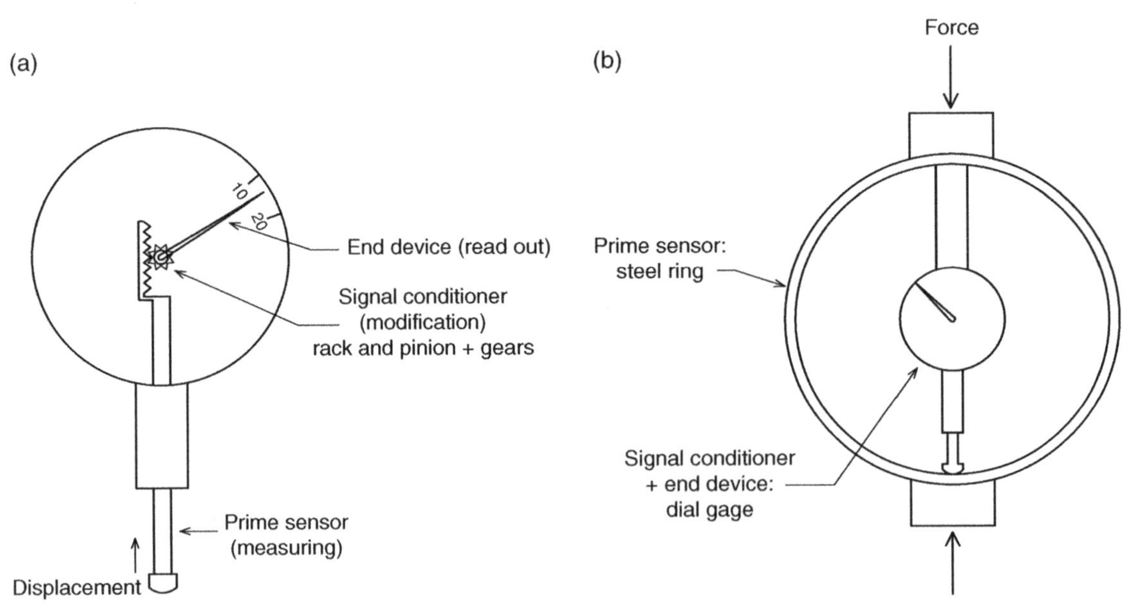

Figure 4.6 Examples of measurement instruments: (a) dial gage; and (b) proving ring.

measuring rapidly changing pressures, deformations, or forces that occur in for example a cyclic triaxial test. Neither are they suitable for automatic logging of data in static or cyclic triaxial tests.

4.4 Electrical instrument operation principles

In addition to mechanical devices such as dial gages, electrical devices are commonly employed as end devices in instruments used for triaxial testing to modify the measurements into electrical signals that may be read manually or by computer. The operating principles and use of the most common types of electrical instruments are briefly reviewed below.

4.4.1 Strain gage

The basic operating principle of a strain gage is that a metal wire exhibits a change in electrical resistance with change in strain. The electrical resistance of a metal wire can be expressed as:

$$R = \rho \cdot \frac{L}{A} \tag{4.1}$$

in which

R = electrical resistance
ρ = resistivity of metal
L = length of wire
A = cross-sectional area of wire

The variation in electrical resistance then occurs according to:

$$dR = \frac{\partial R}{\partial \rho} \cdot d\rho + \frac{\partial R}{\partial L} \cdot dL + \frac{\partial R}{\partial A} \cdot dA \tag{4.2}$$

Performing the derivations as indicated produces:

$$dR = \frac{L}{A} \cdot d\rho + \frac{\rho}{A} \cdot dL - \rho \cdot L \cdot A^{-2} \cdot dA \tag{4.3}$$

and division by R from Eq. (4.1) results in:

$$\frac{dR}{R} = \frac{d\rho}{\rho} + \frac{dL}{L} - \frac{dA}{A} \tag{4.4}$$

Strain sensitivity S_r is defined as the unit change in resistance per unit change in strain:

$$S_r = \frac{dR / R}{dL / L} \tag{4.5}$$

Using Eq. (4.4) produces:

$$S_r = \frac{d\rho / \rho}{dL / L} - \frac{dA / A}{dL / L} + 1 \tag{4.6}$$

For an elastic material the second term is:

$$\frac{dA / A}{dL / L} = \frac{2 \cdot \varepsilon_3}{\varepsilon_1} = -2 \cdot \nu \tag{4.7}$$

in which ν is Poisson's ratio for the metal wire. The strain sensitivity is therefore:

$$S_r = \frac{d\rho / \rho}{dL / L} + (1 + 2 \cdot \nu) \tag{4.8}$$

The strain sensitivity is known as the gage factor.

For conventional strain gages the gage factor is in the range from 1.9 to 2.5. Semiconductor strain gages have gage factors of about 100, that is they are much more sensitive than conventional strain gages. They therefore produce a larger signal for a given strain than conventional strain gages. Semiconductor strain gages are made of piezo-resistive materials that can change resistivity (ρ) with applied stress (tension, compression, hydraulic pressure, shear stress). However, these materials have increased temperature sensitivity and nonlinearity.

Since the material resistivity ρ is temperature dependent for most materials used for strain gages, these devices are temperature dependent. The effect of temperature changes may be reduced by using the gages in half or full bridge circuits, as shown in Fig. 4.7. The principle of a Wheatstone bridge is often used in force and pressure transducers, and it may be used in clip gages for measurement of deformation. Note that opposite strain gages in a Wheatstone bridge are in compression and tension in order that the electrical output is enhanced rather than cancelled.

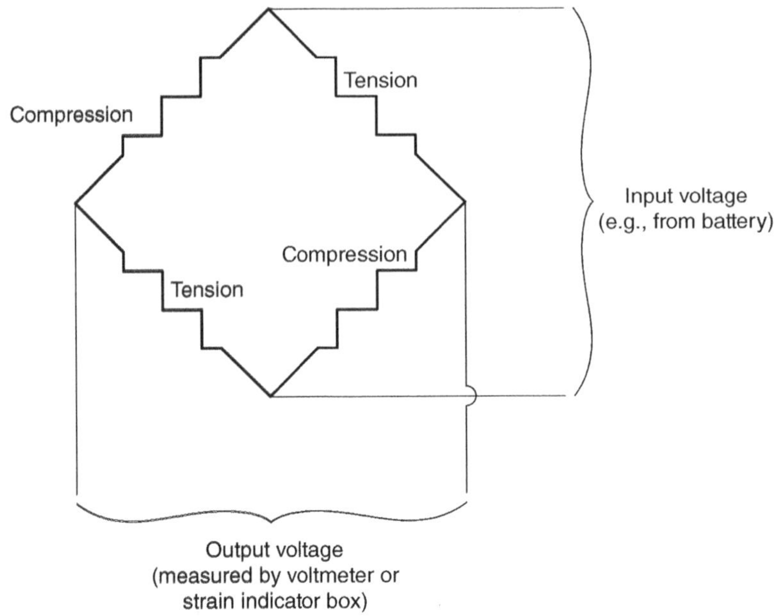

Figure 4.7 Full bridge circuit: Wheatstone bridge.

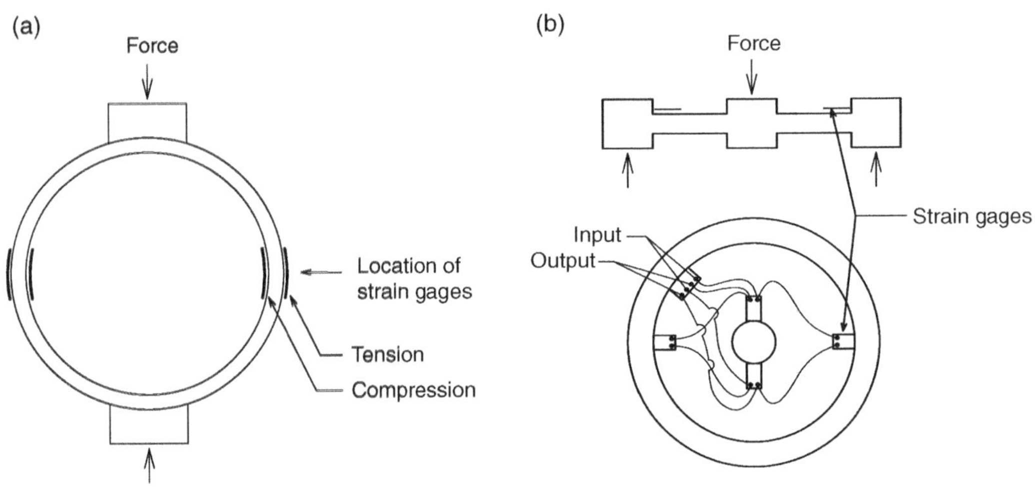

Figure 4.8 Force transducers with strain gages mounted to form Wheatstone bridges: (a) conventional proving ring; and (b) diaphragm load cell.

Force transducers

Figure 4.8 shows examples of employing strain gages in force transducers. Strain gages may be mounted on a proving ring as indicated in Fig. 4.8(a), such that the vertical load applied to a triaxial specimen may be recorded both manually (using the dial gage) and from the electrical signal (using a voltmeter, strain indicator box, or computer). Figure 4.8(b) shows a diaphragm load cell with strain gages mounted near the center and near the outer rim. The wiring diagram follows the principle of the Wheatstone bridge shown in Fig. 4.7. The principles and details of design of diaphragm load cells are given in Appendix B.

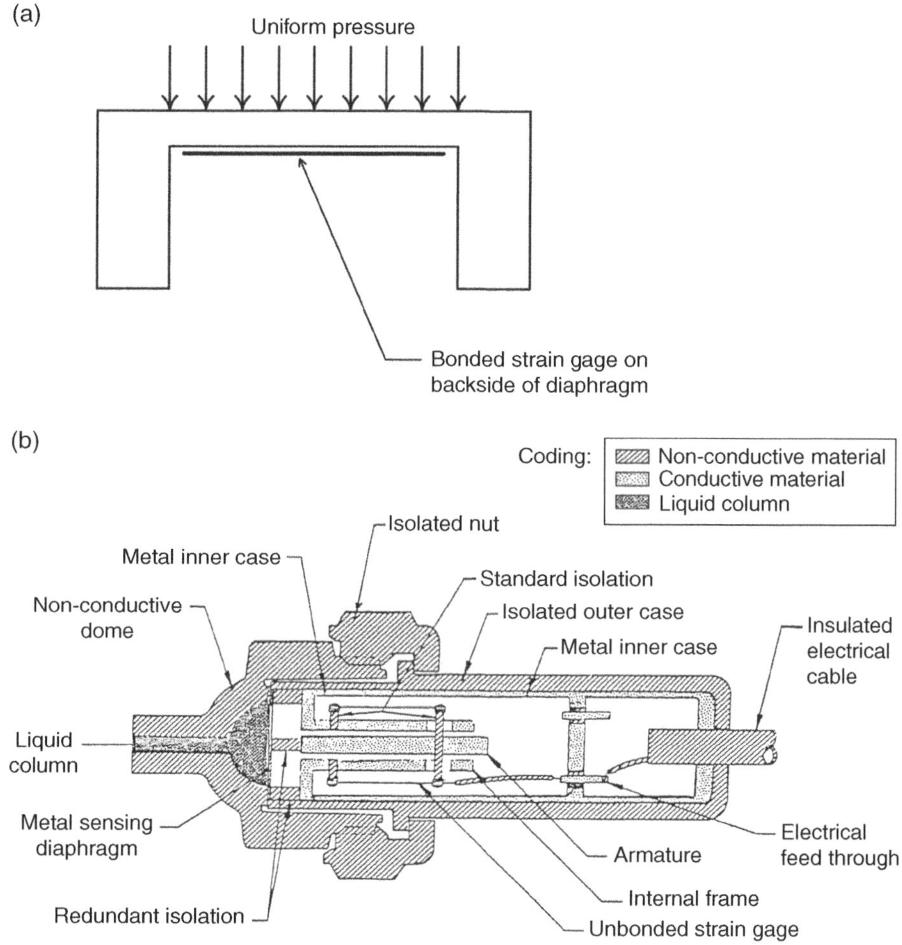

Figure 4.9 Pressure transducers employing strain gages for transduction: transducers with (a) bonded strain gages; and (b) unbonded strain gages.

Pressure transducers

Pressure transducers may employ one of two types of strain gages. Bonded strain gages are glued or cemented to the inside of the diaphragm as indicated in Fig. 4.9(a). Unbonded strain gages are wires whose ends are attached to plates and frames and connected to the diaphragm as shown in Fig. 4.9(b). Pressurizing the diaphragm causes it to deflect and this generates an electrical signal that is measured as explained above.

Linear deformation measurement devices

Several types of devices have been developed on the basis of the bending of relatively weak beams outfitted with strain gages to measure the deformations between particular points. Here they are referred to as clip gages, but some of them have been called local deformation transducers (LDTs) and cantilever gages (Tatsuoka 1988; Goto *et al.* 1991; Hoque *et al.* 1997).

4.4.2 Linear variable differential transformer

A linear variable differential transformer (LVDT) is used to measure displacements. Figure 4.10(a) shows that an LVDT is an electromagnetic transducer consisting of three coils enclosing a movable iron core.

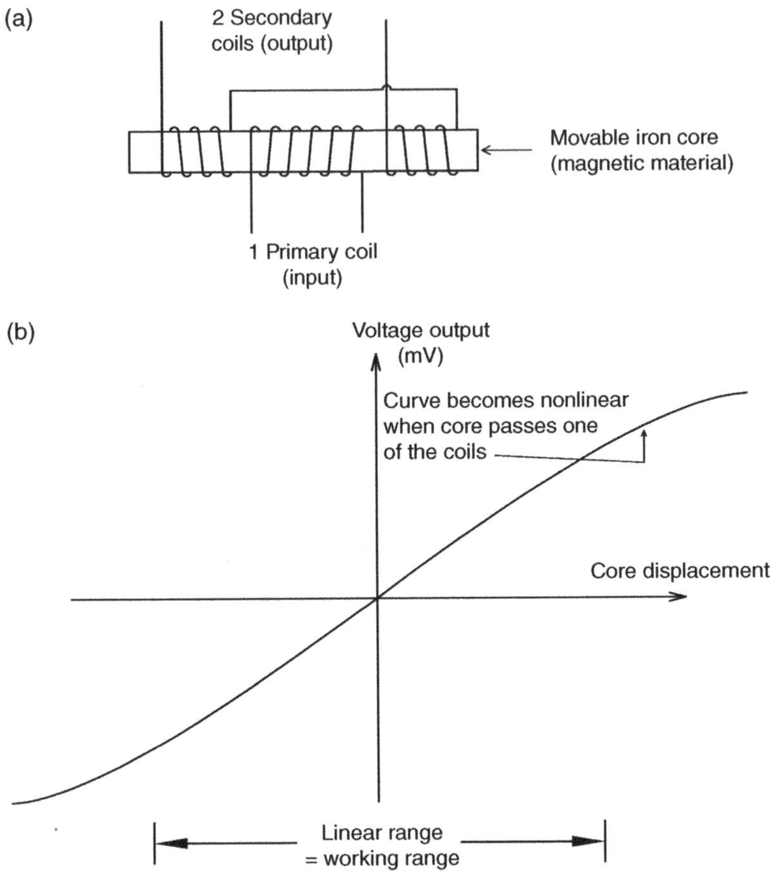

Figure 4.10 (a) Principle of operation of an LVDT and (b) schematic calibration curve.

The input to the centrally located primary coil is an alternating current (AC) that induces voltage differences in the two outer, secondary coils. When the core is in the center or the null position, identical voltages are induced in the secondary coils and zero output is obtained if the two secondary coils are connected in phase opposition. When the core is moved away from the null position, a greater voltage is induced in the secondary coil towards which the core is moved. Consequently, a *differential* voltage output is obtained.

A schematic calibration curve for an LVDT is shown in Fig. 4.10(b). When the core is near the center, the relation between the voltage output and the core displacement is nearly perfectly linear. The relation becomes nonlinear when the core passes one of the secondary coils. The *linear* range, which is the working range for the LVDT, may be *varied* with the input frequency of the AC.

Whereas all LVDTs operate on the basis of AC input, these devices may be purchased with built-in transformers such that direct current (DC) can be input and output. However, such transformers do not need to be part of the LVDTs; they can be located away from the basic AC-LVDT. The linear range of displacement for LVDTs typically varies from 0.25 to 250 mm. The physical sizes of the AC-LVDTs are somewhat proportional to their rated range. DC-LVDTs (sometimes referred to as DCDTs) have minimum sizes larger than the corresponding AC-LVDTs, because they have a built-in transformer.

The sensitivity of an LVDT varies with the input frequency and it may be necessary to specify the input voltage and frequency when quoting a particular value of sensitivity. The sensitivity is expressed as mV/0.025 mm displacement/V input.

Notably the force to move the core in an LVDT depends on the weight of the core itself. It can be as small as 0.3 g. Submersible LVDTs are commercially available, so they may be mounted directly on the triaxial specimen in the cell water.

In addition to deformations, LVDTs may be adapted in primary sensor devices to measure force (in a proving ring), pressure, velocity, and acceleration.

4.4.3 Proximity gage

A proximity gage is a non-contacting displacement transducer which senses the displacement between the sensor and an electrically conducting surface. A proximity gage is most often employed to measure radial deformations of triaxial specimens. It offers the advantages of being free of clamp restraints and of effects of fluid response time as experienced with volume change measurement devices. Figure 4.11 shows that an active coil, activated by an AC (say at 1 MHz; Khan and Hoag 1979), emanates a magnetic flux, which induces eddy currents in a flat, perpendicular, electrically conducting surface. The eddy currents in turn create alternating electric currents in the inactive coil. The magnitude of the eddy currents depends on the distance to the target. As the target moves closer to the sensor, the magnitude of the eddy currents increases and the impedance (the total effective resistance) in the electric circuit increases. The variations in impedance are converted to a DC voltage that is proportional to the distance to the target. The target may consist of a piece of aluminum foil, and due to the high frequency of the eddy currents, they penetrate only a few thousandths of a centimeter into the conductive surface. A target that has a thickness of 0.02 mm is sufficient to produce a stable output.

The proximity gage can operate in gas, water, or oil, and it is insensitive to pressure. Displacements as small as 0.00025 mm (0.00001 in.) can be resolved (Cole 1978). Proximity gages can be used to measure vertical deformations, but are most often fashioned to measure radial deformations.

4.4.4 Reluctance gage

Another magnetic phenomenon (in addition to inductance and eddy currents) is reluctance, which is resistance to magnetic flow, that is the opposition by a magnetic material to magnetic flux. This principle is used in pressure transducers in which the deflection of the diaphragm changes the reluctance of the electric circuit.

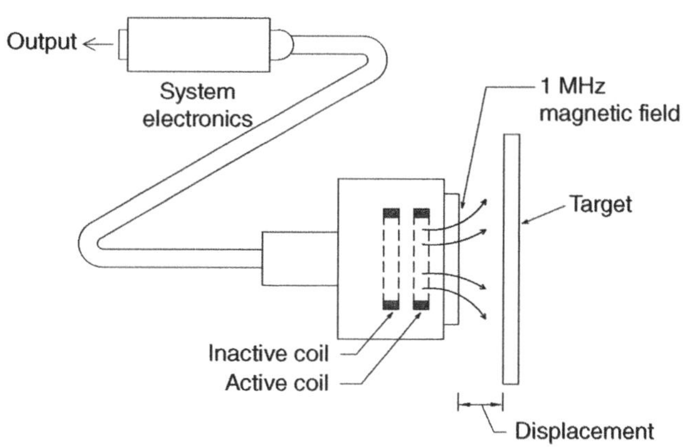

Figure 4.11 A proximity gage is a non-contacting displacement transducer. Reproduced from Khan and Hoag 1979 by permission of Canadian Science Publishing.

Pressure transducers based on the reluctance principle have very high output signals, but they require AC voltage for excitation.

4.4.5 Electrolytic liquid level

Burland and Symes (1982) and Symes and Burland (1984) describe the use of electrolytic levels to measure axial displacement. The electrolytic device consists of a sealed glass capsule with three internal, co-planar electrodes that are partially submerged in an electrolyte. The resistance changes between the central and the outer electrodes as the capsule tilts. When the tilt sensitive capsule is mounted between two points on the surface of a specimen, it can be used to measure local torsional deformations of a hollow cylinder torsion shear specimen. When mounted in a triangular arrangement that converts change in tilt to change in height, as shown in Fig. 4.12, the electrolyte level may be used to measure axial deformation between two points on a triaxial specimen. The linear range is within ±10°.

4.4.6 Hall effect technique

The Hall effect was employed in a device developed by Clayton and Khatrush (1986), who explain that "if a metallic or semiconductor plate, through which current is flowing, is placed in a magnetic field where flux lines are directed perpendicularly to both the plate and the current flow, the charge carriers (i.e., electrons) will be deflected so that a voltage is produced across the plate in a direction normal to the current flow. This voltage is known as the Hall voltage." Figure 4.13 shows the setup for this effect: two small permanent magnets with a gap between them, one with the north pole pointing upwards and the other with the south pole pointing upwards, create the magnetic field. The metallic or semiconductor plate is then passed over this field, as shown in Fig. 4.13, and a voltage will be produced across the plate. As the plate is moved relative to the magnets, the output voltage will change linearly within a range. The Hall effect has been used in setups to measure axial as well as radial deformations.

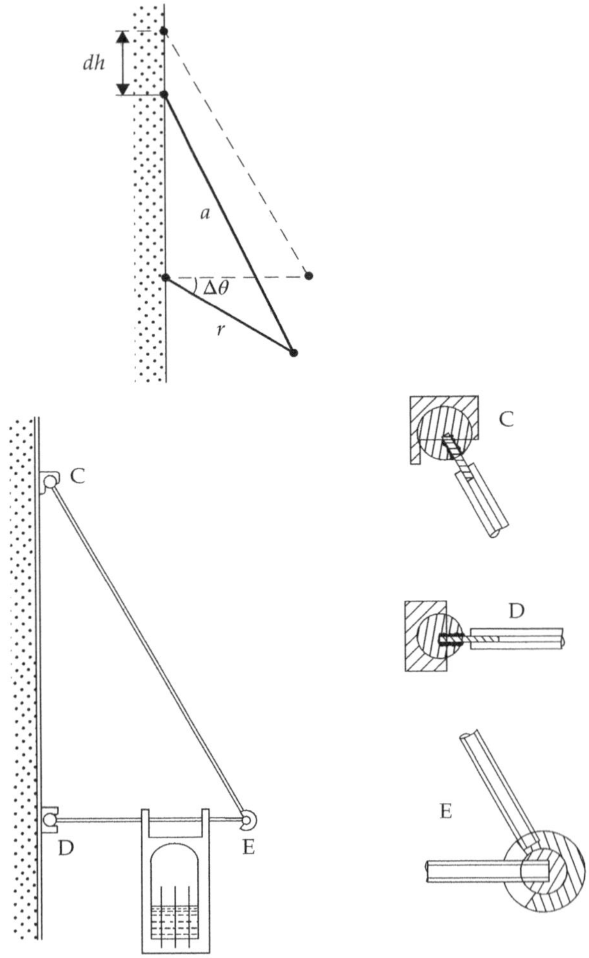

Figure 4.12 Electrolytic level for measurement of axial strain. Reproduced from Symes and Burland 1984 by permission of ASTM International.

4.4.7 Elastomer gage

Safaqah and Riemer (2007) describe a very flexible elastomer gage, that is a strain gage that may be attached to the inside of the rubber membrane and thus follow the specimen during its deformation. Small strains from 0.0005 to over 10% with excellent resolution may be measured with very little adverse effect on the deformation process. The sensing element in this gage consists of a string of electrically conducting liquid alloy contained in a thin capillary tube inside a polyurethane gage body formed as a 1 mm thick and 5 mm wide strip

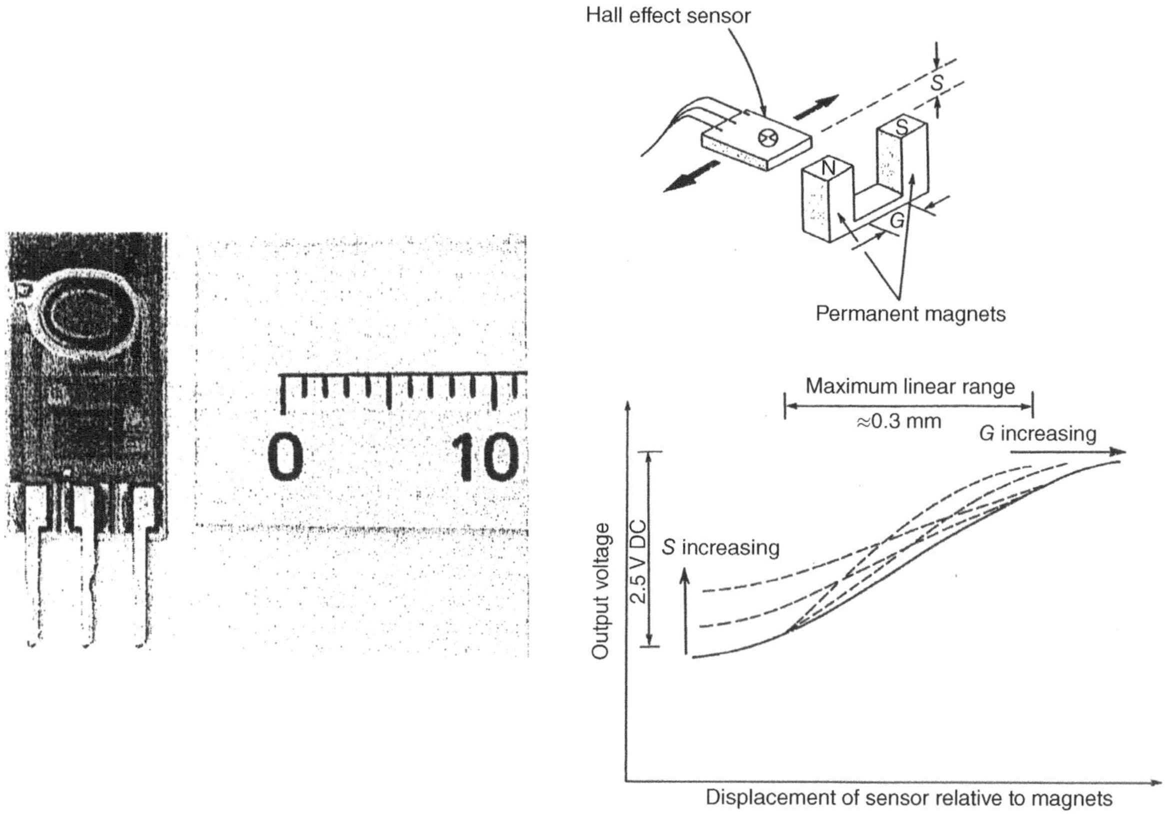

Figure 4.13 Hall effect semiconductor. Reproduced from Clayton and Khatrush 1986 by permission of Geotechnique.

with variable lengths. Nominal lengths of 6–65 mm have been fabricated. The elastomer gage works in extension, but it may be prestressed to accommodate compressive deformation. The operating principles of this device follow those of conventional strain gages, that is it is used in a Wheatstone bridge, activated by a voltage excitation and delivers a signal to be measured. The advantages and limitations are similar to those for conventional strain gages. The elastomer gage may be used to measure axial and radial deformations and it may be inclined to measure shear strains in torsion shear tests.

4.4.8 Capacitance technique

Altschaeffl and Mishu (1970) proposed to employ the change in capacitance between two coaxial cylinders to indicate the radial deformation of a soil specimen surrounded by a thin-walled metal ring capacitor. Considerable influence of the cell pressure and soil type as well as the nonlinear response required substantial calibration efforts. This technique appears to be somewhat cumbersome, and it does not appear to have found much use in triaxial testing.

4.5 Instrument measurement uncertainty

The physical quantities required to interpret a triaxial test are measured by force and pressure transducers, dial gages, clip gages, LVDTs and other types of sensors. The uncertainties with which these instruments measure forces, pressures, and geometric changes to specimen dimensions are discussed below.

4.5.1 Accuracy, precision, and resolution

The measurements by any instrument are described in terms of accuracy and precision. Accuracy is synonymous with degree of correctness, that is it is a measure of how close the measured mean value is to the actual value; accuracy also represents a measure of bias. Precision is synonymous with reproducibility and repeatability, and it expresses the scatter in the measurements. Resolution describes the smallest measurable unit, and it is a measure of readability; for a digital display it is one digit of change in the last digit (Dunnicliff 1988; Germaine and Ladd 1988). Typically, the accuracy is several times larger (possibly in the order of 10 times) than the resolution. The three concepts are illustrated by the four bull's eyes in Fig. 4.14, in which the centers represent the true value of a measured quantity. The measurements shown in Fig. 4.14(a) represent low precision and moderate accuracy (= correctness), while those in Fig. 4.14(b) indicate low accuracy, but high precision (= repeatability). Figure 4.14(c) shows low resolution, low precision, but high accuracy (= correctness), and Fig. 4.14(d) represents both accurate and precise measurements.

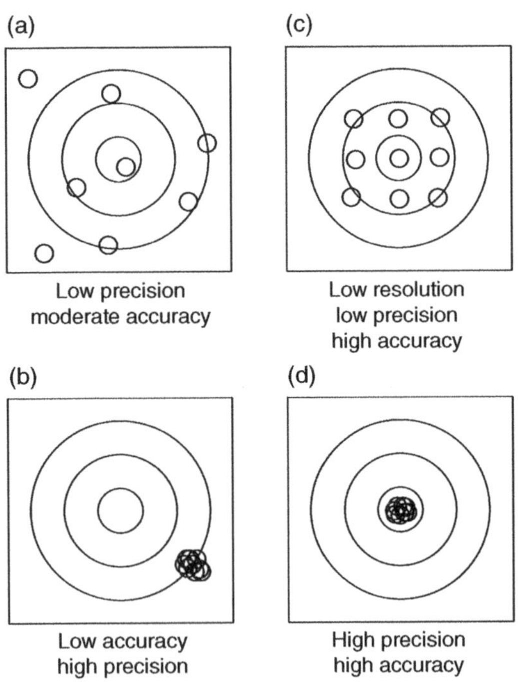

Figure 4.14 Illustration of basic terminology used to evaluate errors. Reproduced from Germaine and Ladd 1988 by permission of ASTM International.

4.5.2 Measurement uncertainty in triaxial tests

The uncertainty in the friction angle obtained due to inaccuracies in measurements may be estimated and provides background for evaluating the accuracy and precision required from the measuring instruments. The friction angle from a drained triaxial test on soil without effective cohesion is calculated from:

$$\sin\varphi = \frac{\sigma_1 - \sigma_3}{\sigma_1 + \sigma_3} = \frac{(\sigma_1 - \sigma_3)}{(\sigma_1 - \sigma_3) + 2\cdot\sigma_3} = \frac{\frac{F}{A}}{\frac{F}{A} + 2\cdot\sigma_{cell}} = \frac{F}{F + 2\cdot A\cdot\sigma_{cell}} \tag{4.9}$$

The uncertainty in $\sin\varphi$ may be calculated from the propagation of error analysis and assumes the variables – in this case F, ΔA, and $\Delta\sigma_{cell}$ – are uncorrelated. With instrumentation error this assumption is usually true, and:

$$\Delta\sin\varphi = \sqrt{\left(\frac{\partial\sin\varphi}{\partial F}\cdot\Delta F\right)^2 + \left(\frac{\partial\sin\varphi}{\partial A}\cdot\Delta A\right)^2 + \left(\frac{\partial\sin\varphi}{\partial\sigma_{cell}}\cdot\Delta\sigma_{cell}\right)^2} \tag{4.10}$$

where

$$\frac{\partial\sin\varphi}{\partial F} = \frac{1\cdot(F + 2A\cdot\sigma_{cell}) - F\cdot 1}{(F + 2A\cdot\sigma_{cell})^2} = \frac{2A\cdot\sigma_{cell}}{(F + 2A\cdot\sigma_{cell})^2} \tag{4.11}$$

$$\frac{\partial\sin\varphi}{\partial A} = \frac{2F\cdot\sigma_{cell}}{(F + 2A\cdot\sigma_{cell})^2} \tag{4.12}$$

$$\frac{\partial\sin\varphi}{\partial\sigma_{cell}} = \frac{2AF}{(F + 2A\cdot\sigma_{cell})^2} \tag{4.13}$$

and

ΔF = estimated error in the deviator force
ΔA = estimated error in cross-sectional area of the specimen
$\Delta \sigma_{cell}$ = estimated error in cell pressure

Since

$$\Delta \sin\varphi = \Delta\varphi \cdot \cos\varphi \tag{4.14}$$

the error in the friction angle may be calculated from:

$$\Delta\varphi = \frac{\Delta \sin\varphi}{\cos\varphi} = \frac{180°}{\pi} \cdot \frac{1}{\cos\varphi} \cdot \frac{2}{\left(F + 2A \cdot \sigma_{cell}\right)^2} \cdot \sqrt{\left(A \cdot \sigma_{cell} \cdot \Delta F\right)^2 + \left(F \cdot \sigma_{cell} \cdot \Delta A\right)^2 + \left(A \cdot F \cdot \Delta\sigma_{cell}\right)^2} \tag{4.15}$$

or

$$\Delta\varphi = \frac{360°}{\pi \cdot \cos\varphi} \cdot \frac{F \cdot A \cdot \sigma_{cell}}{\left(F + 2A \cdot \sigma_{cell}\right)^2} \cdot \sqrt{\left(\frac{\Delta F}{F}\right)^2 + \left(\frac{\Delta A}{A}\right)^2 + \left(\frac{\Delta\sigma_{cell}}{\sigma_{cell}}\right)^2} \tag{4.16}$$

A balanced error, that is an error that is distributed equally on the three quantities, requires that the % error be the same for F, A, and σ_{cell}. To estimate the error with which the cross-sectional area may be measured, the errors in measurements of axial and volumetric strains are required. For the design of volume change devices, the precision with which the volume change is measured should be the same as that with which the axial strain is measured, because both strains appear on an equal basis in the equation for the cross-sectional area (Lade 1988b; see also Section 4.8.1). The premise for this calculation is that the precision of the axial deformation measurements equals the smallest division on a dial gage measuring in millimeters = 0.01 mm. Thus, with equal errors on volumetric and axial strains, $\Delta\varepsilon_v = \Delta\varepsilon_1$, the following estimate of the error on the cross-sectional area may be made. The cross-sectional area is calculated from:

$$A = A_0 \cdot \frac{1-\varepsilon_v}{1-\varepsilon_1} \tag{4.17}$$

and the error may be expressed as:

$$\Delta A = \sqrt{\left(\frac{\partial A}{\partial \varepsilon_v} \cdot \Delta\varepsilon_v\right)^2 + \left(\frac{\partial A}{\partial \varepsilon_1} \cdot \Delta\varepsilon_1\right)^2} \tag{4.18}$$

where

$$\frac{\partial A}{\partial \varepsilon_v} = A_0 \cdot \frac{-1}{1-\varepsilon_1} \tag{4.19}$$

$$\frac{\partial A}{\partial \varepsilon_1} = A_0 \cdot \frac{(1-\varepsilon_1)\cdot 0 - (1-\varepsilon_v)(-1)}{(1-\varepsilon_1)^2} = A_0 \cdot \frac{1-\varepsilon_v}{(1-\varepsilon_1)^2} \tag{4.20}$$

Substitution of Eqs (4.19) and (4.20) into Eq. (4.18) produces the following error in the cross-sectional area:

$$\Delta A = \sqrt{\left(A_0 \cdot \frac{-1}{(1-\varepsilon_1)} \cdot \Delta\varepsilon_v\right)^2 + \left(A_0 \frac{1-\varepsilon_v}{(1-\varepsilon_1)^2} \cdot \Delta\varepsilon_1\right)^2} \tag{4.21}$$

From which the % error may be expressed:

$$\frac{\Delta A}{A_0} = \sqrt{\left(\frac{-(1-\varepsilon_1)}{(1-\varepsilon_1)^2} \cdot \Delta\varepsilon_v\right)^2 + \left(\frac{1-\varepsilon_v}{(1-\varepsilon_1)^2} \cdot \Delta\varepsilon_1\right)^2} \tag{4.22}$$

and for equal errors on the volumetric and axial strains, $\Delta\varepsilon_v = \Delta\varepsilon_1$:

$$\frac{\Delta A}{A_0} = \frac{\Delta\varepsilon_1}{(1-\varepsilon_1)^2} \cdot \sqrt{(1-\varepsilon_1)^2 + (1-\varepsilon_v)^2} \tag{4.23}$$

To evaluate the maximum error from this expression, it is assumed that the soil dilates 20% at 35% axial strain and the % errors in axial strain are the same as those used for evaluation of volume change devices (see Section 4.8.1): $\Delta\varepsilon_1$ = 0.004% for a large specimen with D = 10.00 cm (4.00 in.) and with H/D = 2.5, and $\Delta\varepsilon_1$ = 0.011% for a small specimen with D = 3.56 cm (1.40 in.) and H/D = 2.5. For these conditions the errors in cross-sectional area become 0.010% for large specimens and 0.027% for small specimens.

Table 4.1 Transducer minimum performance characteristics (after Silver 1979)

Property	Load transducer	Displacement transducer	Pore pressure transducer
Minimum sensitivity, mV/V	2	0.2 mV per 0.025 mm/V for AC LVDTs, or 5 mV per 0.025 mm/V for DC LVDTs [a]	2
Nonlinearity, % of full scale	±0.25	—	±0.5
Hysteresis, % of full scale	±0.25	0.0	±0.5
Repeatability, % of full scale	±0.10	±0.1	±0.5
Thermal effects on zero shift, % of full scale per °C (°F)	±0.005 (±0.0025)	—	±0.02 (±0.01)
Thermal effects on sensitivity, % of full scale per °C (°F)	±0.005 (±0.0025)	—	±0.02 (±0.01)
Max. deflection at full rated value, mm (in.)	±0.125 (±0.005)	—	—
Volume change characteristics, cm^3/kPa ($in.^3$/psi)	—	—	2.5×10^{-6} (1.0×10^{-6})

[a] LVDTs, unlike strain gages, cannot be supplied with meaningful calibration data, because system sensitivity is a function of excitation frequency, cable loading, amplifier phase characteristics, and other factors. It is a practical necessity to calibrate each LVDT-cable instrument after installation by making use of a known standard.

These errors, due to uncertainties on measurements of changes in specimen dimensions, are calculated for individual tests, but they do not represent the differences in geometric changes from test to test due to natural scatter in the behavior of the soil being tested. This scatter accounts for much larger differences than the uncertainties due to measurements in geometry of the individual specimens.

Requiring a balanced error, that is an error that is distributed equally on the three quantities in Eq. (4.16), requires that the % error be the same for F, A, and σ_{cell}. However, it is unrealistic to require force and pressure transducers to perform with the precision obtained from the measurements of changes in specimen dimensions. The strains measured in triaxial tests inadvertently vary from test to test, because the scatter on measured strains is much larger than the errors expressed by $\Delta\varepsilon_1$ and $\Delta\varepsilon_v$. This compensates for the uncertainties in transducer measurements, which are larger, but remain constant from test to test. Even with very accurate and precise force and pressure transducer measurements, the natural scatter in the soil, even for laboratory prepared specimens of sand and clay, are such that some differences will always be observed in the measured stress–strain, volume change or pore pressure measurements.

Table 4.1 suggests minimum performance characteristics for transducers that measure loads, pressures, and deformations. The terminology used in this table is reviewed in the following section.

4.6 Instrument performance characteristics

Force and pressure transducers, clip gages, LVDTs, and other types of sensors may possess many desirable features including high measurement accuracy and precision, good long-term stability and good frequency response characteristics. Their performance is measured by a number of characteristics as reviewed below.

4.6.1 Excitation

Alternating current or DC may be used for the operation of strain gages. Linear variable differential transformers operate in principle on AC, but separate or on-board transformers allow operation by DC power.

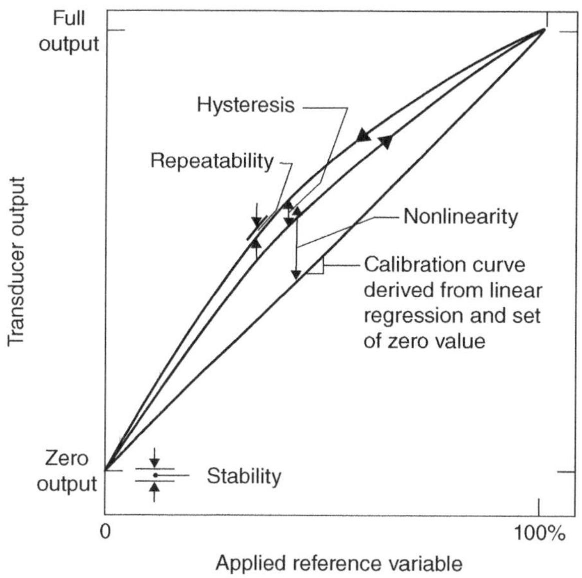

Figure 4.15 Parameters used to quantify the performance of a sensor. Reproduced from Germaine and Ladd 1988 by permission of ASTM International.

4.6.2 Zero shift

Zero shift refers to the change or drift in the output at zero load with time. The issue of stability of the read-out at zero load, shown in Fig. 4.15 depends on the type of construction and environmental factors such as temperature, humidity, and so on.

4.6.3 Sensitivity

The sensitivity of a force or pressure transducer is the full scale output or the change in output signal produced by an increase in applied force or pressure – measured in mV/V (output voltage per volt input).

4.6.4 Thermal effects on zero shift and sensitivity

Transducers can be operated within certain temperature limits such as −50°C to +150°C (−60°F to +300°F). Within these temperature limits the electrical output may change at zero pressure or force, and the sensitivity may change. Both are measured as a percentage of the full scale response per °C (or °F).

4.6.5 Natural frequency

The natural frequency of a transducer is the resonant frequency of the strain gage–diaphragm or proving ring combination, which in turn is the upper limit for pressure or force fluctuations – measured in Hz (cycles/s).

4.6.6 Nonlinearity

Figure 4.15 indicates that nonlinearity is the deviation between the calibration curve and a straight line drawn between the point of zero pressure or force output (applied reference variable) and the point of full scale output – measured as a percentage of the full scale response.

4.6.7 Hysteresis

Figure 4.15 shows that hysteresis is the difference in output at a particular pressure or force when the readings are taken under increasing and decreasing loading – measured as a percentage of the full scale response.

4.6.8 Repeatability

The repeatability, indicated in Fig. 4.15, is the maximum difference occurring between two readings taken at a particular pressure or force – measured as a percentage of the full scale response.

4.6.9 Range

The range of a transducer should be selected to be just a little higher than the magnitude of the pressure or force to be measured. This ensures that the accuracy of the readings is in the order of those quoted for the transducer. Overload capacity and overload protection should be considered together with the range of the transducer.

4.6.10 Overload capacity

Overload capacity is the ability to accept extra load without breaking. It is achieved by designing the device for higher loads than the nominal load for which the device is acquired.

4.6.11 Overload protection

Overload protection is achieved by mechanically stopping the deflection of the prime sensor, thus avoiding its breakage or destruction.

4.6.12 Volumetric flexibility of pressure transducers

The volumetric flexibility is the change of volume per unit change in pressure due to deflection of the diaphragm in the transducer. To obtain reliable measurements of pore water pressure, the pressure transducer should have as small flexibility as possible while it still provides accurate measurements of pressure.

4.7 Measurement of linear deformations

Linear deformations of a triaxial specimen in the axial and radial directions may be measured directly by dial gages, and these produce reliable measurements. However, there are several disadvantages associated with using dial gages, including their requirement for manual datalogging, and they are therefore unsuitable for automated testing, continuous datalogging, and testing at high deformation rates. Much more suitable are electrical devices, which can measure accurately and with high frequency.

4.7.1 Inside and outside measurements

Burland (1989) proposed that working strain levels in soils are typically less than 0.1% in well-designed structures. To reach this level of accuracy, an internal measurement system may be employed to avoid uncertainties in interpretations from apparatus compliance, tilting of specimen and imperfect platen-specimen bedding.

For conventional static triaxial testing, measurement of axial deformation and volume changes outside the cell is most convenient and may result in sufficient accuracy for many practical purposes. The radial deformations and strains as well as the cross-sectional area may be deduced from these measurements. However, a number of sources of errors are present in such measurements, and those associated with the outside measurements of axial deformation are listed in Table 4.2 and indicated in Fig. 4.16 (and previously in Figs 3.33 and 3.34). Including

Table 4.2 Sources of errors in conventional deformation measurements (modified after Scholey *et al.* 1995; Yimsiri and Soga 2002)

Type of error	Caused by
Seating errors	• Gaps closing between piston or internal load cell and specimen cap • Gaps closing between end plates and porous stones
Alignment errors	• Porous stones of nonuniform thickness • Non-verticality and eccentricity of loading ram • Non-horizontality of end platen surfaces • Tilt of specimen
Bedding errors	Surface irregularities and poor fit at the interface between the specimen and the porous stones
Compliance errors	• The tie-rods extend and cause relative displacement of the top plate of the triaxial cell with respect to the piston • Deflection of the internal load cell • The lubricant and rubber sheets are compressed in systems using lubricated ends • The porous paper at specimen ends is compressed
Nonuniform strain errors	Nonuniform deformations in specimen due to end restraint

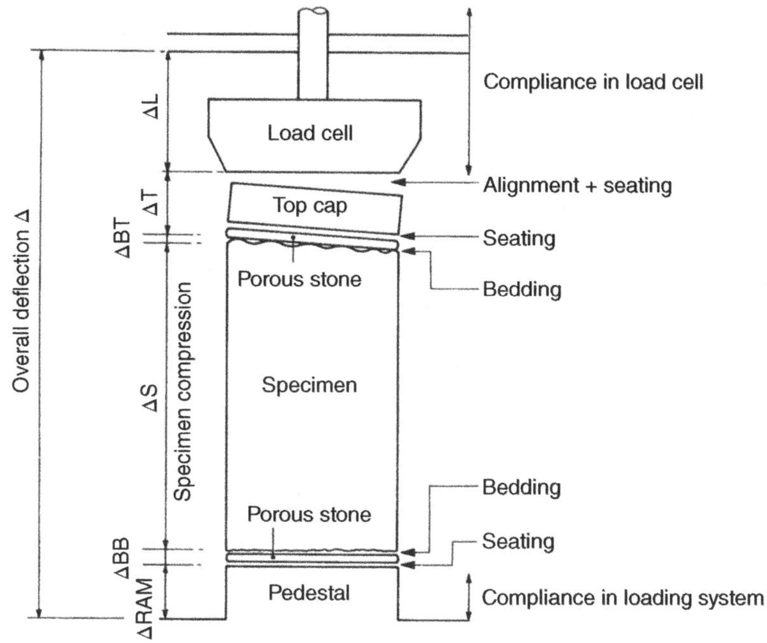

Figure 4.16 Sources of errors in external axial deformation measurements. Reproduced from Baldi *et al.* 1988 by permission of ASTM International.

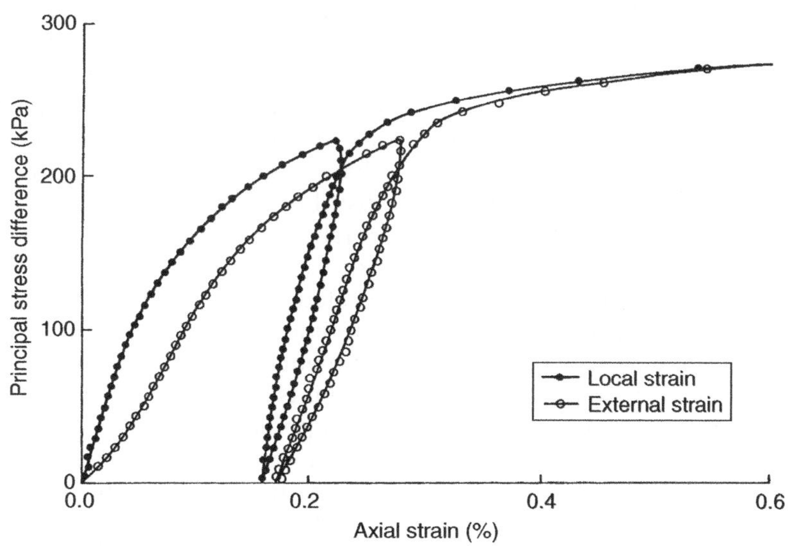

Figure 4.17 Comparison of local and external strains. Reproduced from Clayton and Khatrush 1986 by permission of Geotechnique.

these errors in the measurements often results in too low soil stiffness, both at the beginning of the test and during unloading–reloading cycles, as exemplified in Fig. 4.17. The significance of the errors included in the measurements is more pronounced for stiffer materials.

More accurate measurements may be obtained inside the triaxial cell by avoiding the errors

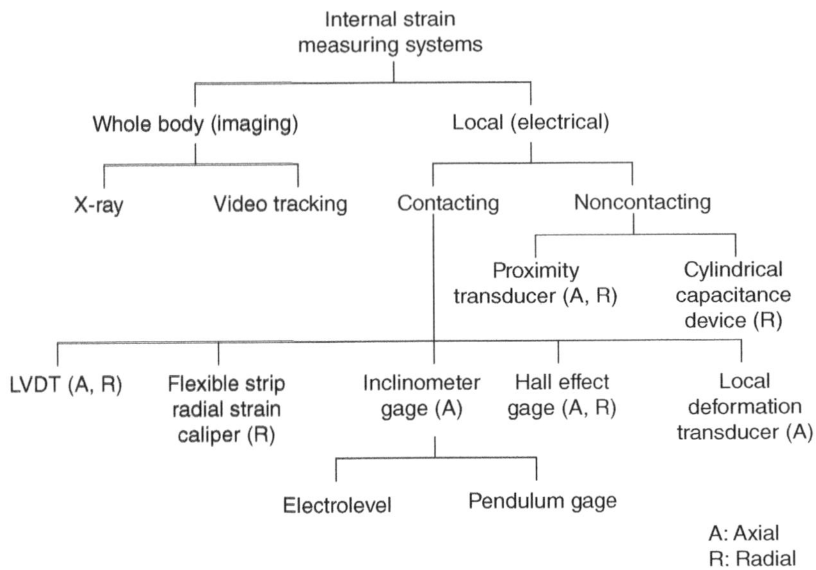

Figure 4.18 Overview of possible internal strain measurement systems. Reproduced from Scholey *et al.* 1995 by permission of ASTM International.

due to seating, alignment, bedding and compliance shown in Fig. 4.16. Measurements directly between the cap and base still include some of the errors due to seating, bedding, and compliance between the specimen and the end plates (which may be outfitted with lubricated ends and/or the filter stones), and this may not be sufficiently accurate for some purposes.

To completely avoid all these errors, it is advisable to measure directly on the triaxial specimen, but away from the end plates, where end friction may skew the measurements. Measuring the deformation of the middle portion is therefore recommended, and to obtain accurate measurements of stiffness requires accurate measurements of small deformations over as long a gage length as possible. Figure 4.18 shows an overview of the possible internal strain measuring systems.

4.7.2 Recommended gage length

Due to the effects of the end plates, it has been recommended to measure over the middle third of the specimen (Kirkpatrick and Belshaw 1968; Kirkpatrick and Younger 1970), because the stress and strain conditions are most uniform and least affected by end restraint in this section. However, a longer section of the specimen may be employed for axial deformation measurements. Lubricated ends may be used to ensure that the specimen deforms as a right cylinder. If end restraints were present, the barreling of the specimen may be approximated by a parabolic shape, as indicated in Fig. 2.8, and the axial strain distribution may be assumed to follow a similar parabolic shape. The average lateral deformation of the parabolic shape occurs at points located 20% of the height from the ends of the specimen. Therefore, to obtain the best measurements possible, even in the presence of small amounts of friction on the end plates, measurement points should be located at the points that are most likely to correspond to the average change in diameter. This also provides a gage length that is almost twice as long as one-third of the height mentioned above.

4.7.3 Operational requirements

A number of different design and operation principles have been employed in designing linear deformation measurement devices. A detailed survey of these devices was published by Scholey *et al.* (1995). The most common principles of linear measurements are embodied in strain gage and LVDT systems with various methods of converting the primary signal to measurable electronic responses. In studying these and other systems, some desirable operational requirements for the ideal system emerge [modified after Scholey *et al.* (1995)]:

1. Highly accurate and capable of resolving very small deformations resulting in strain accuracies better than $\pm 10^{-3}\%$.
2. Measuring systems should be simple to install and operate.
3. Instrumentation must not interfere with the soil behavior.
4. Measuring systems should be able to accommodate coupled axial and radial deformation without loss of accuracy.
5. Deformation measurements should ideally be made locally over the central section (over the central one-third or middle 60%, see discussion in Section 4.7.2) of the specimen so that effects of end restraint are minimized.
6. Instrumentation must be capable of operating in the range of cell pressures required for the triaxial testing.
7. Instruments must be submersible if required to operate in cell fluid.
8. For cyclic systems, instruments must have low hysteresis and rapid response.
9. The instrument should be inexpensive.

Individual measurement devices, whose principles of operation are more uncommon, may have limitations that are not anticipated and discussed above.

Electrical deformation measurement devices for mounting internally in the triaxial cell have been developed based on the principles of operation reviewed in Section 4.4. While some of these devices may provide very accurate measurements, many of them have limited range and may not be suitable for recording over the entire range of deformation experienced by the specimen as it is loaded from the hydrostatic stress state to well beyond peak failure. The operation principles are employed in devices that may be used to measure axial as well as radial deformations.

4.7.4 Electric wires

The wires that carry electric power and output signals from the instruments enter into the cell through the base (or less conveniently through the top plate) and they must be completely sealed so as not to lose cell fluid at the entrance. The wires must also be flexible so that they do not interfere with the specimen deformation as well as the measurement of this deformation. In addition, the wire shielding must provide complete electric seals and they should not compress and collapse under increasing cell pressures. Finding and repairing electric leaks in the measurement systems can be one of the most time-consuming endeavors involved in triaxial testing. Thus, high quality electric wiring is important for successful performance of the test.

Gage systems attached to and carried by the specimen are referred to as floating systems, while those attached to the base are referred to as fixed systems. The interference of stiff wires with the measurements is most critical for the floating-type systems, while the systems fixed to the base are relatively insensitive to the wire stiffness.

4.7.5 Clip gages

The linear deformation of a triaxial specimen may be measured by a clip gage (Lade and Duncan 1973; El-Ruwayih 1976), as shown in Fig. 4.19 or local deformation transducer (LDT) mounted directly on the specimen, as indicated in Fig. 4.20 (Tatsuoka 1988; Goto *et al.* 1991; Hoque *et al.* 1997). A clip gage consists of a thin band of metal (e.g., beryllium copper, phosphor bronze, or spring steel) on which one or two pairs of strain gages are glued on the inside and the outside surfaces. If one pair of strain gages is used on the clip gage, then two dummy gages are mounted on a metal block outside the triaxial cell and together they

Figure 4.19 Cylindrical specimen with mounted collars. Reproduced from Kolymbas and Wu 1989 by permission of ASTM International.

are connected to form a Wheatstone bridge according to the wiring diagram in Fig. 4.21. If two pairs of strain gages are employed, then the four gages are wired to form a Wheatstone bridge on the metal strip inside the triaxial cell, as indicated in Fig. 4.22. If the triaxial cell is filled with water, the gages and wires must be waterproofed.

The metal band with the strain gages may be shaped to fit around the specimen, between the cap and base, or between shoes or hinges attached to the specimen surface, as shown in Fig. 4.20. A hinge for attachment of an LDT for measuring local, lateral deformations is shown in Fig. 4.23 (Hoque *et al.* 1997). Alternatively, two cantilever-type gages attached to a stationary support, as seen in Fig. 4.24, may be employed (Yimsiri *et al.* 2005).

Clip gages or LDTs may be used to measure very small deformations, and they may be employed in pressurized cell water for long periods of time. Further details and evaluation of LDTs have been given by Goto *et al.* (1991), Hoque *et al.* (1997), and Yimsiri *et al.* (2005).

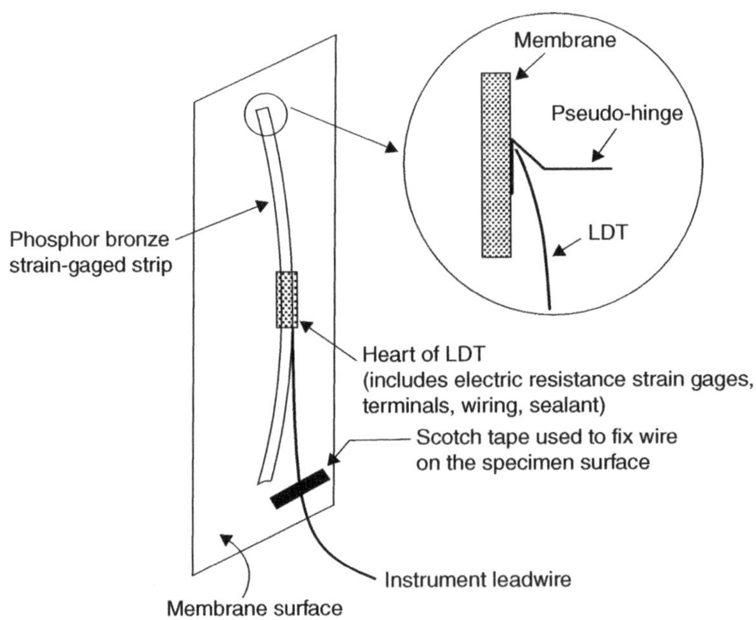

Figure 4.20 An LDT at working condition to measure vertical strains in a triaxial specimen. Reproduced from Hogue *et al.* 1997 by permission of ASTM International.

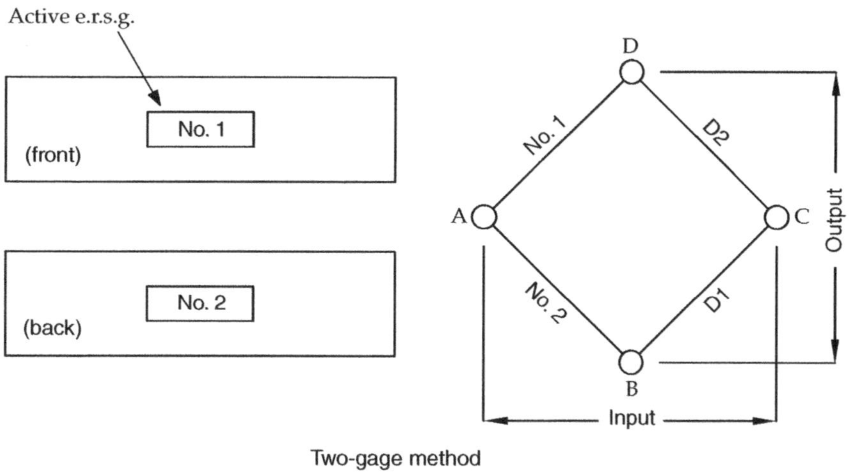

Figure 4.21 An LDT with a two strain gage setup. e.r.s.g., electrical resistance strain gage. Reproduced from Hogue *et al.* 1997 by permission of ASTM International.

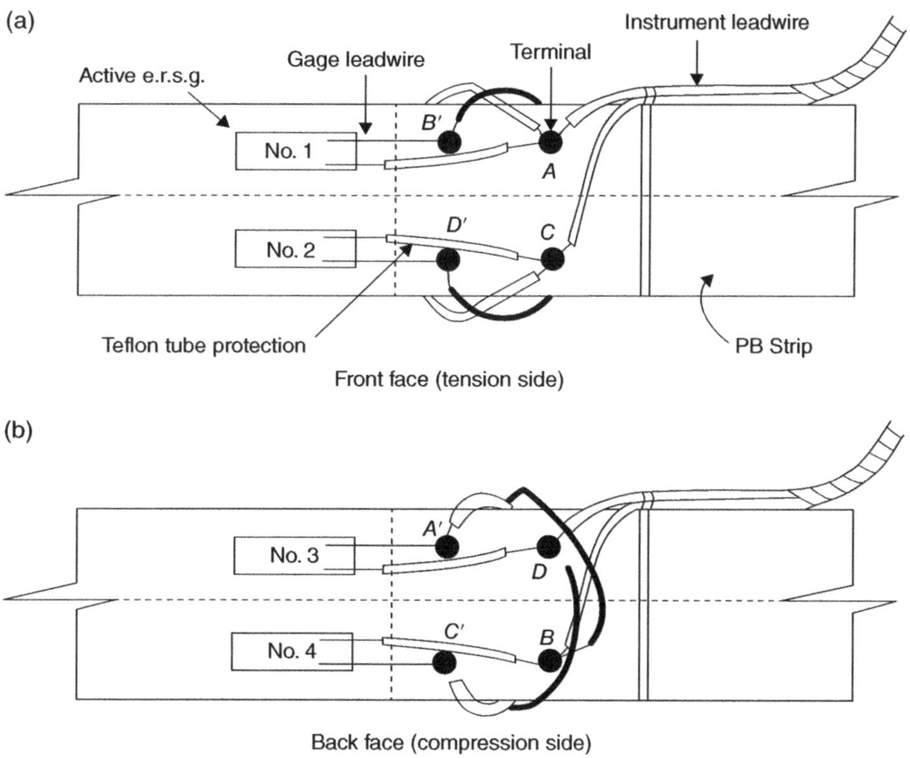

Figure 4.22 Details of the internal connections at the heart of a four-gage type LDT with (a) front face and (b) back face. Reproduced from Hogue *et al.* 1997 by permission of ASTM International.

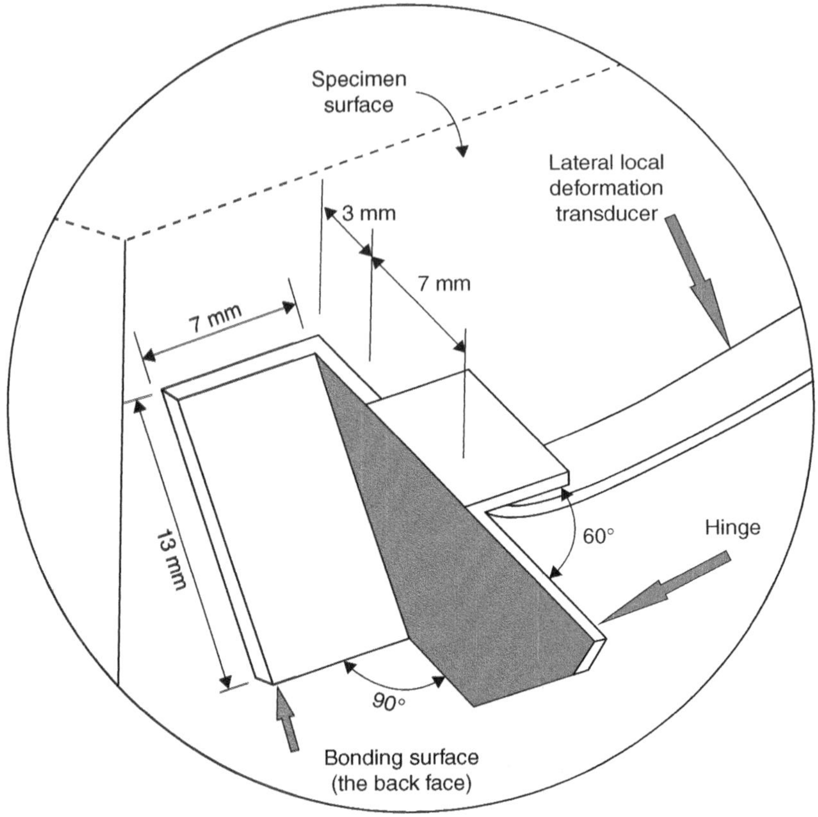

Figure 4.23 Details of the hinge used for an LDT to measure local lateral strains. Reproduced from Hogue *et al.* 1997 by permission of ASTM International.

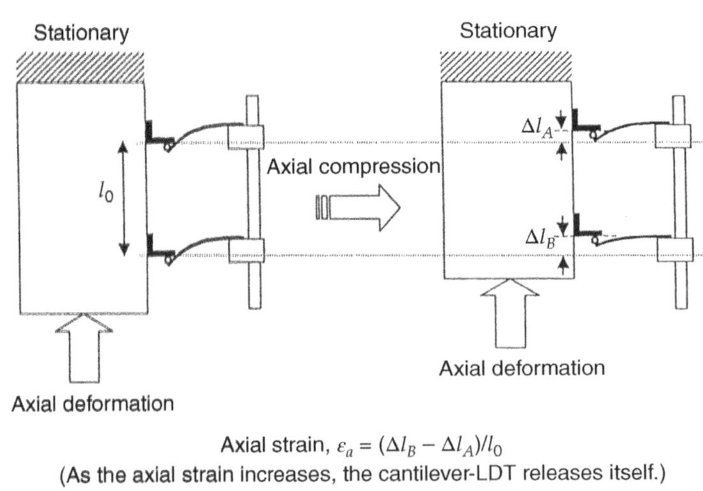

Figure 4.24 Cantilever LDTs to measure the axial deformation by releasing themselves. Reproduced from Yimsiri *et al.* 2005 by permission of ASTM International.

Advantages and limitations

Yimsiri and Soga (2002) summarize the advantages of clip gages (LDTs) as simplicity, low cost, good stability, good accuracy and resolution in the order of $(1–3)\times10^{-3}$. Limitations are considered to be the small working range for LDTs, the nonlinear calibration relation for some LDTs, and limited robustness.

4.7.6 Linear variable differential transformer setup

Linear variable differential transformers are often held and attached locally to the cylindrical specimen by a pair of spring-loaded collars, as seen in Fig. 4.25. Such collars were originally developed for lateral deformation measurements (Bishop and Henkel 1962), but they allow mounting of a pair of axial LVDTs and their associated cores on opposite sides of the specimen (Brown and Snaith 1974). Depending on the wiring of the LVDTs, they may be used to determine the average axial deformation as well as the tilt of the cap, if any (by individual measurements from the two LVDTs), or to determine the average axial deformation directly (by combining the signals from the two LVDTs).

The spring-loaded collars cause the shoes to be pressed against the sides of the specimen. If necessary, fixing pins that go through the rubber membrane may be attached under or be used in place of the shoes to provide clear demarcation of the beginning and end of the gage length over which measurements are taken. This is exemplified in Fig. 4.26, which shows floating-type mountings with pins.

As originally proposed by Bishop and Henkel (1962), the collars may be holding devices such as LVDTs, shown in Fig. 4.27, for measurement of the lateral deformation of the specimen. These are mounted across the gap with the spring, and they measure twice the change in diameter.

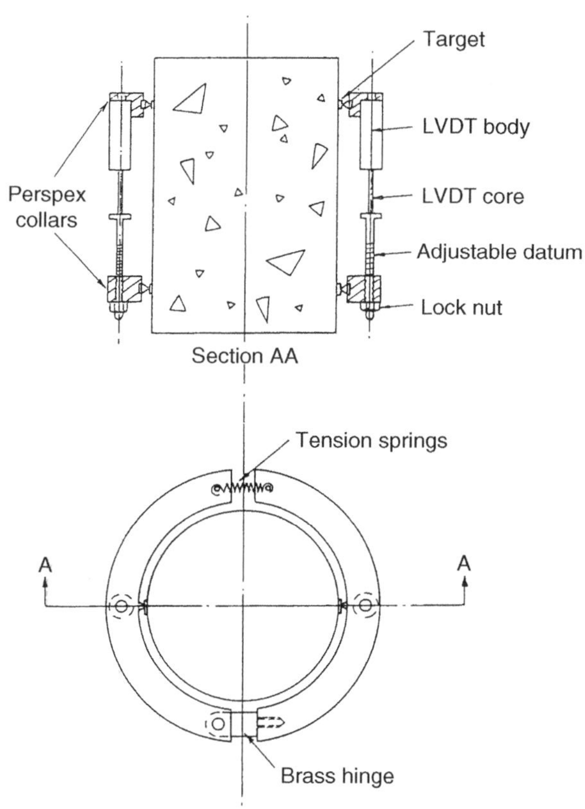

Figure 4.25 Longitudinal strain collars used for bituminous bound materials. Reproduced from Brown and Snaith 1974 by permission of Geotechnique.

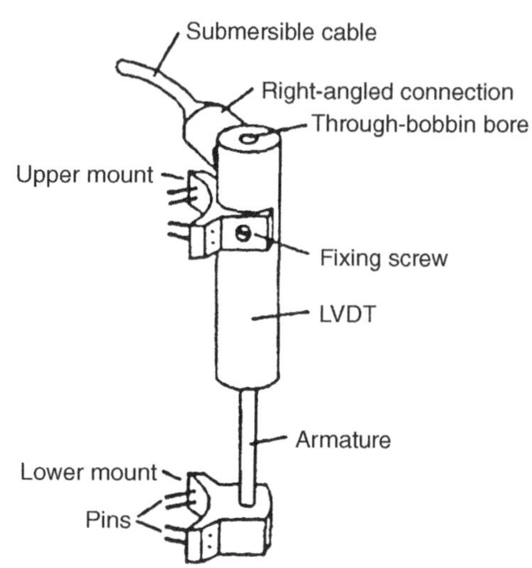

Figure 4.26 Floating-type LVDT attached to the specimen by pins through the membrane. Reproduced from Cuccovillo and Coop 1997 by permission of Geotechnique.

Figure 4.27 Strain collars attached to a specimen of dense bitumen macadam. Reproduced from Brown and Snaith 1974 by permission of Geotechnique.

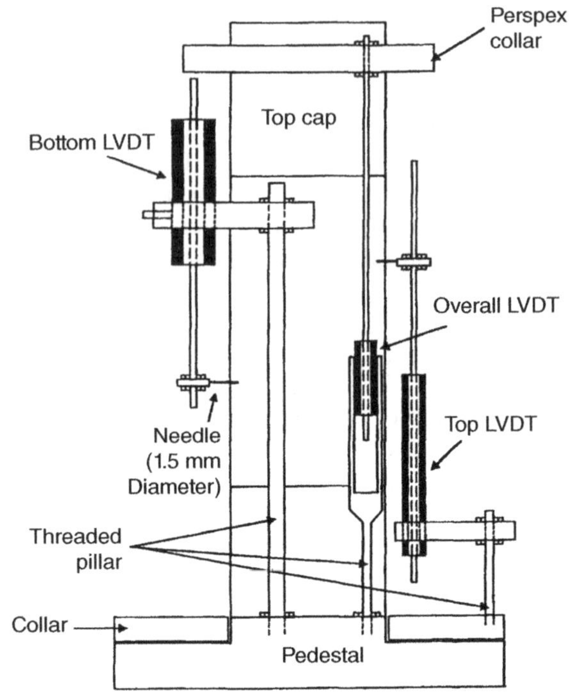

Figure 4.28 Fixed LVDT support system. Reproduced from Costa-Filho 1985 by permission of ASTM International.

Alternatively, the heavier LVDT coil housing may be mounted onto rods or pillars screwed into the base plate of the triaxial cell, as seen in Fig. 4.28. These LVDTs measure the displacement of two individual fix points at which the cores are attached. The difference between these displacements constitutes the deformation between the two fix points, which mark the gage length over which the deformation occurs. This setup also avoids all errors due to seating, alignment, bedding, and compliances as discussed above.

Still another alternative consists of measuring the deformation between the specimen cap and the triaxial base plate by a single LVDT (or possibly two), as also shown in Fig. 4.28. This method may be found to be sturdier than the collar setup, but it includes the errors associated with the interfaces at the cap and base, as discussed above.

Advantages and limitations

Yimsiri and Soga (2002) summarize the advantages of LVDTs as follows: good resolution in the order of 2×10^{-5} –1.4×10^{-3}%; good stability; and linear calibration relation. The limitations consist of requirement for non-conductive cell fluid (air or silicone oil, unless submersible LVDTs are used), high cost, large size (especially if AC–DC converters are built-in), and prone to jamming of the rod with the core, especially if the specimen bulges near and after failure.

4.7.7 Proximity gage setup

Proximity gages are most suitable and were originally devised to measure radial deformations. Proximity sensors are typically held by rods or rigid brackets attached to the triaxial cell base, as seen in Fig. 4.29, and they may be

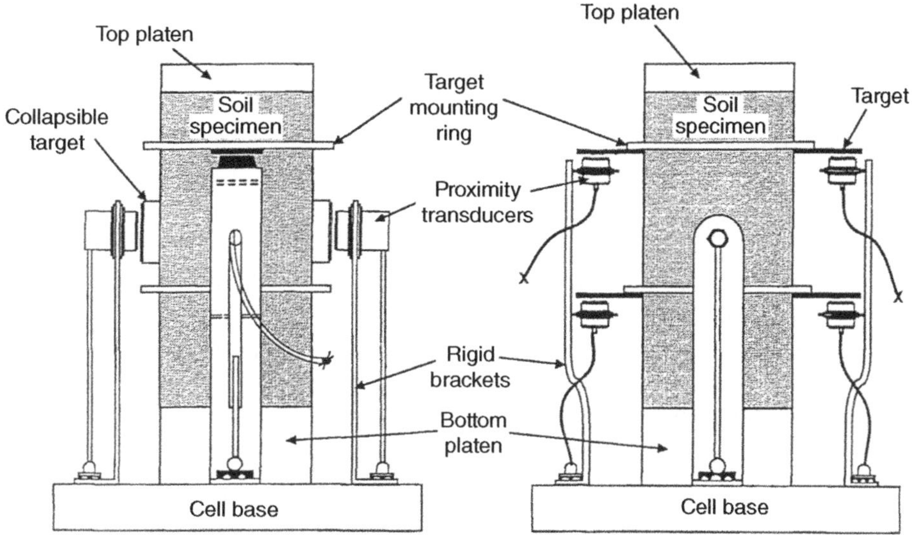

Figure 4.29 Arrangement of proximity transducers for deformation measurement. Reproduced from Hird and Yung 1989 by permission of ASTM International.

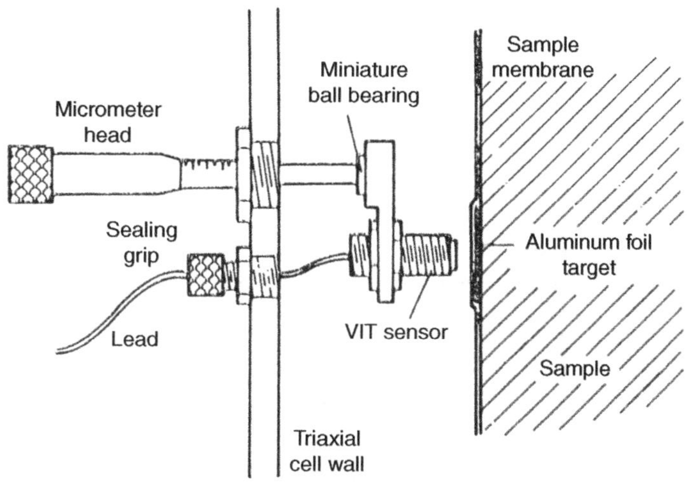

Figure 4.30 Details of proximity sensor mounting. Reproduced from Cole 1978 by permission of Canadian Science Publishing.

arranged to measure axial as well as radial deformations (Hird and Yung 1989). Alternatively, it may be attached to the cell wall, as seen in Fig. 4.30. The latter setup allows direct measurements by adjustment of the micrometer screw (and maintaining the same electric output reading) or it may allow larger measurement range by adjustment as the specimen increases in diameter.

Advantages and limitations

Yimsiri and Soga (2002) consider the advantages of the proximity transducers to be their linear calibration relation, good resolution in the order of 0.001%, and good accuracy in the order of 0.008%. The limitations are potential difficulties in setup procedures including the careful positioning of the target, high cost, and unless special precautions are made they are not water submersible and they are susceptible to pressure changes.

4.7.8 Inclinometer gages

Two types of inclinometer gages have been devised: the electrolytic liquid level type; and the pendulum type. Both involve sensing the inclination of a small capsule and mechanically translating the change in inclination into a measure of deformation. The mechanical mounting and the principles of operation of both types of inclinometer gages are shown in Fig. 4.31. The principle of operation of the electrolytic level type was explained in Section 4.4.

Ackerly *et al.* (1987) devised a pendulum inclinometer consisting of a heavy metal bob attached to the end of a strain-gaged strip of spring steel. This configuration is contained inside a stainless steel capsule, as shown in Fig. 4.32. The electrical signal from this device indicates the inclination of the capsule as it is

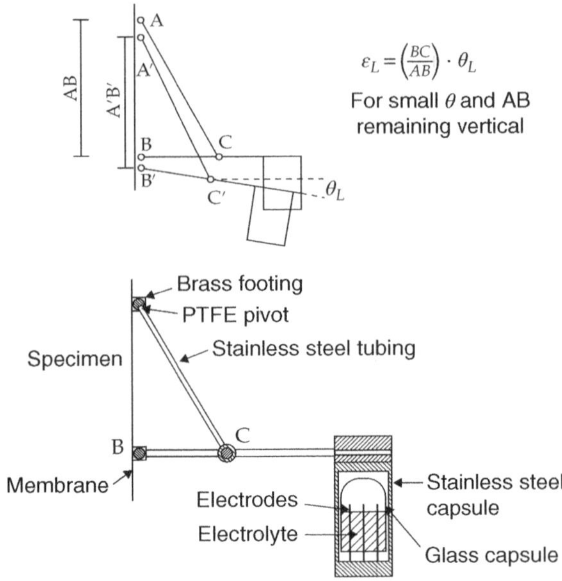

Figure 4.31 Operating principle of inclinometer level. PTFE, polytetrafluoroethylene (Teflon). Reproduced from Jardine *et al.* 1984 by permission of Geotechnique.

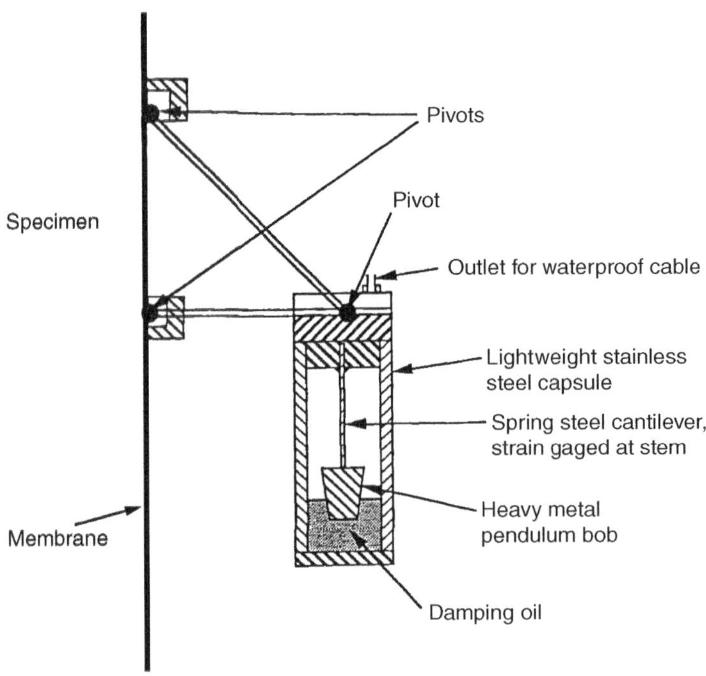

Figure 4.32 Pendulum inclinometer (Reproduced from Ackerly *et al.* 1987 by permission of Geotechnique).

tilted, thereby causing the strain gages to respond to the bending of the cantilever arm.

Advantages and limitations

While the inclinometer gages are very accurate, they are somewhat cumbersome to work with, and they may not be able to track large strains. Much care has to be exercised to make them work correctly.

4.7.9 Hall effect gage

The Hall effect was employed in the device shown in Fig. 4.33 which requires a DC voltage supply between 8 V and 16 V and produces a DC output voltage in the linear range of 2–3 V. Little, if any, amplification is therefore necessary, and this light-weight gage can measure with a resolution of 0.001 mm. It works equally well in air and in pressurized water. This type of gage may be used to measure axial deformation, as shown in Fig. 4.33, and it may be used to measure radial deformation by attachment to a collar (Clayton *et al.* 1989) in place of the LVDT, as shown in Fig. 4.34.

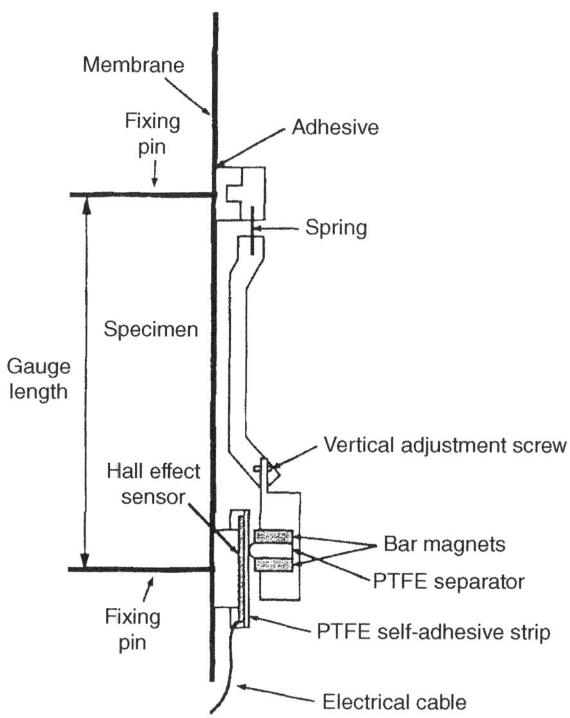

Figure 4.33 Hall effect gage for axial strain measurement (after Clayton and Khatrush 1986).

Advantages and limitations

While the Hall effect gage is very accurate, they are somewhat cumbersome to work with. Much care has to be exercised to make them work correctly.

4.7.10 X-ray technique

Arthur *et al.* (1964, 1977b), Balasubramaniam (1976), and Raju and Deman (1976) used X-ray techniques to follow embedded lead shot to determine the local displacements in triaxial specimens of kaolin clay and sand, respectively. This method requires insertion of a grid of lead shot markers to follow individual points on the X-ray radiographs, and this can obviously only be done in reconstituted specimens.

Advantages and limitations

This technique also requires low deformation rates and taking X-rays at discrete time intervals, and developing and analyzing the picture for each individual data set. This method does not provide very accurate measurements, but it has been used to study the uniformity of deformations inside triaxial and plane strain specimens (Arthur *et al.* 1964; Kirkpatrick and Belshaw 1968; Kirkpatrick and Younger 1970). Expense and safety requirements surrounding this technique are also limitations to its use.

4.7.11 Video tracking and high-speed photography

Tracking of grid points imprinted on the surface of triaxial specimens has been used under conditions where it has been difficult to impossible

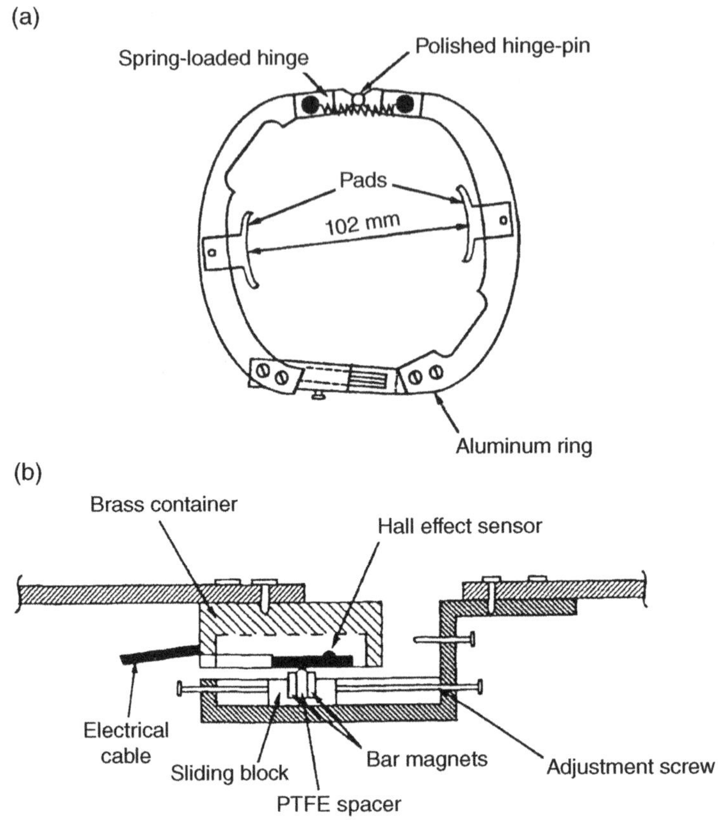

Figure 4.34 Hall effect gage for radial strain measurement: (a) plan view of collar; and (b) Hall effect sensor. Reproduced from Clayton *et al.* 1989 by permission of ASTM International.

to use other methods. Examples include video tracking for experimental validation of discrete element simulations (Hryciw *et al.* 1997), triaxial testing to study the mechanics of granular materials at low stresses and under microgravity conditions in the space shuttle (Macari *et al.* 1997; Sture *et al.* 1998) and studies of sand behavior at very high strain rates (Abrantes and Yamamuro 2002; Yamamuro *et al.* 2011). The video recordings and high-speed films require digital image processing for determination and analyses of both linear and volumetric strains. Gachet *et al.* (2007) used automated digital imaging to track both axial and radial deformations to obtain volumetric changes. They explained the details and accuracy of the image processing.

Advantages and limitations

These methods are very laborious and expensive and they do not provide very accurate measurements.

4.7.12 Optical deformation measurements

Lade and Liu (1998) employed two telescopes to obtain local measurements of the axial and radial deformations between the round heads of four pins inserted through the rubber membrane located at 20% of the specimen height from the lubricated end plates, shown in Fig. 4.35(a). The telescopes were mounted on positioning stages with attached micrometer screws outside the triaxial cell, as shown in

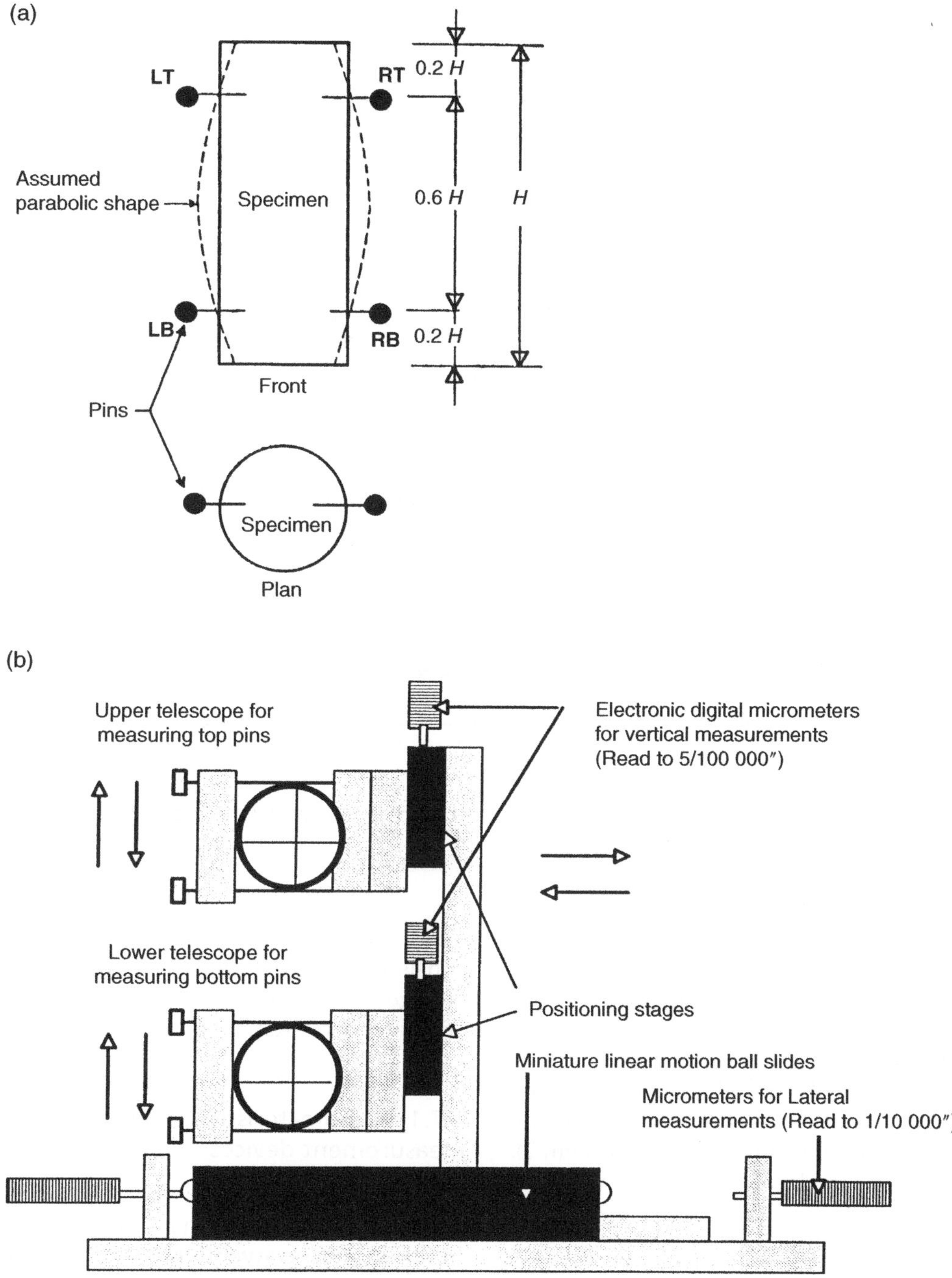

Figure 4.35 (a) Pins located on specimen and (b) telescopes mounted on positioning stages with attached micrometer screws outside the triaxial cell to observe pins through a special window to see the specimen without distortion. Reproduced from Lade and Liu 1998 by permission of ASCE.

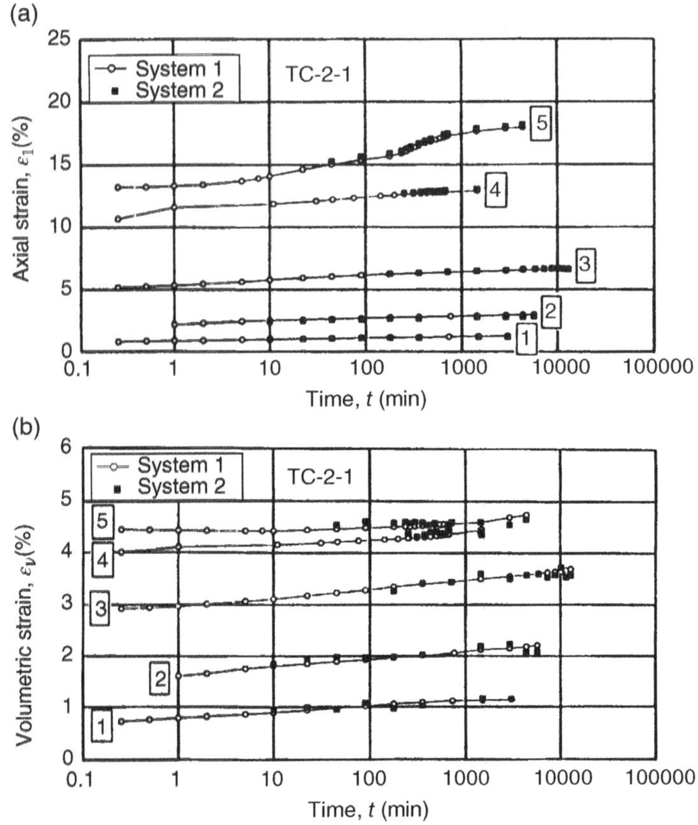

Figure 4.36 Comparison between outside measurements by a dial gage and volume change measurements (system 1) and inside measurements from the pin locations (system 2). The two systems track each other very well. Reproduced from Lade and Liu 1998 by permission of ASCE.

Fig. 4.35(b). They required adjustment for each set of measurements. The triaxial cell wall was supplied with a special window to see the specimen without distortion, as shown in Fig. 3.28. Figure 4.36 shows a comparison between outside measurements by a dial gage and volume change measurements (system 1) and inside measurements from the pin locations (system 2). The two systems track each other very well.

Advantages and limitations

This optical technique was employed for creep testing by Lade and Liu (1998), but sufficient time was not available to adjust the micrometer screws and obtain readings until the 2-min reading and beyond. It was observed that the accuracy of the measurements using this system and a carefully adjusted conventional system consisting of an external dial gage and a volume change device produced essentially equally accurate measurements.

4.7.13 Characteristics of linear deformation measurement devices

Table 4.3 provides a summary of the characteristics of linear deformation measurement devices. In addition to conventional dial gages used to measure outside the triaxial cell, the devices most commonly employed are those involving clip gages, LVDTs, and proximity transducers. Table 4.4 gives a more detailed summary of the characteristics of these three types of measurement systems.

Table 4.3 Summary of internal strain gage characteristics (after Scholey *et al.* 1995)

	Whole Body		Local (Electrical)							
			Contacting						Noncontacting	
					Inclinometer Gage					
Parameter	X-Ray	Video Tracking	LVDT	Flexible-Strip Radial-Strain Caliper	Electrolevel	Pendulum Gage	Hall Effect Gage	Local Deformation Transducer	Proximity Transducer	Cylindrical Capacitance Device
Direction of measurement	Axial, radial	Axial, radial	Axial, radial	Radial	Axial	Axial	Axial, radial	Axial	Axial, radial	Radial
Operating principle	Discrete X-ray images of deformation of lead-shot grid	Tracking of tracer particles, digital imaging	Inductance	Foil strain gages measure flexible strip deformation	Deformation causes tilt of electrolevel	Strain-gaged pendulum	Hall-effect semiconducting sensor	Bending strain in elastic metal strip	Eddy-current loss principle	Capacitance between two cylindrical capacitors
Output characteristics	X-ray images	Digitized images	Linear	Linear	Nonlinear	Nonlinear	Linear over limited range	Nonlinear	Linear	Non-linear
Commercial availability	No	No	LVDTs-yes	No	Yes	Yes	Yes	No	Yes	No
Relative cost	High	High	Moderate	Net known	Low	Low	Low	Low	Moderate	Moderate
Resolution	Poor	Poor	±0.0001–0.001%	±0.005%	±0.001%	±0.0005%	±0.002%	±0.0001%	2.5×10^{-4} mm (R)	±0.01%
Range	Pre-failure	Pre-failure?	Limited by core-body friction/ misalignment	Not Known	35%	Post Failure	1.5 mm (R) 2.5 – 7 mm (A)	±1.5 mm	Limited 2.5 mm (R)	Not Known
Accuracy	±0.1–0.2 mm	1%		Not Known	±0.002%			±0.0001%		Not Known
Specimen restrictions	Reconstituted	Reconstituted	None	Stiffer than flexible strip	None	None	None	None	None	Not Known
Cell fluid restrictions	Consider X-ray retardation		Nonconductor	None	None	None	None	None	Not water	Good dielectric properties
Temperature stability	–	–	Stable	Compensated	±3°C	±3°C	Compensated	Compensated	Compensated	±3°C
Pressure stability	–	–	Stable	Not known	Stable	Stable	Stable	Not Known	Stable	Calibration required
Primary publications	Many	Macari-Pasqualino *et al.*	Many	Holubec and Finn El-Ruwayih	Burland and Symes Jardine *et al.*	Ackerly *et al.*	Clayton and Khatrush	Tatsuoka, Goto *et al.*	Many	Mishu
Year introduced	1960s	1993	1970s	1969, 1976	1982, 1984	1987	1986	1988, 1991	1970s	1966
Applications	Research	Research	Research and commercial	Probably redundant	Research and Commercial	Research and Commercial	Research and Commercial	Research and Commercial	Research and Commercial	Research

Table 4.4 Characteristics of clip gage, LVDT and proximity transducer measurement systems (modified after Yimsiri and Soga 2002)

(a) Clip gage and local deformation transducer measurement systems

Type	Strain direction	Specimen dimensions	Resolution (%)	Accuracy (%)	Working range (%)	Cell fluid	Reference
Floating	Axial	$D = 7$ cm $H = 15$ cm	0.0007	—	2		Goto *et al.* (1991)
Fixed	Axial	$D = 5$ cm $H = 10$ cm	0.0012	0.003–0.004	5		Yimsiri (2001) (as quoted by Yimsiri and Soga 2002)
Floating	Radial	$D = 70$ mm $H = 150$ mm	0.0007	0.003	~28		Lo Presti *et al.* (1995b)
Floating	Radial	$D = 50$ mm $H = 100$ mm	0.0002	—	~8		Zlatovic and Szavits-Nossan (1999)

(b) LVDT systems for axial deformation measurements

Type	Specimen dimensions	Resolution (%)	Accuracy (%)	Working range (%)	Cell fluid	Reference
Fixed	$D = 75$ mm $H = 150$ mm	—	—	13	Silicone oil	Brown *et al.* (1980)
Fixed	$D = 38$ mm $H = 76$ mm	~0.0014	0.01	5	Transformer oil	Costa-Filho (1985)
Fixed	$D = 70$ mm $H = 140$ mm	~0.0003	~0.001	~2.2	Water	Lo Presti *et al.* (1995a)
Floating	$D = 150$ mm $H = 225$ mm	~0.00015	—	—	Air and transformer oil	Brown and Snaith (1974)
Floating	$D = 200–230$ mm $H = 400–580$ mm	~0.001	—	~5	Water	Nataatmadja and Parkin (1990)
Floating	$D = 50$ and 60 mm $H = 100$ and 200 mm	~0.00002	—	~10	Water/oil	Cuccovillo and Coop (1997)
Floating	$D = 100$ mm $H = 200$ mm	0.00001	0.0001	~5	Water	Heymann (1998) (as quoted by Yimsiri and Soga 2002)
Floating	$D = 38$ and 100 mm $H = 76$ and 200 mm	0.0001	—	—	Water	Cabarkapa *et al.* (1999)
Floating	$D = 35$ mm $H = 75$ mm	0.0001	—	~2.5	Silicone oil	Santagata *et al.* (1999)

(c) LVDT systems for radial deformation measurements

Type	Specimen dimensions	Resolution (%)	Accuracy (%)	Working range (%)	Cell fluid	Reference
Lateral strain calliper	$D = 100$ mm	0.005	—	—	Water	Menzies (1976)
Floating	Area = 35.6·35.6 mm^2	0.003	—	~14	Transformer oil	Yuen *et al.* (1978)
Radial strain belt	$D = 100$ mm	0.0001	—	5	Water	Kuwano *et al.* (2000)

(d) Proximity transducer measurement systems

Type	Strain direction	Specimen dimensions	Resolution (%)	Accuracy (%)	Working range (%)	Cell fluid	Reference
	Axial	$D = 100$ mm $H = 200$ mm	0.001	0.008	~5		Hird and Yung (1989)
	Radial (hollow cylinder)	ID = 203 mm OD = 254 mm $H = 254$ mm	0.004	—	—		Hight *et al.* (1983)
	Radial	$D = 100$ mm $H = 200$ mm	0.001	—	~10		Hird and Yung (1989)
	Radial		0.00024	—	—		Shibuya *et al.* (1994)

ID, inside diameter; OD, outside diameter.

4.8 Measurement of volume changes

Measurement of the change in volume of a soil element is uniquely related to soil mechanics, where it plays a significant role in characterization of the soil behavior. No other branch of engineering places as much emphasis on accurate determination of volume change as does soil mechanics. The development of volume change measurement devices has therefore been left entirely to the soil mechanicians' initiative and inventiveness.

Measurement of volume changes of triaxial specimens is most often done by measuring the volume of water expelled from or sucked into a saturated specimen. Volume changes of dry or partly saturated specimens may be obtained by measuring the volume changes occurring in the triaxial cell, as discussed in Section 4.8.3. This requires the cell to be completely filled with water (or other fluid), and appropriate corrections are required for piston intrusion into the cell and for volume changes of the cell (if any). The volume change may also be measured in a separate cell sitting around the specimen inside the triaxial cell, as reviewed in Section 4.8.4. Finally, the volume change may be obtained from measured axial and lateral deformations.

The primary requirements for a volume change device are the volume capacity and the accuracy of measurements. Large volume capacity and high accuracy appear to be mutually exclusive in many designs presented in the literature. Several principles have been applied in the development of volume change devices, and many different types have been designed for soil testing. Some of these devices appear to involve quite complex methods of operation and equipment that is sensitive and difficult to work with on a routine testing basis. Thus, sturdiness and ease of operation are some of the desirable requirements for a volume change device.

4.8.1 Requirements for volume change devices

The requirements for a volume change device may be divided into two parts: those relating to the capabilities and those pertaining to the operation of the device.

Volume capacity, accuracy, and resolution

To evaluate the requirements for a volume change device for triaxial testing, a small study of volume capacity and accuracy is shown in Table 4.5 (Lade 1988b). A large specimen with diameter of 10.0 cm (approximately 4.0 in.) and a small specimen with diameter of 3.56 cm (1.40 in.) are chosen to represent the extreme sizes used for testing on a routine basis. A medium specimen

Table 4.5 Design requirements for volume change devices in terms of volume capacity and resolution (after Lade 1988b)

Quantity	Large specimen	Medium specimen	Small specimen
Diameter (cm)	10.0	7.10	3.56
Height for $H = 2.5 \cdot D$ (cm)	25.0	17.8	8.90
Cross-sectional area (cm²)	78.54	39.73	9.95
Volume (cm³)	1963.5	706.3	88.6
Desired volumetric strain capacity: ε_{vmax} (%) [a]	20	20	20
Volume capacity: $\Delta V = \varepsilon_{vmax} \cdot V/100$ (cm³)	~400	~150	~20
Resolution of axial deformation measurement: ΔH (cm)	0.001	0.001	0.001
Axial strain resolution: $\Delta\varepsilon_1 = (\Delta H/H) \cdot 100$ %	0.004	0.006	0.011
Volume strain resolution: $\Delta\varepsilon_1 = \Delta\varepsilon_v$ (%) [b]	0.004	0.006	0.011
Volume change resolution: $\Delta V = \Delta\varepsilon_v \cdot V/100$ (cm³)	0.080	0.040	0.010

[a] The premise of calculation is that the maximum observable volume change occurs for dense sand at $e_{min} = 0.50$ dilating during shear at low confining pressure to $e_{max} = 0.80$ resulting in $\varepsilon_v = 20\%$.

[b] The premise of calculation is that the resolution of axial deformation measurements equals the smallest division on a dial gage measuring in millimeters (= 0.01 mm = 0.001 cm).

with diameter of 7.1 cm (2.8 in.) is included for comparison. The desired volume capacity is calculated on the premise that the maximum observable volume change occurs during shearing of an initially dense sand with a minimum void ratio of $e_{min} = 0.50$. Shearing at low confining pressure causes the sand to dilate, and at the end of the test the specimen has reached a void ratio of $e_{max} = 0.80$. This change in void ratio corresponds to a volumetric strain of 20%. A similar magnitude of volume change may also be obtained for an initially loose specimen compressed isotropically from low to high confining pressures. Consolidation and shearing of soft clay specimens may also result in total volume changes reaching that order of magnitude. To measure a volumetric strain of 20% in a large specimen with height equal to 2.5 times the diameter requires a device with a volume capacity of approximately 400 cm^3, whereas the small specimen requires a capacity of 20 cm^3. The volumetric strain of 20% is a rather large amount that only occurs in rare cases. More typical values are less than 10%. However, the study of volume capacity provides an appreciation of the necessary requirements to accommodate large volume changes.

The desired resolution of volume change measurements is calculated on the premise that the volumetric strain should be as accurate and precise as the axial strain. The reason is that they enter on equal basis in calculation of the cross-sectional area of the triaxial specimen, and they are also being used side by side in calculations of the lateral strain. The resolution of measurements of axial deformation is assumed to equal the smallest division on a conventional dial gage measuring in millimeters. Thus, the resolution of measurements is assumed to be 0.01 mm = 0.001 cm. Table 4.5 shows that the corresponding resolution on volume change measurements ranges from 0.080 cm^3 for the large specimen to 0.010 cm^3 for the small specimen. Note that these values do not depend on the H/D ratio of the triaxial specimen. If increased resolution is desirable, as for example for testing of very stiff materials, it is easy to scale the values listed in Table 4.5 to obtain corresponding resolutions of volumetric and axial deformation measurements.

The capabilities of many existing volume change devices can be varied by changing the details of the design to accommodate the requirements for the large specimen as well as those for the small specimen. Some of these exhibit adequate precision, and they may be reset relatively quickly to achieve large volume capacities. However, the combination of maximum accuracy and resolution (0.010 cm^3) and maximum capacity (400 cm^3) has only been obtained in one existing volume change device (Chan and Duncan 1967), but this device is not suitable for automatic datalogging.

Operational requirements

A number of different design and operational principles have been employed in building volume change devices. A detailed survey of these devices was published by Alva-Hurtado and Selig (1981). The most common principles of volume change measurements are embodied in buret systems with various methods of measurements of the meniscus level, gravimetric systems in which the change in weight of water expelled or sucked into the specimen is measured, servomechanism systems, mercury pot systems, and rolling diaphragm with displacement transducer systems. In studying these systems, some desirable operational requirements for a volume change device emerge (Lade 1988b; Oswell *et al.* 1989; Tatsuoka 1989):

1. Measurement of volume changes should be performed with minimum interference with the testing process (see Section 4.2). Notably, the change in water level in the device should be minimized or counteracted to avoid changing back pressure and therefore the effective confining pressure. This is most important for drained tests at low confining pressures.
2. The device should produce a linear relation between volume change and measured response. This facilitates easy evaluation of the volume change during the test as well as simple calculation after the test.

3. Application of back pressure should be possible so that saturation of the triaxial specimen can be achieved (see Chapter 6).
4. Changes in back pressure should result in minimal changes in measured response and in calibration. Small changes are unavoidable because of the flexibility of drainage lines, fittings, valves, and so forth, unless the entire device and the connecting tubes are exposed to the back pressure. However, expanding buret, stretching diaphragms, and so forth, which are part of the device, may result in excessive changes in measured response and in calibration caused by changes in back pressure.
5. The device should exhibit good time and temperature stability to avoid incorrect measurements in long-term tests. Parts, such as plastic burets and diaphragms, may exhibit creep deformations when exposed to differential pressures.
6. Diffusion of air from the device into the specimen should be avoided, especially for long-term tests, that is, tests requiring more than 1 day to perform.
7. The device should be designed to avoid loss of water caused by leaks, evaporation, or diffusion through diaphragms.
8. The device should include a minimum number of moving parts, which may "hang up" or provide frictional or variable resistance to movement, thus leading to unsatisfactory operation.
9. The device should not exhibit excessive hysteresis if the flow direction is changed. This hysteresis may be large for long, small bore tubes. Tatsuoka (1988, 1989) discussed this point on the basis of experiments (Tatsuoka 1981; Pradhan *et al.* 1986, 1989), and he suggested to remedy the problem by attaching a short cylinder with an inside diameter (say 2 cm) at the bottom of the buret to reduce the hysteresis in the system. Further details are given by Lade (1989).
10. The device should be easy to set up, prepare for testing (de-air, set initial reading, if required, and so forth), and to operate reliably.
11. It should be possible to measure the response by automatic datalogging *and* by visual reading. A number of problems (from incorrect electric wire connections to malfunctioning of the datalogging equipment) may occur before and during a test, especially at the beginning of automation of triaxial equipment. Measurement redundancy provides a means for continuing the test and obtaining correct measurements despite malfunctioning electric equipment.
12. The device should be sturdy and measurements should be unaffected by vibrations and insensitive to placement in slightly out of plumb position.
13. Measurements should be unaffected by contamination (e.g., by oil), corrosion, and the presence of soil particles, which may inadvertently have entered the device, such that it is unnecessary to clean and reconstitute the device before each test.
14. The device should be easy to disassemble and service in case of breakdown or cleaning of individual components.
15. The device should be simple, inexpensive and easy to construct. However, automatic datalogging requires an electrical device (force, pressure, or displacement transducer) to sense the volume change, modify the signal, and provide an electrical signal that can be measured. The electrical transducer is most often the most expensive part of the volume change device.
16. Oswell *et al.* (1989) point out that for special applications, the volume change device should be suitable for high pressures and high temperatures.

Individual volume change devices, whose principles of operation are more uncommon, may have limitations that are not anticipated and discussed above.

4.8.2 Measurements from saturated specimens

The most common volume change devices measure the amount of water expelled from or sucked into a saturated triaxial specimen.

Examples of buret-type, weighing-type and piston-type devices are given below.

Buret-type volume change devices

Figure 4.37(a) shows a buret-type volume change device designed by Chan and Duncan (1967). It works with a 4-way valve that provides infinite capacity by reversing the fluid flow direction in the buret. Half of the clear, stiff propylene tube is filled with kerosene colored with a dye to make it clearly visible. The movement of one of the interfaces between water and colored kerosene indicates the volume change. For a 1/4 in. tube, and an assumed reading precision of 0.5 mm along the tube, the accuracy on the volume change is about 0.006 cm^3. This device combines the necessary accuracy for small specimens with the capacity for large specimens. It can be used for any size specimen, it is inexpensive to construct, a back pressure can be applied, and it is sturdy. Thus, this device has all the attributes of a good volume change device. However, it is operated manually and read visually, and this device does not lend itself to easy modification for automatic datalogging.

Small modifications to the device shown in Fig. 4.37(a) are indicated in Fig. 4.37(b). A T-fitting has been included at the top of the propylene tube for easy filling (or refilling) of the tube with colored kerosene or with red transmission oil. In addition, a 3-way valve is installed to allow measurement of cell pressure and pore water pressure with the same pressure transducer. The arrangement and use of the valves in the volume change device are further discussed below in Sections 4.10 and 4.16.

Figure 4.38(a) shows another buret-type volume change device designed for automatic datalogging. This device consists of a simple buret that can be read manually. A differential pressure transducer is located at the bottom of the buret for measurement of the height of the water column. A back pressure can be applied on top of the water column and on the backside of the diaphragm in the differential pressure

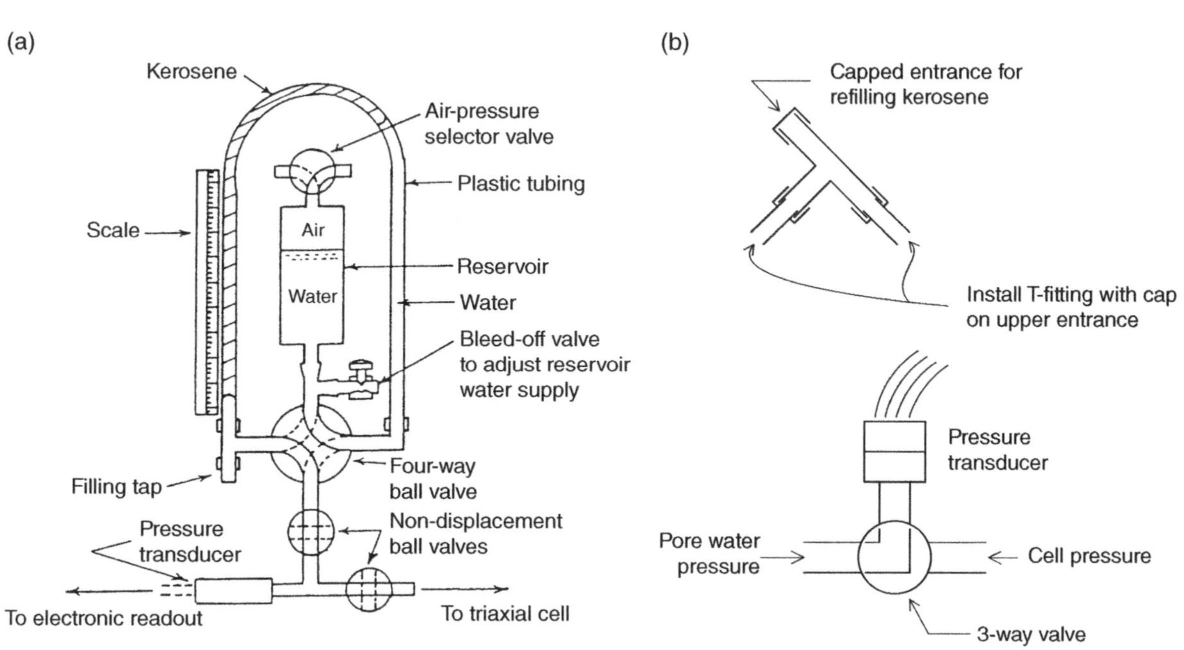

Figure 4.37 (a) Buret system for volume change and pressure measurements. Reproduced from Chan and Duncan 1967 by permission of ASTM International. (b) Small modifications to the device consisting of installing a T-fitting at the top and a 3-way valve to measure back pressure and cell pressure with the same pressure transducer.

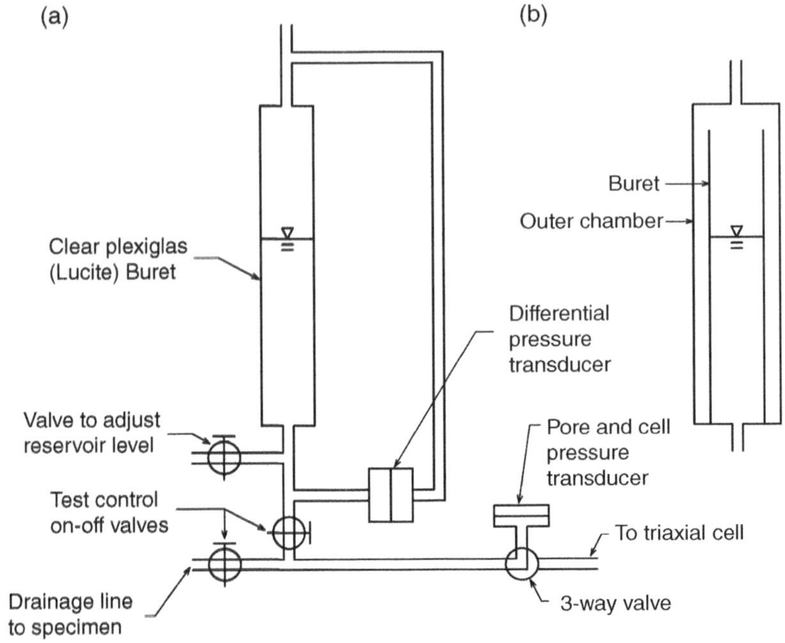

Figure 4.38 (a) Simple buret type device designed for automatic logging of volume change and pressure data and (b) alternate buret with outer chamber (after Bishop and Donald 1961).

transducer. This transducer has a range corresponding to a water column height of 35 cm, and it measures with an accuracy corresponding to a change in water column height of 0.1 mm/mV. By choosing the cross-section of the buret correctly the requirements to volume capacity and accuracy listed in Table 4.5 can be approximately met. Thus, burets with inside diameters of 3.18 cm (1.25 in.), 2.20 cm (0.875 in.), and 1.27 cm (0.50 in.) meet the accuracy requirements and have volume capacities of approximately 275, 135, and 45 cm^3, respectively. By comparing these capacities with the desired values listed in Table 4.5, it is seen that a large specimen whose total volume change exceeds about 14% would require readjustment of the water level. This may be done through the valve at the bottom of the buret.

This design may be improved by enclosing the buret in an outer chamber, as proposed by Bishop and Donald (1961) and by Tatsuoka (1981) and shown in Fig. 4.38(b). Thus, the buret is not exposed to a differential pressure, which may otherwise result in changing calibration due to changing back pressure. Figure 4.38(a) shows that attached to the volume change device is another pressure transducer for measuring pore water pressure as well as cell pressure. This is done through a 3-way valve that allows only one of the pressures to be connected to the transducer at a time. This also allows for easy determination of the specimen *B*-value using just one pressure transducer. This volume change device is easy to construct, the major portion of the cost is associated with the two pressure transducers, a back pressure can be applied, it is sturdy, it can be visually read, and it is designed for automatic datalogging.

Further modifications to the design reviewed above are indicated in the volume change device shown in Fig. 4.39 (Lade 1988b). Photographs of the device are shown in Fig. 4.40. The device consists of four graduated burets connected through a 5-way valve to a differential pressure transducer. Each of the tubes (glass or transparent polycarbonate) is provided with millimeter scales for visual reading of the volume change. The four glass tubes have inside diameters of

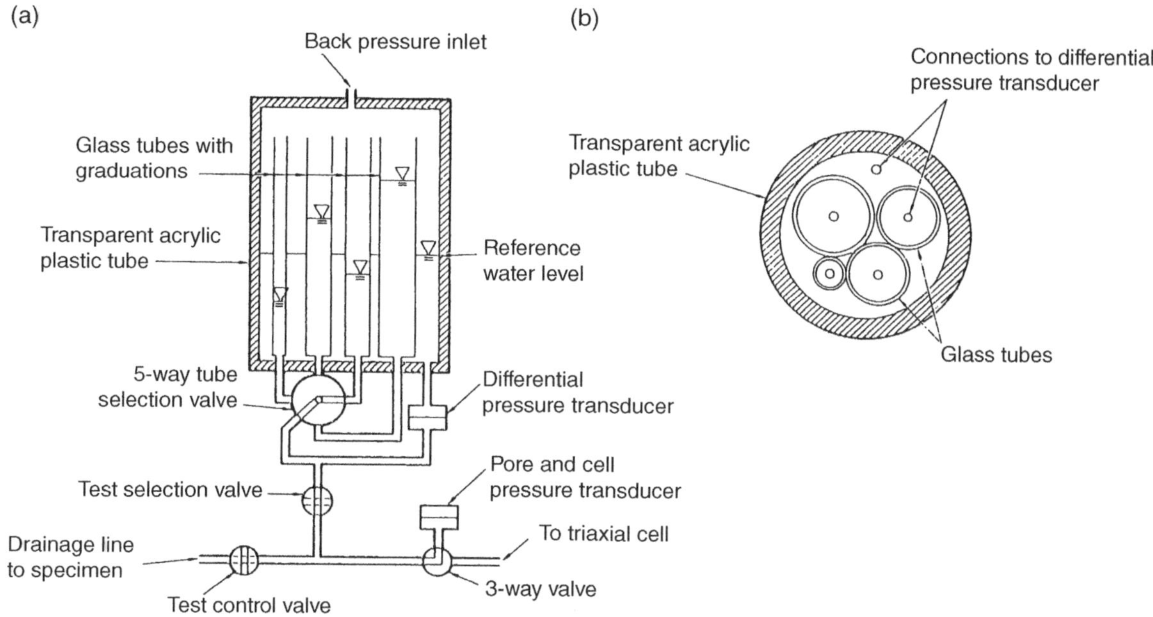

Figure 4.39 (a) Schematic diagram of buret type volume change device for automatic datalogging and (b) actual lay-out of tube assembly. Reproduced from Lade 1988b by permission of ASTM International.

25, 19, 19, and 10 mm, and their effective height is 39 cm. Their volume capacities are 190, 110, 110, and 30 cm^3, respectively. By comparing these capacities with the desired values listed in Table 4.5, it is seen that the smallest tube provides sufficient capacity for the small specimen, and the combined volume capacity in the large and the two medium sized tubes exceeds 400 cm^3, which is sufficient for a large specimen. If it is assumed that the water meniscus can be read visually with an accuracy of 0.5 mm, then the volume change can be determined with accuracies of 0.25, 0.14, 0.14, and 0.04 cm^3 in the four tubes. These values are two to four times larger than the desired values listed in Table 4.5. However, visual reading of volume changes is intended as an alternative method to be employed only if the automatic datalogging malfunctions. The total volume change may still be recorded correctly by visual reading. Only the accuracy of individual readings is affected if visual measurements are required.

The four glass or polycarbonate tubes shown in Fig. 4.39 are enclosed in a chamber, as proposed by Bishop and Donald (1961), consisting of a 6.4 mm (0.25 in.) thick, transparent acrylic plastic tube and two stainless steel end plates held together with three tie-rods. The acrylic plastic tube is 43 cm tall, thus allowing application of back pressure inside and outside the glass or polycarbonate tubes without exposing the tubes to differential pressure, which may otherwise lead to changing calibration caused by the back pressure, and possibly to breakage in the case of glass tubes at high back pressures. The inside diameter of the acrylic tube is 57 mm (2.25 in.), that is barely sufficient to contain the four tubes. Using a design stress of 750 psi (5170 kPa) to avoid stress crazing in the acrylic plastic allows application of back pressures up to 170 psi (1170 kPa = 11.6 atm). Increasing the wall thickness of the acrylic plastic tube or exchanging it with a polycarbonate tube would allow application of higher back pressures.

The water from the triaxial specimen is led through the common port in the 5-way valve to one of the four glass tubes selected for measurement. The differential pressure transducer is located below the 5-way valve and serves to measure the water column height in the selected tube.

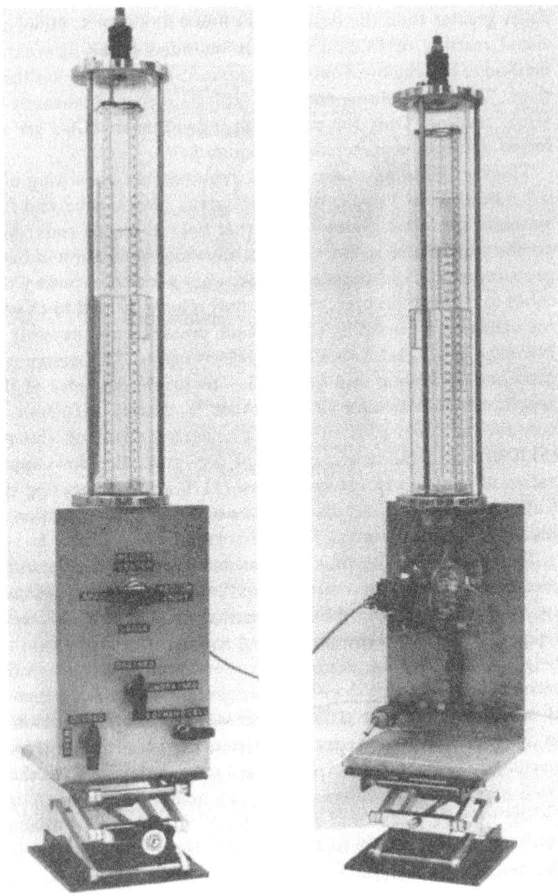

Figure 4.40 Photographs of buret type volume change device seen from (a) the front and (b) the rear. Reproduced from Lade 1988b by permission of ASTM International.

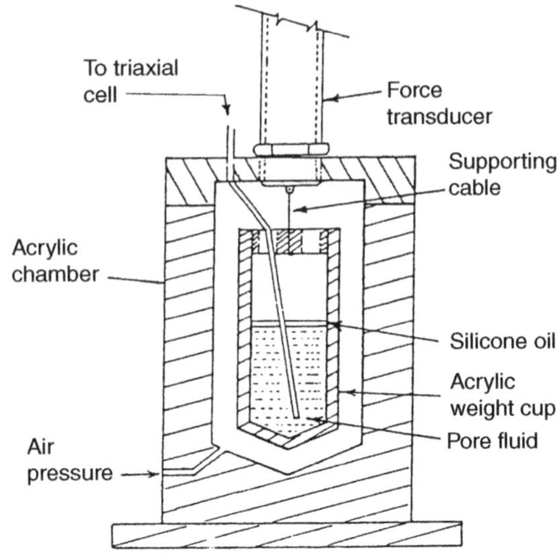

Figure 4.41 Volume change device in which water is weighed by a force transducer. Reproduced from Mitchell and Burn 1971 by permission of Canadian Science Publishing.

The backside of the diaphragm in the transducer is connected with the bottom of the chamber, which is filled halfway with water. The back pressure also acts on the water in the chamber, and the transducer therefore measures the difference between water levels in the selected tube and in the chamber. Both negative and positive pressures can be measured depending on the relative location of the water level in the selected tube. A layer of silicone oil may be floated on each of the water surfaces in the device to reduce the long-term effect of evaporation.

The transducer used in the device described above is a variable reluctance differential pressure transducer. This transducer is sturdy and allows for overpressures in either direction of 200%. It has a range of ±0.5 psi (±3.45 kPa) corresponding to a water column height of ±35 cm, and it measures with an accuracy better than 0.1 mm of water per millivolt (mV) at an excitation of 6 V. This corresponds to accuracies of 0.049, 0.028, 0.028, and 0.008 cm^3 in the four tubes. These accuracies are better than those listed in Table 4.5.

Finally, the volume change device is placed on a laboratory jack, as shown in Fig. 4.40. This is to be able to perform drained triaxial tests at low effective confining pressures which may be excessively affected by the change in water level. This could result in changing the effective confining pressure by up to 39 cm of water (3.8 kPa). This effect may be counteracted by adjusting the height of the laboratory jack periodically to maintain approximately constant water level in the selected measuring tube relative to the triaxial specimen.

Weighing-type volume change devices

Figure 4.41 shows another principle of operation. In this device, designed by Mitchell and Burn (1971), the volume change is determined

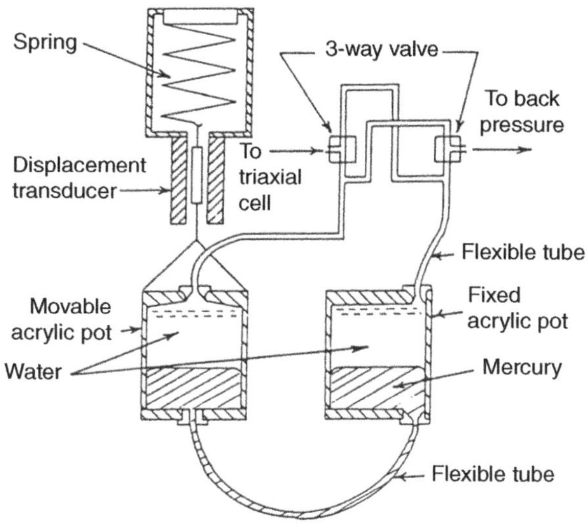

Figure 4.42 Volume change measurements by a mercury pot system. Reproduced from Darley 1973 by permission of Geotechnique.

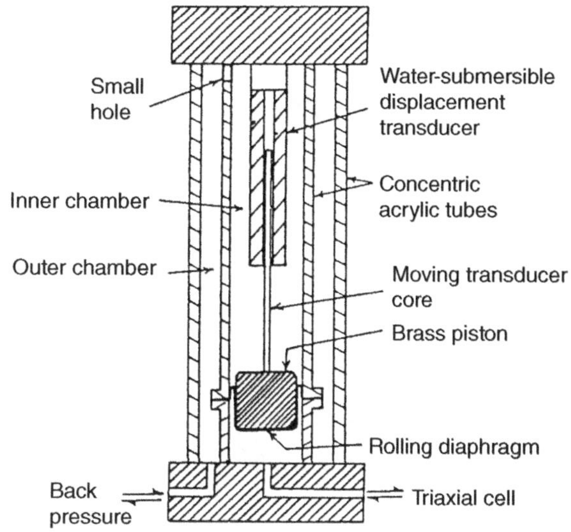

Figure 4.43 Volume change device in which water level is measured by an internal displacement transducer. Reproduced from Menzies 1975 by permission of Geotechnique.

by weighing the water expelled from or sucked into a cup freely hanging from a sensitive force transducer. The cup is contained inside a chamber that allows application of back pressure to the specimen.

Figure 4.42 shows a schematic diagram of a mercury pot system in which the water from the specimen displaces mercury from one pot to the other. One of the mercury pots is suspended in a spring and its elongation is measured by an LVDT. By displacing mercury, the weight of the water is magnified 12.55 times (= 13.55 – 1.00). This system was developed by Darley (1973) as a modification to a system suggested by Rowland (1972). Back pressure can also be applied to this type of device.

Piston-type volume change devices

Menzies (1975) developed a volume change device in which the water is confined like in a loading piston, that is within a tube and a rolling diaphragm, as seen in Fig. 4.43. A brass piston applies a small differential pressure across the convolute of the diaphragm resulting in a low friction rolling action. A displacement transducer inside the device measures the vertical displacement and thereby the volume change of the specimen. A back pressure can be applied in the device.

The volume change device shown in Fig. 4.44 was designed by Johnston and Chiu (1982), and it operates as a double acting piston in which the water is alternately pushed into or expelled from one of the two chambers. By installing a 4-way valve between the two chambers, Lade (1988b) increased the volume capacity to the volume of the water reservoir seen in Fig. 4.45(a). Figure 4.46 shows photographs of the device. The two chambers are located at the closed ends of a stainless steel tube, in which water is confined by two rolling diaphragms separated by a lightweight hollow aluminum piston. The rolling diaphragms are clamped between flanges near the ends of the tube. The aluminum piston is provided with two guiding Teflon bushings with sufficient clearance inside the steel tube to allow for smooth axial piston movement. The frictionless movement of the piston is monitored by a displacement transducer and a dial gage, both mounted externally on the steel tube. An arm, rigidly attached to the piston, protrudes through a longitudinal slot in the steel tube and provides the connection between the

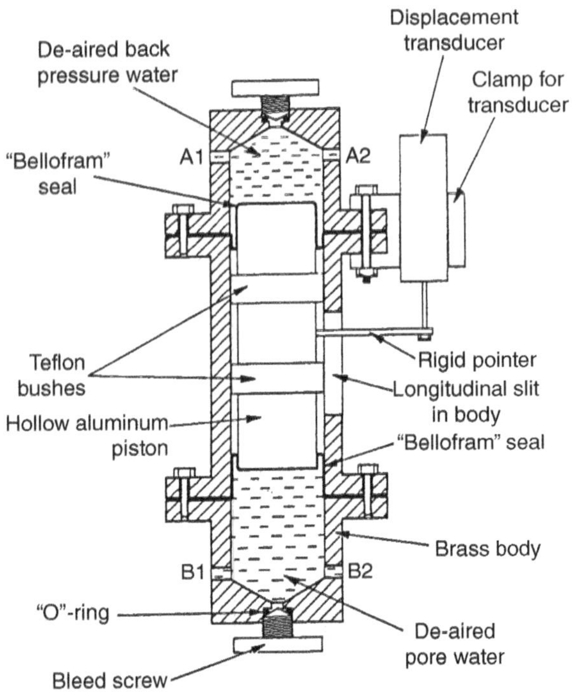

Figure 4.44 Volume change device in which water level is measured by an external displacement transducer. Reproduced from Johnston and Chiu 1982 by permission of Elsevier.

piston and the measuring devices. The dial gage is used for visual readings, and the displacement transducer is employed for automatic datalogging.

The inside diameter is the same along the tube, the two rolling diaphragms are identical and the piston has the same diameter in both ends. Therefore, for any amount of water being pushed out of one chamber, an equal amount is sucked into the other chamber. A back pressure must be applied to the chamber, and this pressure will act with equal magnitude on the other chamber. Johnston and Chiu (1982) reported that this device has excellent accuracy, but relatively small volume capacity.

However, by combining the principle of the double acting piston device with the principle of reversible flow embodied in the device presented by Chan and Duncan (1967), it is possible to obtain a combination of excellent accuracy and infinite volume capacity. Thus, a 4-way valve is used to connect the two chambers alternately to the drainage line from the triaxial specimen and to a large reservoir of water to which the back pressure can be applied.

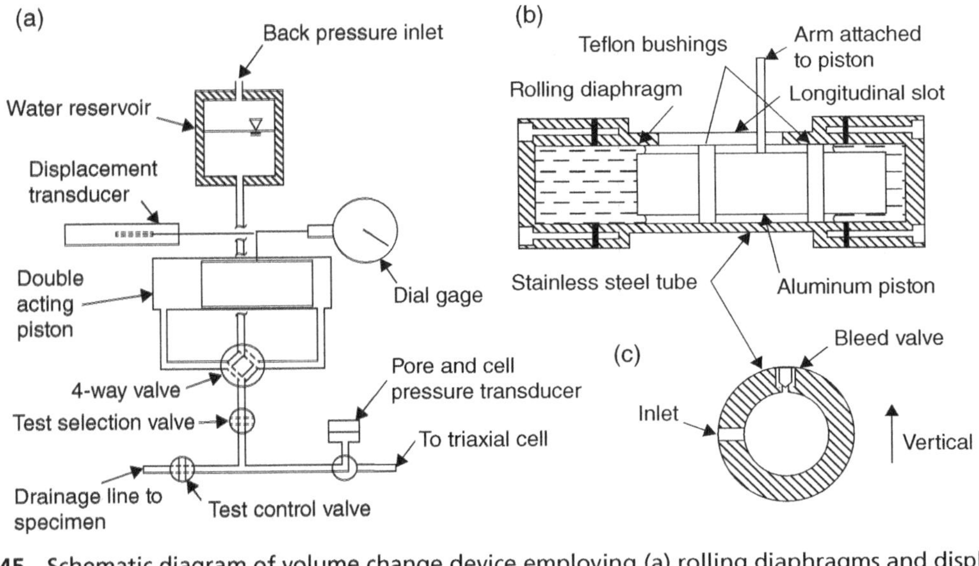

Figure 4.45 Schematic diagram of volume change device employing (a) rolling diaphragms and displacement transducer for automatic datalogging, and (b) details and (c) end view of double acting piston. Reproduced from Lade 1988b by permission of ASTM International.

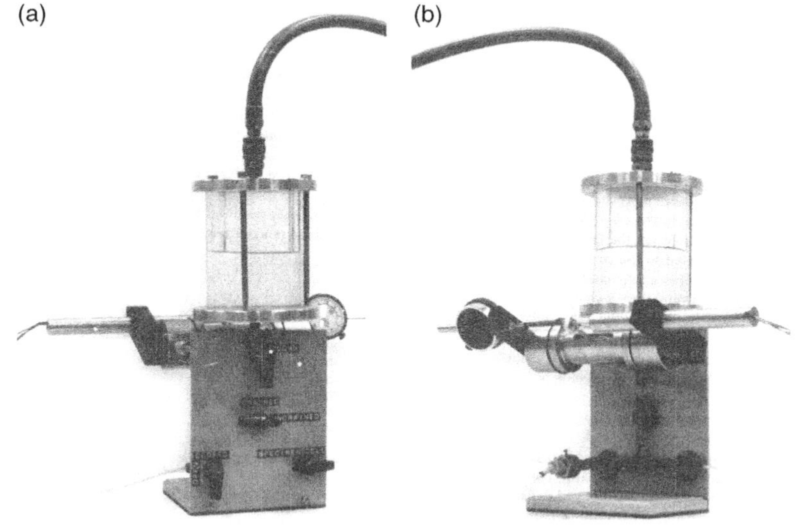

Figure 4.46 Photographs of volume change device with double acting piston seen from (a) the front and (b) the rear. Reproduced from Lade 1988b by permission of ASTM International.

The device shown in Figs 4.45 and 4.46 is prepared for testing by flushing de-aired water through the chambers, one at a time, and out through the bleed valves located on top of the horizontal steel tube. Once the device is de-aired, it is ready for testing. To check whether the device is de-aired before a test, the back pressure is varied, and the 4-way valve is turned. The displacement transducer and the dial gage should not respond to this operation. If additional de-airing is required, this may be done by opening one bleed valve and applying a small back pressure to the chamber in the other end, thus pushing the piston towards the open bleed valve and forcing air out. The procedure is repeated for the other chamber by turning the 4-way valve.

The accuracy of the volume change device is a function of the effective diameter of the piston with the enveloping rolling diaphragm. In the device shown in Figs 4.45 and 4.46 the effective diameter is 28 mm. Using visual readings from the dial gage with an accuracy of 0.01 mm results in a volume change accuracy of 0.006 cm^3. This value is smaller than the accuracy required for testing small specimens, as indicated in Table 4.5. Using a displacement transducer with better accuracy than the dial gage it is possible to obtain an accuracy of approximately 0.001 cm^3, the same value presented by Johnston and Chiu (1982). With a piston stroke of 38 mm, the volume capacity corresponding to piston movement in one direction is about 23 cm^3. However, by reversing the flow through the 4-way valve, an infinite capacity can in principle be obtained. In practice, the volume capacity depends on the size of the reservoir located on top of the device. In the present design the reservoir volume is approximately 785 cm^3. This reservoir consists of a 10-cm tall acrylic plastic tube with inside diameter of 10.0 cm (4.0 in.) and wall thickness of 12.4 mm. (0.5 in.) contained between two stainless steel end plates connected with three tie-rods. The acrylic plastic tube may be reinforced with hose clamps to allow high back pressures to be applied. The design of this reservoir is such that only small changes in water level occur for large volume changes. This reduces the effect of water level changes on the effective confining pressures, thus minimizing the interference with the testing process.

The maximum back pressure that may be applied to the device may be limited by the strength of the rolling diaphragms. These can withstand differential pressures of up to 500 psi

(3450 kPa), thus exceeding most requirements for back pressure. Oswell *et al.* (1989) constructed a similar volume change device, but they measured the position of the piston by an LVDT located at one end of the device. This allowed them to completely enclose the piston and the rolling diaphragms in a tight compartment, thus being able to apply back pressures up to 10 MPa by pressurizing the compartment at the desired back pressure (say 10 MPa) minus 0.1 MPa.

The requirements for volume capacity and accuracy listed in Table 4.5 are exceeded by the volume change device described above, and the operational requirements are also fulfilled with the exception of one: diffusion of water through the rolling diaphragms was measured by closing the drainage valve and applying a back pressure of 30 psi (207 kPa). Over a period of 1 week, the stability was measured to be 0.01 $cm^3/24$ h.

Other principles of volume change measurements

Sharpe (1978) directed the water flowing into or out of the specimen into a reservoir which displaced paraffin into a buret that contained an electrolyte (sodium chloride solution). The electrolyte then flowed into a second buret which was wrapped in aluminum foil and acted as a variable capacitor from which the volume change could be determined.

Test control and pressure measurements

Some of the volume change devices presented above are provided with non-displacement shut-off valves for test control and with a pressure transducer for measuring pore water and cell pressures.

One shut-off valve, located immediately below the volume change devices, shown for example in Figs 4.37(a) and 4.38, is employed to select the drainage condition during shearing. This test selection valve is open during the consolidation phase of the triaxial test, and it remains open during shearing in a drained test. For undrained tests, this valve is closed to prevent further drainage during shearing. The other shut-off valve is employed to control the initiation of drainage during the consolidation phase of the triaxial test. Thus, the cell pressure and the back pressure are set, finely adjusted, and measured with the test control valve closed. Consolidation begins when the valve is opened.

The pressure transducer is connected to the common port of a non-displacement 3-way valve, as shown in Fig. 4.37(b). This setup enables measurement of cell pressure, back pressure, and pore water pressure using only one pressure transducer. By turning the valve 180°, the transducer can be connected to either the cell pressure line or to the back pressure/pore water pressure line. During an undrained test, the test control valve is open, and the 3-way valve is turned to measure pore water pressures.

This setup also provides for easy measurement of the *B*-value for saturation check on the triaxial specimen. Whereas the 3-way valve is normally a non-displacement valve, very small changes in volume may be registered as a consequence of turning the valve. This small volume change may be caused by the reduced deflection of the diaphragm in the pressure transducer when the pressure changes from the higher cell pressure to the lower pore water pressure. These small volume changes may affect the pore pressures during a *B*-value test, especially for stiff soils. To overcome this problem, the *B*-value test for an isotropically consolidated specimen with free cap (no piston uplift force is applied to the specimen) may be performed as follows:

1. Set the 3-way valve and measure the cell pressure.
2. Turn the 3-way valve and measure the back pressure/pore water pressure.
3. Shut off the volume change device through which the back pressure is applied.
4. With the 3-way remaining in position to measure the pore pressure, increase the cell pressure by the desired amount (the approximate cell pressure may be monitored on a Bourdon gage connected to the triaxial cell, or similar).

5. Measure the increase in pore pressure.
6. Turn the 3-way valve and measure the exact increase in cell pressure. If the pressure transducer provides a linear response, it is not necessary to convert the readings to actual pressures. The *B*-value may be calculated directly from the readings in the usual manner (see Chapter 6).

The setup described here provides all the necessary measurements and controls for consolidated- drained as well as consolidated-undrained triaxial tests. No additional devices or valves are required for performance of these types of tests. The volume change devices outfitted with pressure transducer and valves as presented above may also be useful for performance of other types of tests such as plane strain, true triaxial and torsion shear tests.

Digital pressure/volume controller

The pressure/volume controller described in Section 3.4.4 is devised to control either pressure or volume through the liquid applied from the device to the components of the triaxial cell, such as (1) the triaxial specimen in which either the back pressure or the specimen volume is controlled while the other is measured, (2) the triaxial cell in which the pressure is controlled, and (3) the axial loading device in which the pressure is controlled.

The working principle of the digital pressure/volume controller is shown in Fig. 4.47 and described by Menzies (1988). The liquid in the hydraulic piston (de-aired water) is pressurized by a piston that is pushed or pulled by a stepper motor through a ball-screw that guides the piston rod linearly a certain amount for each turn. The stepper motor is outfitted with a gear so as to be able to advance or retract the piston at different rates, and a pressure transducer measures the liquid pressure produced by the stepper motor action with feedback to the digital controller. The digital controller responds to the measured pressure so as to increase, decrease or maintain constant output pressure, as desired, and it measures the volume of fluid pushed into or retracted from the specimen by the number of turns by the stepper motor multiplied by an appropriate calibration factor. Very accurate volume measurements may be obtained by such systems. The volumetric capacity depends on the piston diameter and travel (e.g., 200 cm^3 and 1000 cm^3), and resolution down to 0.001 cm^3/step of the stepper motor may be obtained (Menzies 1988). The pressures generated may be resolved to 0.2 kPa and controlled to 0.5 kPa and varied over a wide range up to 64 MPa (Menzies 1988).

4.8.3 Measurements from a triaxial cell

For cases where access to measuring the volume change directly by the expelled or imbibed amount of water from the specimen is not possible, measurements of specimen volume

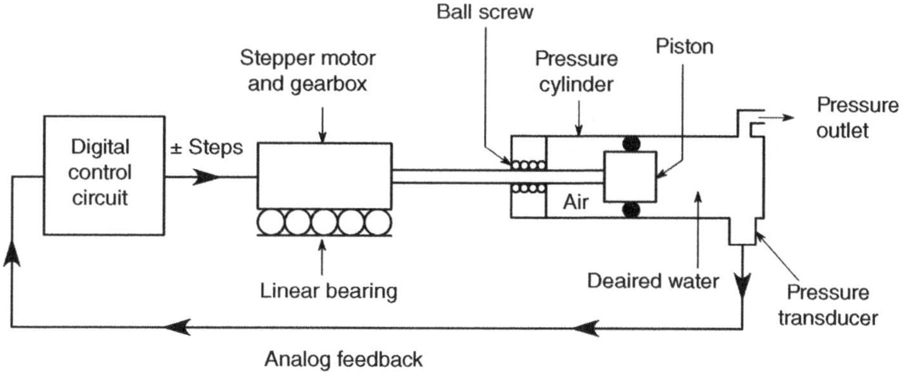

Figure 4.47 Working principle of the digital pressure/volume controller (after Menzies 1988).

changes may be performed by measuring the complementary volume change that occurs in the triaxial cell, or the volume change may be calculated on the basis of axial and radial deformation measurements using the methods discussed in the previous section. The volume change from the triaxial cell represents the mirror image of the volume change of the specimen. This technique may be necessary, for example (1) for tests on frozen soils, (2) for undrained tests on dense, saturated soils that tends to dilate in which volume changes occur after the pore water cavitates, (3) for measuring volume changes in tests on dry, and (4) for partly saturated soils, such as in triaxial tests on compacted soils.

For this purpose the volume change device is connected with the inlet to the triaxial cell through which the cell pressure is applied, and the cell pressure is directed through the volume change device as though it were a back pressure. Application of this technique requires that (1) the cell is completely filled with water, (2) the change in volume of the cell due to changes in cell pressure can be accounted for, (3) the cell does not change volume with time, and (4) it necessitates a correction for piston intrusion into the cell.

Completely filling the triaxial cell with water without trapping air bubbles may be done by applying the CO_2-method for specimen saturation on the cell. Thus, gaseous CO_2 is slowly let into the cell from the bottom, thereby pushing the air (which is heavier than gaseous CO_2) out through the top port of the cell. This is followed by letting water in from the bottom thereby pushing the gaseous CO_2 out through the top port, which is then closed after a suitable amount of water has passed through. Any small amount of gaseous CO_2 left in the cell will dissolve in the water filling the cell. The CO_2-method of water saturation is further explained in Section 6.6.2.

Any triaxial cell will change volume due to a change in cell pressure. If this volume change is elastic in nature, and therefore repeatable, and if it can be reliably quantified by suitable calibration, then it is possible to perform tests with varying cell pressure and calculate the change in cell volume. This may be possible only for cells made of metals such as aluminum and stainless steel. Triaxial cells with cell walls made of acrylic plastic will creep and change volume with pressure and time in a relatively unpredictable fashion, as indicated schematically in Fig. 4.48 so they cannot be relied upon to possess predictable volume changes.

Figure 4.48 also shows that correction for piston intrusion consists of volume adjustments calculated from the cross-sectional area of the piston and the measured axial deformation of the specimen, which represents the length of the piston intrusion.

Inner cylinder triaxial cell

Most of the problems reviewed above and resulting in required corrections when measuring volume changes from the triaxial cell may be circumvented by using an internal cylindrical cell wall sitting around the specimen, as shown in Fig. 4.49. The extra internal wall is acted upon inside and outside by the same pressure and consequently there is no volume change due to changes in pressure or due to creep with time. It is essential to avoid air bubbles in the water and to maintain full saturation in the internal cell to measure the volume change confidently and correctly. The water level of the inner cell is measured by the differential pressure transducer using the principle explained for the volume change device in Fig. 4.38. To increase the change in height of water level and therefore the resolution of the volume change measurements, the upper portion of the wall may be narrower around the piston than the lower portion, as shown in Fig. 4.49. The correction for piston intrusion into the water in the internal cell is similar to that indicated in Fig. 4.48.

Double wall triaxial cell

Wheeler (1988) proposed to employ a double wall triaxial cell in which the inner cell is completely saturated and enclosed in the outer cell. The inner cell is sealed to its own top plate and to the common base plate, and the same pressure is acting in the inner and outer cells. This setup

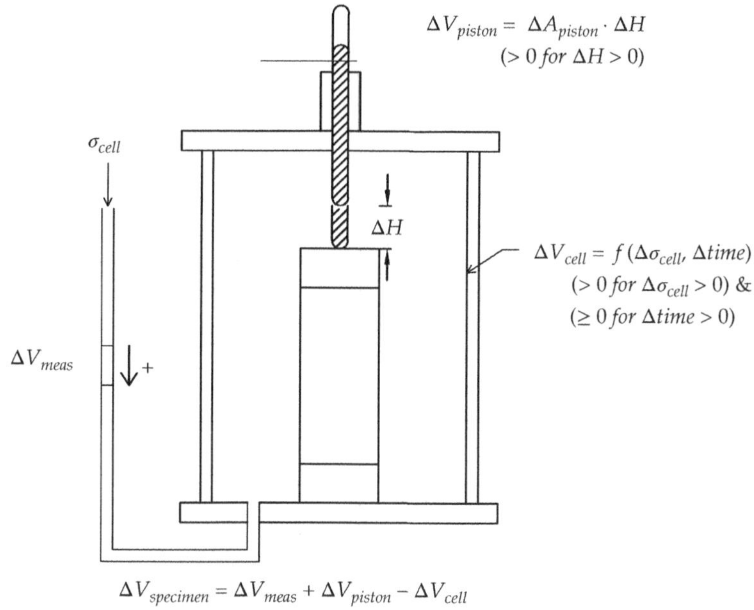

Figure 4.48 Errors in measurements of specimen volume change from a triaxial cell due to piston intrusion and cell wall deformation.

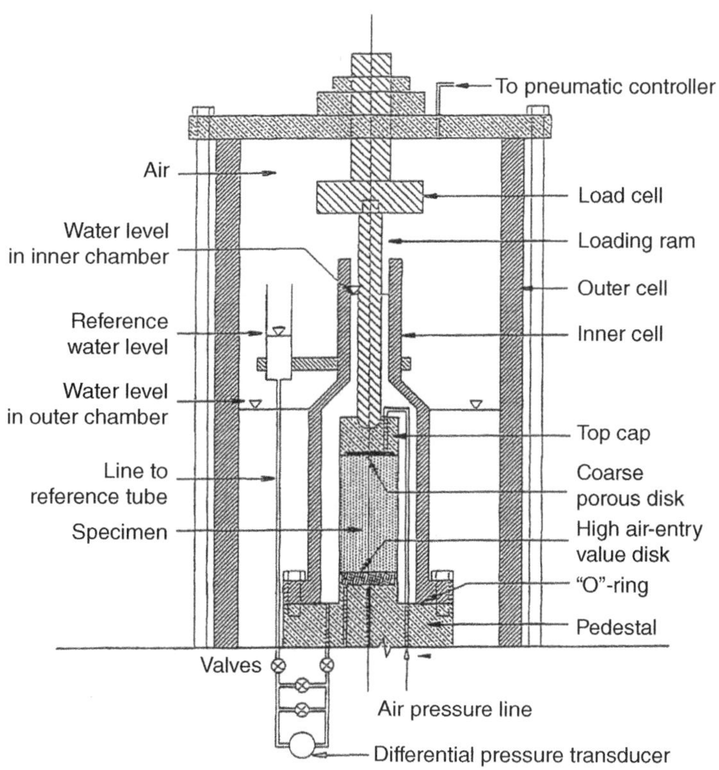

Figure 4.49 Volume change measurements from an inner cylinder triaxial device. Reproduced from Ng *et al.* 2002 by permission of Canadian Science Publishing.

avoids volume change of the inner cell, and the volume change of the specimen is then measured from the water leaving or entering the inner cell using a conventional volume change device as reviewed in Section 4.8.2. Correction for piston intrusion in the inner cell is still required. Both Yin (2003) and Sivakumar *et al.* (2006) evaluated the accuracies of double wall triaxial cells and showed that essentially the same volume changes were obtained from measurements from the inner cells as from the fully saturated specimens. Thus, negligible errors were experienced from the inner cell measurements during consolidation as well as during shearing of the specimens.

Comparison of inner cell and double wall techniques

A comparison of the two techniques was carried out by Laloui *et al.* (2006) as shown in Table 4.6 for one inner cell device and two double wall devices.

Temperature effects

Changing ambient temperature during the period of testing may affect the volume changes inferred from the volume of the confining water in the triaxial cell. Temperature fluctuations in the order of ±1°C (±2°F) typically occur in the laboratory from day to night unless special precautions are made. Thus, for tests lasting more than a few hours, it may be necessary to consider corrections to the volume changes for temperature variations. This correction is specific to the equipment employed, and calibration of the equipment may be required. The details of such calibration were demonstrated and discussed by Stewart and Wong (1985) and by Leong *et al.* (2004).

4.8.4 Measurements from dry and partly saturated specimens

Measurement of air volume change

It is possible to measure the volume change of dry and some partly saturated specimens by measuring the amount of air expelled or sucked into the specimen. To measure the volume change correctly, it is important that all air in the specimen is freely accessible and can flow in and out of the specimen. However, if the specimen contains air bubbles in water, the specimen

Table 4.6 Characteristics of inner cylinder and double wall triaxial cells (modified after Laloui *et al.* 2006 by permission of Elsevier)

	Inner cylinder	Double wall	
Reference	Ng *et al.* (2002)	Wheeler (1988)	Yin (2003)
Instantaneous deformation from 0 to 500 kPa	500 mm^3 (0.5 %)	700 mm^3 (1.75 %)	400 mm^3 (0.25 %)
Adsorption or creep of the cell or membrane at 200 kPa	80 mm^3 per week (0.09 %)	150 mm^3 per week (0.125 %)	—
Temperature variations	40 mm^3/°C (0.05 %/°C) Daily variations: about 2°C	70 mm^3/°C (0.07 %/°C)	—
Accuracy of the piston displacement calibration	Very good	α = 70 mm^3 β = ±0.07 %	—
Accuracy of the apparent volume change calibration	Good	α = 530 to 950 mm^3 β = ±0.53 to 0.95 %	—
Accuracy of the volume measurement system	Differential pressure transducer precision: α =31.4 mm^3 β = ±0.04 %	Buret precision: α = 10 mm^3 β = ±0.025 %	Automatic volume meter precision
Total best possible accuracy	α =31.4 mm^3 β = ±0.04 %	α = 600 to 1020 mm^3 β = ±0.66 to 1.13 %	—

α and β are the absolute errors in volume change and volumetric strain, respectively.

may change volume and the bubbles may contract or expand with resulting higher or lower pressures inside the air bubbles, but with only little air coming out or going into the specimen. The amount of air measured from the specimen is therefore not representative of the specimen volume change. For example, compacted soils may have much of the air present in inaccessible bubbles. Nevertheless, well on the dry side of the optimum water content, the water may sit only at the grain contact points and the soil may be sufficiently permeable to air that the expelled or imbibed air is representative of the volume change of the specimen.

Since air is very compressible, it is important to maintain the air pressure and the temperature constant while measuring the volume change. Bishop and Henkel (1962) devised a relatively simple constant pressure air system, as shown in Fig. 4.50. In this apparatus the volume of air coming from the triaxial specimen is measured by the change in mercury level, which is adjusted to maintain constant reading on the oil manometer, thus maintaining atmospheric pressure inside the closed system of the specimen and air volume change device. Air absorption by oil and mercury are both negligible, so errors are minimized.

A simpler version of this apparatus is indicated in Fig. 4.51. In this device water is replacing the mercury and the oil. Water absorbs only a little air and since the water is maintained at atmospheric pressure and is generally in equilibrium with the surrounding atmospheric pressure and temperature, this replacement may be acceptable for relatively

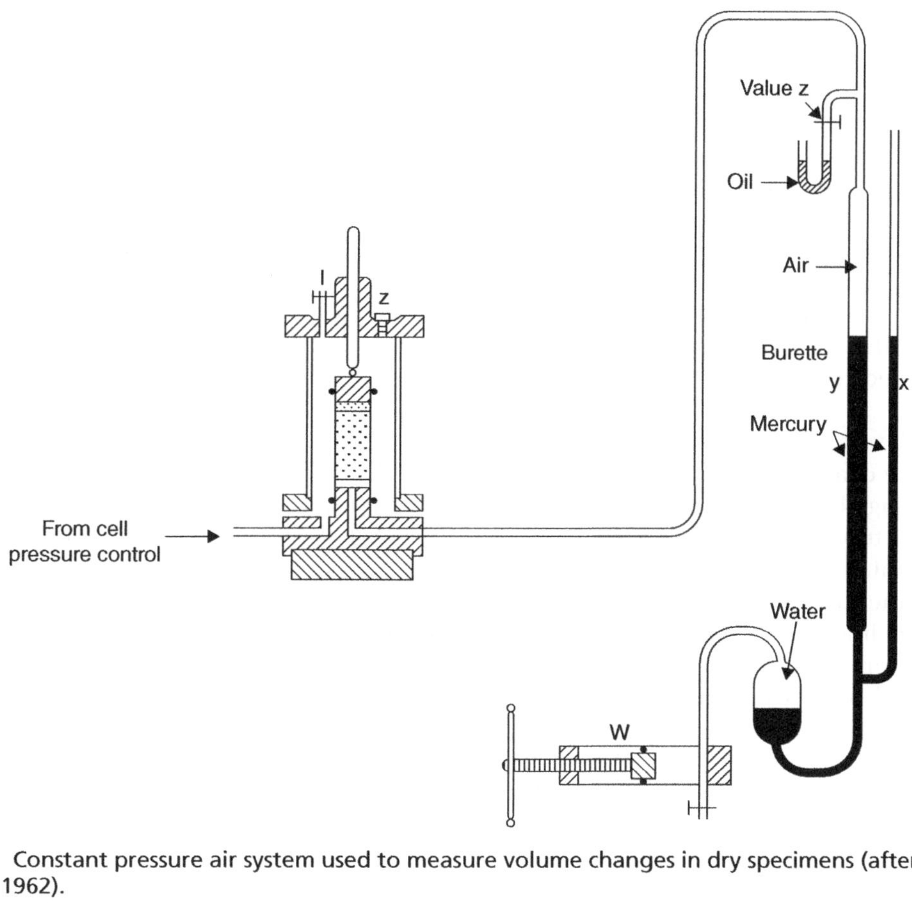

Figure 4.50 Constant pressure air system used to measure volume changes in dry specimens (after Bishop and Henkel 1962).

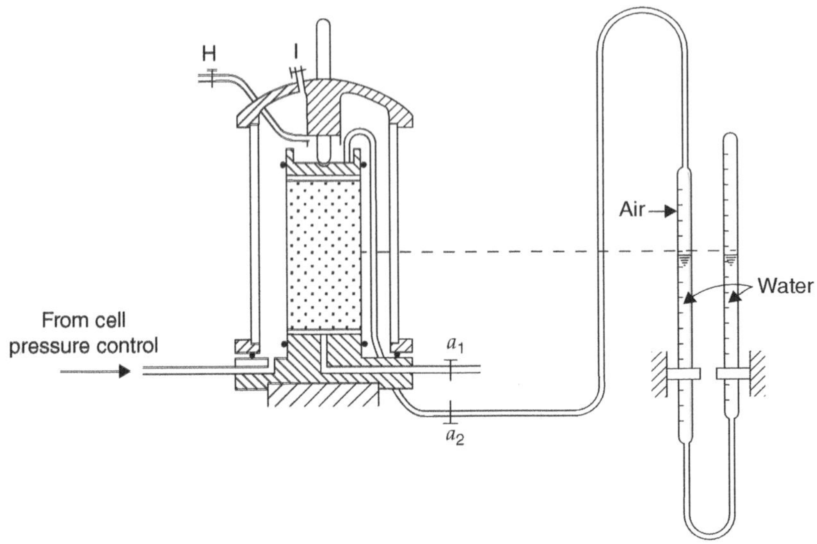

Figure 4.51 Measurement of the volumes of both air and water expelled from a partly saturated specimen (after Bishop and Henkel 1962).

short-term experiments on dry or partly saturated soils. The volume change is measured by the change in water level in the left hand buret while maintaining atmospheric pressure inside the closed system (specimen and device) by lowering or elevating the right hand buret. An automated version of this apparatus was described by Laudahn *et al.* (2005) in which photoelectric sensors and computer control combined with a pressure/volume controller were employed to maintain constant pore air pressure.

Another device that may be used for measuring volume changes of dry sand specimens is shown in Fig. 4.52. A 1/4 in. outside diameter propylene tube is curled up in a snail shape lying horizontally and attached to a board. One end of the tube is connected to the specimen drainage line and the other end is open to atmospheric pressure. In the case where this device is used for a vacuum triaxial test, the open end is attached to a vacuum regulator that applies the vacuum to the specimen. The inside of the tube has been made water repellent by flushing it with water repellent fluid, and just before the test is performed, a slug of water is carefully let into the tube through a shut-off

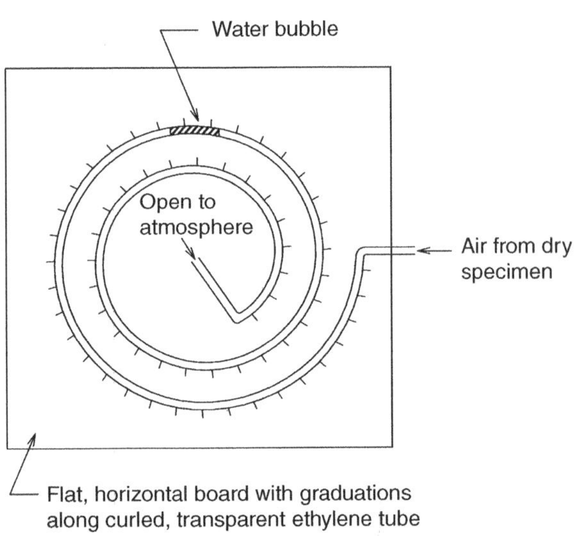

Figure 4.52 Horizontal snail with slug of water to measure air volume change of dry or nearly dry specimen.

valve connected to a small water reservoir. A scale is marked on the board along the curled tube and an appropriate volume calibration of the tube is used to determine the volume change. The water repellent coating will reduce the drag on the water slug due to the menisci that set up at both ends. The movement of the water slug can be followed on the scale along the tube and will indicate the specimen volume change.

Measurement of air and water volume changes

The volume changes of the air and water of a partly saturated specimen may be measured by separating the air from the water and measuring the two fluids separately using appropriate devices as discussed above. For this purpose a standard porous stone is used at the cap and a high air entry porous stone is employed at the base. Alternatively, the volume change of a partly saturated specimen may be measured by digital pressure–volume controllers as shown in Fig. 4.53. The total volume change is obtained by adding the measurements from the two devices.

Photography and image processing

Using the methods discussed for measurement of linear deformations, including video tracking and photography, the volume change has also been determined from image processing by Rifa'i *et al.* (2002) and Gachet *et al.* (2007).

Other principles of volume change measurements

Romero *et al.* (1997) determined the volume change using a laser technique in which the radial deformations were measured on two

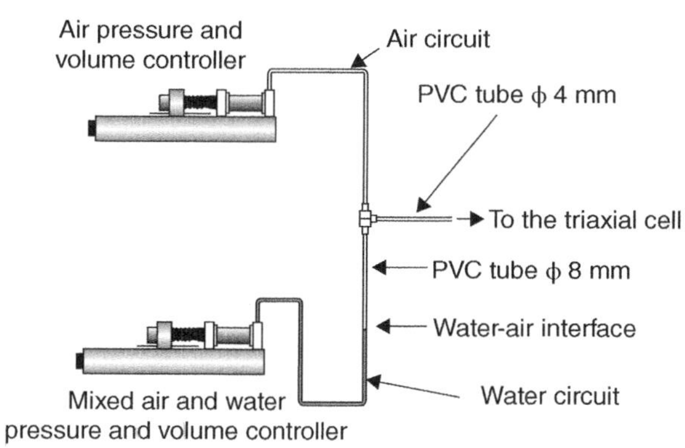

Figure 4.53 Device with air volume controller and mixed air–water volume controller. Reproduced from Laloui *et al.* 2006 by permission of Elsevier.

diametrically opposite sides of the specimen by sweeping with the laser over the entire specimen height by a non-contacting laser system mounted outside the triaxial cell.

Comparison of various volume change measurement methods

Laloui *et al.* (2006) compared three different techniques of measuring the volume change of partly saturated specimens: (a) cell liquid measurements; (b) air–water volume measurements; and (c) direct measurements on the specimens. Table 4.7 shows the comparison of these three principal methods of measurements.

4.9 Measurement of axial load

4.9.1 Mechanical force transducers

Load cells are used to measure the axial force transmitted to the specimen in a triaxial test. When the load cell is properly calibrated and maintained, it provides accurate and reliable measurements of the axial load. Different design principles have been employed with the bonded strain gage load cells offering accuracies from 0.03 to 0.25% of full scale response. Strain gage load cells are relative simple and inexpensive and they may be fabricated "in house" to fit a particular application with excellent results.

Other principles of measuring axial loads include using a proving ring. A proving ring must deform to measure a load, which may become an issue with regard to stiffness of the load measuring system. For this reason two nested proving rings may be used, such that when the outer (larger and more flexible) ring has deformed beyond some specified amount, the inner (smaller and stiffer) one will engage.

A hydraulic load cell may be employed in which the force to be measured is applied to a piston that causes a change in the pressure in the fluid of the internal cell, which is measured. To reduce friction along the piston, the hydraulic load cell may be provided with a rolling diaphragm. The pressure in the fluid increases

Table 4.7 Comparison of three methods of volume change measurements for partly saturated soils (modified after Laloui *et al.* 2006 by permission of Elsevier)

Type of device	Advantages	Limitations	Absolute errors on ΔV (α) and ε_v (β)
		Method (a): Cell liquid measurements	
Standard triaxial cell (a1)	Use of standard cell, without modifications	Indirect method, involving long calibration process	$\alpha = \pm 0.45$ cm^3 $\beta = \pm 0.22\%$
Inner cylinder (a2)	Minimizes or strongly decreases the undesired volumetric changes observed with (a1) as the confining pressure is imposed on both sides of the inner wall	Indirect method, involving calibration process	$\alpha = \pm 0.21$ cm^3 $\beta = \pm 0.08\%$ Bishop and Donald (1961): $V_{spec} = 100$ cm^3: $\alpha = \pm 0.1$ cm^3 $\beta = \pm 0.1\%$
Double walled cell (a3)	Same as (a2) Enables continuous measurements	Indirect method, involving calibration process	For specimens of 100 cm^3: $\alpha = \pm 0.6$ to 1.02 cm^3 $\beta = \pm 0.6$ to 1.0 % depending on the cell. The average global accuracy is believed to be better
		Method (b): Air–water volume measurements	
Air filled controller (b1)	Direct measurement or imposition of the volume of air	Air volume is strongly influenced by temperature and atmospheric pressure. Undetectable air leakage	$\alpha = \pm 2.2$ cm^3 $\beta = \pm 1.1\%$ +Continuous air leakage of 2–3 cm^3/day
Mixed air–water filled controller (b2)	Same as (b1) Minimizes the air volume and the possible errors	Same as (b1), but less important	$\alpha = \pm 2.2$ cm^3 $\beta = \pm 0.11\%$ +Continuous air leakage of 0.2 cm^3/day
		Method (c): Direct measurements on the specimens	
Hall effect captor with radial strain measurements (c1)	Direct measurement on specimen Enables continuous measurements	Conceived for small strain measurements Problems of accuracy for barrel-shaped specimens equipped with only one radial strain gage. Mounting or sealing transducer on the specimen is quite delicate and requires an initially fairly rigid specimen	—
Laser technique (c2)	Direct, continuous, non-contacting measurements Measurement of entire specimen profile Possible measurement all around specimen	High costs and long calibration process	Estimate based on Romero *et al.* (1997): $\beta = \pm 0.007\%$
Image processing (c3)	Direct, non-contacting measurements Measurement of entire specimen profile Computer controlled calibration process	Not valid for asymmetric specimen when using only one camera	$\alpha = \pm 0.25$ cm^3 $\beta = \pm 0.1\%$

linearly with the applied force and it is measured with a pressure transducer. Pressure transducers are discussed in Section 4.10. However, the electrical strain gage load cell is most often employed.

4.9.2 Operating principle of strain gage load cells

Electrical strain gages are bonded to the primary sensor (see Section 4.4.1) which deforms when a load is applied. The load cell therefore stretches the strain gages, which consequently change resistances. Four strain gages connected in a Wheatstone bridge are most often employed to obtain maximum sensitivity and temperature compensation. Two of the gages are in tension and the other two are in compression, as shown in Fig. 4.7. When a force is applied to the load cell the strain gages change their electrical resistance in proportion to the force, as explained in Section 4.4.1. Application of a constant excitation voltage, as shown in Fig. 4.7, produces an electrical signal in proportion to the force applied to the load cell.

4.9.3 Primary sensors

The primary sensor to which the four strain gages are attached may deform in bending, shear or direct compression or tension. Figure 4.54 shows different designs of axial force cells. The action of the bending beam, the proving ring and the rotationally symmetric diaphragm or pancake design are indicated in Fig. 4.54(a), (b), and (c), respectively. The strain gages are

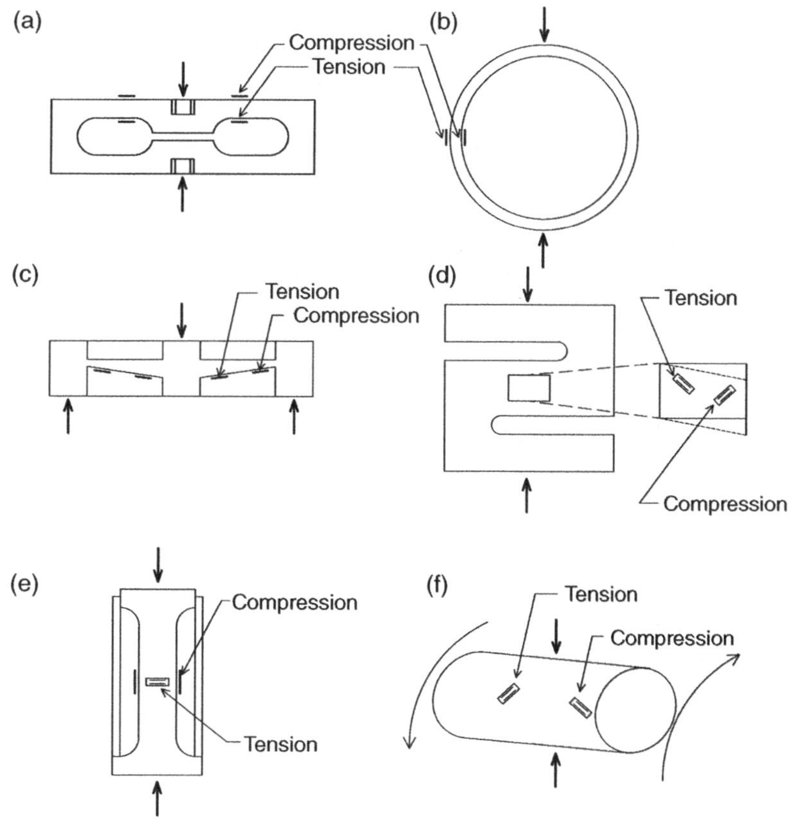

Figure 4.54 Load cell designs: (a) bending beam; (b) proving ring; (c) rotationally symmetric diaphragm or pancake; (d) S beam; (e) direct stress or column/canister; and (f) helical.

attached, two and two, at the points of tension and compression and connected to form a Wheatstone bridge.

Figure 4.54(d) shows a shear beam design in which the strain gages are attached at 45° inclination on opposite sides of a thin web that deforms in uniform shear under an applied force. At these inclinations the strains are in compression or in tension as indicated in the diagram. The shear beam design is employed in an "S" beam load cell used for compression or tension, as indicated in Fig. 4.54(d).

The direct stress or column/canister load cell shown in Fig. 4.54(e) employs two compression strain gages in the longitudinal direction and two tension strain gages mounted in the transverse direction. The cross-section of the column may be square, circular, or circular with sections on which to mount the strain gages. The column with the strain gages and wiring may be protected inside a tubular sleeve, as shown in Fig. 4.54(e).

The helical spring load cell shown in Fig. 4.54(f) handles eccentric loads better than the direct compression load cell, because it relies on the spring action in which the torsional moment in the coil balances the axial force. Off-axis loading has little effect on the spring compression and load eccentricity is therefore not important for measuring the axial force. Strain gages are mounted on the spring as indicated in Fig. 4.54(f).

Each of these types of load cells may be outfitted with central holes with threads for attachment of the piston so that both axial compression and tension may be applied and measured, as discussed in Section 3.3.

The most convenient load cells for triaxial testing are the diaphragm or pancake load cell and the "S" beam load cell. The former type of load cell is suitable for building into the cap or placing below the base, while the "S" beam load cell is not suitable for building into the cap or placing under the base.

4.9.4 Fabrication of diaphragm load cells

While most types of load cells are commercially available, the diaphragm type load cell may be fabricated "in house" to fit anywhere in the triaxial setup, such as outside the cell, inside the cell above the cap, in the cap or under the base, as discussed in Section 3.1.1. Appendix B provides design specifications and charts for custom design of diaphragm load cells.

Materials for load cells are most often stainless steel, aluminum, and beryllium copper. The latter has a very linear stress–strain relation and is often employed in high quality load cells and other measurement devices such as clip gages. The properties of these metals are given in Appendix B.

4.9.5 Load capacity and overload protection

When choosing a load cell, two conditions should be considered when the capacity of the load cell is determined. On one hand the capacity should be matched to the maximum load anticipated in the given situation, such that the full range of the load cell is used, thus producing the most accurate measurements possible with the given equipment. On the other hand, the load to be measured may be higher than anticipated, so some additional capacity should be added. While some load cells are designed with overload capacity, that is they have extra capacity and thus may be loaded beyond their nominal capacity without damage, it is prudent to arrange for capacity beyond the maximum load anticipated in a given situation. It is recommended that the anticipated maximum load should be at around 75% of the nominal capacity of the load cell.

To avoid damage to the load cell it may be possible to design it with overload protection. While it is difficult to arrange for overload protection for "S" beam load cells, the diaphragm load cells may be protected against overload in compression and/or in tension. This is done by calculating the deflection at the maximum load and then preventing the diaphragm from deflecting beyond this distance. The details of the corresponding designs are discussed in Appendix B.

4.10 Measurement of pressure

The pressures to be measured in a triaxial test include the cell pressure and the pore pressure.

4.10.1 Measurement of cell pressure

The cell pressure may be supplied by regulated air pressure, which may then be converted to fluid pressure before entering the triaxial cell. Because of the large amount of pressurized air or fluid available, the system may be considered to be open, and the pressure measuring device may be chosen without consideration to the volume required to deform the primary sensor. Thus, a U-tube manometer or a Bourdon gage, which require considerable volumes of air or fluid to be activated, may be used to measure the pressure. However, the cell pressure may also more conveniently be measured by a pressure transducer, as discussed below.

4.10.2 Measurement of pore pressure

Measuring the pore pressure in an undrained test is difficult, because the volume change required to activate the primary sensor must be as small as possible to minimize the resulting change in pore pressure. If essentially incompressible water fills the pores of the specimen, the pore pressure will drop as some of the water comes out of the specimen to activate the pressure measuring device; however, the drop in pore pressure also depends on the soil compressibility, as discussed in detail in Section 6.4. It is possible to measure pore pressure using a manometer or a Bourdon gage by involving a null indicator in the process, as shown in Fig. 4.55 (Bishop and Henkel 1962). This requires another control device such as a screw control, as also shown in Fig. 4.55(a), and it requires the constant attention of an operator to adjust the screw control and for reading the pressure gage. While such a system may be automated by modern methods, as indicated by Laudahn *et al.* (2005), it is not often used today, because it may be replaced with a closed (dead end) electrical pressure transducer, which automatically maintains undrained conditions while the pore pressure is measured.

Thus, a pressure transducer may be used to accurately measure pore pressure without disturbing its magnitude significantly. Very small volume changes are required to activate the primary sensor for a given change in pressure. In this regard the volume flexibility is the relevant measure of the pressure transducer. The stiffer the pressure transducer the better, but the resolution of the response also decreases with increasing stiffness. Since volume flexibility is rarely an important attribute for a pressure transducer, information regarding this property is not often provided by manufacturers. This is because pressure transducers have many applications in industry beyond those in soil mechanics.

The primary sensor in most pressure transducers is a diaphragm that deflects in response to a differential pressure between its two sides. Thus, all transducers are, in principle, differential transducers. A reference pressure is applied to one side and the pressure to be measured is applied to the other side. Thus, the reference pressure may be atmospheric pressure in which case the pressure is referred to as gage pressure, that is the gage pressure is defined relative to atmospheric conditions. This is accomplished by leaving the low pressure side open to the atmosphere. The reference pressure may also be a vacuum, in which case the measured pressure is referred to as absolute pressure. This may be achieved by employing a sealed vacuum reference on the low pressure side.

The designation of the pressure transducer is often given by the reference pressure (gage or absolute) indicated by the pressure units followed by a "g" (e.g., "psig" which means pounds per square inch gage) or by an "a" (e.g., "psia" which means pounds per square inch absolute).

If the transducer is such that two different pressures can actively be directed to the two sides of the diaphragm, then such a transducer may be used to measure the effective pressure directly by applying the pore pressure on one side (as the reference pressure) and the cell pressure on the other side.

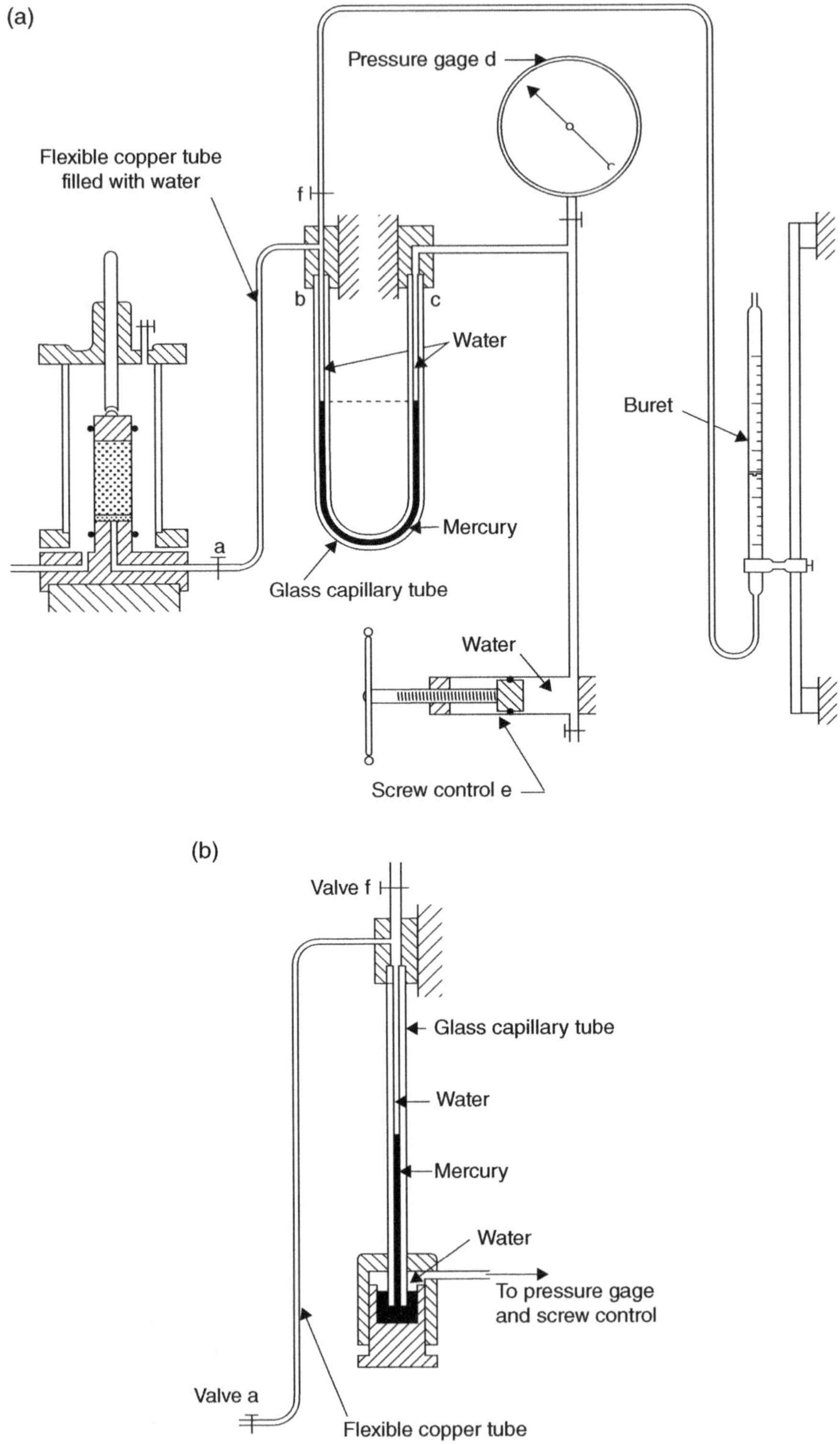

Figure 4.55 (a) Original arrangement for null method of pore pressure measurement and (b) modified null indicator for pore pressure measurement (after Bishop and Henkel 1962).

4.10.3 Operating principles of pressure transducers

Pressure transducers differ in the operating principles employed to measure the deflection of the diaphragm. The most common principles used in pressure transducers employed for triaxial testing are reviewed below.

Strain gage pressure transducers

Strain gage-type pressure transducers are widely employed in triaxial testing. Either a circular bonded foil strain gage already connected in a full Wheatstone bridge is glued to the backside of the diaphragm, as shown in Fig. 4.9(a), or it is connected with unbonded strain gages, as shown in Fig. 4.9(b).

Reluctance pressure gages

The reluctance principle is used in pressure transducers in which the deflection of the diaphragm changes the reluctance, that is the resistance to magnetic flow in the electric circuit. Pressure transducers based on the reluctance principle have very high output signals, but they require AC voltage for excitation.

4.10.4 Fabrication of pressure transducers

Because of the complications involved and the precision required of pressure transducers, they are difficult to manufacture "in house." In particular, the attachment of strain gages to the inside of diaphragms present an obstacle to producing high quality pressure transducers.

4.10.5 Pressure capacity and overpressure protection

Similar to load cells, two conditions should be considered when the capacity of the pressure transducer is determined. On one hand the capacity should be matched to the maximum pressure anticipated in the given situation, such that the full range of the pressure transducer is used, thus producing the most accurate measurements possible with the given equipment. On the other hand, the pressure to be measured may be higher than anticipated, so some additional capacity should be added. While some pressure transducers are designed with overpressure protection, that is they have extra capacity, typically in the order of 50–200%, and thus may be pressurized beyond their nominal capacity without damage, it may be prudent to arrange for capacity beyond the maximum pressure anticipated in a given situation. On the other hand, since the cell pressure is usually the independent variable, that is it is determined a priori and it represents the highest pressure to which the transducer will be exposed, it may be possible to employ pressures near 100% of the nominal capacity of the pressure transducer.

4.11 Specifications for instruments

The instruments, whose principles of operation are reviewed above, are those most often employed in equipment for triaxial testing of soils. Additional transducers and instruments with still other operational principles may sometimes be required for specific test procedures. However, any further instrumentation will not be reviewed here.

To provide an idea about the specifications for electrical instruments, which may be required to assure adequate quality of the triaxial test data, proposed minimum performance characteristics for transducers that can be purchased on a regular basis are listed in Table 4.1. Specifications are listed for load transducers, displacement transducers (LVDTs) and pore pressure transducers. Similar proposed specifications are listed in Table 4.5 for volume change devices. The values given in Tables 4.3 and 4.4 are those that would be desirable under ideal conditions. They may not be necessary for all triaxial testing, and they may be found to be somewhat restrictive and may limit the choice and availability of instruments from different manufacturers. However, these specifications do provide guidelines regarding the capabilities and the orders of magnitude of accuracy that are available and may be obtained from instrument manufacturers.

4.12 Factors in the selection of instruments

A number of factors play a role in selection of measuring instruments. A list of considerations is given below with a few comments:

1. Physical quantity to me measured? (e.g., force, pressure, deformation)
2. Nature of the measurement? (static, dynamic, long term stability of instrument)
3. Effect of instrument on physical process? (maximum signal for minimum interference)
4. Environment for instrument? (water, humidity, pressure, temperature, vibrations, shock, magnetic field to disturb measurements, can calibration be checked after installation?)
5. Limits of instrument? (maximum range, sensitivity, etc. – see Section 4.5)
6. Compatibility with existing equipment? (space for equipment, is read-out equipment available for LVDT?, etc.)
7. Cost, availability, warranty, service facilities nearby?
8. Past experience?

4.13 Measurement redundancy

Because instruments may fail during a test, it is prudent to have redundant capabilities on as many instrument stations as possible. Mechanical devices such as dial gages, Bourdon tube pressure gages, and proving rings are sturdy, and they perform steadily over long periods of time. However, they require manual datalogging. Electrical devices are often more accurate, more convenient to apply, easier to read when provided with digital read-out devices, and they can be used for automatic datalogging. However, electrical instruments are often more sensitive to mechanical handling and other environmental effects (water, shock, etc.), and they are consequently more liable to malfunction during a test than mechanical devices. In addition, their functioning depends on their correct connection to read-out devices such as voltmeters, dataloggers, and computers, which may themselves malfunction. Thus, many more devices are required to perform correctly when advanced measurement systems are used. Further, some of the instruments may be located inside the triaxial cell where they cannot be reached, repositioned and repaired during the test.

It is therefore prudent to consider redundant instrumentation of the physical processes to be measured. For example, an LVDT may work in parallel with a dial gage, as shown in Fig. 4.56. Similarly, the dial gage may remain in the proving ring after strain gages or an LVDT have been attached to it. Instruments may be applied inside the triaxial cell to measure vertical and horizontal deformations. Redundant measurements may then be made by using conventional measurement systems such as a vertical deformation dial gage or LVDT and a volume change device outside the cell. These may all be electrical devices.

It is also important to note and record the time occasionally during a triaxial test, because the vertical deformation may be double checked on the basis of the deformation rate employed, and any leak in the membrane developing during the test may be calibrated at the end of the test and this calibration may be used to correct

Figure 4.56 Dial gage with LVDT mounted on its back side. Reproduced from Berre 1982 by permission of ASTM International.

the volume change after the leak developed. Thus, there are many ways of making redundant measurements.

4.14 Calibration of instruments

Primary standards against which to calibrate linear deformation devices, load cells and pressure transducers are provided by the National Institute of Standards and Technology (NIST) in Gaithersburg, MD, USA and this Institute will calibrate such devices for a fee. Standards calibrated by NIST may then be employed as secondary standards against which laboratory measurement devices may be calibrated. However, given the scatter in properties of soil specimens either from the field or produced in the laboratory, such calibration may not be required for most practical purposes of triaxial testing.

The calibration process relates the output magnitude of a measuring system to the magnitude of a known input. This requires known secondary standards with known accuracies. The calibration relationship is then inverted to express the unknown input magnitude (e.g., in newtons, N) as a function of the system output magnitude (e.g., in millivolts, mV). If the instrument has a linear calibration relationship, for example in N/mV, then this calibration constant is employed by multiplying the load measured in millivolts by the calibration constant to obtain the load in newtons.

It is recommended to repeat the calibration procedure several times and using the average response from these several calibrations to obtain the best possible calibration constant for the instrument. The calibration should also include values measured during both increasing and decreasing inputs to check for any hysteresis.

The calibrations of most instruments are linear and a single calibration constant (e.g., in N/mV for a load cell) may therefore be established as indicated above. It is important to know the range within which the calibration relationship remains linear, because exceeding this range would result in a nonlinear relationship and incorrect calculation of the load outside this range.

Nevertheless, it may be possible to calculate the load outside the linear range of the instrument, but this requires that the actual output values, not only the change in these values, are known. Using the actual output referenced to the null setting, the displacement may be determined by simply following the nonlinear relationship. However, using any instrument in the nonlinear range should only be done in exceptional cases, for example in a case where the load to failure is much larger than anticipated and resetting or exchanging the instrument is not possible without interrupting the experiment.

Methods of calibrating devices for measurement of linear deformation, volume change, load, and pressure are reviewed below.

4.14.1 Calibration of linear deformation devices

Calibration of linear deformation devices such as clip gages, LVDTs and other devices reviewed in Section 4.7 are best performed in a calibration setup as shown in Fig. 4.57. In this calibration bench one part of the deformation measuring device (the coil of the LVDT shown in Fig. 4.57) is held by one post and the other part (the core for the LVDT) is attached to the end of a micrometer screw held by another post. Moving the micrometer screw in increments of known amounts thereby pushing or pulling the core

Figure 4.57 Setup for calibration of a pair of LVDTs using micrometer screws.

relative to the coil and measuring the resulting changes in output from the LVDT produces a relationship between the input (the displacement of the core relative to the coil) and the output (the change in electric signal, e.g., in mV).

The distance between the two posts on the calibration bench shown in Fig. 4.57 may be changed to accommodate LVDTs with different ranges by moving one of the posts to a different position as indicated by the screw holes in the base.

The calibration bench may also be used to calibrate clip gages, proximity gages, inclinometer gages, and Hall effect gages by appropriately modifying the posts to hold these devices and by placing the bench in a vertical position when necessary.

4.14.2 Calibration of volume change devices

Volume change devices may best be calibrated by letting the device expel water into a small beaker, which is placed on a sensitive scale, as shown in Fig. 4.58. Since 1 g of water has a volume of 1 cm^3, the volume change is most accurately determined by weighing the amount of water expelled. The relation between the visible response or the output from the volume change device and the amount of volume change measured by the expelled water is then established. If a more accurate calibration is required, the relationship between weight and volume of water as dependent on the temperature (given in tables in most undergraduate textbooks) may be used. Linear relationships, which allow single calibration constants to be established, are obtained from most volume change devices.

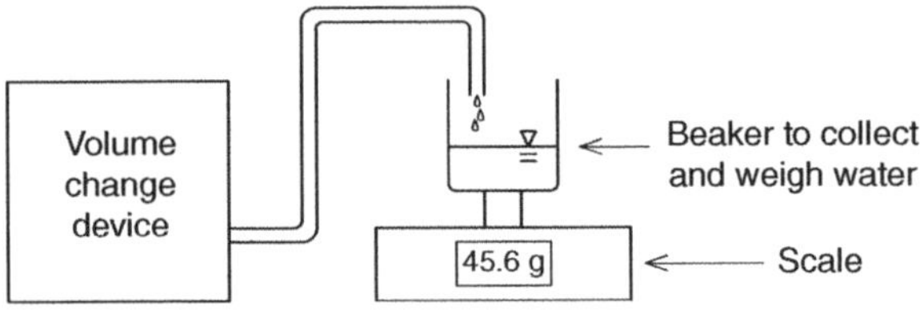

Figure 4.58 Calibration of volume change device by measuring weight of expelled water.

4.14.3 Calibration of axial load devices

The most accurate method of calibrating a proving ring or an electrical load cell is by using deadweights. For this purpose a yoke may be employed as shown in Fig. 4.59. Beginning with zero load on the device followed by load increments corresponding to the available weights in the laboratory and recording the output produces a relation between load and response from which a calibration relation can be determined. The weight of each deadweight should be known with a certain accuracy.

While this method is the most reliable, it is relatively limited due to the limit of the loads that may be applied to the device. Higher loads may be achieved by using a hydraulic loading machine which has been calibrated to a certain standard. Such a secondary standard may be the best available device for producing higher loads by which to calibrate proving rings and load cells.

4.14.4 Calibration of pressure gages and transducers

Pressure gages and pressure transducers may be calibrated by a deadweight tester, as shown in Fig. 4.60 or by connection to a secondary laboratory standard calibration device, as indicated in Fig. 4.61. While the secondary standard may be less accurate than the deadweight tester or the primary standard provided by and traceable to NIST, it provides a more convenient means of calibrating other instruments.

The deadweight tester consists of a piston with known cross-sectional area, A, that fits into a cylinder reservoir filled with hydraulic oil. The piston is outfitted with a platform on which known deadweights, W, are placed to create a known pressure, P, in the reservoir ($P = W/A$). The pressure transducer to be calibrated is attached to a branch from the fluid reservoir. The platform is then loaded up in increments with deadweights thus creating known pressures in the fluid and the corresponding responses from the pressure transducer are recorded. The piston is made to fit precisely into

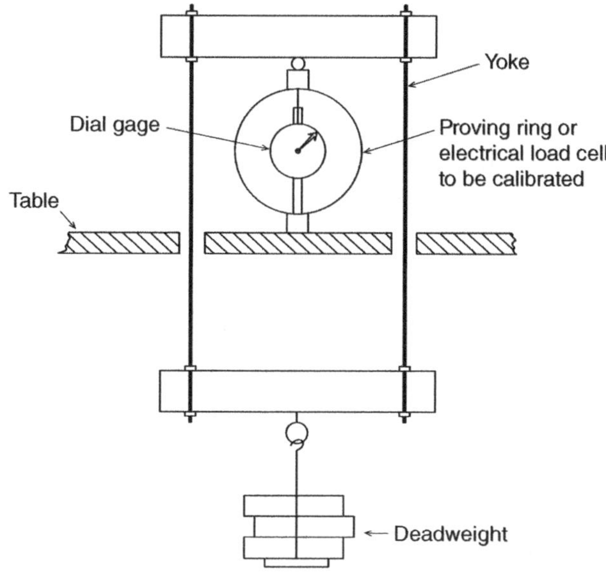

Figure 4.59 Calibration of load measuring device by deadweights.

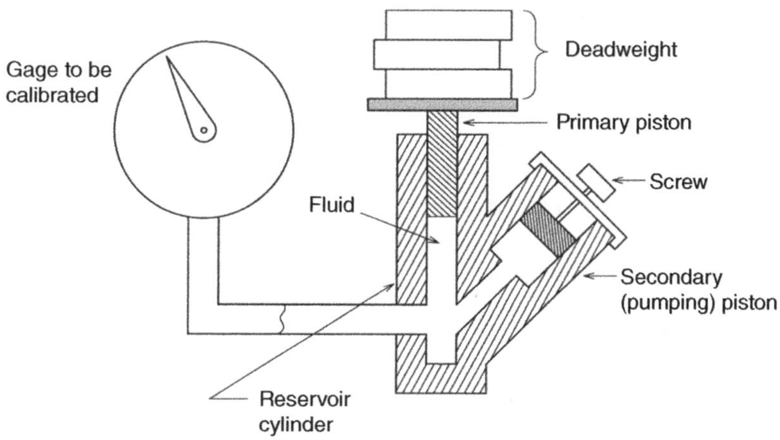

Figure 4.60 Calibration of pressure gage by connection to a deadweight tester.

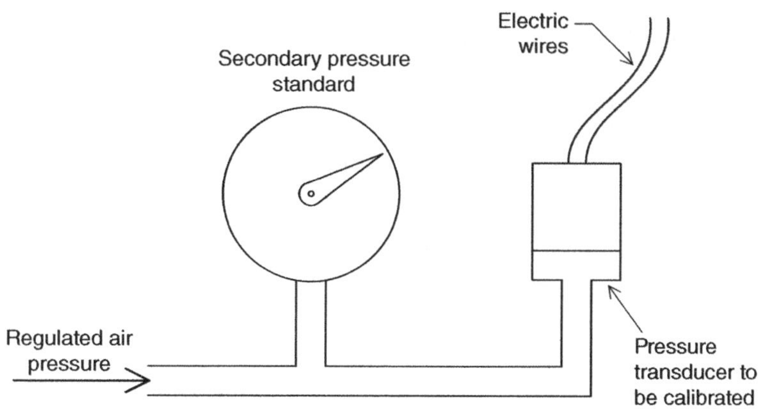

Figure 4.61 Calibration of pressure gage or pressure transducer by connection to a common pressure source and secondary pressure standard.

the bushing with very small tolerance, thus allowing a small amount of oil to seep past it and provide lubrication. To reduce vertical friction around the piston to zero, the piston may be rotated around its vertical axis.

Once a secondary pressure standard is available, it may be connected to a common pressure source and with parallel attachment to the pressure device to be calibrated, as shown in Fig. 4.61. The pressure source may be the air pressure house line or a bottle with compressed gas. The secondary pressure standard may be a previously calibrated pressure gage. By regulating the pressure in increments and recording the simultaneous responses from the secondary pressure standard and from the pressure transducer, a calibration relationship may be established.

Note that it is of primary importance that the calibrations of load cells and proving rings correspond with the calibrations of pressure transducers and gages, such that pressures determined from load measuring devices and cross-sectional areas are consistent with the pressures measured by pressure transducers and gages.

4.15 Data acquisition

Recording of data may be done manually, by a datalogger, or by a computer. Manual recording of instrument measurements is increasingly being replaced by automatic datalogging by a computer from electrical instruments. Figure 4.62 shows a schematic diagram of a data acquisition system suitable for triaxial testing. There are several requirements for satisfactory performance of such a system. These have been discussed by Silver (1979).

4.15.1 Manual datalogging

It is most practical to record the measured data in a sequence that makes it convenient to perform the calculations afterwards. Thus, for a drained triaxial test performed with constant confining pressure, the linear deformations and the volume changes should be recorded first, because they enter into calculation of the cross-sectional area, which in turn is used to calculate the deviator stress. Thus, it is practical to record the measurements as indicated in the following sequence: ΔH, ΔV, F, from which the following quantities are calculated: ε_1, ε_v, A, $(\sigma_1-\sigma_3)$, σ_1/σ_3. For an undrained triaxial test, the most practical sequence of recordings is as follows: ΔH, F, Δu, and the important quantities are calculated in the following sequence: ε_1, A, $(\sigma_1-\sigma_3)$, σ_3', σ_1'/σ_3'.

In addition, all data sheets should have room for notes where anomalies, strange occurrences, and procedural errors can be recorded. This is true for the computerized systems as well.

4.15.2 Computer datalogging

Datalogging may also be done automatically by a computer. This requires the instrumentation to provide electrical signals from each of the four key components of the triaxial setup. The four key components are: (1) the axial loading frame with stepper motor speed measurements: (2) the triaxial cell with cell pressure measurements; (3) the axial deformation measurements; and (4) the pore pressure/volume change measurements. Each of these components may be monitored continuously and provide input for calculation of stress–strain and volume change or pore pressure relations.

4.16 Test control

Test control can often be done with minimal modifications of the existing equipment. It requires control of axial load, cell and pore pressures and/or volume changes, as well as axial deformations.

4.16.1 Control of load, pressure, and deformations

The types of electrical measuring devices reviewed above may be used for control. This requires a loading frame or a Bishop–Wesley

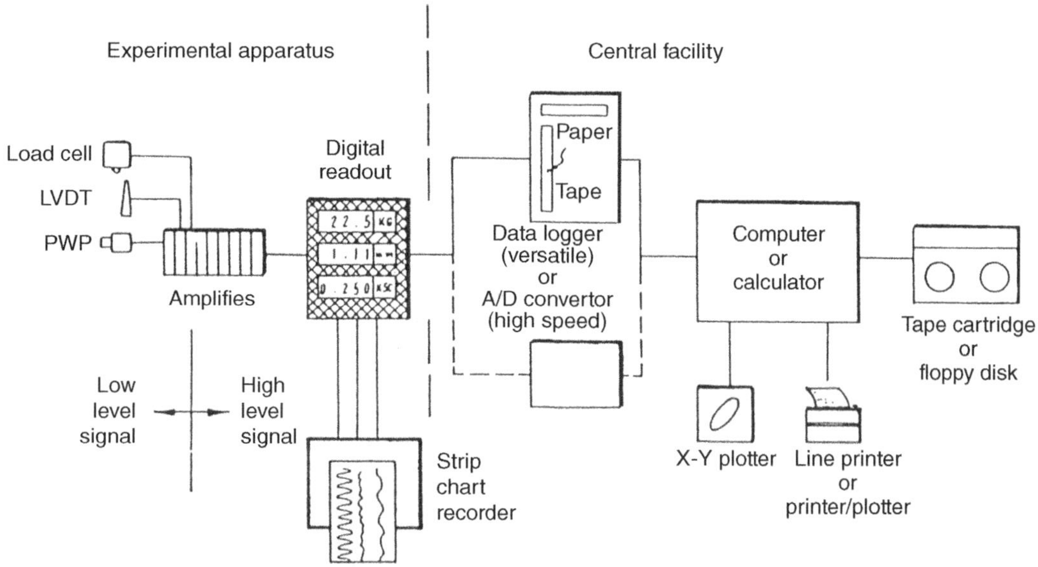

Figure 4.62 Schematic diagram of data acquisition system suitable for triaxial testing. Reproduced from Silver 1979 by permission of ASTM International.

triaxial setup, a triaxial cell with cell pressure, two axes of motion control, instrumentation, signal conditioning, and a computer. Figure 4.63 shows the process control loop and Fig. 4.64 indicates the loading system employed for a high pressure triaxial compression system, as indicated by Yamamuro and Lade (1993b). Note that the loading system in Fig. 4.64 is similar to the Bishop–Wesley system, shown in Fig. 3.47.

The two axes of control consist of hydraulic cylinders that are actuated by stepping motors through reduction gears and ball screw jacks. The whole system is close-looped controlled by a computer, which operates an analog-to-digital converter for data acquisition and a stepping motor controller to control the stepping motors, all supervised and run by custom-control programs. All instrumentation signals are conditioned through amplifiers and active filters before entering the computer.

The first control axis regulates either the true vertical stress or strain by feeding hydraulic fluid into the axial loading frame's hydraulic cylinder, which raises or lowers the table on which the triaxial cell is sitting. The second control axis operates the confining pressure by feeding hydraulic fluid into the high-pressure triaxial cell. The high-pressure triaxial cell and instrumentation can test the specimens in either compression or extension employing confining pressures of up to 100 MPa. Custom control software can be developed to perform a wide variety of tests under both true strain and true stress control.

Similar systems of control have been described by Li *et al.* (1988), Ampuda and Tatsuoka (1989), and Sheahan *et al.* (1990).

4.16.2 Principles of control systems

The computer program to control the triaxial test may be a commercial program, such as LabVIEW^R from National Instruments, which works with an on-screen setup that allows both datalogging and test control. The computations of adjustments to current stresses and strains are based on the virtual stiffness, that is the estimated additional pressure or axial force generated by each motor step. If the virtual stiffness matches the physical stiffness, the difference is

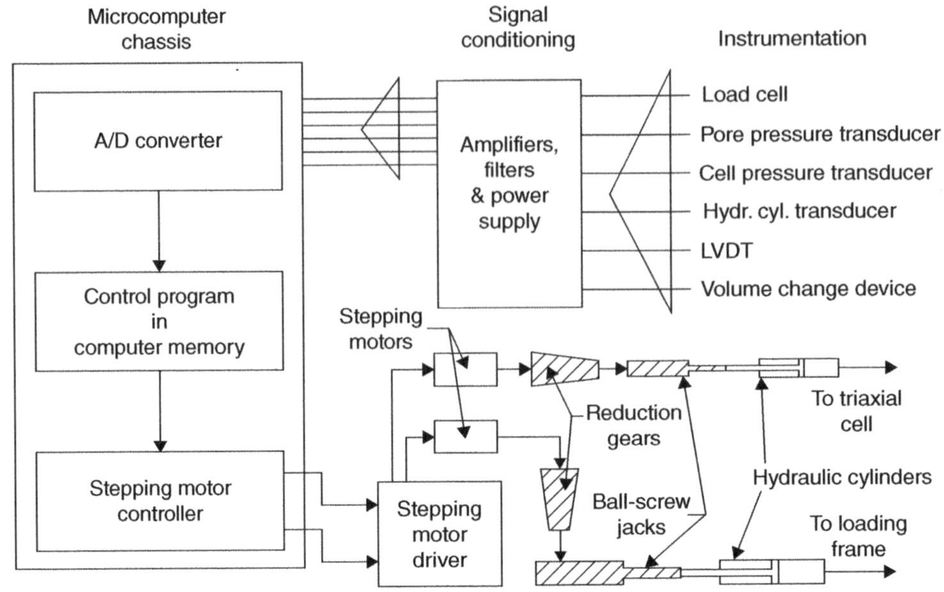

Figure 4.63 Process control loop for automatic control of a triaxial test. Reproduced from Yamamuro and Lade 1993a by permission of ASTM International.

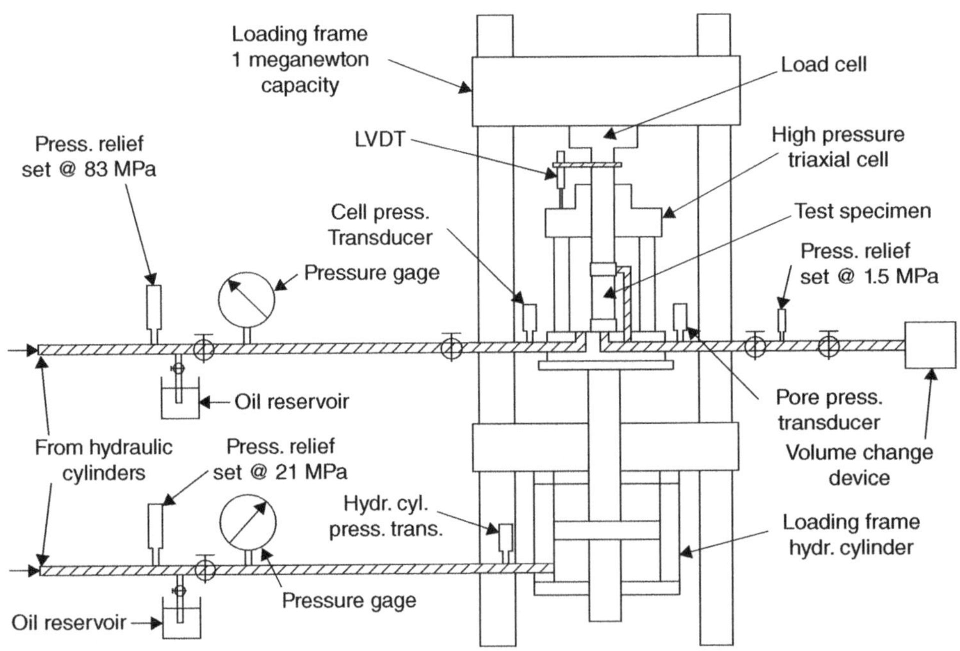

Figure 4.64 Loading system employed for a high pressure triaxial compression and extension test (after Yamamuro and Lade 1993a).

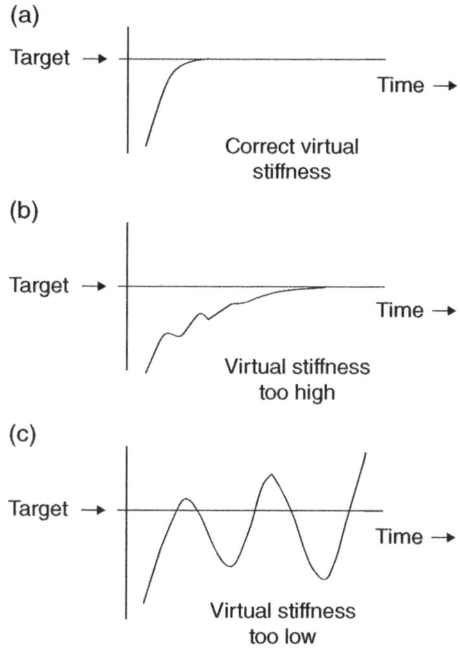

Figure 4.65 Computations of adjustments to current stresses and strains are based on the virtual stiffness: (a) if the virtual stiffness matches the physical stiffness, the difference is closed rapidly and without overshooting the target; (b) if the virtual stiffness is too high, that is the system is less responsive than assumed in the control algorithm, the controlled component behaves slowly and may take several cycles to eliminate the difference; and (c) if the virtual stiffness is too low, that is the system is more responsive than assumed, then the component will continually overshoot the desired value and become unstable. Reproduced from Sheahan and Germaine 1992 by permission of ASTM International.

closed rapidly and without overshooting the target, as shown in Fig. 4.65(a). If the virtual stiffness is too high, that is the system is less responsive than assumed in the control algorithm, the controlled component behaves slowly and may take several cycles to eliminate the difference, as indicated in Fig. 4.65(b). If the virtual stiffness is too low, that is the system is more responsive than assumed, then the component will continually overshoot the desired value and become unstable, as shown in Fig. 4.65(c). It is therefore necessary to ensure that the virtual stiffness is greater than the stiffest response in the physical system.

5 Preparation of Triaxial Specimens

5.1 Intact specimens

Dealing with intact specimens, their sampling using thin-walled tubes, the sealing of the soil inside the tube using wax and tube caps, selecting representative specimens, extruding the soil cylinder, and trimming the specimen to fit the triaxial cell are issues that have all been dealt with and have been described well by Germaine and Germaine (2009).

5.1.1 Storage of samples

Soil samples are usually contained in tubes of steel, brass, or polyvinyl and they are sealed at the ends to avoid any loss of water which would allow the soil to dry out and to avoid oxidation at the clay surface causing aging. These Shelby tubes are contained in a room with controlled humidity (at 90% humidity) and temperature (8–9°C) to further help the soil contain its in situ water content (La Rochelle *et al.* 1986).

Earlier sealing techniques employed regular paraffin wax, but this was too brittle and insufficient to protect the samples. They allowed fissures to appear and brownish and yellowish coloration would form along these fissures indicating oxidation of the clay (Lessard and Mitchell 1985). Mixing the paraffin with beeswax makes a better sealing material for the ends.

The sealing technique described by La Rochelle *et al.* (1986) appears to be optimal, because it was shown that the water contents, the Atterberg limits and the pH of two sensitive Canadian clays did not change over a period of 8 years. According to this technique, the large diameter specimens (20 cm) were extruded from the sampler and sliced into lengths of 12.5 cm or more depending on the size of the laboratory specimen desired. These were then sealed as follows: a wax compound consisting of 50% paraffin wax and 50% vaseline was kept in a warming pot at a temperature of 60–65°C. Thick plywood boards (25 cm square) were prepared by painting them with a layer of wax compound, placing a plastic sheet and painting it with another layer of wax compound. The plastic sheet is first dipped into the wax compound and smoothed out along the board to avoid trapping any air pockets. The sliced clay specimen is then placed by sliding it onto the board to prevent it from catching air bubbles at the base of the sample.

The sample is then turned upside down and the same procedure is now applied to the other side, thus preventing any part of the sample from being without protection from air intrusion. Two layers

Triaxial Testing of Soils, First Edition. Poul V. Lade.

of plastic sheets immersed in wax compound cover the ends of the sample. The exposed cylindrical faces of the sample are then painted with wax compound and covered with two sheets with wax compound between them. Visual inspection is carried out to avoid any trapping of air bubbles between the sample and the plastic sheets.

Transportation of the sealed samples to the humidity room occurs on a 10 cm thick rubber foam layer to avoid any vibration and therefore disturbance of the sensitive clay.

While these 20 cm diameter samples may be considered to be block samples, the sealing technique described here may also be expected to work well for larger block samples. Such larger samples may be carved out of the ground, enclosed in the system of plastic sheets dipped in the wax compound and surrounded by a box, which is then placed in the humidity room.

Figure 5.1 shows a comparison of stress–strain relations obtained from different sampling methods, and it is clear that the block samples represent the most undisturbed sampling techniques, and the other sampling techniques produce less satisfactory results.

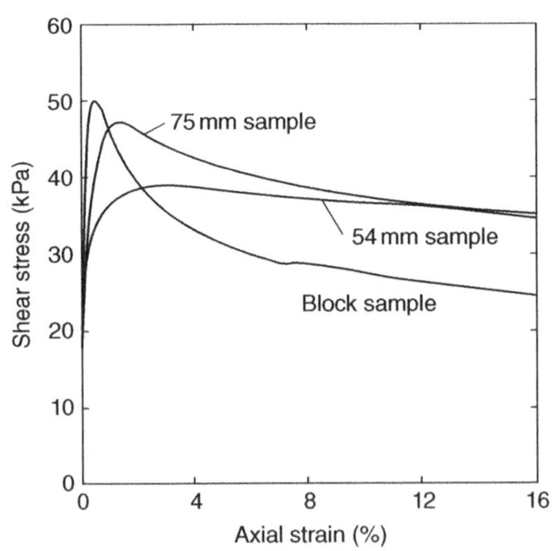

Figure 5.1 Example of triaxial test results on a block sample versus 75 mm and 54 mm samples on Onsøy clay (after Lunne *et al.* 1997, as quoted by Karlsrud and Hernandez-Martinez 2013). Reproduced from Karlsrud and Hernandez-Martinez 2013 by permission of Canadian Science Publishing.

5.1.2 Sample inspection and documentation

The minor geologic details that constitute soil fabric may seem insignificant, but often control the engineering behavior of the soil mass. Undrained shear strength, permeability, rate of consolidation, compressibility, and sensitivity are the properties most affected. To insure that laboratory samples represent field conditions, visual inspection, radiography, and microscopic examination can be used to identify fabric. Then the proper size and orientation of the specimen can be chosen to include critical geologic details. It may be possible to use fabric to decipher the geologic history of the soil and therefore deduce the engineering behavior. Failure may result if fabric is not recognized or not accounted for in sampling and testing.

Visual inspection

The advantage of samples is that they can be split lengthwise and photographed in various phases of drying. Upon drying, the specimen will often break along fractures, peds, inclusions, or coarser layers (Rowe 1972). The surface can also be wire brushed when dry to reveal any layering or varves which may have been obscured by smear on the surface as the sample was split.

From the vertical samples of San Francisco Bay Mud cored in thin-walled steel tubes 30 cm in diameter and 30 cm in length, the soil fabric and stratigraphy were determined by a drying technique, as follows: six slices were cut from the perimeter of each core, as shown in Fig. 5.2, placed on a table in the laboratory and allowed to dry. As the water dries out, the specimen slice becomes lighter in color and it may reveal any layering present. When the six parts of the core were laid side-by-side, the silt and shell layers could be traced from piece to piece. The resulting pattern is a sine curve, as shown in Fig. 5.2, because the silt layers are slightly dipping. Figure 5.2 illustrates the core and the inclined layer in three dimensions. When slices are taken from the perimeter of the core and laid flat, the dipping layer is viewed from many different

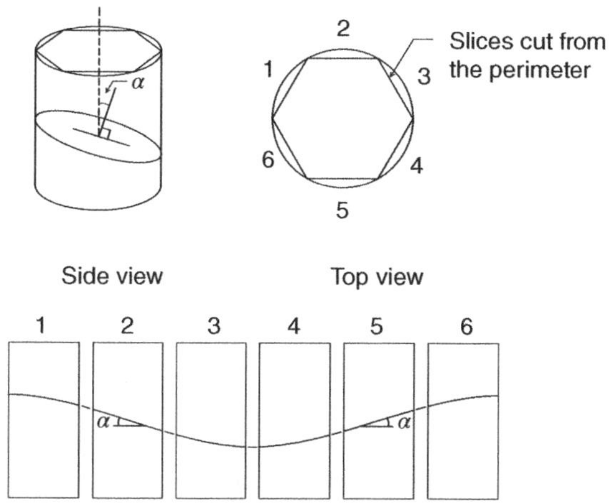

Figure 5.2 Six slices cut from the perimeter of a cylindrical core of San Francisco Bay Mud, placed on a table in the laboratory and allowed to dry. The steepest part of the sine curve represents the true inclination of the layer in the field.

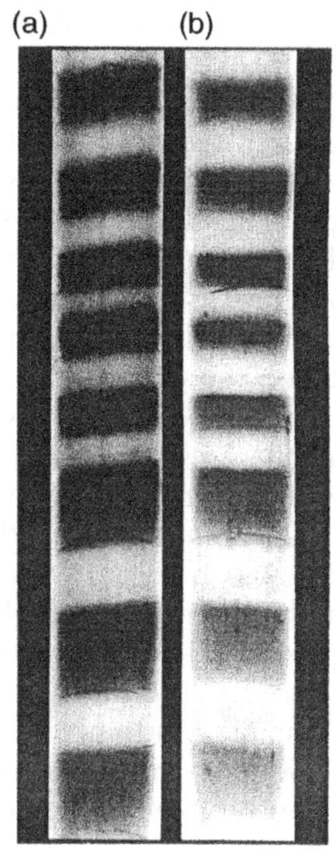

Figure 5.3 X-rays of samples taken from borings 200 ft (61 m) apart: (a) sample from elevation 561 ft (161 m); and (b) sample from elevation 556 ft (159 m). Soil width is 2 in. (5 cm). The X-rays show the extent of the disturbed zone before extrusion. Reproduced from Kenney and Chan 1972 by permission of Canadian Science Publishing.

angles. The steepest part of the sine curve represents the true inclination. For the seven cores where fabric was continuous, dip angles measured consistently between 6° and 7°. The purpose was to identify any layers in the seemingly homogeneous mud and to determine if these layers were horizontal or inclined.

Fabric was identifiable in 7 of the 12 San Francisco Bay Mud cores inspected. The most traceable features on the seven specimens were thin, light colored bands, possibly silt layers, within the dark gray mud. Also present were shells, but they were not forming continuous layers. Both of these features were most conspicuous when the sample was rewetted and allowed to dry partially. The silt layers tend to dry faster and therefore stand out with a lighter color.

Radiography

More subtle details of fabric can be revealed by the use of radiographs (X-ray images). Minor changes in density, delicate fractures, filled root, animal burrows, and very fine stratification may not be detected by visual examination. These can be seen and permanently recorded with radiographs. X-rays can resolve fine details that are only partially visible in reflected light, and can reveal particles such as shells and nodules beneath the surface and can detect structures not evident in reflected light. They also confirm whether a sediment that appears homogeneous is actually so. Samples of stiff clays of clay shales can be broken in the tube. X-ray images may be used to determine where to cut the tubes, especially where obtaining samples has been difficult.

X-ray images are taken with the samples still in the tubes, so they are non-destructive, and non-consuming (Kenney and Chan 1972). Samples are not smeared as they would be by extrusion. Preliminary examination of fabric and screening for sample damage can be done easily before samples are chosen for laboratory tests. Figure 5.3 shows X-rays of soil specimens in the sampling tube that reveals the degree of disturbance before extrusion.

Radiographs are produced when radiation is passed through an object and a shadow image is recorded on sensitive film. Objects which absorb radiation appear light because less radiation reaches the film. A stone would appear as a light spot and wood or ice as a dark spot. Tonal variations can be interpreted in terms of porosity and water content, because soil particles absorb more radiation than air or water. For example, in a varved deposit, the dense silt would appear light and the clay with the larger water content would appear darker. Cracks appear as dark lines.

Radiographs also have the advantage that the X-rays can be oriented parallel to inclined layering, slip surfaces, and fractures to get clearer pictures. If the joints, slickensides, or complex layering are randomly oriented, some features will appear masked or distorted. Radiographs can be overlapped to avoid distortion at the edges.

Microfabrics

Fabrics which control engineering properties may be at the microscopic scale, visible only by the polarizing microscope, X-ray diffractometer, or electron microscope (Barden 1972).Two types of microfabrics are possible in natural clays. Clay particles are shaped like thin, flat plates or rods (Mitchell 1956). If attractive forces are greater than repulsive forces, then the clay particles will flocculate. This occurs in solutions with high salt content, since interparticle repulsive forces decrease with increasing electrolyte concentration and increasing cation valence. In a flocculated structure, the particles are randomly oriented so that positively charged edges are attracted to negative particle surfaces.

The clay particles can also form a dispersed structure in which they are in a parallel arrangement. Dispersed clay occupies a smaller volume for the same weight and its properties are distributed more uniformly. Clays deposited in fresh water remain dispersed and settle at a slower velocity with a more parallel orientation. When pressure is applied to flocculated clay there is more shifting or rearrangement of particles than for dispersed clay, and the rearrangement is towards a more parallel arrangement. This microstructure affects certain engineering properties of the soil.

Based on the previous discussion it is clear that soil fabric has a great influence on the engineering properties of the soil. Therefore it is essential that samples include the minor geologic details which control soil behavior. Choosing the location, quality, and size of the sample as well as appropriate drilling technique to minimize disturbance are important for obtaining data relevant to the soil in its natural state (Rowe 1972).

Figure 5.4 shows how the fabric governs the size of the specimen required for testing. Depending on the spacing of the dominant features of the deposit, a small sample may not represent the mass. Table 5.1 lists the size of the sample necessary depending on the clay and type of fabric present. For example, the orientation and size of samples in fissured clays depend on fissure geometry and whether the fissures are empty or filled with sand or silt. Rowe (1959) described problems in defining the coefficient of consolidation, c_v, in varved clays where different size specimens give different values of c_v. In such clays the consolidation rate is dominated by the thickness of the clay layers, which is the same for all samples, whereas c_v is computed from the varying sample size. For measuring undrained strength or consolidation rate, small 37–76 mm diameter samples are not appropriate. Rowe (1970) recommends that the 260 mm diameter size be used as a standard "large sample." Although the cost of drilling may be greater, fewer samples are needed. Large samples of sensitive clays may experience less disturbance.

5.1.3 Ejection of specimens

Samples may be ejected from the Shelby tubes using an ejector with a sample trough to support the ejected sample. Care should be taken to avoid squeezing water out of the samples during this operation.

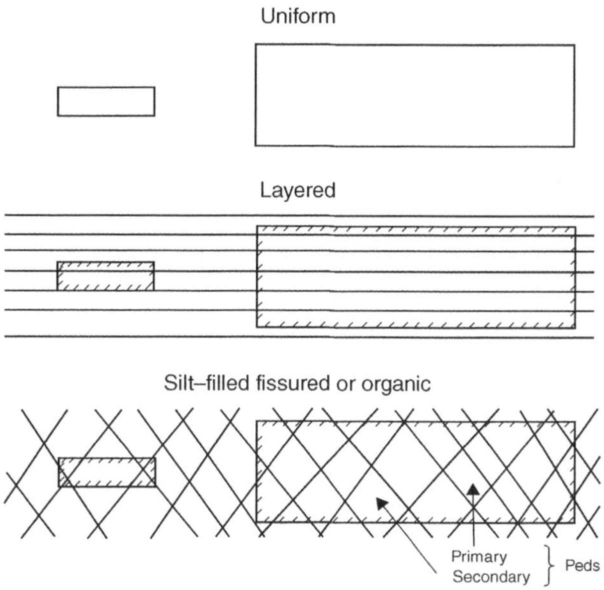

Figure 5.4 Size of the specimen required for testing is governed by the soil fabric. Reproduced from Rowe 1972 by permission of Geotechnique.

5.1.4 Trimming of specimens

The cylindrical specimen just extruded from the sampling tube is placed on a trimming device as shown in Fig. 5.5. The trimming is then performed using a wire saw and a straight edge as shown in the pictures. The trimming device has stiff edges along which the wire saw may be guided while the trimmings are cut off. The straight edge is used for final adjustment of the specimen shape. The outer disturbed zones are trimmed from the sample. Figure 5.5 shows two types of trimming devices: the one used in Fig. 5.5(a) and (b) has a vertical rod that can be positioned such that the desired specimen diameter is achieved by holding the wire saw and the straight edge against the two vertical rods. The other device, shown in the background of Fig. 5.5(a) and (b) has two translating plates against which the wire saw and straight edge can be held. The length of the specimen may be adjusted by surrounding the specimen by plastic wrap (to avoid adherence to the cradle) and then placing it in a cradle that allows cutting off a slice, as shown in Fig. 5.6.

Before trimming of specimens, the sample tube may be cut open to see the extent of the zone of disturbance due to intrusion sampling or the specimen may be X-rayed in the tube to see the disturbed zone before extrusion, as shown in Fig. 5.3.

Specimens of peat may be carved using an electric knife rather than a knife or a wire saw, which may rip the fabric of the organic materials.

During trimming of clay specimens to the size required for the triaxial cell, it may be possible to perform stratigraphic studies. In this context, problems may be encountered with pebbles or small rocks, shells, wood pieces, organic matter or nodules sticking out of the side of the specimen, and this may make fitting a rubber latex membrane around the specimen difficult. In such cases it may be beneficial to remove the small rock and fill the hole with plaster of Paris or hydrostone with a smooth surface to fit the radius of the specimen. Similarly, holes in the surface of the soil specimen may be filled before the membrane is fitted around the specimen.

Table 5.1 Minimum sizes of specimens from thin-walled piston samples of natural clay deposits, except deposits too weak, too strong, too variable, or too gravelly. Reproduced from Rowe 1972 by permission of Geotechnique

Clay type	Macro fabric	Mass hydraulic conductivity, k (m/s)	Parameter	Specimen diameter (mm)
Non-fissured	None	10^{-10}	c_u, c', φ'	37
Sensitivity <5			m_v, c_v	75
	Pedal, silt, sand layers,	10^{-9}–10^{-6}	c_u	100–250
	inclusions		c', φ'	37
			m_v, c_v	75
				250
	Organic veins			
	Sand layers >2 mm at	10^{-6}–10^{-5}	c', φ'	37
	<0.2 m spacing		m_v	75
Sensitivity >5	Cemented with any above		c_u, c', φ', m_v, c_v	50–250
Fissured	Plain fissures	10^{-10}	c_u	250
			c', φ'	100
			m_v, c_v	75
	Silt or sand filled fissures	10^{-9}–10^{-6}	c_u, c_v	250
			c', φ'	100
			m_v	75
Jointed	Open joints		φ'	100
Pre-existing slip			c_r, φ_r	150 or remolded

c_u, undrained cohesion; c', effective cohesion; c_r, residual cohesion; c_v, coefficient of consolidation; m_v, coefficient of compressibility; φ', effective friction angle; φ_r, residual friction angle.

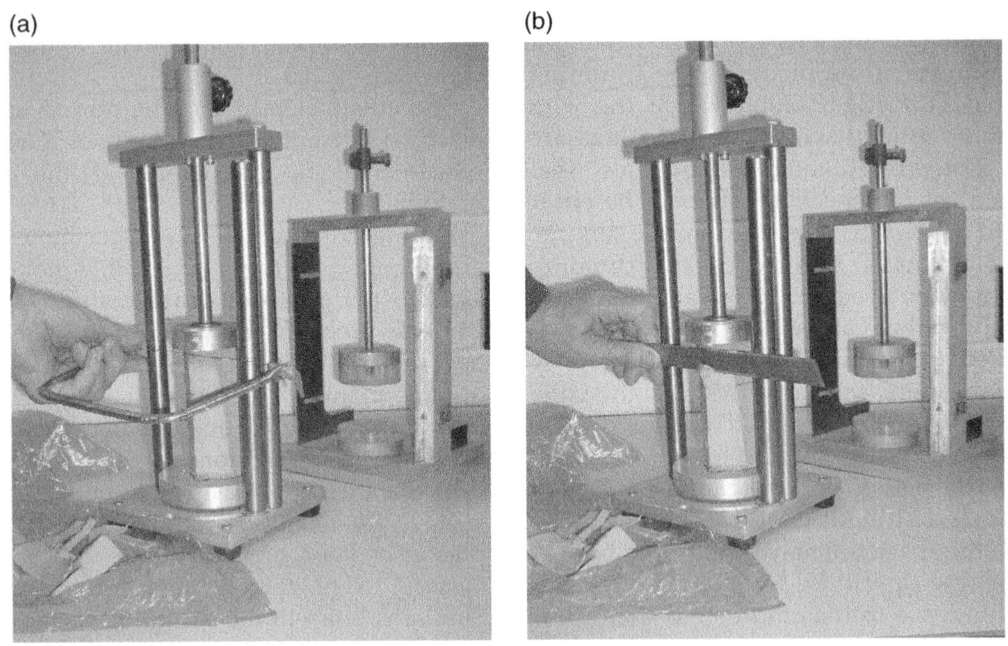

Figure 5.5 Trimming device with (a) wire saw and (b) straight edge for trimming the sample into a cylindrical specimen for testing.

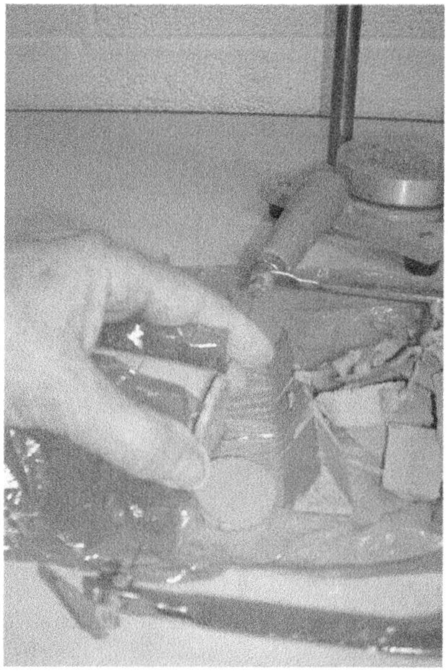

Figure 5.6 Specimen enclosed in plastic wrap to avoid sticking to cradle while being trimmed to the right length.

Alternatively, a rubber latex membrane may be built around the specimen. For this purpose, the water in the specimen is first sealed off by a uniform spray of diluted rubber cement with a coloring agent (e.g., red) and letting it dry. The coloring agent allows the rubber cement (which otherwise is clear when dry) to be visible so that it may be checked that the entire surface has been covered. Now the water from the specimen is sealed off from the water in the fluid latex rubber, which is to be applied on top of the rubber cement. One or two layers of fluid latex rubber is painted or sprayed on the specimen and allowed to dry after each application. This membrane is then sealed to the cap and base with O-rings.

5.1.5 Freezing technique to produce intact samples of granular materials

Frozen samples may be extracted from the ground to produce intact specimens of granular materials. This technique is reviewed in Section 10.5. Frozen specimens may be reduced in diameter using a lathe to trim the outer layers of frozen sand. This requires a cold room with freezing temperatures. This topic is beyond the scope of this book.

5.2 Laboratory preparation of specimens

In addition to intact specimens from the field, specimens may be fabricated in the laboratory by different methods, as reviewed below. This is necessary, especially for clean sand deposits, which cannot be sampled without disturbance, unless they are frozen in situ, cored, shaped into appropriate specimens in frozen condition, and tested after thawing inside the triaxial apparatus.

5.2.1 Slurry consolidation of clay

Specimens made from reconstituted clay may be carved from large blocks created in the laboratory by mixing clay powder with water and consolidating the slurry in a tank by the desired pressure. Clay powders for studies of the behavior of reconstituted clay are readily available from various vendors. Alternatively, clay powder may be created by drying of clay lumps and crushing these lumps to form fine powder.

Experiments show that the water content at which the clay slurry is sufficiently liquid and flows easily out of the mixing bowl is two to three times the liquid limit (LL). Often a water content of two times the LL is sufficient. The required quantity of (distilled) water is weighed in a stainless steel bowl and the clay powder, corresponding to a mixture with two times the LL, is weighed in a separate container. The clay powder is then slowly sifted through a No. 10 sieve (to break up any lumps) onto the water surface, where it is wetted and sinks into the water. Sufficient time must be allowed for the clay particles to sink into the water before more clay powder is sifted onto the water surface. After all clay powder has been mixed with the water, additional mixing may be achieved by hand to produce a smooth slurry without any lumps. A vacuum may be applied to the bowl to

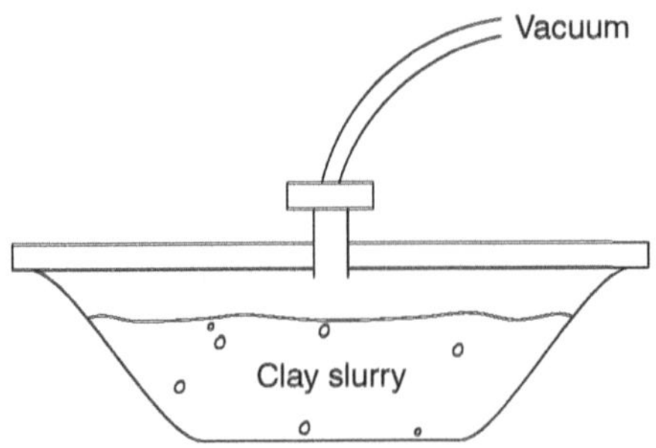

Figure 5.7 Application of vacuum to clay slurry to remove air bubbles.

extract any air that may have entered the water with the clay particles, as shown in Fig. 5.7.

The clay slurry is then poured into a large (double drainage) consolidometer, shown in Fig. 5.8, and consolidated at the desired pressure. The drains may consist of disks of porous plastic that fit inside the cylindrical consolidometer. Filter paper may be placed on the side facing the clay slurry. It is prudent to initially let the clay slurry consolidate under a rather low pressure to let it gain some strength to avoid it flowing past the drains with the filter papers. Following the small strength gain, the pressure can be increased to higher levels. It may be necessary to consolidate the clay in two or three pressure increments to avoid any slurry circumventing the drains.

The design of the large consolidometer shown in Fig. 5.8 is particularly practical, because it is relatively simple to extrude the clay cake from the cylindrical tank by releasing the pressure, removing the top plate and then pressurizing the space below the piston again to slowly move the cake up and slide it off on the lower porous plastic drain. The clay specimen should not be allowed drainage during the unloading phase, because the material will imbibe water and swell thus becoming very soft. At the end of unloading the porous plastic end drains can easily be removed, because they are flexible and can be pried off.

The large consolidation tank may have any dimensions that will produce the desired sample size. The tank shown in Fig. 5.8 produces a cake with dimensions of 30.5 cm (12 in.) in diameter

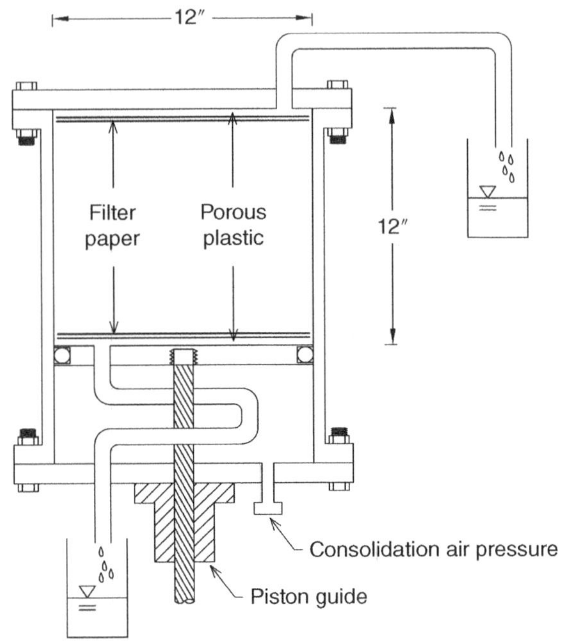

Figure 5.8 Large consolidometer tank for preparation of cylindrical blocks of reconstituted clay.

and 10–18 cm (4–7 in.) in height depending on the LL for the clay.

Sheehan and Krizek (1971) describe the preparation of homogeneous soil samples by slurry consolidation.

Preparation of clay specimens with identifiable fabric

Krizek *et al.* (1975) and Prashant and Penemadu (2007) describe how to prepare specimens with predetermined and identifiable fabric. Techniques are described in which the clay–water system is controlled along with the isotropic or anisotropic consolidation stress path and the magnitude of the consolidation stresses. Anisotropic consolidation tends to induce preferred particle orientation, whereas isotropic consolidation tends to produce basically random orientation of clay particles. Krizek *et al.* (1975) found that anisotropically consolidated samples from dispersed slurries exhibited greater preferred particle orientation than those from flocculated slurries and the particle orientation was enhanced with increasing major principal stress.

5.2.2 Air pluviation of sand

Pluviation of sand appears to best simulate the sedimentation leading to formation of sand deposits in nature, whether in air or in water. The main factors controlling the behavior of sands are the void ratio and the sand structure. Air pluviation is used to create sand specimens directly inside a membrane held on a forming jacket. Specimens reconstituted by air pluviation (AP) may be formed by a sand rainer, as shown in Fig. 5.9. It may consist of, Fig. 5.9(a), a container with a shutter at the bottom, a diffuser screen, and a long tube with an equivalent inside diameter to the split mold or forming jacket that holds the specimen membrane. Alternatively, Fig. 5.9(b), the sand may be dispensed by a spoon to rain slowly (a few grains at a time) through two diffuser screens (No. 4 U.S. sieve), oriented at 45° to each other, and a long tube. This tube should be made of material that does not generate static electricity, because this may influence the uniformity of the raining process. A cardboard tube is preferable. The tube is placed on top of the split mold extension, which rests on the split mold holding the membrane. A variation in drop height is used to create specimens of different densities (Miura and Toki 1982; Vaid and Negussey 1984b, 1988; Rad and Tumay 1987). To form a specimen of uniform density, Vaid and Negussey (1984b) suggested that the fall height should remain constant relative to the top of the specimen as it is formed.

Different sand densities may also be achieved by changing the rate at which the sand is poured through the tube. Shutters with different numbers of holes and different hole sizes may be used to control the deposition rate. As this rate is decreased, the density of the specimen is increased, since the sand grains

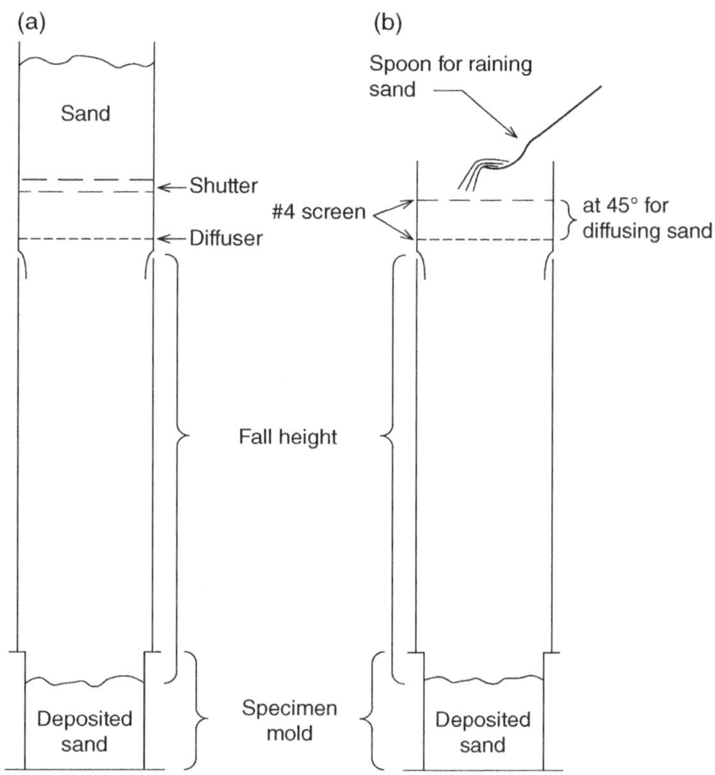

Figure 5.9 Sand rainer setup consisting of (a) sand container, shutter, diffuser, and long tube and (b) spoon for slow dispensing of sand and two #4 screens at 45° to each other.

have adequate time and depositional energy to achieve a dense configuration. Faster pouring rates and lower drop height tend to promote arching of particle structures that lock them into a looser state, thus preventing the sand from reaching higher densities. Miura and Toki (1982), Vaid and Negussey (1984b), Rad and Tumay (1987), and Kuerbis *et al.* (1988) reported similar findings.

While there is interaction between the deposition rate, controlled by the shutter, and the fall height from the lowest screen to the top surface of the sand specimen (typically zero to 70 cm, with up to 200 cm used; Kolbuszewski 1948), the actual void ratios or relative densities created by the sand rainer also depends on the sand itself. Thus, experimentation may be necessary for a particular sand to achieve the desired void ratio.

Following the air pluviation, the specimen is confined by placing the cap on top of the horizontal specimen surface, flipping the membrane up around the cap and sealing it with one or two O-rings. A small vacuum is now applied to the drainage lines to create an effective confining pressure on the specimen, and the forming mold can be removed. The vacuum may be applied through a bubble chamber, which will indicate whether a leak is present in the membrane. Fig. 5.10 shows the setup required for this operation. The bubble chamber consists of a transparent bottle to which the regulated vacuum is applied as indicated. The short line with the vacuum goes through a rubber stopper in the bottle neck, while the line that goes to the specimen dips down in water at the bottom of the bottle. If air leaks through a hole in the membrane, the bubbles in the water will continue until the leak is stopped. If no leak is present, then the bubbles will stop after the air is evacuated from the specimen. The hole in the membrane may be repaired by painting the outside of the membrane with fluid latex rubber. When the repair is finished, the bubble chamber is tilted so water cannot be sucked into the specimen when the vacuum is released.

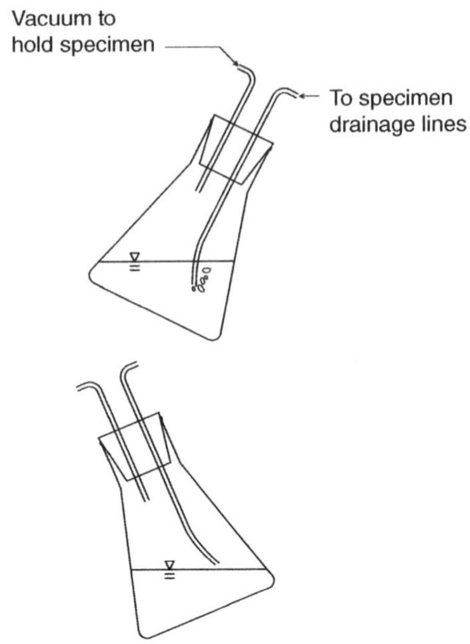

Figure 5.10 "Bubble chamber" with vacuum applied to specimen drawn through water to indicate a leak in the latex rubber membrane. A transparent bottle is used with a little water through which the vacuum can be applied to the dry specimen to provide confining pressure.

Preparation of sand specimens with identifiable fabric

Dry or wet pluviation may be used to prepare specimens with cross-anisotropic fabrics and this may have considerable influence on the stress–strain and strength behavior of sand. Specimens may be prepared by pluviation with strong preferred particle orientation as in natural deposits. These specimens may be temporarily frozen to facilitate their installation and desired orientation in the testing apparatus. Several studies have indicated the importance of cross-anisotropy on the observed behavior, for example Oda (1972a, 1972b, 1981), Oda and Koishikawa (1977), Oda et al. (1978), Yamada and Ishihara (1979), and Ochiai and Lade (1983).

Ochiai and Lade (1983) performed true triaxial tests on cross-anisotropic sand deposits prepared by uniformly graded Cambria sand with particle sizes between No. 10 and No. 20 U.S.

sieves (2.00–0.84 mm) with maximum and minimum void ratios of 0.80 and 0.51, respectively. These grain sizes were chosen so as to be able to determine the three principal dimensions, that is length, width, and height using two microscopes. Results based on 250 particles are shown in Fig. 5.11(a). The results are presented as length to height, L/H, ratios and length to width, L/W, ratios and they indicate that the sand grains were somewhat long and flat. For Cambria sand these ratios vary from 1.4 to 2.0 and they are typical of values for natural sands. While the cross-anisotropic behavior depends on the directions of the sand grain contacts, these are difficult to determine, and they are replaced by the L/H and L/W ratios.

Specimen preparation

Cubical specimens with side lengths of 76 mm were prepared by pouring and shaking sand grains in several layers in the cavity created by the specially designed mold. The saturated specimen was then frozen in the mold, which was designed to avoid any expansion or disturbance of the sand structure during freezing. The four sides and the top plate of the mold consisted of varnished wood and the bottom plate was made of copper. The mold could be completely disassembled to remove the frozen specimen. Because copper has a much higher heat conductivity than wood, freezing of the specimen proceeded upwards from the bottom plate and pushed excess water out through a hole in the top plate. To further enhance the process of freezing from the bottom, the mold was placed on a large, solid piece of aluminum (which also has high heat conductivity) inside the freezer.

A specimen was prepared in the mold in 10 layers. Each layer consisted of approximately 80 g of sand, which was poured into de-aired water in the mold. The mold was then placed on a vibrator and shaken for 1 min by horizontal movements evenly distributed in all directions.

Specimens prepared by this method had void ratios of 0.53–0.54 corresponding to relative densities of 90–93%.

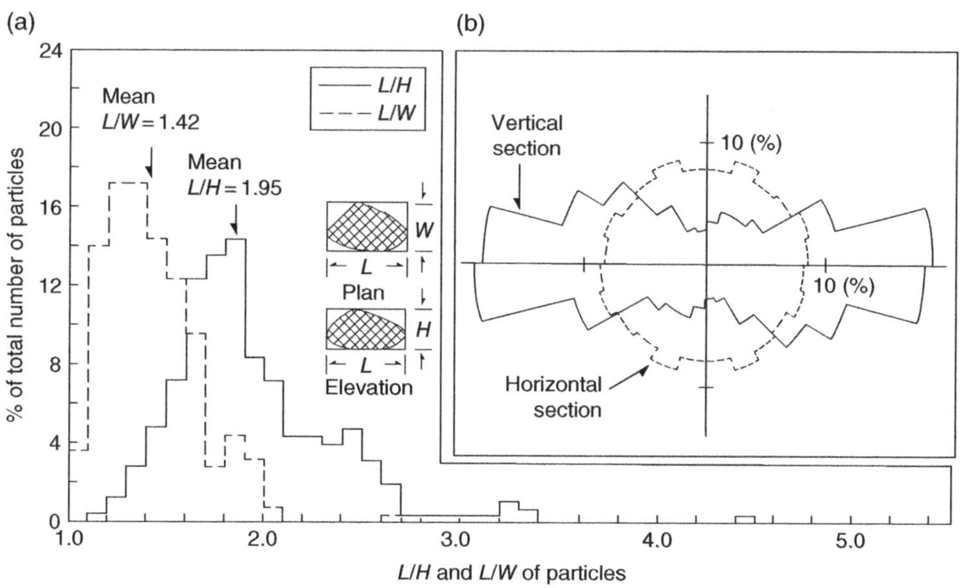

Figure 5.11 (a) Grain shape distributions and (b) Rose diagrams of particle long axis orientations for specimens of Cambria sand. Reproduced from Ochiai and Lade 1983 by permission of ASCE.

Fabric characterization

To examine the fabric of a specimen, photographs were taken of horizontal and vertical sections through central regions of the specimen. This was accomplished by melting part of the frozen specimen. The central region was used to avoid effects of side walls which may locally have influenced the fabric. Measurements of orientation were made on photographic enlargements. The orientations of apparent long axes in horizontal and vertical sections of a specimen are shown on the rose diagram in Fig. 5.11(b). In this study the orientation of each particle was assigned to one of the 15° intervals between 0° and 180°. Figure 5.11(b) shows that the particles in the specimens prepared by the method described above had strong preferred orientations in the vertical section, but almost completely random orientations in the horizontal section.

To compare the intensity of fabric anisotropy with previously obtained intensities, the mean vector direction, α, and vector length, L, defined by Curry (1956) and used by Oda and his co-workers (1972, 1977, 1978), were calculated according to:

$$\bar{\alpha} = \frac{1}{2} \cdot \arctan \frac{\Sigma n \cdot \sin 2\alpha}{\Sigma n \cdot \cos 2\alpha} (°) \tag{5.1}$$

$$L = \frac{100}{\Sigma n} \cdot \sqrt{(\Sigma n \cdot \sin 2\alpha)^2 + (\Sigma n \cdot \cos 2\alpha)^2} \ (\%) \tag{5.2}$$

in which α is the orientation of the apparent long axis relative to a reference axis, and n is the number of particles at α. The value of L varies from 0 to 100%. L = 0% corresponds to completely random orientation of particle long axes, whereas L = 100% corresponds to all long axes having exactly the same direction.

For the vertical section of a specimen of Cambria sand $\bar{\alpha}$ = 3.5°. This means that the apparent long axes in the specimen were preferably parallel to horizontal, which was used as the reference axis. The vector length was calculated to be L = 39.5% for the vertical section. Compared with values of L obtained in previous studies of natural sand deposits by Oda and Koishikawa (1977) and Oda *et al.* (1978), this value of vector length corresponds to a high degree of preferred particle orientation. Corresponding values of $\bar{\alpha}$ and L for the horizontal section of the Cambria sand specimen were 11.9° and 0.9%, respectively. Thus, an almost completely random orientation of particles in the horizontal direction was obtained. The specimen fabric is therefore of the cross-anisotropic type with a vertical axis of rotational symmetry and horizontal planes of isotropy. Ochiai and Lade (1983) explain the testing of such specimens and the results obtained from triaxial compression, plane strain and true triaxial tests on Cambria sand.

Isotropic specimens

Note that the particle contact angles will always favor the vertical direction when the pluviation method is applied, and isotropic specimens cannot be created by pluviation of particles. It is not known how to systematically and repeatedly create isotropic specimens of granular materials.

5.2.3 Depositional techniques for silty sand

Silts have properties between those of clays and those of sands. Clay specimens will stand up due to sustained suction and small menisci at the specimen surface, while sand specimens will fall apart due to lack of sufficiently small menisci. In comparison, silt specimens may not be able to stand alone for an extended period of time, because the menisci at the specimen surface are broken and the suction developed in fine-grained soils is lost. It is therefore most prudent to form silt specimens inside a membrane held on a forming jacket as done for sand specimens.

Silty sands behave differently depending on the amount of silt and the location of the silt particles, either between and separating the coarser sand grains or in the voids formed by the coarser sand grains. Different deposition methods have been developed to try to capture the structure created in silty sands that are prone to liquefaction.

Specimens may be reconstituted by several different techniques: dry funnel deposition (DFD); water sedimentation (WS); slurry deposition (SD); mixed dry deposition (MDD); and air pluviation (AP). Specimens formed using wet depositional methods (water sedimentation and slurry deposition) already are saturated prior to assembly of the triaxial cell. However, specimens formed from an initially dry state are first flushed with gaseous CO_2 for 30 min prior to water saturation, which is then achieved by slowly percolating de-aired water through the specimen from the bottom drain. Past research has shown that significant volume changes may occur during the saturation stage of specimen preparation, particularly for loose silty sands (Sladen and Handford 1987). Settlement during saturation may be measured by monitoring the change in volume of the cell water using a calibrated buret. During this stage, the change in height of the specimen is measured using a dial gauge. A back pressure may be used to ensure full saturation and to prevent cavitation of the pore water during undrained shearing.

Dry funnel deposition

Specimens prepared using DFD are formed by initially placing the spout of a funnel on the bottom of a split mold. The sand–silt mixture is placed into the funnel, which is then slowly raised along the specimen axis of symmetry as shown in Fig. 5.12(a). This allows the sand to be deposited in a low-energy state without any drop height. This technique is commonly used for testing silty sands (Ishihara 1993; Lade and Yamamuro 1997; Yamamuro and Lade 1997; Zlatovic and Ishihara 1997; Yamamuro and Covert 2001). To achieve higher densities, the split mold is gently tapped in a symmetrical pattern. This method of creating specimens is referred to as tapped funnel deposition (TFD). Denser specimens may also be prepared by raising the funnel more quickly (although still without a drop height) prior to tapping. This reduces the tapping required to achieve a desired density. This technique is referred to as fast funnel deposition (FFD).

Water sedimentation

Although some WS techniques involve raining dry soil through water (Tatsuoka *et al.* 1986; Ishihara 1993; Zlatovic and Ishihara 1997), it has been experimentally determined that saturation is best achieved by depositing the soil in a fully saturated state. Backpressure greater than 100 kPa to fully saturate the specimen may be undesirable due to the increased effects of piston uplift at low effective confining pressures (25 kPa). Therefore, the specimen is placed in a 2000 ml volumetric flask half filled with water, boiled for approximately 30 min, and then the remainder of the flask is filled with de-aired water. The completely saturated mixture is allowed to cool overnight.

Prior to placement of the sand specimen, the split mold and drain lines are filled with de-aired water. The flask is capped with a thin foam plastic disk coated with latex rubber and then repeatedly rotated to evenly mix the sand sample. The flask is then inverted and lowered to the bottom of the split mold, at which point the disk is removed. Similar to DFD, the sand specimen is formed by slowly raising the flask along its axis of symmetry, allowing for zero fall height of the sand, as shown in Fig. 5.12(b). As the soil flows slowly out of the flask into the split mold, an equal volume of water from the mold is pulled back into the flask. This is caused by the suction created within the flask as the sand empties from it. After deposition is completed, the disk is carefully placed back over the mouth of the flask, which is then removed. Any soil left in the flask after deposition is dried and weighed to determine the exact specimen weight. Similar water sedimentation techniques have been used by Lee and Seed (1967), Finn *et al.* (1971), Mulilis *et al.* (1977), Vaid and Negussey (1984b), and Vaid and Thomas (1995).

While the drop height and therefore the terminal velocity of the individual sand grains to some degree controls the void ratio produced in air pluviated sand, the terminal velocity of sand grains in water is reached in about 0.2 cm (Vaid and Negussey 1988) and control of drop height

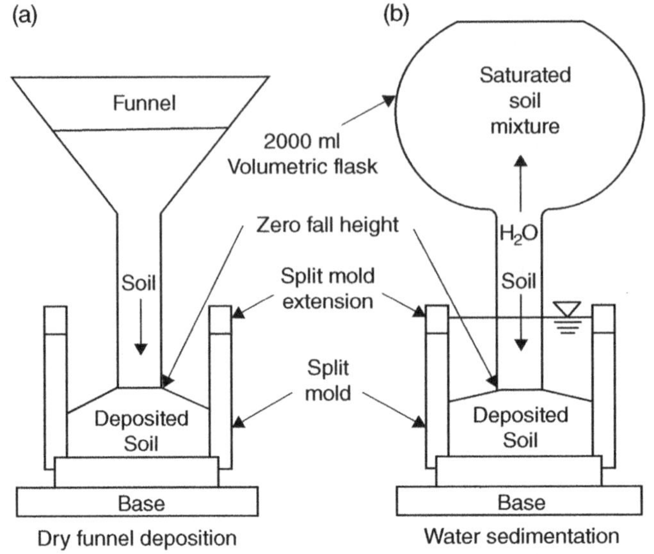

Figure 5.12 Schematic diagram showing: (a) DFD or TFD; and (b) WS. Reproduced from Wood *et al.* 2008 by permission of Canadian Science Publishing.

in water pluviation is therefore ineffective in obtaining different void ratios.

Most of these studies used vibration as a means to densify the specimens while the specimen was still in the mold. However, with silty sands this method of densification has been observed to trigger clearly visible small sand boils, that is liquefaction (Wood *et al.* 2008).

Slurry deposition

The SD method is similar to that presented by Kuerbis and Vaid (1988). The method was modified by Wood *et al.* (2008) to accommodate a different sized specimen, as well as lubricated and enlarged end platens. To ensure full saturation the sand sample is boiled the night before testing, similar to the WS technique. Rather than depositing the soil directly into the split mold, it is first placed into a mixing container. The mixing container, shown in Fig. 5.13(a), consists of a galvanized steel pipe 101 mm in diameter and 220 mm in length that has been lathed down to a 1 mm wall thickness. A 0.3 mm thick latex rubber membrane seals the bottom end and is held in place by an O-ring. Through-bolts hold the tube in place between two aluminum disks,

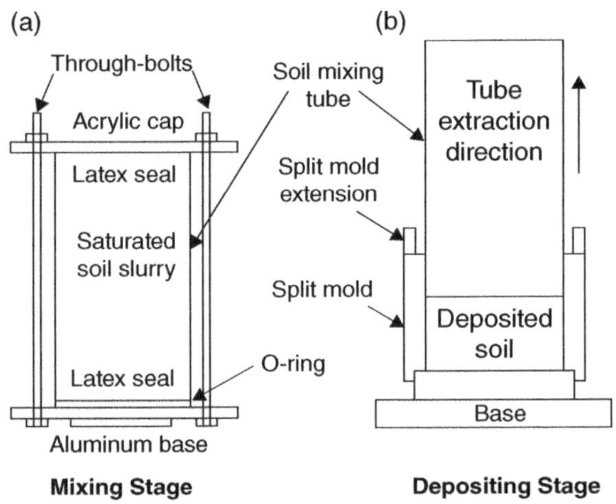

Figure 5.13 Schematic diagram showing the SD method. Reproduced from Wood *et al.* 2008 by permission of Canadian Science Publishing.

compressing the latex rubber and sealing the bottom. After the mixing container is filled with de-aired water, the flask containing the boiled soil mixture is inverted and lowered through a hole in the center of the top disk. The cap of the flask is removed and the water in the mixing container is exchanged with the soil in the flask, similar to the WS method. When all of the soil is deposited, the top disk is replaced with a solid acrylic disk affixed with another latex rubber membrane to provide a watertight seal.

The mixing container is subjected to a series of slow rotations to create a uniform distribution of the base sand and silt. After sufficient mixing of the soil (generally about 1 h), the top disk and through-bolts are removed. The triaxial cell base assembly, including the split mold, is then inverted and lowered onto the mixing container. The final step in the mixing stage involves rotating the triaxial cell base assembly (now containing the mixing container) back to its upright position. After a mold extension is placed on the split mold, the mixer is removed slowly, leaving the soil specimen in the split mold, as shown in Fig. 5.13(b). The top platen of the specimen is then attached. Care should be taken to ensure that all of the drain lines are saturated prior to assembly to obtain full saturation and an acceptable *B*-value (at least 0.99 in all cases of loose silty sand).

Mixed dry deposition

Mixed dry deposition is very similar to the SD method, except that the soil is deposited in a dry state. A funnel is used to deposit the dry silty sand into the soil mixing container. After a period of mixing (about 1 h) the triaxial cell base assembly, which includes the split mold, is inverted and placed over the mixing tube. The final rotation of the mixing stage involves returning the triaxial base to its upright position. The mixing tube is carefully extracted, and the specimen is then saturated exactly as in the DFD method.

Moist tamping

Moist tamping has been used in many studies (Ladd 1974; Mulilis *et al.* 1977; Tatsuoka *et al.* 1986; Sladen and Handford 1987; Ibrahim and Kagawa 1991; Ishihara 1993; Pitman *et al.* 1994; Zlatovic and Ishihara 1997; Jang and Frost 1998). Its ability to produce very loose specimens generally is considered beneficial, particularly in liquefaction testing, where contractive specimens are desirable. However, Casagrande (1975) noticed that sand formed in a moist state forms a "bulked, honeycomb structure" that is likely to liquefy upon saturation. Vaid and Thomas (1995) discouraged the modeling of loose water-deposited sands by moist tamping, recognizing that the results appear to produce an unrealistically volumetrically contractive response. Jang and Frost (1998) observed that specimens created using moist tamping were less homogeneous than air pluviated specimens.

The method of rodding is a variation of moist tamping in which a flat-ended rod is used to compact the specimen. Thus, this is not different from the moist tamping technique described above. If rodding is employed with soft particles, such as found in calcareous sand, caution should be exercised to prevent undue breakage of conglomerate particles.

Comparison of depositional techniques for silty sands

Careful evaluations of the depositional techniques were performed in terms of the uniformity of the silty sand specimens and their undrained behavior by Wood *et al.* (2008) and in terms of the microstructures that were achieved by Yamamuro *et al.* (2008). Table 5.2 shows the silt distribution in specimens of non-plastic silt and Nevada 50/200 sand created by five different deposition methods. The silt used by Wood *et al.* (2008) was a darker color than the base sand, and photographs of partially saturated specimens that retained their shapes after disassembly indicated the distributions of silt and sand. The purpose of this analysis was to investigate the extent to which silt content could vary within different regions of the specimens formed by different depositional methods. Table 5.2 shows the silt distributions obtained from the sieve analyses. The inset diagrams in Table 5.2 show typical visual patterns of silt and sand distribution that were observed for each method. The observed trends were repeatable, although the magnitudes varied slightly. However, the same patterns generally were observed, regardless of specimen density.

Table 5.2 Silt distribution in specimens of non-plastic silt and Nevada 50/200 sand created by five different deposition methods. Reproduced from Wood *et al.* 2008 by permission of Canadian Science Publishing

Deposition method	Total silt content (%)	Specimen region	Silt distribution (%)	Visual observations
Tapped funnel deposition (TFD)	18.4	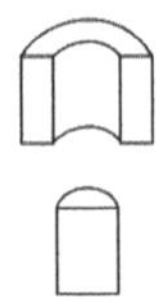	Shell: 18.9; core: 17.7	Less silt in core region
Water sedimentation (WS)	18.4	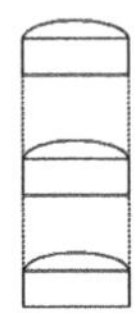	Top: 21.8; middle: 17.8; bottom: 16.3	Silt bands visible near edges of specimen
Slurry deposition (SD)	20.0	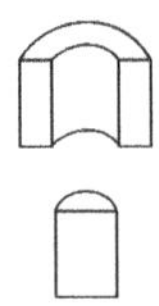	Shell: 19.3; core: 20.1	Silt appears evenly distributed
Mixed dry deposition (MDD)	20.0	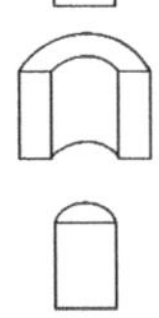	Shell: 19.3; core: 21.0	Silt appears evenly distributed
Air pluviation (AP)	9.6	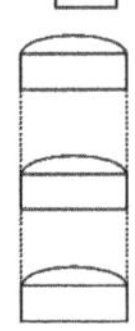	Top: 9.5; middle: 10.0; bottom: 9.2	Slightly inclined layers across specimen

Dry funnel deposition

Table 5.2 indicates that TFD specimens do not show distinct layering between dark and light areas. However, the core region of the specimens appeared slightly lighter, as shown in the inset diagram, suggesting that more silt resided toward the outer region (shell) of the specimen. A small degree of anisotropy was also evident in the form of very thin layers inclined toward the central vertical axis. This pattern appeared to correspond with the direction of sand flowing away from the bottom of the funnel during deposition and it was approximately inclined at the angle of repose. The grain size distribution suggests that the base gradation remained constant between the core and shell of the specimen. Sieve analysis detected slightly more silt within the outer region of the specimen.

Water sedimentation

The WS specimens tended to exhibit distinct layering within their outer regions. This variation did not appear to continue through the entire diameter of the specimen. Visual examination did not suggest that there was a vertical variation of silt content within the specimen. However, sieve analyses consistently revealed that the silt content gradually increased from the bottom of the specimen to its top. The bottom third of the specimen contained 16.3% silt, while the top third contained 21.8% silt. When the soil is deposited into the split mold, the larger grains fall faster through the water, even though the soil is thoroughly mixed in the flask. Due to the upward flow of water from the split mold into the flask during soil deposition, some silt grains deposited within the vicinity of the mouth of the flask were carried back inside. The grain size distribution curves suggest that this effect was only observed with the silt particles and not the base sand.

Slurry deposition

The grain size distribution of the base sand within each section appeared similar. However, the core region contained slightly more silt. Sieve analysis performed by dividing the specimen into three horizontal layers produced almost identical grain size distributions. Slurry deposition specimens appeared to be visually homogeneous when thoroughly mixed.

Mixed dry deposition

The results for MDD specimens are similar to those with SD specimens. They indicate that the core region of the MDD specimen also contained slightly more silt. Particle segregation between the base sand and silt was not visually observed provided that the sand sample was thoroughly mixed prior to deposition. However, there did appear to be a small degree of variation between the grain size distributions of the base sand obtained from the inner and outer regions of the specimen. The mixing method may deposit coarser-grained particles toward the outside of the specimen in slightly greater proportions.

Air pluviation

The gradation of the base sand generally did not vary within AP specimens. No pattern was observed with regard to silt content variation between layers. Very thin layering was observed in each AP specimen. The layering was generally inclined at a small angle across the specimen. There were no visually observed areas of great contrast as with WS specimens. The observed layers within AP specimens spanned across the entire diameter, unlike those observed in WS specimens.

Overview of volumetric behavior trends for silty sands

An overview of the trends in silty sand structure, stability and volumetric behavior as influenced by the method of reconstitution and the silt content is given in Fig. 5.14. The consequent particle structure, as indicated in the inset diagram, had a decreasing effect on behavior with increasing specimen density. Further discussions of these effects are given by Wood *et al.* (2008) and Yamamuro *et al.* (2008).

5.2.4 Undercompaction

Undercompaction is a method of static compaction devised by Ladd (1978) to reduce soil grain segregation and to obtain specimens with relatively uniform density. The method was developed for reconstitution of triaxial specimens of sand for cyclic strength testing, but it may also be used for static testing of other specimens and soil materials. The method was devised to overcome the nonuniformity produced by compaction in layers in which the compaction of each succeeding layer can further densify the sand below it. Thus, each layer is typically compacted to a lower density than the final desired value by a predetermined amount, which is defined as percent undercompaction U_n. The U_n-value in each layer is varied linearly from the bottom to the top layer. The bottom (first) layer has the maximum U_n-value. Figure 5.15 shows the variation of the percent undercompaction with the layer number. The percent undercompaction in each layer is calculated from:

$$U_n = U_{ni} - \left[\frac{(U_{ni} - U_{nt})}{n_t - 1} \cdot (n-1) \right] \tag{5.3}$$

in which U_{ni} is the percent undercompaction selected for the first layer and U_{nt} is the percent undercompaction for the final (top) layer (usually zero).

The sand specimen is prepared in an internal split mold inside a membrane attached to the base of the triaxial setup, as shown in Fig. 5.16.

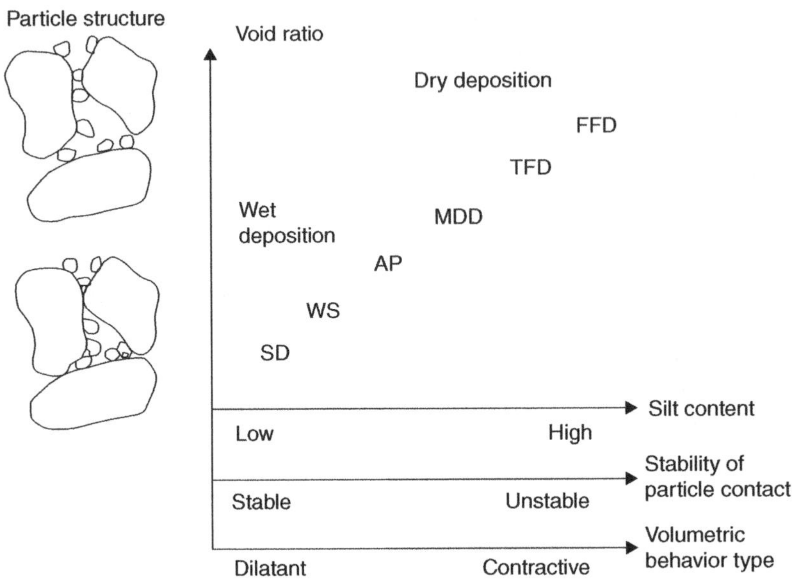

Figure 5.14 Overview of trends in silty sand structure, stability and volumetric behavior as influenced by the method of reconstitution and the silt content. Reproduced from Wood *et al.* 2008 by permission of Canadian Science Publishing.

Alternatively, it may be prepared directly in an external mold as shown in Fig. 5.17. In the latter case, the external split mold is easier to use than a mold from which the specimen needs to be extruded. However, many soils with fines may have sufficient strength due to capillary forces to be extruded and set up on the triaxial base without any significant disturbance to the soil structure.

Procedure

The following procedure for undercompaction is provided by Ladd (1978):

1. Adjust the water content of the air-dried material so that the initial degree of saturation of the compacted material will be between 20% and 70%. Oven-drying the material is not recommended. The lower the percentage of fines in the material, the lower the degree of saturation required. A degree of saturation greater than 70% can be used if water does not bleed from the specimen during compaction. The material should be mixed with water about 16 h before use.
2. Determine the average water content of the prepared material using a minimum of two determinations.
3. Assemble and check all necessary equipment to be used in preparing the test specimen. Determine the inside diameter and the height of the mold to within ±0.02 mm (±0.001 in.) and calculate the volume based on these measurements. If an internal split mold is used, correct the diameter measurement for the average thickness of the rubber membrane.
4. Select the number of layers to be used in the preparation of the specimen. The maximum thickness of the layers should not exceed 25 mm (1 in.) for specimens having diameters less than 102 mm (4 in.). Typically, the required number of layers increases as the required dry unit weight increases. Layers having a thickness of about 12 mm (0.5 in.) are recommended.
5. Determine the total wet weight of material required for sample preparation:

$$W_T = \gamma_d \cdot (1 + w) \cdot V_{spec} \tag{5.4}$$

in which γ_d is the dry unit weight, w is the water content, and V_{spec} is the volume of the specimen.

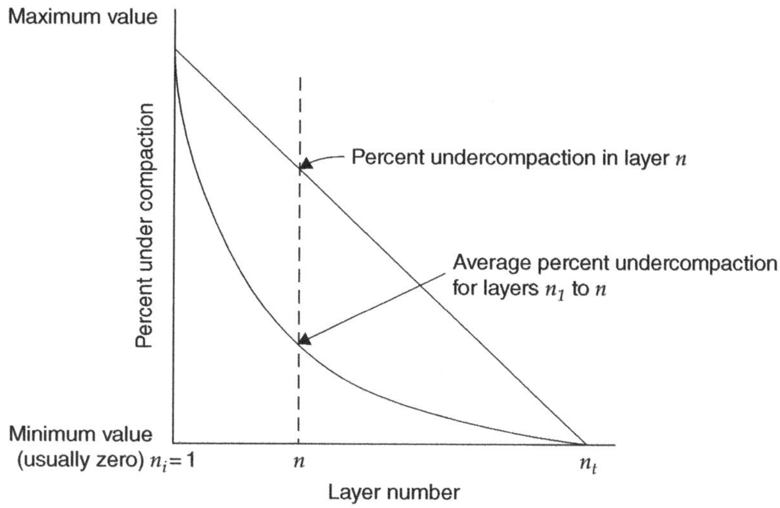

Where: A. Percent undercompaction in layer being considered, U_n

$$U_n = U_{ni} - \left[\frac{(U_{ni} - U_{nt})}{n_t - 1} \times (n-1)\right]$$

B. Average percent undercompaction for layers compacted, $\bar{U}_n$

$$\bar{U}_n = \frac{U_n}{n}$$

U_{ni} = Percent undercompaction selected for first layer

U_{nt} = Percent undercompaction selected for first layer (usually zero)

n = Number of layer being considered

n_i = First (initial) layer

n_t = Total number of layers (final layer)

Figure 5.15 Variation of the percent undercompaction with the layer number in undercompaction procedure. Reproduced from Ladd 1978 by permission of ASTM International.

6. Determine the moist weight of material required for each layer:

$$W_L = W_T / n_t \qquad (5.5)$$

in which W_T is the total weight and n_t is the total number of layers in the specimen.

7. For the first layer to be compacted, select a value of U_{ni}. Typically, this value ranges between zero for the preparation of dense specimens to about 15% for the preparation of very loose specimens. For the preparation of very dense specimens, it has been found that negative values are sometimes required. Each subsequent layer receives a lesser percentage of undercompaction, conforming to the relationship in Fig. 5.15.

The correct (optimum) value of percent undercompaction may be determined experimentally by one of the following methods:

a) Run a series of cyclic triaxial strength tests with the same effective consolidation stress and applied stress ratio, but with different U_{ni}-values, to determine the optimum value.
b) Observe the behavior of the specimen during cyclic loading. Excessive necking or bulging in a layer or layers, either at the top or at the bottom of the specimen, indicates a specimen with inappropriate value of U_{ni}.
c) Observe the behavior of the specimen during unconsolidated-undrained loading. Non-uniform vertical strains indicate an inappropriate value of U_{ni}.

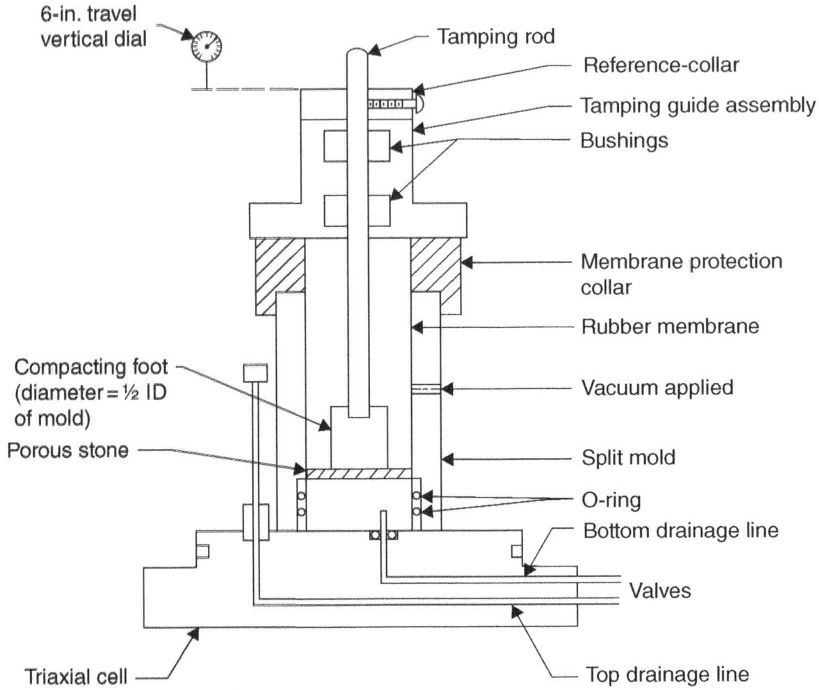

Figure 5.16 Preparation of sand specimen in an internal split mold inside a membrane attached to the base of the triaxial setup. Reproduced from Ladd 1978 by permission of ASTM International.

d) Observe the fabric of the specimen. A honeycomb structure at either the top or the bottom of the specimen indicates an inappropriate value of U_{ni}.

e) Measure the dry unit weight of the prepared test specimen as a function of its height. A dry unit weight not uniform with height indicates an inappropriate value of U_{ni}.

8. Calculate the required height of the specimen at the top of the nth layer:

$$h_n = \frac{h_t}{n_t} \cdot \left[(n-1) + \left(1 + \frac{U_n}{100}\right) \right] \tag{5.6}$$

in which h_t is the total height of the specimen, n_t is the total number of layers, n is the number of the layer being considered, and U_n is the percent undercompaction for the layer being considered.

9. Weigh the amount of material required for the layer, as determined in step 6, and place it into a closed container. If each layer requires a weight greater than about 80 g, it is usually easier to weigh the amount of material required for each layer and place it into small containers.

10. Adjust the reference collar on the tamping rod to obtain the proper h_n. Weigh (if not done before) the amount of material required for the layer, and place it into the mold. During weighing, care must be taken to lose as little moisture as possible. Using the tamping rod, guided by the tamping rod assembly, compact the surface of the material (after it has been leveled) in a circular pattern starting at the periphery of the mold and working toward the center of the mold. Initially, a light tamping force should be used to distribute and seat the material uniformly in the mold. The force should then be gradually increased until the reference collar uniformly hits the top of the tamping rod guide assembly. For the last few coverages,

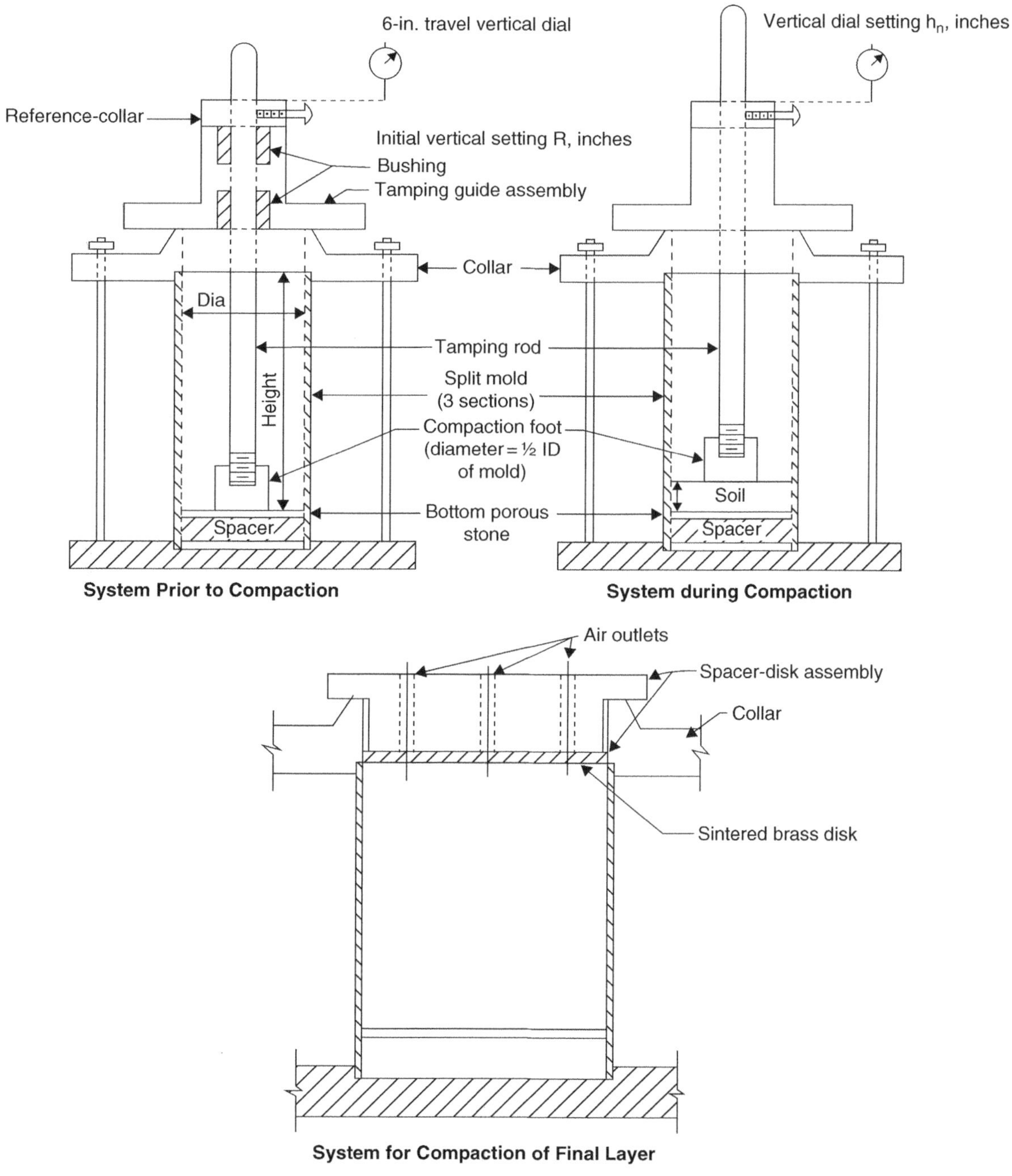

Figure 5.17 Alternative method in which a specimen is prepared directly in an external mold. Reproduced from Ladd 1978 by permission of ASTM International.

it may be necessary to hit the tamping rod with a rubber mallet to compact the material into a dense state. Next, scarify the compacted surface to a depth equal to about one-tenth of the thickness of the layer.

11. Repeat steps 9 (if required) and 10 until the last layer is in place. During compaction of the last layer, the tamping rod should be used until the surface of the compacted material is about 0.4 mm (1/64 in.) higher than required. Then, for specimens prepared in an external split mold, place the spacer disk assembly into position and lightly strike it with a rubber mallet until it is seated. For specimens prepared with the internal split mold, place the top cap and the porous stone directly on the specimen. The top cap should be attached to the loading piston, which, in turn, should be guided by the bushing located in the top of the triaxial cell. Then lightly strike the loading piston with a rubber mallet until the compacted material reaches the prescribed height. The procedure ensures that there is proper alignment and seating at the top cap in relation to the specimen and the loading mechanism of the triaxial cell.

5.2.5 Compaction of clayey soils

Compaction of clayey soils is performed in the field to increase the dry density, increase the strength, decrease the compressibility, and to control the permeability and the volume change due to swelling and shrinkage. These properties are in turn controlled by the structure of the clay particles, flocculated or dispersed, which are influenced by the interparticle forces. The clay structure varies with the compaction water content, compaction effort and compaction method, as indicated in Fig. 5.18. Three main methods of compaction may be employed: static, vibratory, and kneading. There is little difference in strength on the dry side, but on the wet side of the optimum water content and at the same water content, they produce increasing strength in the sequence of kneading, vibratory and static compaction, as shown in Fig. 5.19. A comprehensive review of the behavior of compacted clayey soils was presented by Seed *et al.* (1960).

Determination of the dry density and the optimum water content for use in the field may take the form of specifying the relative compaction based on ASTM D698 (2014) Standard Test Methods for Laboratory Compaction Characteristics of Soil Using Standard Effort [12 400 ft-lb/ft^3 (600 kN-m/m^3)], or ASTM D1557 (2014) Standard Test Methods for Laboratory Compaction Characteristics of Soil Using Modified Effort [56 000 ft-lb/ft^3 (2700 kN-m/m^3)] or other preferred procedures from the relevant literature and organizations.

Soil preparation

Soils that are used for compaction are first air-dried and crushed and then sieved through a No. 4 U.S. sieve, unless special circumstances warrant the presence of the larger particles. A sufficient amount of air-dry soil is then weighed off for the required number of specimens, in addition to an allowance for water content determination and specimen trimming waste. The air-dry soil is then mixed with the appropriate amount of water using a spray bottle for easy distribution to produce the desired water content for the test specimens. The wet soil sample is mixed for 5 min and stored in a polyethylene bag and cured for 24 h prior to compaction. Prior to specimen compaction, the wet soil is again thoroughly mixed by hand.

Static compaction

The soil structure in specimens that are statically compacted simulate the soil structure produced in the field by smooth steel drums or rubber tire rollers. Static compaction is performed by placing an amount of wet soil in a specimen mold

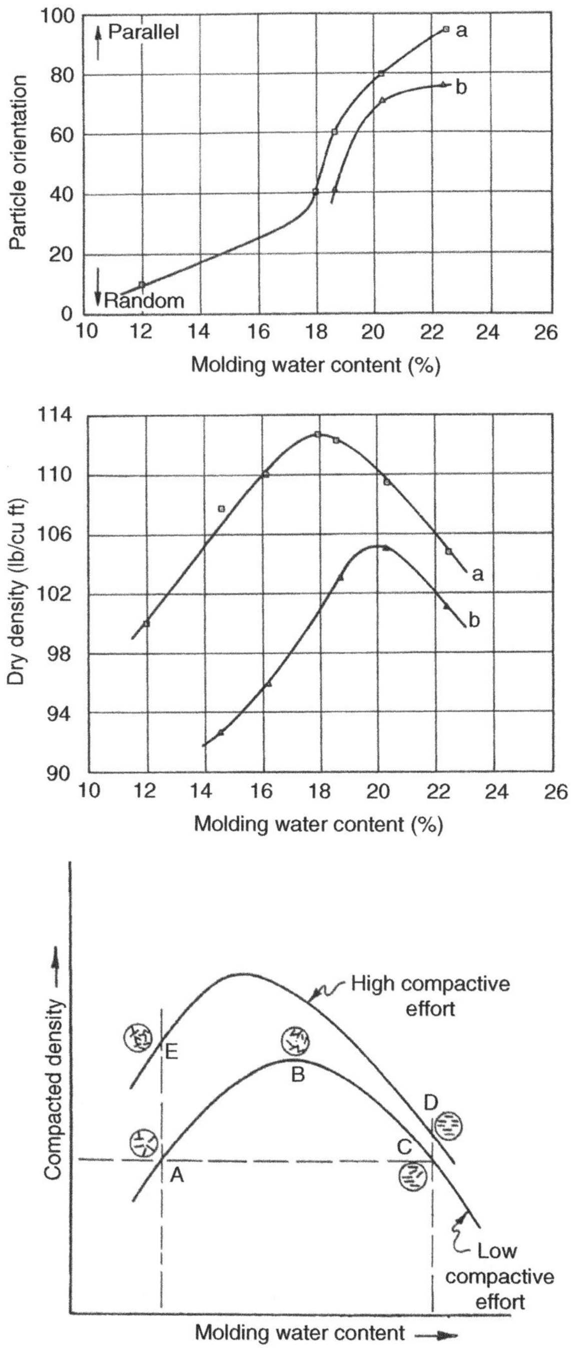

Figure 5.18 Influence of molding water content on particle orientation for compacted samples of Boston Blue clay. Reproduced from Seed *et al.* 1960 by permission of ASCE.

and compressing it between two pistons to the desired static pressure. The cylindrical mold may consist of a length of steel tube with the diameter and length of the desired specimen. A mold with an extension collar may be required to contain the soil. The compacted soil specimen is then extruded from the cylindrical mold. Alternatively a split mold may be used for ease of removal of the compacted specimen. The soil may be compressed in one layer or in several layers. If more than one layer is used, the soil interface should be scarified before addition of the following layer of soil. The ends of the specimen are carefully trimmed to produce square and smooth ends while still in the compaction mold.

Vibratory compaction

Vibratory compaction results in structure and properties between those from the static compaction and kneading compaction, as seen in Fig. 5.19.

Kneading compaction

Kneading compaction simulates the compaction obtained by a sheepsfoot roller in which large shear strains are imparted to the soil and a dispersed structure is produced. Kneading compaction is achieved in the laboratory by compacting the soil by for example a Harvard miniature compaction tamper. This tamper consists of an adjustable spring-loaded circular piston with a diameter of approximately 0.5 in. Alternatively, a pressurized piston with a similar diameter that outputs a constant force upon release from the end against which it is pressurized may be employed. Several layers are used for preparation of the soil specimen. The output force at the end of the piston and the number of tamps per layer and the number of layers in the specimen are adjusted according to the desired compactive effort. The compactive effort may be calculated as the compactive force per tamp multiplied by the number of tamps per layer multiplied

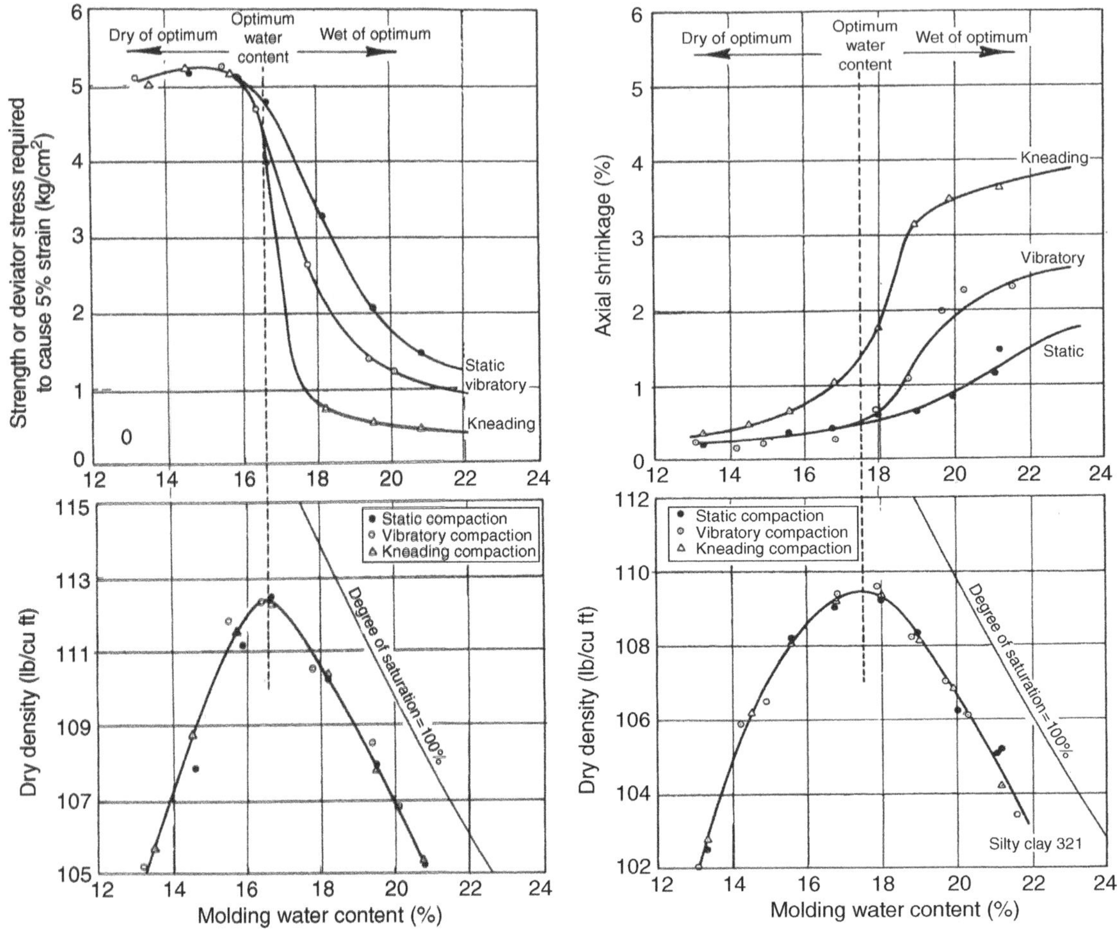

Figure 5.19 Influence of method of compaction (kneading, vibratory, and static) on strength and shrinkage of silty clay. Reproduced from Seed *et al.* 1960 by permission of ASCE.

by the number of layers in the specimen divided by the volume of the specimen. Thus, the compactive effort is measured in force per volume (lbf/ft^3 or kN/m^3).

5.2.6 Compaction of soils with oversize particles

Soils with oversize particles such as rockfill cannot be compacted in the conventional mold used for the standard or the modified compaction efforts, because the mold is too small for the larger particles. A compaction procedure for rockfill was proposed by Rolston and Lade (2009) on the basis of a method devised for determination of the maximum density of a mixture of soil, gravel, cobbles, and boulders. The procedure is straightforward and involves compaction of rocks (gravel, cobbles, and boulders) separately and compaction of the soil fraction in the usual manner. Only one compaction test is necessary for the rocks, because the minimum void ratio for rocks is insensitive to water content. Based on the maximum dry density of the soil and the minimum void ratio of the rocks, the maximum density of the rockfill (consisting of soil and rocks) is calculated according to two different formulas depending on the relative contents of soil and rocks. The method is relatively insensitive to the grain size used for splitting the rockfill into soil and rocks portions, assuming the rockfill follows the "maximum density grading curve" or is shallower, that is is more well-graded than the "maximum density grading curve." This ensures that the smaller particles fit in the voids between the larger particles. Compaction density of soil with considerable amounts of oversize material can then be determined mathematically by calculating the volume of oversize material on the basis of its weight and specific gravity and then include it with the compaction density for the soil alone.

If the grain size distribution curve is shallower than that for the "maximum density grading curve" (Fuller and Thompson 1907), which is often the case for rockfill, then the distinction between soil particles and rock particles is relatively unimportant, because the smaller particles fit in the voids between the larger particles.

Mathematical formulas were given by Rolston and Lade (2009) to calculate the maximum dry density of the rockfill. These formulas depend on the relative amounts of soil and rocks in the total sample.

5.2.7 Extrusion and storage

Following soil preparation, the specimen may be extruded from the mold in which it was compacted and stored under conditions similar to those used for intact specimens.

5.2.8 Effects of specimen aging

Due to thixotropic action after compaction of clayey specimens, they may gain strength with time (Mitchell 1960). To simulate this aging effect under conditions of minimal interference from the outside, the specimens may be enclosed between the cap and base and surrounded by a rubber membrane sealed to the end plates with O-rings, which keeps the interaction between outside and inside water and air to a minimum. These specimens may be placed in a water or oil bath that maintains the same temperature and prevents any interaction with the surroundings. After appropriate time in this condition, the specimens may be taken out and tested to find the effects of thixotropy on the behavior of the compacted clayey soil.

5.3 Measurement of specimen dimensions

5.3.1 Compacted specimens

For specimens compacted in an external split mold, the specimen should be removed from the split mold (using extreme caution to prevent disturbance) to obtain its weight, height, and diameter. The weight should be determined to the nearest 0.01 g. However, for specimens weighing more than 1000 g, measuring to the nearest 0.1 g is adequate. The height and diameter should be determined to the nearest 0.02 mm (0.001 in.) using a dial comparator. The dial gage contact points on these instruments should have a flat surface with a minimum diameter of about 5 mm (1/4 in.).

For specimens compacted in an internal split mold, the initial weight cannot be directly checked. Therefore, the oven-dry weight of the specimen should be checked after the test. However, the height and diameter of the compacted specimen should be measured after a slight vacuum is applied and the mold is removed. A Pi Tape is recommended for measuring the diameter.

5.4 Specimen installation

Irrespective of the method used to prepare a soil specimen for triaxial testing outside the triaxial cell, by trimming of intact samples, by compaction of clayey soil, or other method that allows the specimen to be free standing without collapse, the specimen is now ready to be installed in the triaxial apparatus. Other specimens prepared as reviewed above are made directly on the base of the triaxial cell, and need no further preparation detail.

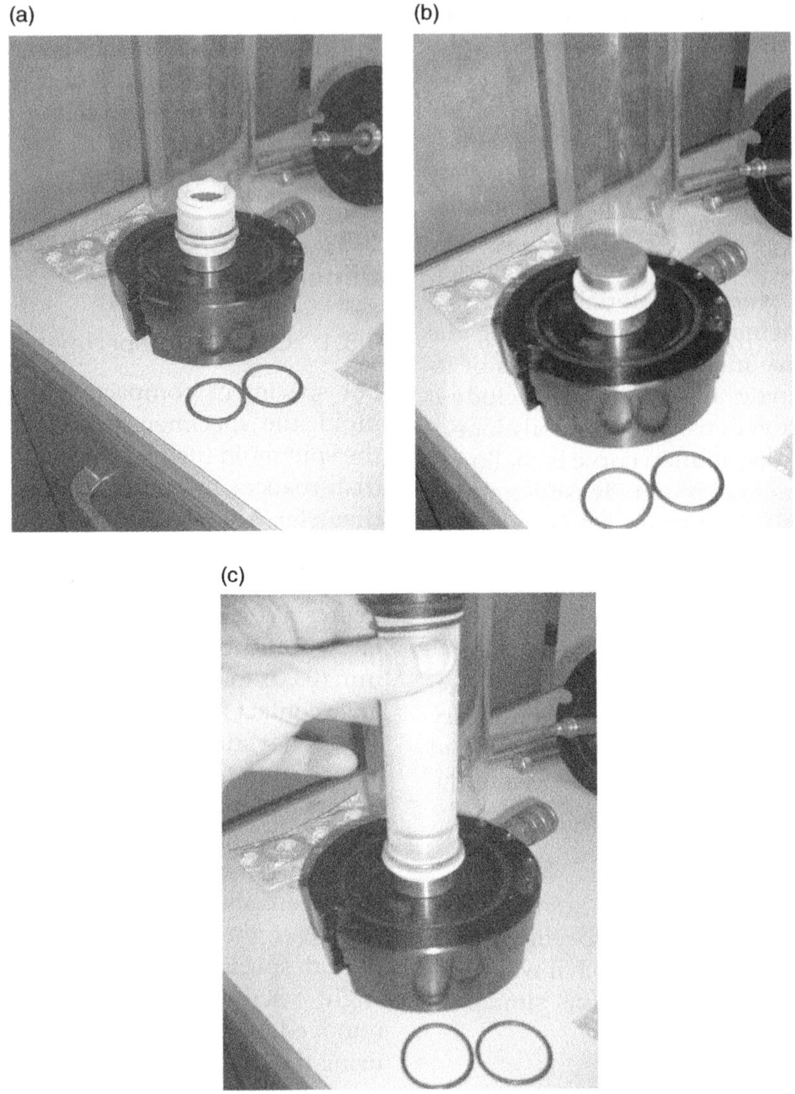

Figure 5.20 (a, b) Prophylactics placed around base and sealed with O-rings and (c) rolled up around the specimen and sealed with O-rings around the cap.

5.4.1 Fully saturated clay specimen

For installation of a fully saturated clay specimen prepared by trimming, the base should be flushed with de-aired water before the specimen is placed directly on the base plate or the filter stone to avoid trapping air between the base and the specimen. Thus, water is pushed out from under the specimen during the installation. This is followed by positioning of the rubber membrane(s), either by a membrane stretcher reaching over both ends of the specimen assembly, or by rolling the thin membranes (prophylactics) initially positioned around the base, as shown in Fig. 5.20(a, b), and Fig.5.20(c),

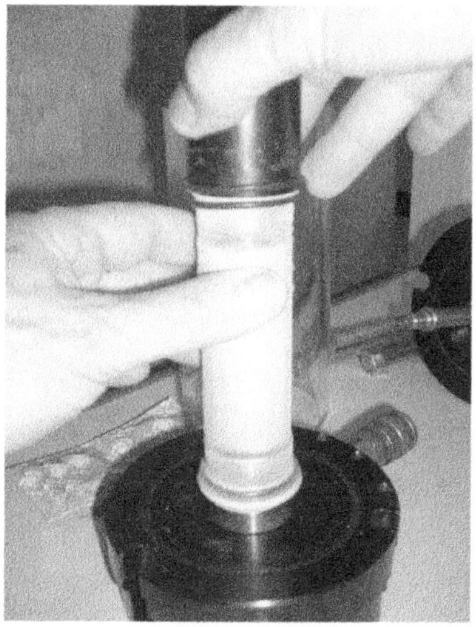

Figure 5.21 O-ring being placed on the membrane around the cap by flipping it off the O-ring stretcher by using the ring finger.

rolled up around the specimen. Because these soil specimens are sensitive to disturbance due to positioning of the upper O-rings, these are positioned with the help of an O-ring stretcher. This is a tube that fits over the diameter of the cap. The O-rings are flipped off onto the cap by the ring finger on one hand while using the fingers on the other hand to prevent the O-rings from falling down on the soil specimen, as shown in Fig. 5.21.

5.4.2 Unsaturated clayey soil specimen

The unsaturated clayey soil specimen is placed on the high air entry filter stone and the membrane is positioned around the specimen, as described above. The connection between the water in the specimen and the water in the filter stone is checked before testing is initiated.

6 Specimen Saturation

6.1 Reasons for saturation

Saturation of soil specimens is necessary to provide reliable measurements of pore water pressure and volume change. In an undrained triaxial test, the condition of no volume change may be simulated if the specimen is completely saturated. In a drained triaxial test, the volume change of the specimen may be determined by measuring the amount of water entering or leaving the saturated soil. The importance of complete saturation is most pronounced for undrained tests in which the development of pore water pressure depends strongly on the stiffness of the pore fluid.

There are basically two categories of field conditions for which duplication of saturated soil conditions in the laboratory are important:

1. Undisturbed natural soil deposits below the ground water table. This includes soil at the ocean bottom where the pore pressure is high.
2. Compacted or natural soils that have or may become saturated by (a) percolation or (b) compression due to overburden pressure.

Soil specimens from these two categories require testing under fully saturated conditions.

6.2 Reasons for lack of full saturation

Often soil specimens are not completely saturated immediately before testing. Some of the reasons for lack of complete saturation may be:

1. Specimens from natural deposits may have fissures that open up due to sampling and unloading.
2. Specimens from inhomogeneous deposits such as for example varved clay may undergo redistribution of pore water and partial replacement of pore water with intruding air.
3. Specimens of coarse grained soils are unable to form menisci at the specimen surface. This results in air intrusion, disturbance and, in some cases, collapse of the specimen.
4. Reduction of the pore water pressure due to sampling causes release of dissolved gases and consequent lack of full saturation. This effect is particularly pronounced in (a) soils with organic materials in which gases such as methane are formed in situ and (b) soils from the ocean floor in which the change in pore water pressure (from large, positive to negative) is particularly large.
5. Air may be introduced during handling and trimming of undisturbed specimen.

Triaxial Testing of Soils, First Edition. Poul V. Lade.

6. Air may be trapped between specimen and membrane during preparation of specimen setup.
7. Specimens prepared by deposition of granular soils, even when done under water are often not fully saturated.
8. Specimens prepared by compaction at the optimum water content are only 80–95% saturated.

The requirement of full saturation extends to include that portion of the pore pressure measuring equipment that together with the specimen forms a closed system during the undrained test. It should also be noted that an initially saturated specimen may become unsaturated during the test if the soil tends to dilate during shear resulting in decreasing pore pressure. This in turn may cause unintended release of dissolved gases or cavitation of the pore water, which occurs at −1 atm (or more precisely at 17 mm Hg) for clean, de-aired water at room temperature.

6.3 Effects of lack of full saturation

Even a small amount of air will have a large effect on the pore pressures that develop during an undrained test. This is because the volumetric compressibilities of water and air are vastly different. The soil skeleton in a saturated specimen reacts against a very stiff pore fluid and may create large changes in pore pressure. Replacement of some of the stiff water with very compressible gas results in severely reduced pore pressure changes. This is because the tendency for volume change of the soil skeleton is not forcefully resisted by the pore fluid whose compressibility is dominated by the free air. The trend of the pore pressure change (increase or decrease) is not affected, but the magnitude or the pore pressure is directly influenced by the amount of free air in the specimen. Thus, the pore pressure response is buffered by the presence of air.

The lack of full saturation affects the pore pressure which in turn affects the stress–strain relation, the effective stress path, and the undrained strength. The effective strength envelope is not appreciably affected by the lack of full saturation. Figure 6.1 shows the effects of lack of full saturation on the compressive strengths of compacted kaolinite. At low consolidation pressure the compacted clay behaves as an overconsolidated soil that tends to dilate during shear. The corresponding decreases in pore pressures in the partly saturated specimens are too small resulting in smaller effective confining pressures and therefore lower compressive strengths than obtained from the fully saturated specimens.

At high consolidation pressures the compacted clay behaves as a normally consolidated clay that tends to contract during shear.

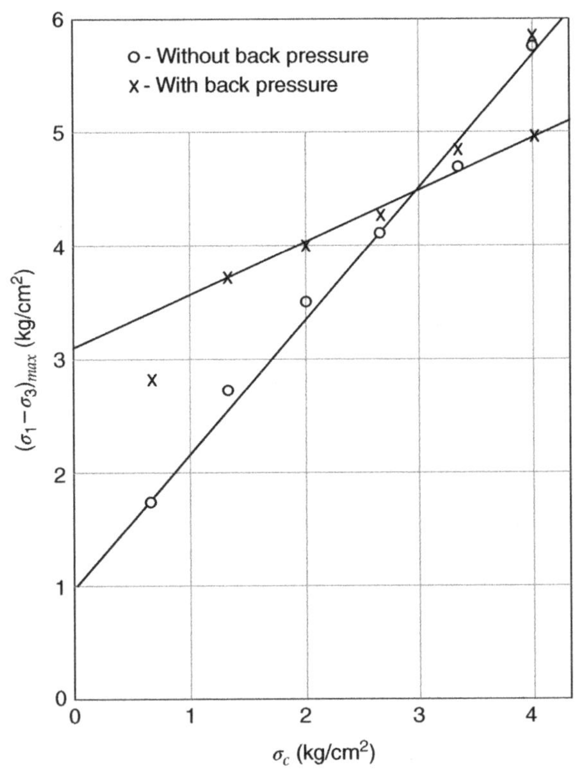

Figure 6.1 Effect of lack of full saturation on the compressive strengths of compacted kaolinite at various consolidation pressures. Reproduced from Allam and Sridharan 1980 by permission of ASTM International.

The resulting increases in pore pressures during undrained shear are too small in the partly saturated specimen, and the resulting effective confining pressures and consequently the compressive strengths are lower than those obtained from the fully saturated specimens. The compressive strength envelopes shown in Fig. 6.1 therefore cross each other.

Since soils that contract during shear have lower undrained strengths when tested in the saturated condition than when tested at partial saturation, it is essential that the triaxial specimens be fully saturated. Even if the soil may not reach full saturation in the field, it is on the side of caution to test such material at full saturation. On the other hand, soils that dilate during shear have higher undrained shear strengths when tested in the saturated condition, and it may be too conservative to employ the lower undrained shear strength from a partly saturated specimen if the soil in the field is always at full saturation.

It is obvious that the degree of saturation plays an important role in the measured soil properties. For tests whose results rely on fully saturated specimens it is of interest to determine the degree of saturation before shearing or before any saturation procedure is attempted.

6.4 *B*-value test

A very sensitive measure of the degree of saturation is obtained by measuring the value of the pore pressure coefficient *B*, as defined by Skempton (1954) (see Section 2.2.3):

$$B = \frac{\Delta u}{\Delta \sigma_3} \tag{6.1}$$

in which $\Delta\sigma_3$ represents an imposed change in the all-around, isotropic cell pressure, and Δu represents the resulting change in pore pressure obtained under the undrained isotropic compression conditions.

A number of factors affect the value of *B*, but it depends strongly on the degree of saturation. Typically, the value of *B* is close to 1.0 for a saturated soil. *B*-values smaller than 1.0 may be indicative of incomplete saturation, but other factors may cause *B* to become lower than unity. For example, leakage between the specimen and the atmosphere results in decreasing pore water pressure and lower *B*-value, whereas a *B*-value greater than unity may suggest leakage from the cell to the specimen. To interpret the value of *B* correctly, the factors that affect *B* are reviewed and discussed below.

6.4.1 Effects of primary factors on *B*-value

For the purpose of the present discussion, the primary factors that affect the *B*-value are considered those associated with the soil itself. These are the soil skeleton, the soil grains, the porosity, the water, the degree of saturation, and the pore pressure itself. Secondary factors are considered to be those imposed by the testing technique. These include the measuring system and the membrane that surrounds the specimen.

The influence of the degree of saturation *S* on the measured *B*-value is always very pronounced. This may be seen from the expression for the *B*-value applicable to undrained conditions in a triaxial specimen.

The expression for the parameter *B* is derived on the basis that the overall volume change of a closed system consisting of a partly saturated specimen must be zero. During a *B*-value test the isotropic stress increase results in a volume change of the soil skeleton, ΔV_{sk}, which must be matched by volume changes of the water, ΔV_w, the soil grains, and of the air, ΔV_a, in the specimen. Thus:

$$\Delta V_{sk} = \Delta V_w + \Delta V_g + \Delta V_a \tag{6.2}$$

The term on the left-hand side is a function of effective stresses and the terms on the right-hand side are functions of pore pressures. These terms may be expressed as:

$$\Delta V_{sk} = C_d \cdot V_0 \cdot \Delta\sigma_3' = C_d \cdot V_0 \cdot (\Delta\sigma_3 - \Delta u) \tag{6.3}$$

in which C_d is the volume compressibility of the soil skeleton and V_0 is the total volume of the specimen.

$$\Delta V_w = C_w \cdot n \cdot V_0 \cdot S \cdot \Delta u \qquad (6.4)$$

in which C_w is the compressibility of water (approximately $4.8 \cdot 10^{-7}$ m²/kN = 4.8 $\cdot 10^{-7}$ vol/vol/kPa) and n is the porosity.

$$\Delta V_g = C_g \cdot (1-n) \cdot V_0 \cdot \Delta u \qquad (6.5)$$

Using Boyle's law for compression of gas (absolute pressure times volume is constant at constant temperature), ΔV_a may be found from:

$$u_1 \cdot V_a = (u_1 + \Delta u)(V_a - \Delta V_a) \qquad (6.6)$$

in which $V_a = n \cdot V_0 \cdot (1-S)$. Thus:

$$\Delta V_a = n \cdot V_0 \cdot \frac{1-S}{u_2} \cdot \Delta u \qquad (6.7)$$

in which u_2 $(= u_1 + \Delta u)$ is the absolute pressure in the pore air after application of an increase of an increment in stress.

Substitution of Eqs (6.3), (6.4), (6.5) and (6.7) into Eq. (6.2), solving for Δu and dividing by $\Delta\sigma_3$ yields:

$$B = \frac{\Delta u}{\Delta \sigma_3} = \frac{C_d}{C_d + n \cdot S \cdot C_w + (1-n) \cdot C_g + n \cdot \dfrac{1-S}{u_2}} \qquad (6.8)$$

This expression indicates that the magnitude of B depends on the following variables: (1) the degree of saturation; (2) the compressibility of the soil skeleton; (3) the porosity of the soil; (4) the compressibility of the soil grains, which is very small compared with those of the water and the soil skeleton; and (5) the absolute pressure of the pore fluid after application of the increment in cell pressure. For specimens with high degrees of saturation, the value of u_2 may be taken as the sum of the absolute pressure in the pore fluid before the pressure increment and the increment in cell pressure. The compression of the air incorporated in Eq. (6.8) follows Boyle's law. The expression is based on the assumption that pore air is not dissolved in the pore water during the B-value test and that the pressure in the air and in the water is the same. These assumptions have been shown to be reasonable for B-value tests on specimens with high degrees of saturation (Black and Lee 1973).

To illustrate the dependence of B on the degree of saturation S, theoretical B-value versus S curves have been calculated for a few representative cases. Four different classes of soils were treated (Black and Lee 1973):

1. *Soft soils*, which are soft normally consolidated clays with for example $e \approx 2.0$ and C_d = $1.45 \cdot 10^{-3}$ m²/kN (=$1.0 \cdot 10^{-2}$ in.²/lb).
2. *Medium soils*, which are compacted silts and clays, and lightly overconsolidated clays with for example $e \approx 0.6$ and $C_d = 1.45 \times 10^{-4}$ m²/kN (=$1.0 \cdot 10^{-3}$ in.²/lb).
3. *Stiff soils*, which are overconsolidated stiff clays, and average sands of most densities with for example $e \approx 0.6$ and $C_d = 1.45 \times 10^{-5}$ m²/kN (=$1.0 \cdot 10^{-4}$ in.²/lb).
4. *Very stiff soils*, which are very dense sands or stiff clays, especially at high confining pressures with for example $e \approx 0.4$ and $C_d = 1.45 \cdot 10^{-6}$ m²/kN (=$1.0 \cdot 10^{-5}$ in.²/lb).

The computed variations in B-value with increasing degree of saturation for these four classes of soil are shown in Fig. 6.2. Since the soil compressibilities for soft and medium soils are greater than the water compressibility, the value of B is for practical purposes equal to unity for a saturated soil, as indicated in Fig. 6.2. Thus, for a fully saturated specimen of soft and medium soils, an increase in confining pressure results in an equal increase in pore water pressure. However, the presence of even very small volumes of undissolved air or gases in the pore water results in much increased compressibility of the pore fluid, and the value of B consequently decreases below unity. The value of B therefore provides a sensitive measure of the degree of saturation, and B-value tests are often used to check the degree of saturation of triaxial specimens (Bishop and Henkel 1962; Wissa and Ladd 1965; Lee *et al.* 1969; Wissa 1969).

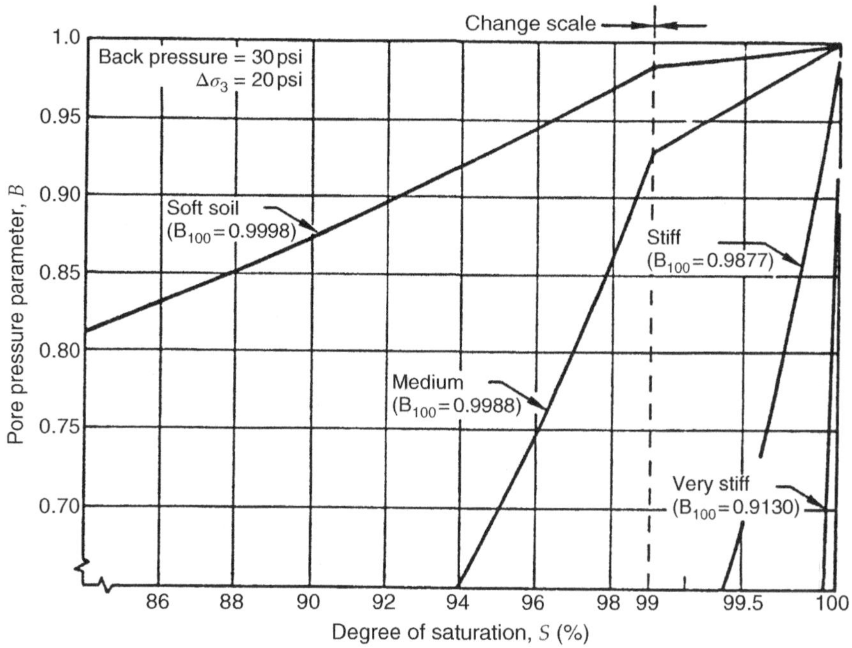

Figure 6.2 Variation of *B*-values with degree of saturation for four classes of soil. Reproduced from Black and Lee 1973 by permission of ASCE.

According to Eq. (6.8) and Fig. 6.2, the value of *B* will also decrease below unity for fully saturated but very stiff soils. If the soil is cemented (Wissa and Ladd 1965) or confined at high pressures (Lee *et al.* 1969), its compressibility is relatively low, and it may approach the same order of magnitude as the compressibility of water. Clay soils consolidated at moderately high pressures and exposed to very small pressure increments exhibit compressibilities that may be smaller than the compressibility of water (Schuurman 1966; Lee *et al.* 1969).

The compressibilities of soil grains are very small (about $0.2 \cdot 10^{-7}$ m^2/kN) (Skempton and Bishop 1954) compared with those of the water and the soil skeleton, and it has been omitted from the pore pressure analysis above. Bishop (1973) showed that for soils with skeleton compressibilities greater than that of water, the influence of the soil grain compressibility is negligible. However, as the compressibility of the skeleton decreases below that of water and approaches that of the soil grains, the difference between *B*-values calculated from Bishop's expression and from Eq. (6.8) becomes significant. Skeleton compressibilities of sufficiently low magnitudes to cause an effect of the grain compressibility on the pore pressure in saturated specimens may be encountered in very stiff soils and in porous and fissured rock.

6.4.2 Effects of secondary factors on *B*-value

The secondary factors affecting the measured *B*-value are those associated with the testing of the soil. Thus, the pore pressure measuring system itself and penetration of the membrane into the voids between the particles in granular soils affect the measured pore pressure.

The pore pressure measuring system, which most often consists of an electrical pressure transducer connected to the specimen through tubings, fittings, and valves, forms a closed system with the specimen. Due to the flexibility of the pressure measuring system (including the compressibility of the fluid filling the system), pore fluid will flow into or out of the pore space of the saturated specimen, thus causing the

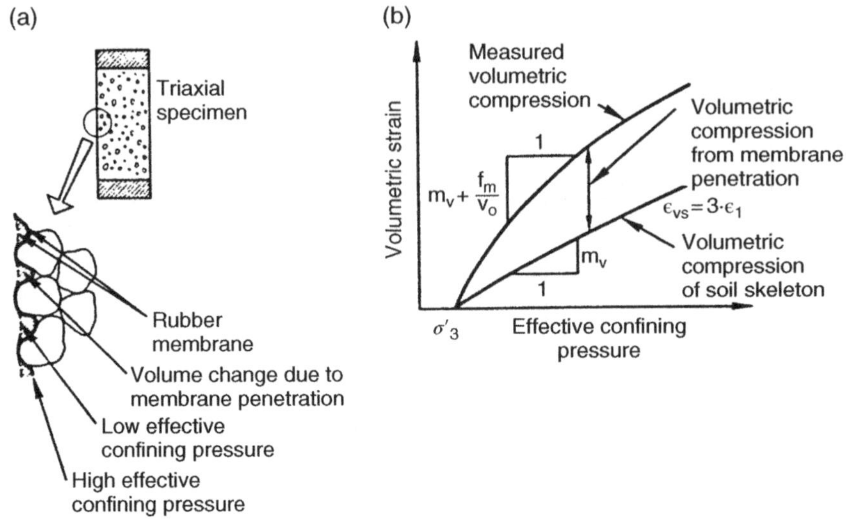

Figure 6.3 (a) Membrane penetration in a triaxial specimen and (b) determination of membrane flexibility from an isotropic compression test. Reproduced from Lade and Hernandez 1977 by permission of ASCE.

specimen to change volume, and the test is therefore not truly undrained. The resulting equilibrium pore pressures will generally differ from the true pore pressures that the measuring system was installed to measure (Bishop and Henkel 1962; Wissa and Ladd 1965; Wissa 1969).

Penetration of the rubber membrane enclosing the triaxial specimen into the voids between the particles in granular soils causes volume changes in the tests with changing effective confining pressures, as shown in Fig. 6.3. In undrained tests membrane penetration will tend to buffer the changes in pore pressure, thus having effects which in some respects are similar to those of air in the specimen.

As before, an expression for the parameter B may be derived on the basis that the overall volume change of the closed system consisting of a partly saturated specimen with surrounding membrane and a pore pressure measuring system must be zero. This requires that the volume changes of the individual components must balance each other, expressed as:

$$\Delta V_{sk} + \Delta V_m = \Delta V_w + \Delta V_g + \Delta V_a + \Delta V_s \quad (6.9)$$

in which ΔV_{sk} is expressed in Eq. (6.3), ΔV_w is expressed in Eq. (6.4), ΔV_g is expressed in Eq. (6.5), and ΔV_a is expressed in Eq. (6.7). ΔV_m is the volume change due to membrane penetration and ΔV_s is the volume change of the pore pressure measuring system. Note again that the terms on the left-hand side of Eq. (6.9) are functions of effective stresses and the terms on the right-hand side are functions of pore pressures.

The component due to membrane penetration is expressed in terms of the flexibility of the membrane, f_m, which is defined as the volume change of the specimen due to membrane penetration, $\Delta\sigma_3'$ (Lade and Hernandez 1977):

$$f_m = \frac{\Delta V_m}{\Delta \sigma_3'} \quad (6.10)$$

from which

$$\Delta V_m = f_m \cdot (\Delta\sigma_3 - \Delta u) \quad (6.11)$$

The flexibility of the pore pressure measuring system is expressed in terms of the coefficient f_s such that:

$$\Delta V_s = f_s \cdot \Delta u \quad (6.12)$$

Substitution of Eqs (6.3), (6.4), (6.5), (6.7), (6.11), and (6.12) into Eq. (6.9), solving for Δu and dividing by $\Delta\sigma_3$ gives (Lade and Hernandez 1977):

$$B = \frac{C_d + \frac{f_m}{V_0}}{C_d + n \cdot S \cdot C_w + (1-n) \cdot C_g + n \cdot \frac{1-S}{u_2} + \frac{f_s}{V_0} + \frac{f_m}{V_0}} \tag{6.13}$$

In this expression for B the effect of the pore pressure measuring system appears only in the denominator and it acts to further reduce the value of B. The term for membrane penetration appears in both the nominator and the denominator and its effect is to stabilize B at unity.

The coefficient f_s represents the total flexibility of the system consisting of pressure transducer, tubings, fittings, valves, and water filling these components. Values of transducer flexibilities reported in the literature range from as low as $0.8 \cdot 10^{-7}$ cm^3/kPa (Wissa 1969) to more typical values of $5 \cdot 10^{-7}$–$100 \cdot 10^{-7}$ cm^3/kPa (Morgan and Moore 1968). The total system flexibility is usually higher due to tubings, valves, and so on.

The effect of the measuring system on the equilibrium pore pressure developed in soft soils with high compressibilities is negligible, even for systems with appreciable flexibility. However, for stiff soils the system flexibility can have a considerable effect on the developed pore pressures. Equation (6.13) shows that the effect of the measuring system may be reduced by using a system with low flexibility or by testing specimens with large volumes. A detailed analysis of these possibilities was presented by Wissa (1969).

The system flexibility for a given pore pressure measuring system may be determined for practical purposes once and for all by closing the control valve that leads to the triaxial specimen and increasing the back pressure by a given amount, observing the volume change for the system and calculating f_s from Eq. (6.12). If it is desirable to include the flexibility of the tube leading to the triaxial specimen, a separate flexibility test may be performed on a length of this type of tube and adding this contribution to that of the volume change/pore pressure measuring device.

The influence of membrane penetration on the measured B-value for specimens of granular soils tends to make evaluation of the degree of saturation more complicated. In this case it is necessary to determine the compressibility of the specimen–membrane system $[(= C_d + f_m/V_0)$, see Fig. 6.3(b)] at the appropriate pressure and substitute the value into the following expression for B for a saturated specimen:

$$B = \frac{\left(C_d + \frac{f_m}{V_0}\right)}{\left(C_d + \frac{f_m}{V_0}\right) + n \cdot C_w + (1-n) \cdot C_g + \frac{f_s}{V_0}} \tag{6.14}$$

The B-value test is then performed as usual, followed by opening the drainage valve while the cell pressure is maintained at its new higher value $(= \sigma_3 + \Delta\sigma_3)$. Granular soils with any significant membrane penetration are usually relatively permeable, and the amount of volume change resulting from opening the drainage valve is used to calculate the appropriate value of the term:

$$\left(C_d + \frac{f_m}{V_0}\right) = \frac{\Delta V_{measured}}{V_0 \cdot \Delta\sigma_3} \tag{6.15}$$

Substitution of Eq. (6.15) into Eq. (6.14) results in a theoretical B-value that may be compared directly with the measured B-value to evaluate the specimen saturation. Note that by determining the term in the nominator as one term, there is no need to make assumptions about isotropy of the soil specimen, as implied in Fig. 6.3. Note also that the membrane flexibility (as well as the soil skeleton compressibility) varies with effective confining pressure and the flexibility determined from an isotropic compression test is particular to the specimen size and geometry used. This is automatically included in the method indicated above. Additional information about the factors that control membrane penetration are presented in Section 9.7.1.

The effect of membrane penetration at low effective confining pressures can easily be much larger than the terms accounting for grain and water compressibility and flexibility of the measuring system. The value of B for a saturated specimen at low effective confining pressures may therefore be closer to unity than indicated by Eq. (6.8). The membrane flexibility decreases with increasing effective confining pressure and the B-value calculated from Eq. (6.13) therefore approaches the value calculated from Eq. (6.8).

6.4.3 Performance of B-value test

The actual performance of a B-value test for determination of the degree of saturation of a triaxial specimen may be done according to different procedures depending on the compressibility of the soil, the occurrence of membrane penetration, the permeability of the soil, and the anticipated degree of saturation.

The B-value test should be done while the specimen is under an isotropic state of stress, and the specimen has fully consolidated under the current effective confining pressure. The B-value test may, in principle, be done at any effective isotropic confining pressure. However, practical considerations indicate that it is best done at a low confining pressure in the early stages of the specimen setup, before the final consolidation pressure is applied, because, in the event of measurement of a low B-value, it may be possible to remedy the low degree of saturation during the following consolidation stage.

The fact that the B-value tests should be done from an isotropic stress state is not obvious from the classical derivation based on elastic properties, whether isotropic or anisotropic properties are assumed. A derivation for an elastic-plastic material shows that the shear-volume coupling terms, which become active when a deviator stress is applied, will cause either a too high or too low B-value depending on whether the change in cell pressure is positive or negative (J.F. Peters, personal communication, 2015), and this B-value is therefore not useful in terms of indicating the degree of saturation of the soil.

In general, the B-value test is performed by closing the drainage valve, increasing the cell pressure by a sufficient amount (say $\Delta\sigma_3 = 20$–70 kPa (3–10 psi)) to allow determination of a reasonably accurate value of B, measuring Δu, and calculating the value of B ($= \Delta u/\Delta\sigma_3$). The cell pressure is usually reduced to its initial value and the drainage valve opened again to return to the state of stress in the specimen before the B-value test.

For soft and medium soils the value of B is unity for a saturated specimen, that is the change in confining pressure causes an equal change in pore pressure. For soft non-saturated soil, the degree of saturation is approximately equal to the measured value of B, as indicated in Fig. 6.2. Increasing the cell pressure in a test on a non-saturated soil causes an increase in effective confining pressure of:

$$\Delta\sigma_3' = \Delta\sigma_3 \cdot (1 - B) \tag{6.16}$$

This in turn causes the specimen to compress, and it may result in overconsolidation of the specimen, as illustrated in Fig. 6.4. The subsequent reduction in confining pressure results in rebounding the specimen, but the void ratio at the desired effective confining pressure is now

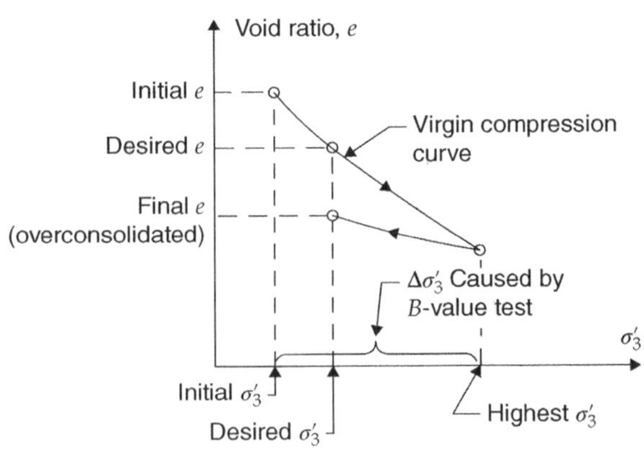

Figure 6.4 Schematic illustration of the behavior of a partially saturated soil during a B-value test.

lower than that corresponding to the virgin compression curve. It is therefore important that the increment in cell pressure used for determination of B be chosen with consideration to the possibility of unintentionally overconsolidating the specimen. The magnitude of $\Delta\sigma_3$ should therefore be small for a specimen whose degree of saturation is anticipated to be low.

Soft soils with low permeability

Soft clays have low permeabilities that prevent immediate response in the measured pore water pressure when the cell pressure is increased. Soft clays exhibit viscous behavior as indicated in Fig. 6.5, which shows a diagram of B-value measurements for triaxial specimens of normally consolidated San Francisco Bay Mud. The diagram indicates that it may take some time for the pore water pressure to respond sufficiently and the B-value to reach unity. In fact, if unity is not reached within a reasonable time, it will never be reached and the specimen is not saturated. If, however, unity is reached within a reasonably short period, then the specimen can be taken to be saturated.

Very stiff soils

For very stiff soils the value of B is significantly less than unity, even for fully saturated specimens, as shown in Fig. 6.2. It is therefore difficult to evaluate the degree of saturation on the basis of a simple B-value test alone. In this case, the B-value for a fully saturated specimen may be calculated from Eq. (6.14) for comparison with the measured value.

Alternatively, the values of B may be determined at several successively higher back pressures while keeping the effective consolidation pressure approximately constant, as shown in Fig. 6.6 (Lee *et al.* 1969; Wissa 1969). A measured B-value that is constant and independent of the magnitude of the back pressure indicates full saturation.

Specimens with applied deviator stress

Figure 6.7 shows the variation in measured B-values for saturated clay and saturated sand specimens at increasing stress ratios (Lade and Kirkgard 1984). For the Bootlegger Cove Clay, a relatively soft clay from Anchorage, AK, USA, the B-values measured initially at the isotropic state of stress were reasonably close to unity.

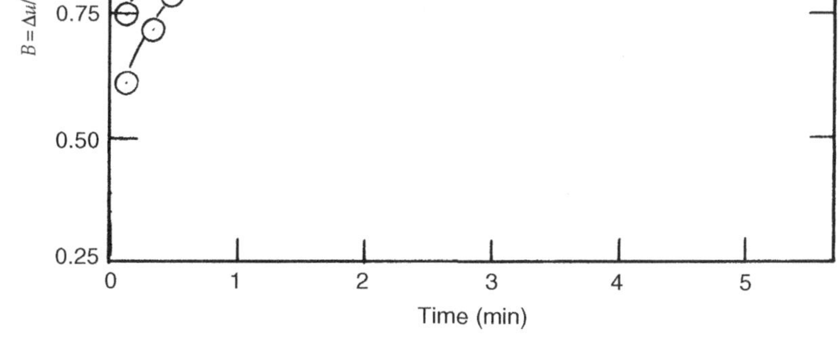

Figure 6.5 Diagram of B-value measurements for triaxial specimens of viscous, normally consolidated San Francisco Bay Mud indicating that it may take some time for the pore water pressure to respond sufficiently and the B-value to reach unity.

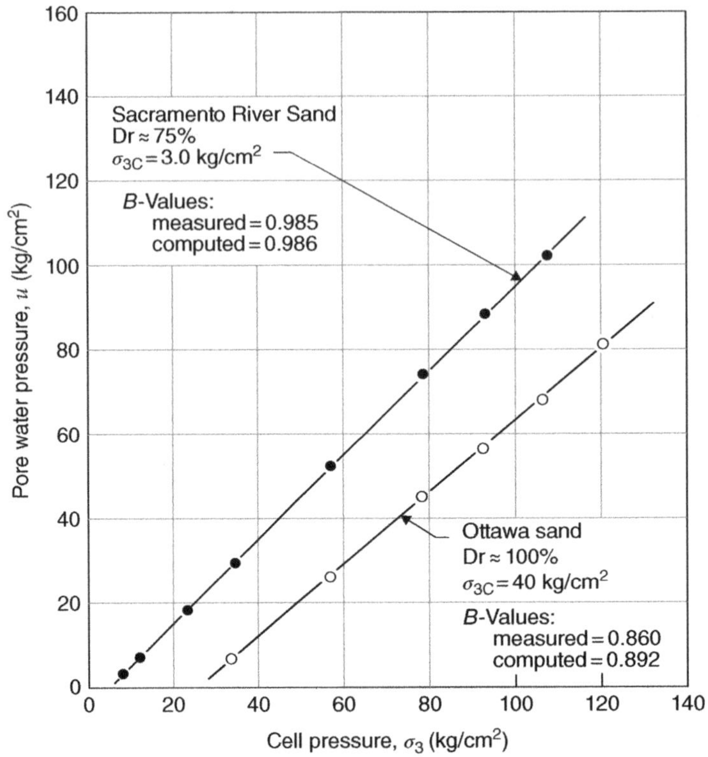

Figure 6.6 Values of *B* may be determined at several successively higher back pressures while keeping the effective consolidation pressure approximately constant (after Lee *et al.* 1969).

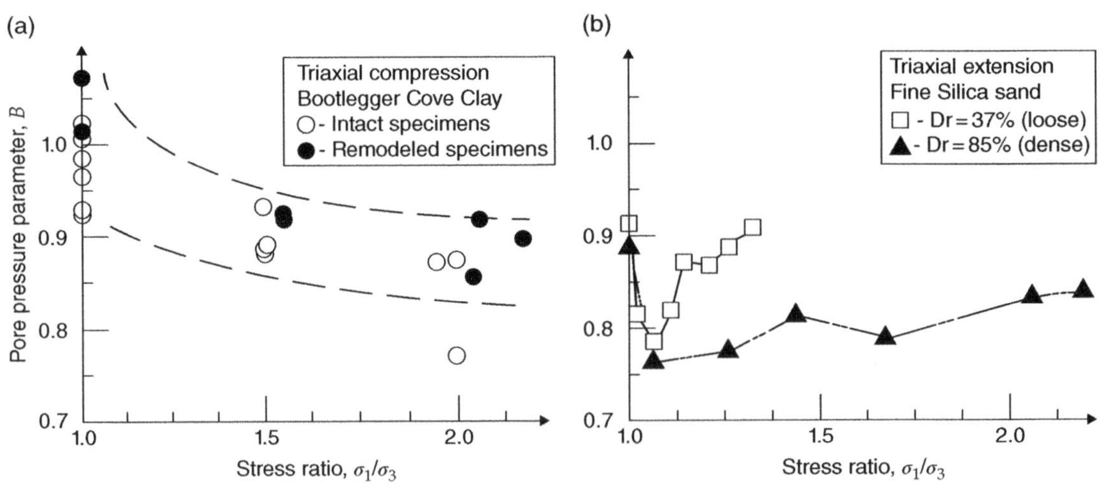

Figure 6.7 Variation of measured *B*-values for (a) saturated clay and (b) saturated sand specimens at increasing stress ratios. Reproduced from Lade and Kirkgard 1984 by permission of ASCE.

Increasing the consolidation stress ratio resulted in increasing *B*-values for each of the specimens tested. The intact clay specimens generally exhibited slightly lower *B*-values, possibly due to the anisotropic structure of the natural clay.

The results of the two triaxial extension tests on saturated, Fine Silica sand also show considerable variation in the measured *B*-values with change in stress ratio. The *B*-values decrease sharply immediately upon increase in stress ratio. Further increase in stress ratio causes the *B*-values to increase again towards the original value measured in the isotropic state of stress. The *B*-values increase rapidly for loose sand whereas much slower increases are observed for the dense sand.

From these examples it may be seen that it can be difficult to employ the *B*-value test as a check on the degree of saturation of a soil specimen. As an example, consider a specimen of Bootlegger Cove Clay initially K_0-consolidated before further testing. A *B*-value test results in $B = 0.90$. Such a low value may be particularly surprising since this clay is relatively soft, and a much higher value approaching unity might have been expected for a soft clay. However, it is difficult to interpret this value of *B* relative to the degree of saturation without further measurements and analyses.

Thus, if anisotropic consolidation is required at the end of the consolidation stage, then it is advisable to determine the *B*-value at an earlier stage when the specimen is still under isotropic pressure.

6.5 Determination of degree of saturation

Once the *B*-value is determined, the degree of saturation, *S*, may be determined from the expression for *B* in Eq. (6.13):

$$S = \frac{\left(C_d + \frac{f_m}{V_0}\right)(1-B) - B\cdot\left(n\cdot\frac{1}{u_2} + (1-n)\cdot C_g + \frac{f_s}{V_0}\right)}{B\cdot n\cdot\left(C_w - \frac{1}{u_2}\right)} \tag{6.17}$$

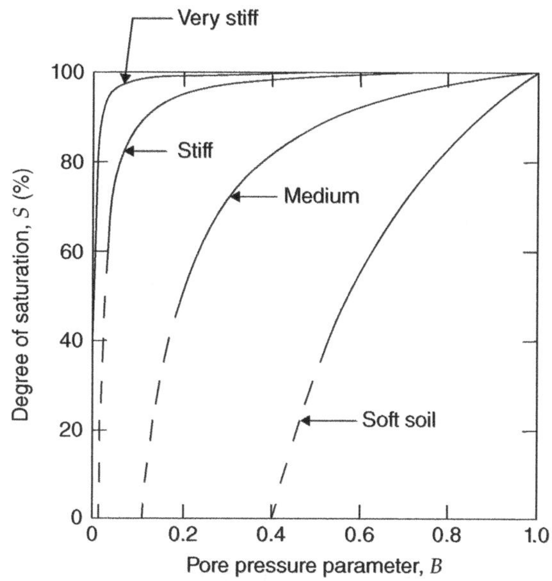

Figure 6.8 Variation of degree of saturation, *S*, with measured *B*-value for four classes of soil for the case of no membrane penetration and negligible influence of the pore pressure measuring system.

Because C_w and C_g are always much smaller than $1/u_2$, this expression simplifies to:

$$S = 1 - \frac{u_2}{n}\cdot\left(C_d + \frac{f_m}{V_0}\right)\cdot\frac{1-B}{B} - \frac{f_s}{V_0} \tag{6.18}$$

The influence of *B* on *S* is shown in Fig. 6.8 for the case of no membrane penetration and negligible influence of the pore pressure measuring system. The four classes of soil used for the diagram in Fig. 6.2 are also employed here. In fact, the diagram in Fig. 6.8 shows the inverse of the relations shown in Fig. 6.2, but over the full range of *B*- and *S*-values. Note that, except for soft and medium soils, the *B*-values are significantly below 1.0 at full saturation, and they reduce very rapidly for *S* less than 100%. If, for example, $B = 0.95$ was accepted as indicating full saturation, it may be seen from Fig. 6.8 [or from Eq. (6.18)] that this would correspond to a degree of saturation of 99.9% for a very stiff soil. However, for a soft soil with the same *B*-value, the degree of saturation would be 96%.

6.6 Methods of saturating triaxial specimens

A very important and sometimes time consuming stage in a triaxial test is the process of saturating a specimen that, after mounting and enclosure in the triaxial cell, is not completely saturated. Several methods are available to increase the degree of saturation and to completely saturate triaxial specimens with water. The methods that have been employed for this purpose are: (1) percolation with water; (2) the CO_2-method; (3) application of back pressure; and (4) the vacuum procedure. Each of these methods are presented below.

6.6.1 Percolation with water

In the percolation method, water is introduced through the bottom drainage line and allowed to seep up through the specimen under a hydraulic gradient. This is intended to flush the air out through the top drainage line thereby saturating the specimen with water, as illustrated in Fig. 6.9. The process may be assisted by applying a vacuum to the specimen causing the volume of air to enlarge and, in part, be sucked out of the specimen. Then water is introduced to the bottom of the specimen while maintaining a vacuum at the top. Further assistance to the percolation process may be provided by applying a pressure to the water at the bottom, thus increasing the hydraulic gradient in the specimen.

This method may be used to increase the degree of saturation from low values in soils that are relatively permeable. However, it does not lend itself to saturate specimens completely on a consistent basis. Lowe and Johnson (1960) reported typical data for cohesive soil specimens prepared at degrees of saturation ranging from less than 70% to about 95%. After persistent attempts at percolation of water, the specimens reached degrees of saturation ranging from about 81 to 100%. Complete saturation of fine grained soils has rarely been achieved by percolation of water. Clean sands with high permeability may be easier to saturate by this method, but even this type of soil may exhibit some resistance to become completely saturated by percolation. Small air bubbles may hang on to sand grains and are difficult to flush out.

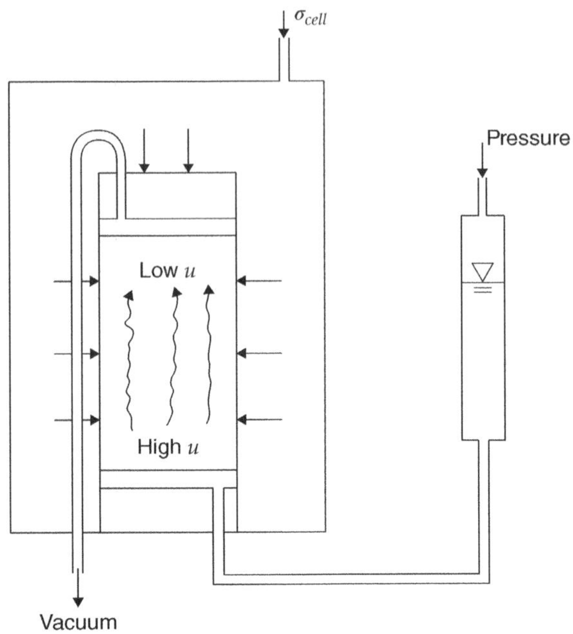

Figure 6.9 Saturation of triaxial specimen by percolation of water, thus flushing air out through the top drainage line.

The process of percolation may require assistance from high differential pressures at the drainage ports to develop significant flow through the specimen. However, the cell pressure, which is essentially constant along the specimen, must be higher than the pressure applied internally at the bottom to maintain a positive effective confining pressure on the specimen. Thus, effective confining pressures of different magnitudes are applied along the height of the specimen with the lowest pressures near the bottom. Further, percolation assisted by differential pressures imposes seepage forces along the axis of the specimen. These effects cause undesirable and uneven consolidation of the specimen. Finally, the process of percolation takes considerable time in fine grained soils.

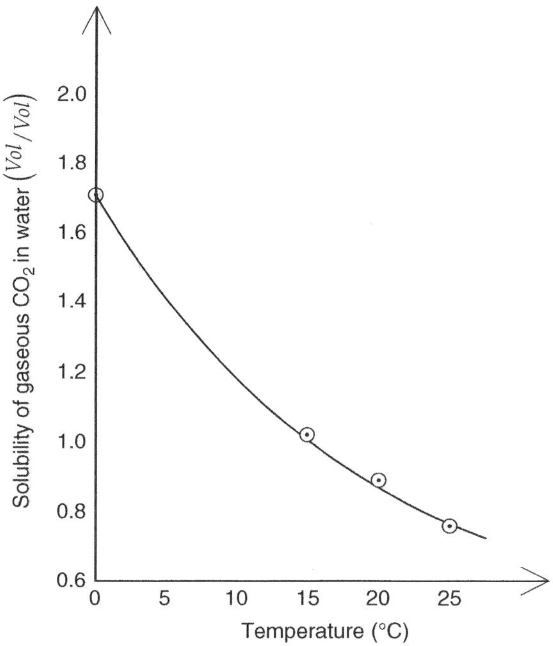

Figure 6.10 Variation of Henry's coefficient of solubility with temperature for CO_2 in water.

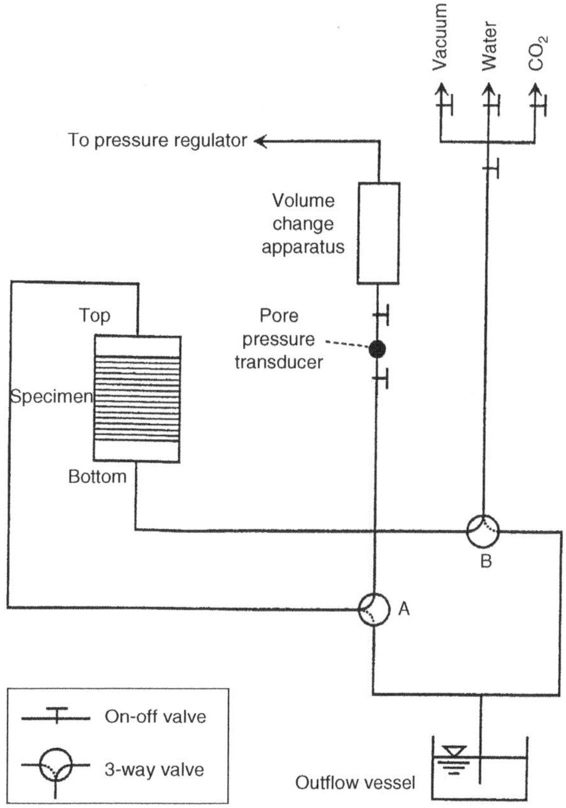

Figure 6.11 A setup suitable for application of the CO_2-method. Reproduced from Lade and Abelev 2005 by permission of Elsevier.

6.6.2 CO_2-method

In the CO_2-method of specimen saturation (Lade 1972; Lade and Duncan 1973), gaseous CO_2 is introduced through the bottom drainage line, thereby pushing the lighter air in the specimen out through the top drainage line. The CO_2 is allowed to seep up through the specimen for about 15 min, ensuring complete replacement of the air, which is lighter than CO_2. De-aired water is then introduced through the bottom drainage line and also allowed to seep slowly up through the specimen, thereby pushing most of the CO_2 out through the top drainage line. The variation of Henry's coefficient of solubility with temperature for CO_2 in water is shown in Fig. 6.10. A volume of gaseous CO_2 can be dissolved in approximately an equal volume of de-aired water, that is Henry's coefficient of solubility is nearly 1.0 at room temperature (i.e., it is highly soluble in water). Any CO_2 left in the specimen will dissolve in the intruding water, which in turn will fill the voids in the specimen. An additional pore volume of water may be allowed to seep up through the specimen to replace any remnants of the CO_2. A setup suitable for application of the CO_2-method is shown in Fig. 6.11. Note that while gaseous CO_2 is being forced through the specimen, it is important to monitor the gas pressure to avoid a reduced effective confining pressure at the bottom of the specimen where the gas is introduced.

The method is suitable to saturate specimens completely with water. Measured *B*-values in initially dry sand specimens saturated by the CO_2-method have been found to be 0.97–1.00 in most cases. The success of this method depends on the ability to replace all air in the specimen with CO_2. Thus, if air in a partly saturated specimen is present as air bubbles surrounded by water, it may not be possible to replace the air by CO_2. However, the method works well for specimens that are initially dry and relatively permeable. The gradients that are required to cause the gaseous CO_2 and the water to flow are small, and the method therefore does not impose any undesirable consolidation pressures on the soil.

6.6.3 Application of back pressure

Saturation of soil specimens by application of elevated back pressure has been widely used. Elevated back pressure may also be employed to simulate the pore water pressure present in the field. This may be particularly important for soil specimens from (1) deposits with organic content and dissolved gases, (2) deposits in the ocean bottom where the pore water pressure is high, and (3) deposits in which cavitation of the pore water during shear is to be avoided in the laboratory tests.

The basic concept in saturation of soil specimens by this method is to apply a sufficiently high pressure to the pore fluid to cause the pore air to dissolve completely into the pore water. The fundamental physical phenomena occurring in the process of saturating a soil specimen under pressure may be described by the following basic laws dealing with gas–liquid systems:

1. *Boyle's law*: For a perfect gas at constant temperature, the product of pressure and volume is constant:

$$P \cdot V = const. \tag{6.19}$$

in which V is the volume of the gas and P is the absolute pressure in the gas.

2. *Henry's law*: At a given temperature, the mass of gas that can be dissolved in a mass of liquid is directly proportional to the pressure. Expressing this law on the basis of volumes, it takes the following form at any pressure:

$$V_{gas,dissolved} = H \cdot V_{liquid} \tag{6.20}$$

in which H is Henry's coefficient of solubility. The value of Henry's coefficient for air in water varies with temperature and at room temperature $H = 0.02$ (vol/vol).

3. *Kelvin's equation*: The pressure inside a gas bubble is higher than the pressure in the surrounding liquid due to the surface tension effect at the air–water interface:

$$P_{gas} = P_{liquid} + \frac{2 \cdot T_s}{r} \tag{6.21}$$

in which T_s is the surface tension of the gas–liquid interface and r is the radius of the air bubble. For an air–water interface $T_s = 7.27 \cdot 10^{-5}$ kN/m.

With regard to the question of full saturation of a triaxial specimen, the effect of surface tension is usually small (Black and Lee 1973; Rad and Clough 1984) and it will be disregarded in the following quantitative developments. Thus, the pore air pressure and the pore water pressure are assumed to be the same.

There are two ways in which saturation is achieved by the back pressure method. First, the increased pressure in the pore water causes the air to compress according to Boyle's law. This results in a greatly increased degree of saturation. Secondly, the compressed air is dissolved in the water, which at the higher pressure can dissolve a larger mass of air according to Henry's law of solubility.

Passive development of back pressure

A partly saturated soil specimen may be saturated by increasing the cell pressure while the specimen is maintained in the undrained condition, that is neither water nor air is allowed to cross the specimen boundaries. The soil skeleton of the partly saturated soil compresses and causes the pore pressure to increase. This allows additional amounts of air to be dissolved in the pore water.

The increase in the pore water pressure, Δu, required to completely saturate the specimen (Bishop and Eldin 1950) may be calculated with reference to the illustrations in Fig. 6.12. The volume of air in the specimen at the initial pore pressure, u_0, is:

$$V_0 = V_a + V_{ad} = V_s \cdot (1 - S_0) \cdot e_0 + H \cdot V_s \cdot S_0 \cdot e_0 \tag{6.22}$$

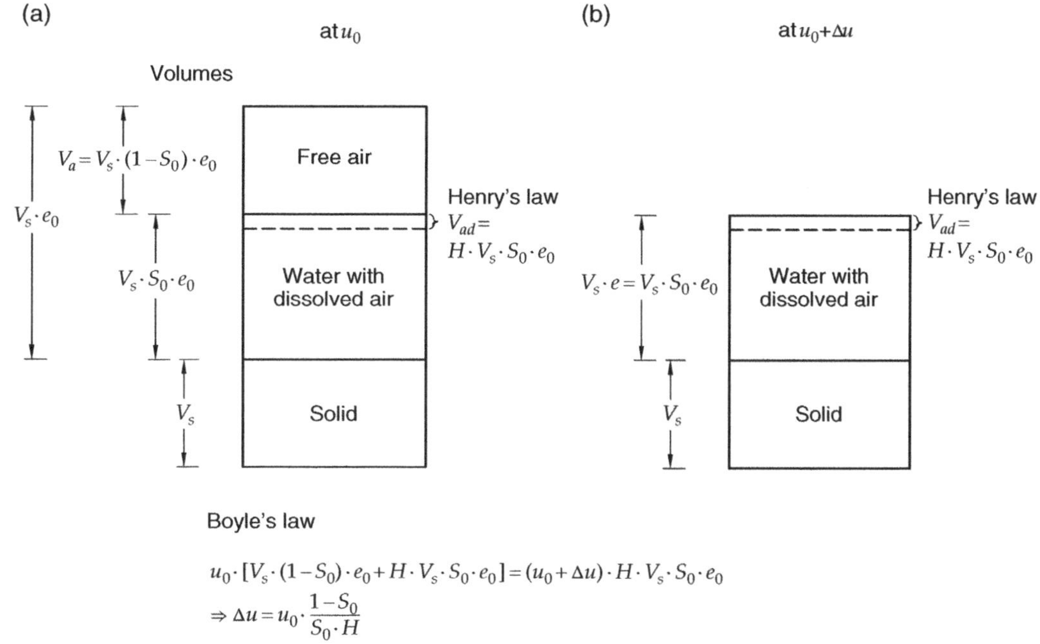

Figure 6.12 Determination of additional pore pressure, Δu, required to produce saturation of a soil specimen by passive development of back pressure. (a) Initial conditions of soil and (b) complete saturation of soil.

The first term in this expression accounts for the free air, and the second term represents the volume occupied by the dissolved air if it were extracted from the water and compressed at the current pore pressure. Complete saturation of the soil occurs when the air is compressed to such a small volume that all air can be dissolved in the pore water at the new higher pressure, $u_0 + \Delta u$.

Expressing the volume of air that can be dissolved at the higher pressure by Henry's law, and using Boyle's law for the initial and final conditions of the air, results in:

$$u_0 \cdot \left[V_s \cdot (1 - S_0) \cdot e_0 + H \cdot V_s \cdot S_0 \cdot e_0\right] = (u_0 + \Delta u) \cdot H \cdot V_s \cdot S_0 \cdot e_0 \quad (6.23)$$

which can be reduced to yield the increment in pore pressure:

$$\Delta u = u_0 \cdot \frac{1 - S_0}{S_0 \cdot H} \quad (6.24)$$

in which u_0 is the initial absolute pore pressure and S_0 is the initial degree of saturation. The total pore pressure or back pressure required in the specimen to cause full saturation is then:

$$u_{back} = u_0 + \Delta u = u_0 \cdot \frac{1 - S_0 \cdot (1 - H)}{S_0 \cdot H} \quad (6.25)$$

in which u_{back} is absolute pressure.

The increments in pore pressure, Δu, required to saturate soil specimens with various initial degrees of saturation at atmospheric pressure are indicated by the upper curve in Fig. 6.13. The cell pressures required to generate these pore pressure increments depend on the compressibility and the initial degree of saturation of the soil specimen.

In this procedure of passive development of back pressure, the soil skeleton compresses to compensate for the space formerly occupied by the air. This simulates the behavior of a relatively impermeable soil in an earth embankment or earth dam during construction. The lower middle section of such an earth dam may become saturated during construction due to compression caused by the overburden

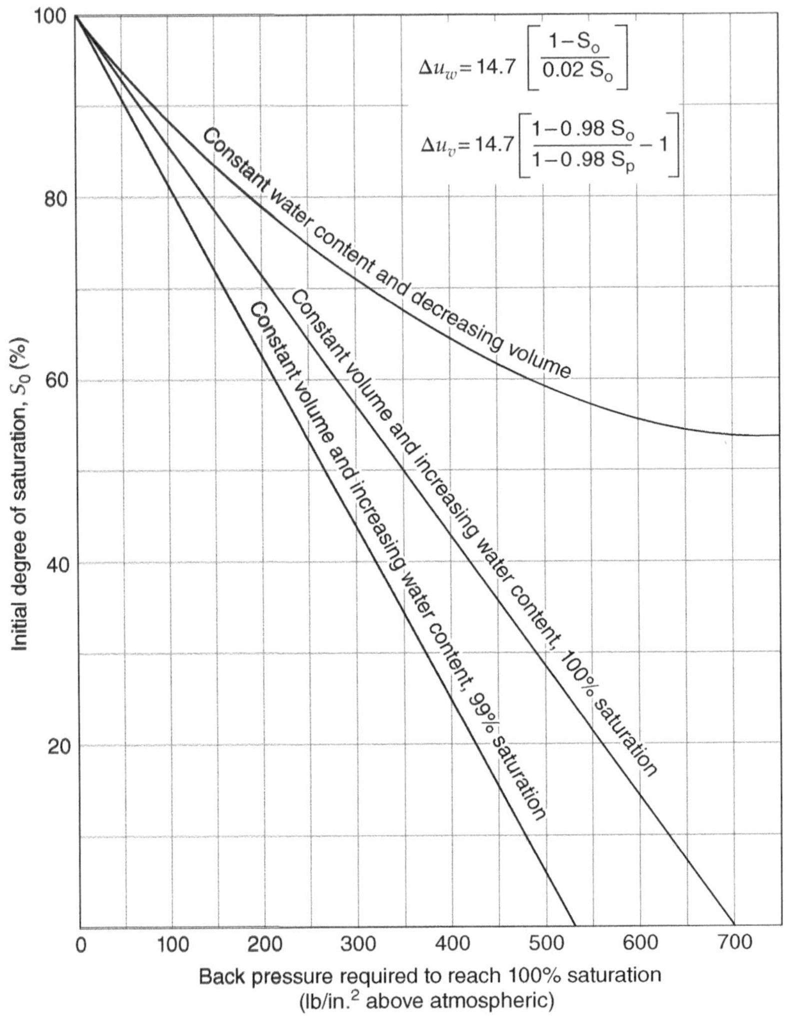

Figure 6.13 Magnitudes of pore pressure increments, Δu, above atmospheric pressure, required to saturate soil specimens with various initial degrees of saturation at atmospheric pressure indicated by the upper curve. Reproduced from Lowe and Johnson 1960 by permission of ASCE.

pressure. The corresponding type of laboratory test is the UU-test in which the soil behaves in a similar fashion (see Chapter 1). The application of this saturation procedure is limited to those conditions described above. It is not suitable for most laboratory testing procedures carried out to simulate field loading conditions in which compression of the soil skeleton and consequent decreases in void ratio and increases in dry density may be untimely and/or undesirable. The procedure involving active application of back pressure may be more suitable for such conditions, as explained below.

Active application of back pressure

In the more conventional method, a back pressure is actively applied to the pore water through a reservoir of water, such as the volume change buret, as shown in Fig. 6.14. The water from the buret is allowed to flow into the

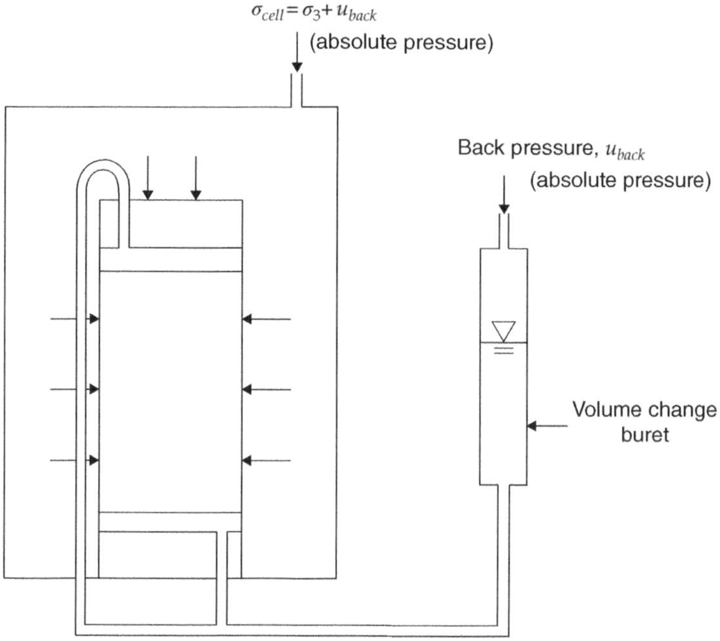

Figure 6.14 Conventional saturation method in which back pressure is actively applied to the pore water through a reservoir of water, such as the volume change buret.

specimen to replace the air being dissolved in the pore water. In this method the soil skeleton can be kept at its initial volume, such that void ratio and dry density remain constant. The axial strain should be monitored during the saturation process, because it is the only variable available to signal disturbance in the state of the specimen. The value of the axial strain should be reported in the test notes. However, volume changes may also be allowed so that equilibrium can be reached under a given effective consolidation pressure. Thus, both saturation and consolidation of the specimen may be achieved at the same time.

The increase in pore pressure, Δu, required to increase the degree of saturation (Lowe and Johnson 1960) may be calculated with reference to the illustrations in Fig. 6.15. The volume of air in the specimen at the initial pore pressure, u_0, is:

$$V_0 = V_a + V_{ad} = V_s \cdot (1 - S_0) \cdot e_0 + H \cdot V_s \cdot S_0 \cdot e_0 \quad (6.26)$$

in which the first term accounts for the free air and the second term represents the volume occupied by the dissolved air if it were extracted from the water and compressed at the current pore pressure. The final degree of saturation, S_f (not necessarily full saturation), is obtained after introducing an additional amount of water into the specimen and raising the pressure to $u_0 + \Delta u$. At equilibrium, the total volume of air in the specimen is:

$$\begin{aligned} V_f = V_{af} + V_{ad1} + V_{ad2} &= V_s \cdot (1 - S_0) \cdot e_0 \\ &+ H \cdot V_s \cdot (S_f - S_0) \cdot e_0 + H \cdot V_s \cdot S_0 \cdot e_0 \end{aligned} \quad (6.27)$$

in which the second and third terms represent the volumes occupied by the dissolved air if it was extracted from the newly introduced and the original water, respectively, and compressed at the new higher pore pressure. Using Boyle's law for the initial and final conditions of the air results in:

$$\begin{aligned} u_0 \cdot &\left[V_s \cdot (1 - S_0) \cdot e_0 + H \cdot V_s \cdot S_0 \cdot e_0\right] \\ &= (u_0 + \Delta u) \cdot \left[V_s \cdot (1 - S_f) \cdot e_0 \right. \\ &\left. + H \cdot V_s \cdot (S_f - S_0) \cdot e_0 + H \cdot V_s \cdot S_0 \cdot e_0\right] \end{aligned} \quad (6.28)$$

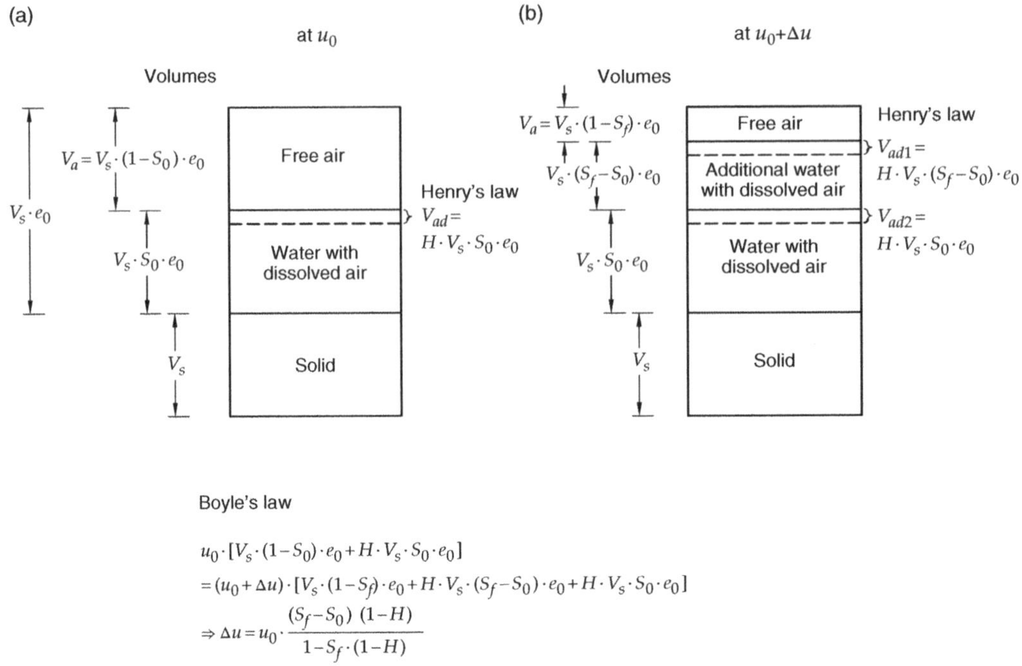

Figure 6.15 Determination of additional pore pressure, Δu, required to produce a final degree of saturation, S_f, of a soil specimen by active application of back pressure: (a) initial conditions of soil; and (b) final condition of soil.

which, after reduction, yields the *additional* pore pressure required to reach the final degree of saturation, S_f:

$$\Delta u = u_0 \cdot \frac{(S_f - S_0)(1 - H)}{1 - S_f \cdot (1 - H)} \qquad (6.29)$$

in which u_0 is the initial absolute pore pressure and S_0 is the initial degree of saturation. This result may also be expressed as the *total* pore pressure or the back pressure to be applied to the specimen to reach S_f:

$$u_{back} = u_0 + \Delta u = u_0 \cdot \frac{1 - S_0 \cdot (1 - H)}{1 - S_f \cdot (1 - H)} \qquad (6.30)$$

The back pressure required to reach full saturation is:

$$u_{back} = u_0 \cdot \frac{1 - S_0 \cdot (1 - H)}{H} \qquad (6.31)$$

The back pressures in these two equations are given as absolute pressures. The increments in pore pressures, Δu, required to reach 100% and 99% saturation for various initial degrees of saturation at atmospheric pressure are shown by the two lower lines in Fig. 6.13.

Soils compacted near their optimum water contents are usually at 80–95% saturation. For these degrees of saturation the upper left corner of Fig. 6.13 applies. Thus, back pressures in the order of 50–150 psi (350–1000 kPa) are usually required to saturate compacted soils.

The method of active application of back pressure is suitable for CD- as well as CU-tests. It can, in principle, be applied to any type of soil. It may be employed in tests in which the actual magnitude of the back pressure does not affect the soil behavior adversely, that is in a manner not conflicting with simulation of the soil behavior in the field.

Back pressure techniques

When a back pressure is applied, an equal pressure is added to the cell pressure to maintain the effective confining pressure constant.

Specimens with high degrees of saturation, such as soft clay sampled below the ground water table, may contain small amounts of air trapped during preparation of the triaxial setup. For such basically saturated soils, back pressures in the order of 200 kPa may be sufficient to dissolve any small amount of air in the pore water. Back pressures of this magnitude may be applied routinely for basically saturated specimens. For such cases the specimen is usually consolidated with a back pressure such that complete saturation is achieved during the same time consolidation of the specimen occurs.

For a partly saturated specimen requiring substantial back pressures, both the cell pressure and the back pressure may have to be applied in increments such that the specimen is not becoming overconsolidated. Although the cell pressure and the back pressure are applied simultaneously, it takes time for the additional water to seep into the specimen and for the dissolution of air into the pore water. The latter process occurs by diffusion and this can be very time consuming (see below). Therefore, the initial result of increasing the cell and back pressures is an increase in the effective confining pressure of:

$$\Delta\sigma_3' = \Delta\sigma_3 \cdot (1-B) \tag{6.32}$$

Applying the total back pressure required for complete saturation may therefore result in excessive compression of the soil, as shown schematically in Fig. 6.16. Applying the back and cell pressures in increments small enough to avoid exceeding the desired consolidation pressure is illustrated in Fig. 6.17. Choosing the pressure increments so small that the maximum effective confining pressure during saturation is somewhat smaller than the desired consolidation pressure enhances the possibility of reaching the desired and correct void ratio at the end of the subsequent consolidation of the specimen.

It may be prudent to check the B-value and recalculate the necessary back pressure following equilibrium after each increment of back pressure. Note that the B-value increases as the specimen becomes saturated. This would allow increasing increments in back and cell pressures during the following stages without exceeding the maximum effective pressure reached during the previous stage. To save time, it may be possible to perform the consolidation during the final stage of the saturation procedure.

Time for saturation by dissolving air

Dissolving air in the pore water is a diffusion process that takes time. However, the time to reach full saturation can be substantially decreased by increasing the back pressure above the value required, as calculated from Eq. (6.31). Based on an experimental study, Black and Lee (1973) developed an expression for the diffusion time, t_d, to reach a final degree of saturation, S_f:

$$t_d = \left\{ \frac{1}{K} \cdot \left[\frac{1-S_0}{1+\frac{(1-H)}{H} \cdot R \cdot (1-S_0)} - (1-S_f) \right] \right\}^{1/x} \tag{6.33}$$

in which R is the ratio of applied back pressure to required back pressure for full saturation, and K and x are coefficients that may be determined from:

$$K = 0.0094 - 0.01 \cdot S_0 \quad \text{for } S_0 < 0.8 \tag{6.34a}$$

$$K = 0.0014 \quad \text{for } S_0 \geq 0.8 \tag{6.34b}$$

and

$$x = 0.085 + 0.133 \cdot S_0 \quad \text{for } 0 \leq S_0 \leq 1.0 \tag{6.35}$$

These expressions were developed from experimental data that showed a fair amount of scatter, especially at low degrees of saturation. However, even with the uncertainty, calculation from Eq. (6.33) should be sufficiently realistic to indicate the general order of magnitude of the time delay for reaching a desired degree of saturation.

Figure 6.18 shows the time required for complete saturation plotted versus the initial degree of saturation for different values of R (= $u_{back}/u_{back,100}$ in this diagram). Because both K and x vary with the initial degree of saturation, the calculated time delays first increase and then

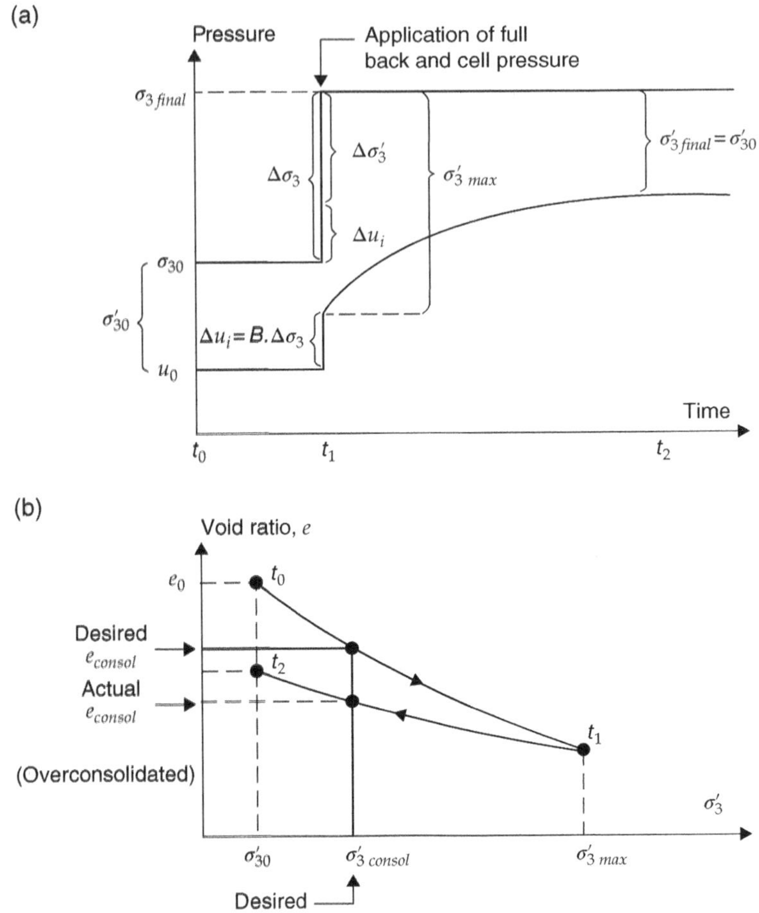

Figure 6.16 Application of full increments of back pressure and cell pressure to a partly saturated specimen resulting in unintended overconsolidation of soil. (a) Sequential application of cell and back pressure with time and (b) effects on effective pressure–void ratio relation.

decrease with increasing value of S_0. According to these relations it can take longer time to saturate a fairly wet specimen than a fairly dry specimen for the same applied *relative* back pressure. However, the actual *required* back pressure would be much larger for the dry specimen. Black and Lee (1973) suggested that the form of the relations shown in Fig. 6.18 may indicate a difference in the nature of pore air at low as compared with high degrees of saturation.

Saturation of soil specimens may at times require very high back pressures to reach full saturation and to decrease the time delay for saturation. Thus, equipment suitable for high pressures as well as considerable time is required. It is, however, possible to reduce the back pressure and the time delay substantially as described below.

6.6.4 Vacuum procedure

According to Eq. (6.31) the back pressure required to saturate a soil specimen is proportional to the initial absolute pressure in the pore fluid. Decreasing the pore pressure by applying a vacuum to the specimen should therefore reduce the back pressure required for saturation. Rad and Clough (1984) performed a study in which specimens of uncemented, fine to

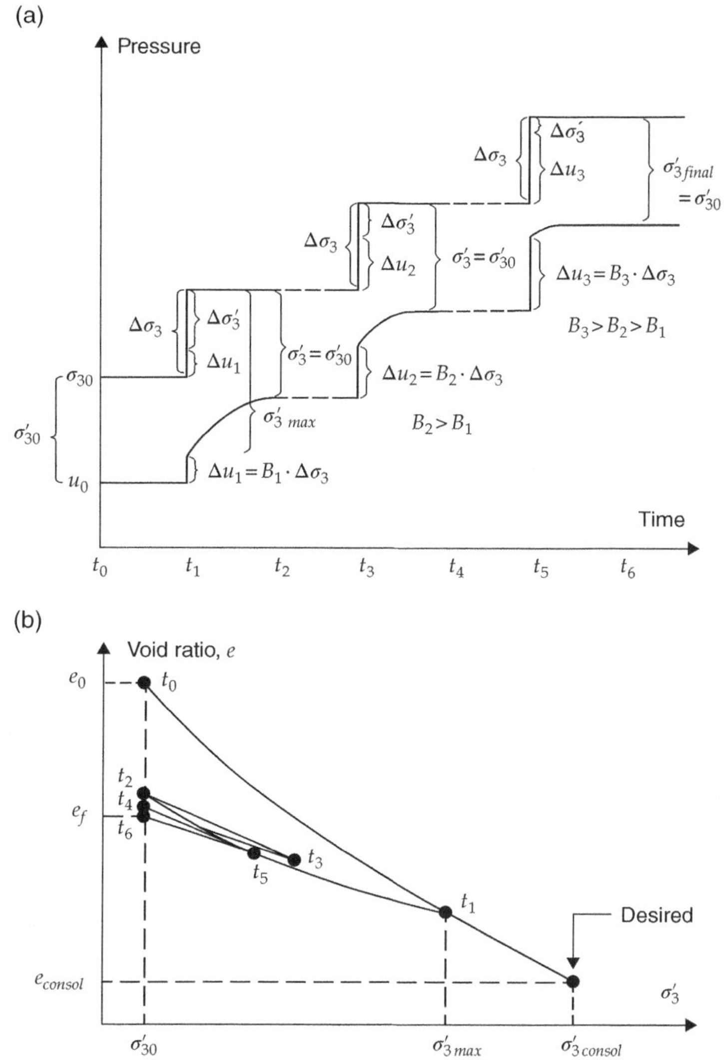

Figure 6.17 Stage saturation consisting of application of three equal increments of back pressure and cell pressure to a partly saturated specimen in preparation for correct consolidation of soil. (a) Sequential application of cell and back pressure with time and (b) effects on effective pressure–void ratio relation.

coarse sands and cemented sands were initially exposed to vacuum of different magnitudes before application of back pressure. They included the effect of surface tension in air bubbles [Eq. (6.21)] present in the partly saturated specimen in the expression for the required back pressure:

$$u_{back} = u_0 \cdot \frac{1 - S_0 \cdot (1 - H)}{H} + \frac{2 \cdot T_s}{H} \cdot \frac{(1 - S_0)}{r_0} \quad (6.36)$$

in which $r_0 = 0.1$ mm was assumed for the initial radius of the air bubbles. Schuurman (1966) suggested that for degrees of saturation greater than 85%, the pore air is present as individual bubbles rather than as a continuous phase. The addition of the surface tension term for such high degrees of saturation therefore has relatively little impact.

Figure 6.19 shows the influence of an initial vacuum on the subsequent required back pressure to reach full saturation, as calculated

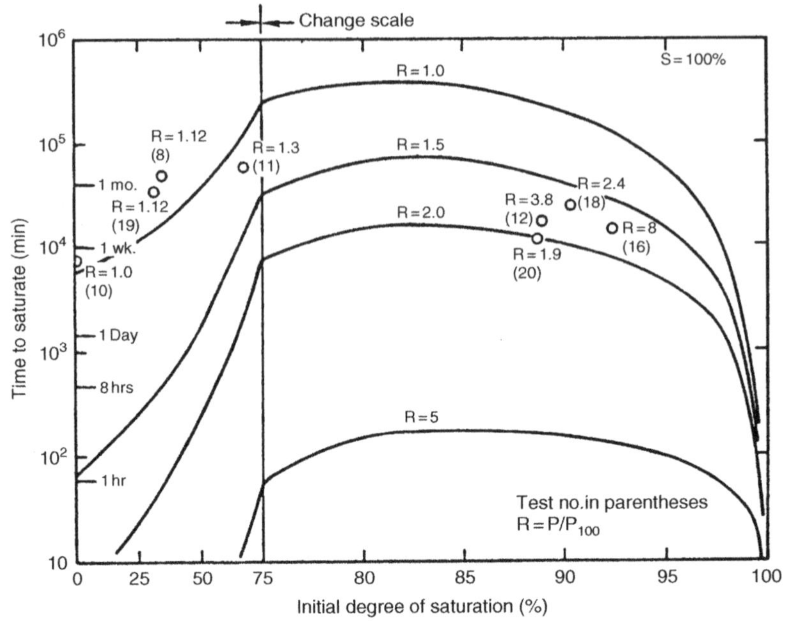

Figure 6.18 Time required for complete saturation plotted versus the initial degree of saturation for different values of $R = u_{back}/u_{back,100}$. Reproduced from Black and Lee 1973 by permission of ASCE.

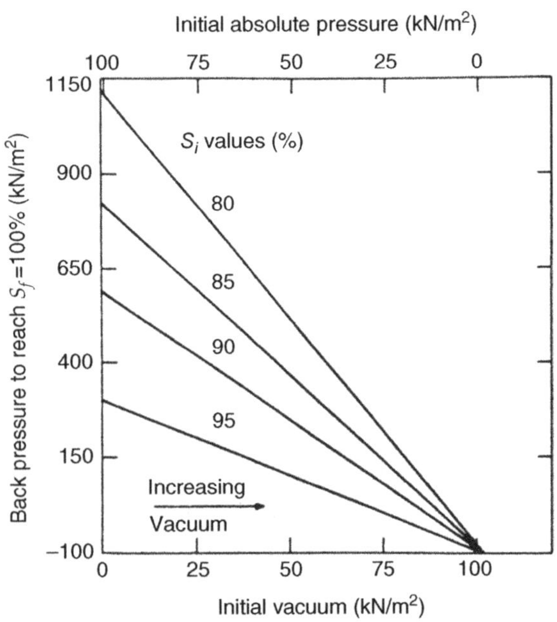

Figure 6.19 Back pressure required to cause complete saturation after initial application of vacuum. Reproduced from Rad and Clough 1984 by permission of ASCE.

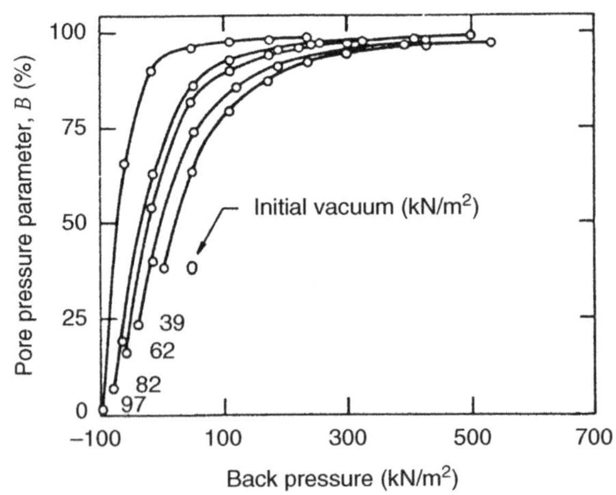

Figure 6.20 Effect of amount of applied vacuum on back pressure values for Monterey No. 3 sand at a relative density of 80%. Reproduced from Rad and Clough 1984 by permission of ASCE.

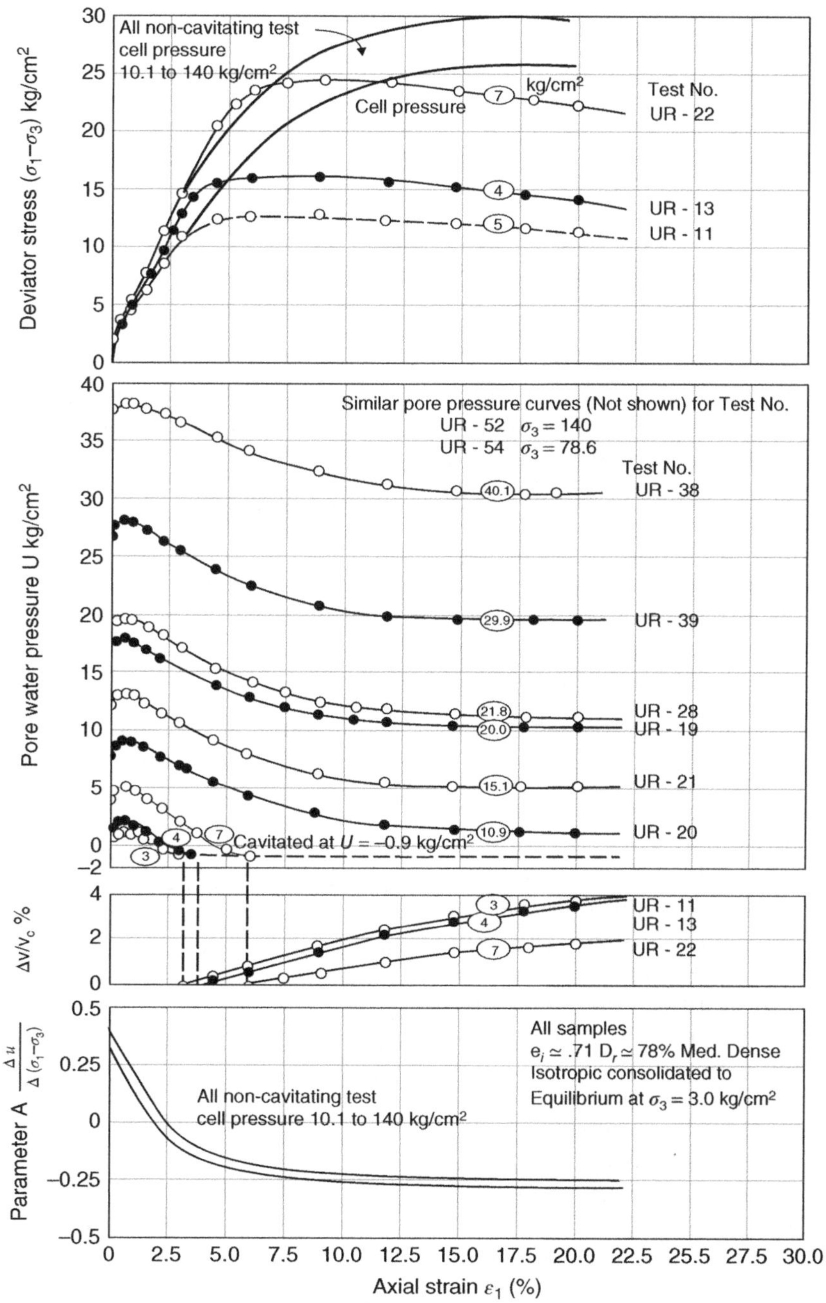

Figure 6.21 Experiments with different back pressure magnitudes indicate negligible effects on the observed behavior (after Lee 1965).

from Eq. (6.36). This diagram shows that for any initial degree of saturation, a lower back pressure is required to obtain full saturation if a vacuum is initially applied to the specimen. Theoretically, if an absolute vacuum (17 mm Hg) is first applied to the specimen, only a nominal absolute back pressure is required to achieve full saturation. However, since water boils at room temperature at an absolute pressure of 17 mm Hg = 2.3 kPa (absolute) = 99.0 kPa below atmospheric pressure (= −99.0 kPa relative to atmospheric pressure), this is the lowest possible vacuum that can be applied to the water in the specimen. At this value, the water releases all dissolved air and water vapor comes out of solution. Release of the vacuum followed by inflow of de-aired water results in full saturation with no required back pressure in excess of atmospheric pressure.

During application of vacuum, a lower vacuum may be applied to the triaxial cell to avoid undesired increases in the effective consolidation pressure. The maximum required length of vacuum application is 5 min for granular materials with any density, particle sizes, and initial degree of saturation.

Figure 6.20 shows an example of the effects of initial application of vacuum to dry specimens and the subsequent back pressure to cause a certain *B*-value. The line corresponding to the initial vacuum of zero represents the case of conventional application of back pressure. This diagram clearly indicates the beneficial effects of the vacuum procedure. Saturated specimens can be obtained at atmospheric pressure by initially exposing the specimen to a sufficiently low, absolute pore pressure. However, also less than the highest possible vacuum promotes saturation of soil specimens when followed by much lower back pressures than usually required.

Figure 6.21 indicates that the effect of magnitude of back pressure is negligible on the measured stress–strain and pore water pressure relations for triaxial compression tests on Sacramento River sand. Thus, the back pressure can be determined freely without any consideration of its effect on the measured behavior.

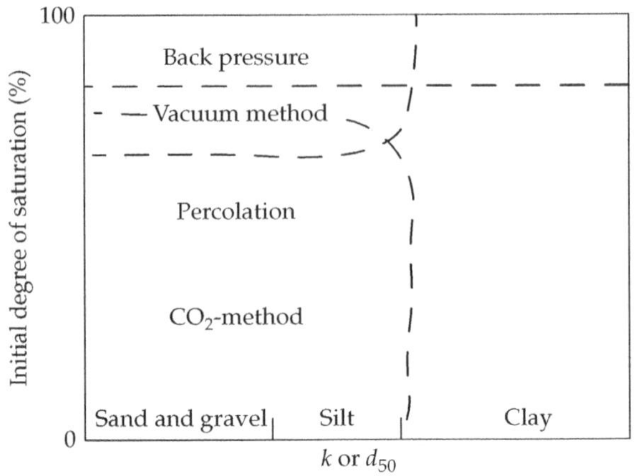

Figure 6.22 Estimate of the soils and the conditions to which the saturation methods appear to be most suitable. The saturation methods overlap in terms of the soils to which they apply. *k*, hydraulic conductivity; d_{50}, mean grain size.

Most effective combination of saturation methods

The most efficient method of specimen saturation includes a combination of the vacuum procedure and a nominal back pressure. Thus, following the vacuum procedure, which may not produce 100% saturation in the first vacuum cycle, another cycle may be employed with resulting increase in *B*-value. Finally, the specimen may be exposed to a nominal back pressure such as 200 kPa, which after resting on the specimen for a short time, may result in 100% saturation. Such a combination of procedures was employed to a partially dried specimen of "dirty" sand, and the specimen was brought to full saturation within a couple of hours.

6.7 Range of application of saturation methods

While the saturation methods reviewed above are clearly most applicable to particular soils, the methods do overlap in terms of the soils to which they apply. Figure 6.22 gives an estimate of the soils and the conditions to which they appear to be most suitable.

7 Testing Stage I: Consolidation

7.1 Objective of consolidation

The objective in the first stage of the triaxial test is to consolidate the specimen to establish a base for evaluation of the test results. All aspects of the soil behavior depend on the previous stress history and the current effective confining pressure acting on the soil. It is therefore required that the initial stresses are applied and sufficient time is allowed for complete consolidation to occur under the applied stresses. These consolidation stresses together with the previous stress history are the bases for evaluating the soil behavior to be obtained during the subsequent shearing stage discussed in Chapter 8. The properties of the soil relating to the rate of consolidation are combined in the coefficient of consolidation, c_v, which may be determined from the time curve obtained during consolidation.

Depending on the size of the soil sample recovered from the field, multiple specimens may be trimmed from the same elevation. Three specimens may be produced for testing over a range of confining pressures to establish the failure envelopes corresponding to the consolidated-undrained or to the consolidated-drained strength envelopes. However, the engineer should not just send requests to the laboratory without considering the testing procedure in relation to the project to which the results are to be applied, and the laboratory technician may not understand the request for complicated testing procedures. Therefore, interaction between the engineer and the laboratory technician is necessary to obtain relevant and satisfactory results of the testing.

7.2 Selection of consolidation stresses

Since the concept of testing is to simulate the field loading conditions as closely as possible, consolidation should ideally occur under the same effective and pore pressures as those present in the field before shearing. Because the effective strength envelope is curved, it is important to perform the tests in the range of stresses to be encountered in the field.

For deep deposits of uniform soil, specimens from shallow depth may be consolidated at higher stresses to simulate the conditions and obtain shear strengths corresponding to deeper regions of the deposit. While this may be a generally acceptable concept, practice indicates that soil deposits are rarely uniform but vary with depth, or they may be sensitive deposits that change character when consolidated to higher

Triaxial Testing of Soils, First Edition. Poul V. Lade.

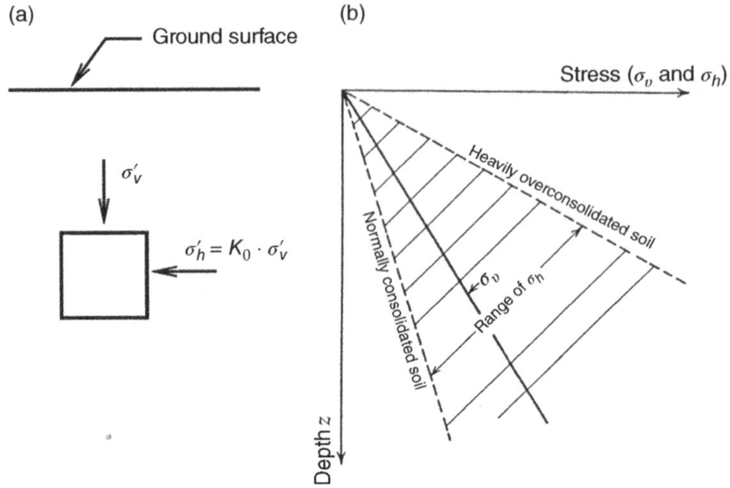

Figure 7.1 (a) Effective stress state under approximately horizontal ground surface and (b) possible range of horizontal stresses. Reproduced from Lambe and Whitman 1979 by permission of John Wiley & Sons.

stresses, which may cause disturbance of the soil fabric. Nevertheless, the behavior of soils may be studied over large ranges of confining pressure as long as the limitations of such studies are realized.

Ideally, it is most desirable to consolidate the specimen to the stresses acting in the ground at the depth from where the specimen has been retrieved, or possibly slightly higher, as explained below. Since these stresses are rarely if ever isotropic, anisotropic stress states should be used for the initial consolidation in the triaxial test. However, it is more difficult to consolidate a specimen anisotropically than isotropically, so isotropic consolidation is often employed in practice.

7.2.1 Anisotropic consolidation

Anisotropic stress ratios corresponding to K_0-stress conditions occur in field deposits with approximately horizontal ground surface, as shown in Fig. 7.1(a). In this case the effective, horizontal stress may be written as:

$$\sigma_h' = K_0 \cdot \sigma_v' \tag{7.1}$$

in which K_0 is the coefficient of earth pressure at rest and σ_v' is the effective, vertical stress. The latter may be calculated with accuracy as the static overburden pressure:

$$\sigma_v' = \sum_{i=1}^{N} (\gamma_i \cdot h_i) - \gamma_w \cdot h_w \tag{7.2}$$

in which h_i is the thickness of the soil strata with mass density γ_i, and h_w and γ_w are the depth below the ground water table and the density of water, respectively. The range of possible effective, horizontal stresses is shown in Fig. 7.1(b).

Normally consolidated soil

For normally consolidated soils the value of K_0 may be obtained from (Jaky 1948; Bishop 1958; Ladd *et al.* 1977):

$$K_0 = 1 - \sin\varphi' \tag{7.3}$$

in which φ' is the effective friction angle obtained from triaxial compression tests. The variation of K_0 calculated from Eq. (7.3) is shown in Fig. 7.2 along with the Mohr circle corresponding to an initial K_0-stress state for a normally consolidated soil.

Stress application

Ideally, the consolidation stresses should be applied to the normally consolidated specimen in such a manner that the state of stress never

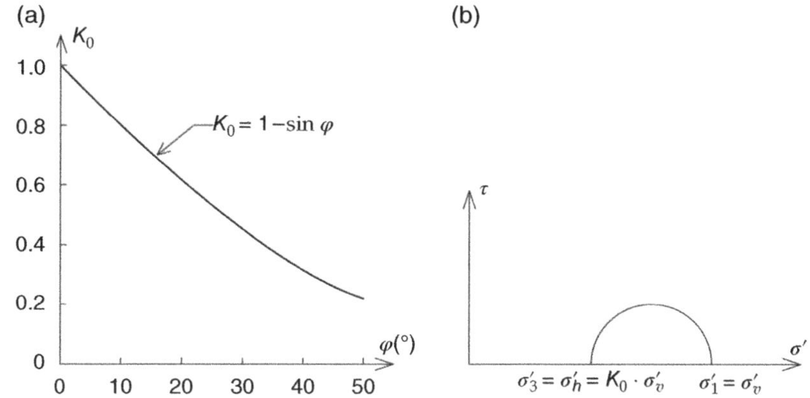

Figure 7.2 (a) Typical variation of K_0 with friction angle and (b) Mohr circle for K_0-stress state for normally consolidated soil.

pierces and causes the yield surface to be pushed out from its location in the ground. The yield surface for a normally consolidated soil is located around the K_0-line as shown in the triaxial plane in Fig. 7.3. Note that the yield surface is located a little further out than the current state of stress in the ground due to creep effects occurring over time. Thus, it is rare to find a truly normally consolidated soil for which the preconsolidation stress state matches the current stresses in the ground. To stay within the yield surface, the effective stress path during reestablishment of the K_0-stress state may simply follow the K_0-line, or it may follow a zigzag path consisting of alternating increments in effective confining pressure and deviator stress. Both types of stress paths are indicated in Fig. 7.3.

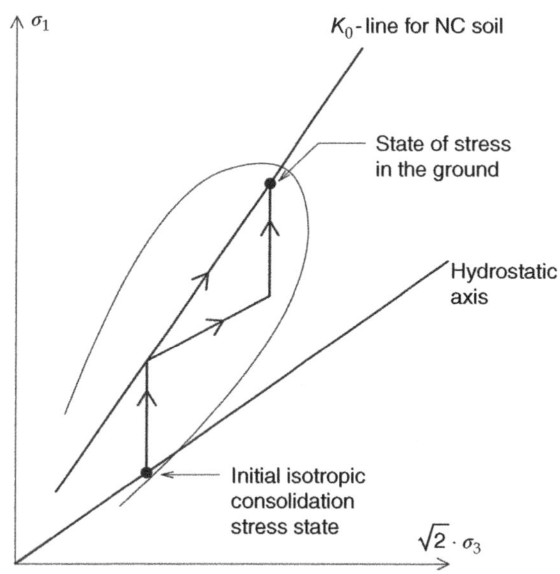

Figure 7.3 Yield surface for a normally consolidated (NC) soil located around the K_0-line. Note that the yield surface is located a little further out than the current state of stress in the ground due to creep effects occurring over time. Reestablishing the K_0-stress state may occur along the K_0-line or in small increment in confining pressure and deviator stress, as indicated.

Overconsolidated soils

For overconsolidated soils the value of K_0 may be estimated from the following approximation formula (Schmidt 1966; Mayne and Kulhawy 1982):

$$K_0 = (1 - \sin\varphi') \cdot OCR^{\sin\varphi'} \qquad (7.4)$$

The relationship between K_0 and ϕ' for various values of the overconsolidation ratio (OCR) are also shown in Fig. 7.4.

Alternatively, the value of K_0 may be estimated from the curves in Fig. 7.5 based on the plasticity index and the OCR for the soil. For K_0-values greater than unity the horizontal stress is greater than the vertical stress. Consolidation under such conditions requires application of a negative deviator stress.

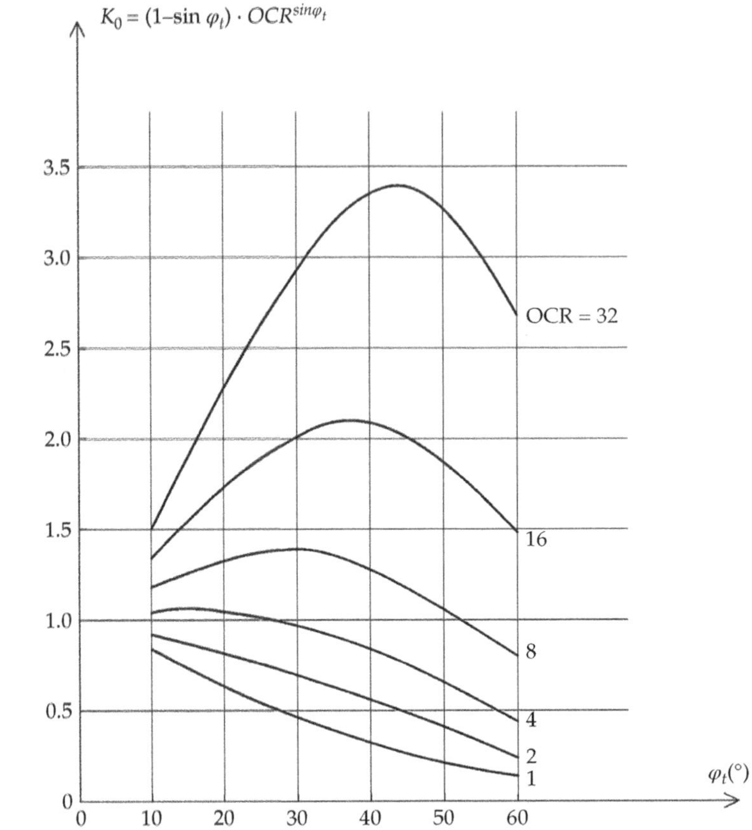

Figure 7.4 The K_0-value may be estimated from curves based on the effective friction angle and overconsolidation ratio (OCR) for the granular soil.

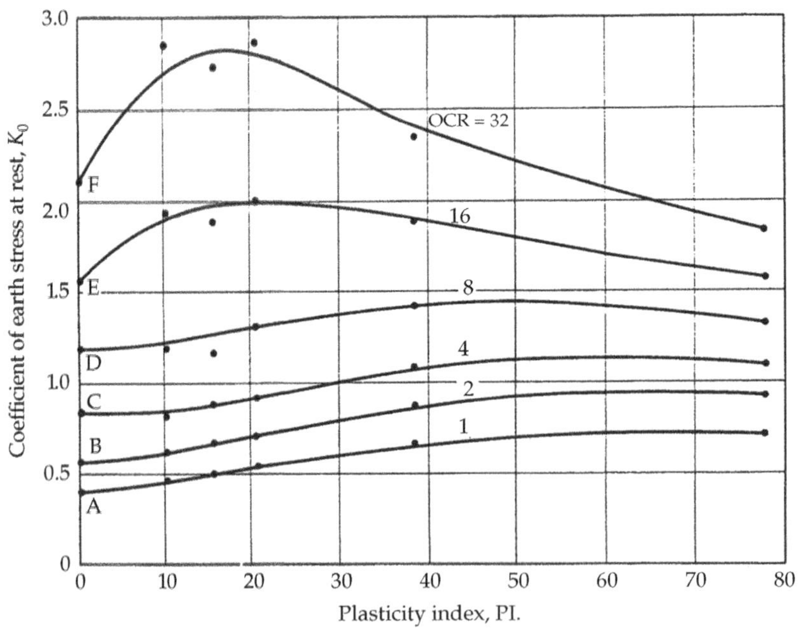

Figure 7.5 The K_0-value may be estimated from the curves based on the plasticity index and the overconsolidation ratio (OCR) for the soil. Reproduced from Brooker and Ireland 1965 by permission of Canadian Science Publishing.

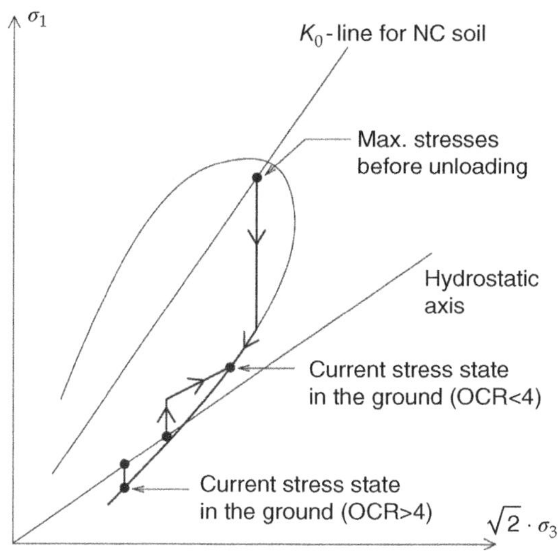

Figure 7.6 Establishing the K_0-stress state for an overconsolidated soil: one stress path follows a zigzag line and leads directly to the current stress state. The other path retraces the stress history, first along the K_0-line followed by K_0-unloading leading to an effective stress path, as shown. NC, normally consolidated; OCR, overconsolidation ratio.

Stress application

For an overconsolidated soil, for which the yield surface is located further away from the current state of stress in the ground, but with relatively low value of K_0, this state of stress may be reestablished by following one of two possible effective stress paths, as shown in Fig. 7.6. One path follows a zigzag line and leads directly to the current stress state. The other path retraces the stress history, first along the K_0-line followed by K_0-unloading leading to an effective stress path as indicated. Alternatively, the loading path may follow the zigzag line shown in Fig. 7.3 followed by the K_0-unloading path shown in Fig. 7.6.

It may be easier to reach the current state of stress for a soil with higher value of K_0, because the isotropic consolidation line is entirely within the yield surface. Consequently, following the isotropic consolidation line up to the value of σ_h' in the field, followed by application of the appropriate deviator stress produces the current state of stress in the ground.

According to Becker (2010), the effect of anisotropic consolidation is more important for the measured deformation parameters than for the measured strength.

It should be noted that consolidating normally consolidated clay under a relatively high stress ratio can be problematic, because the specimen can fail under undrained conditions, even if drainage has been provided. The stress path from an anisotropically consolidated specimen often is directed downward toward the apparent critical state point and is therefore unstable. Pore pressures can be generated faster than they can be dissipated in the test, leading to an undrained failure. This may be avoided by following the isotropic line and then applying a deviator stress under deformation control up to the estimated K_0-state.

7.2.2 Isotropic consolidation

It is most convenient to use isotropic consolidation in the triaxial test. In this case the three principal stresses are equal, and they are applied by the effective cell pressure ($\sigma_1' = \sigma_2' = \sigma_3' = \sigma_{cell}'$). If the piston is attached to the cap (see Chapter 3), a deviator load is required to compensate for the uplift pressure. The magnitude of the deviator load should be $\sigma_{cell} \cdot A_{piston}$, where σ_{cell} is the actual cell pressure used.

The magnitude of the isotropic consolidation pressure may be equal to the minor principal stress acting in the field. However, under such conditions the specimen may be found to swell. Frequently, the magnitude of the isotropic consolidation pressure is set equal to the effective overburden pressure, σ_v'. This is because the vertical, effective stress is often the major principal stress and this stress has been found to play the major role in the compression of soils (Rutledge 1947; Lee and Seed 1967). Choosing the vertical overburden pressure as the consolidation pressure avoids the complication of having to determine the horizontal stress acting in the field (or the value of OCR), and it most often results in compression of the specimen, thus causing it to become, in part, reconstituted, as explained below. For a normally consolidated

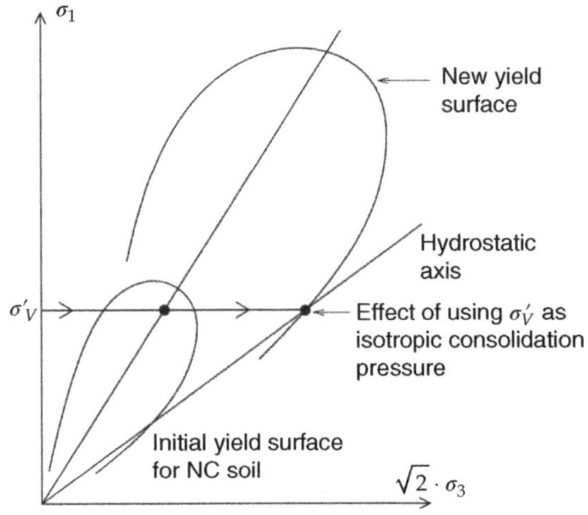

Figure 7.7 Schematic diagram showing the creation of a new yield surface when employing vertical stress as the isotropic cell pressure for a normally consolidated (NC) soil.

soil, employing the vertical stress as the cell pressure causes the yield surface to be pushed out, as shown in Fig. 7.7, while smaller effects are produced for overconsolidated soils. Note, however, that an isotropic state of stress only rarely occurs in the field, and it is for convenience that isotropic consolidation is employed in the performance of triaxial tests.

7.2.3 Effects of sampling

Removing a sample of soil from the ground inadvertently changes the state of stress acting on the soil in the field. The soil undergoes several disturbances and the state of stress before installation in the triaxial apparatus is far removed from that in the ground. Figure 7.8 llustrates the disturbance due to drilling down to the soil sample in the ground, tube sampling, extrusion from the tube, loss of suction in the sample (due to water redistribution and possibly due to water cavitation), and trimming and handling. The resulting loss in effective confining pressure can have severe consequences for the effective stress path, the stress–strain behavior, and the undrained strength in a UU-test, as illustrated in Fig. 7.9. Even if the specimen was sampled "perfectly" (in which only the stresses change from the K_0-state to the isotropic state) and then sheared from an initial isotropic stress state, Fig. 7.10 shows that the measured undrained strength of a normally consolidated clay is not as high as that experienced in the field.

However, it may be possible, at least in part, to reconstitute the specimen by reconsolidating it at a pressure higher than the minor principal stress in the field. This technique, which is described in further detail below, may not work for soils whose grain structure or fabric has been or will be severely or irrecoverably disturbed by reconsolidation to a higher pressure.

7.2.4 SHANSEP for soft clay

In the design method for clay deposits presented by Ladd and Foott (1974), the soil behavior is evaluated according to the concept of SHANSEP (stress history and normalized soil engineering properties). This concept is based on the fact that all aspects of the behavior of clay specimens from the same deposit and with the same OCR can be normalized on the basis of the consolidation pressure, σ_c'. Thus, the stress–strain relation, pore pressure, effective stress path and undrained strength can be normalized for each value of the overconsolidation ratio.

Figures 7.11 and 7.12 show examples of normalized soil behavior for normally consolidated clay and overconsolidated clay. To obtain reliable test data, which may be normalized as shown in the diagrams, it is necessary to consolidate specimens to stresses in excess of those at which they were consolidated in the field. This is done to establish maximum consolidation pressures located on the virgin compression curve, as exemplified by the dotted reconsolidation curve in Fig. 7.13. This is followed by unloading and swelling to create a specimen with known OCR and devoid of any effects of the sampling procedure. Thus, a specimen that has been disturbed during sampling, shown by the dotted line 1–2 in Fig. 7.13, is reconstituted (2–3) and again unloaded (3–4) before it is sheared.

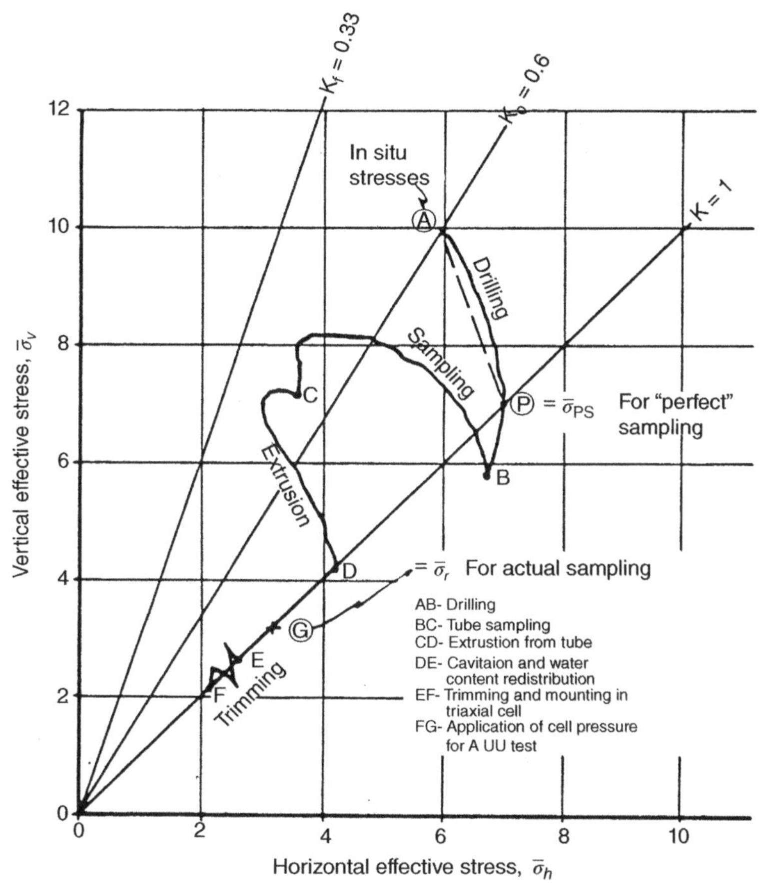

Figure 7.8 Diagram illustrating the disturbance due to drilling down to the soil sample in the ground, tube sampling, extrusion from the tube, loss of suction in the sample (due to water redistribution and possibly due to water cavitation), and trimming and handling. Reproduced from Ladd and Lambe 1963 by permission of ASTM International.

This procedure requires that the specimen be consolidated back to the virgin compression curve, and this necessitates consolidation pressures greater than 1.5 to 2 times the maximum vertical in situ consolidation pressure, σ_{vm}'. To standardize the effect of secondary compression, the last consolidation increment should be left on the specimen for one log cycle of secondary compression.

Ladd and Foott (1974) recommend using the following procedure:

1. Consolidate specimens to approximately 1.5, 2.5 and 4 times the in situ value of σ_{vm}' and determine the normalized undrained shear strength, s_u/σ_{vc}'. If the clay exhibits normalized behavior, the values of s_u/σ_{vc}' will be the same, at least at the two higher pressures. If the s_u/σ_{vc}'-values vary consistently with pressure, then the normalization procedure does not apply to the clay.
2. To obtain the relation between s_u/σ_{vc}' and OCR, use the minimum value of σ_{vc}' resulting in normalized behavior as the laboratory value of σ_{vm}', and perform tests at OCR-values of 2 ± 0.5, 4 ± 1, and 6 ± 2. Compare the results with those shown in Fig. 7.14 to check their reliability. The data points should form a smooth upward concave curve.

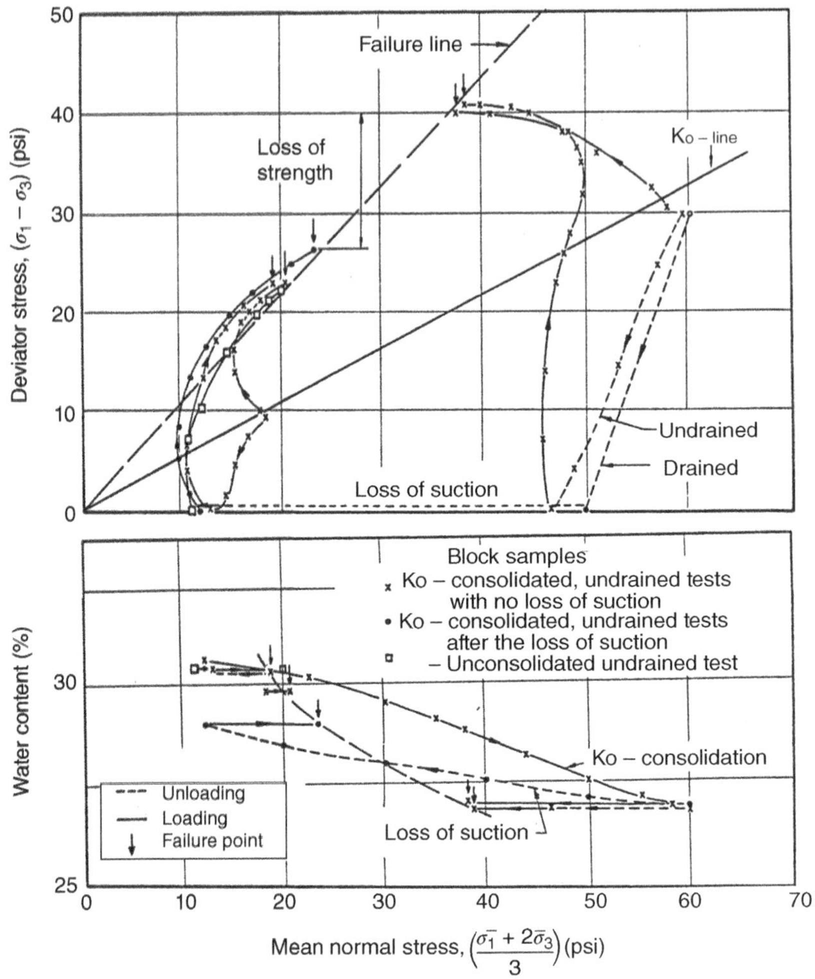

Figure 7.9 Diagram illustrating loss of suction with severe consequences for the effective stress path, the stress–strain behavior, and the undrained strength in a UU-test. Reproduced from Adams and Radhakrishna 1971 by permission of ASTM International.

To use the resulting soil behavior patterns in design, it is necessary to know the reconsolidation pressure in the field. This may be established with good accuracy from oedometer tests.

This method results in much better definition of the soil behavior than that produced directly from unconfined compression tests or UU-tests on "undisturbed" specimens from the field. Note that the reconsolidation pressures required in this procedure are much higher than the in situ stresses to overcome the sample disturbance effects. Reconsolidation to the in situ stresses does not overcome the effects of sampling.

Because the method depends on reconstitution of the soil specimens by reconsolidation to higher pressures, it may not be useful for very sensitive clays, in which the initial fabric is highly structured, or for cemented deposits. The initial structure or the cementation cannot be regenerated. In fact, it may be further broken down by reconsolidation.

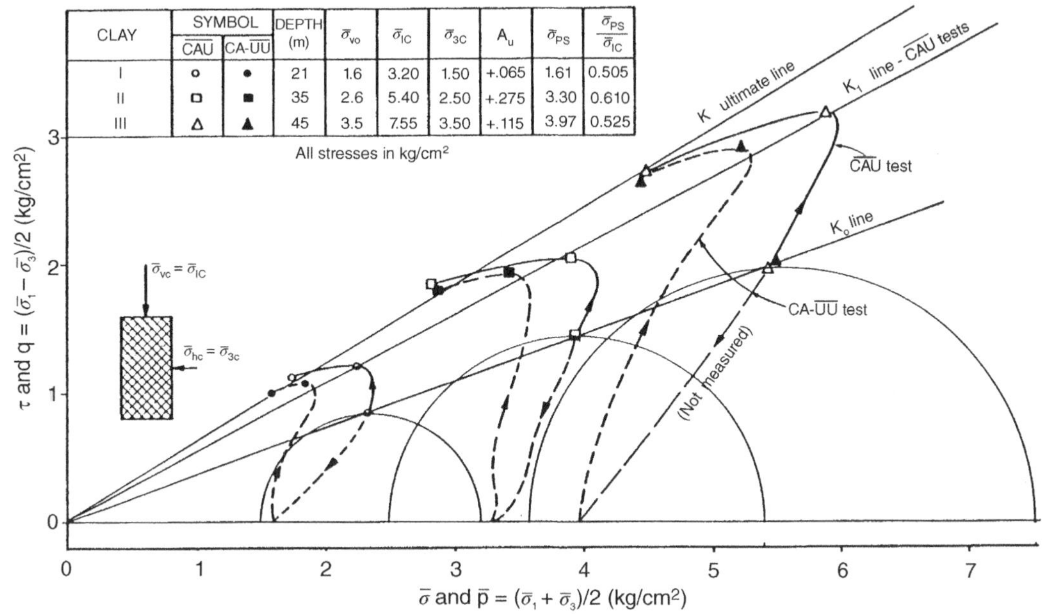

Figure 7.10 Effect of "perfect" sampling on stress paths and undrained strength for normally consolidated Kawasaki clays. Reproduced from Ladd and Lambe 1963 by permission of ASTM International.

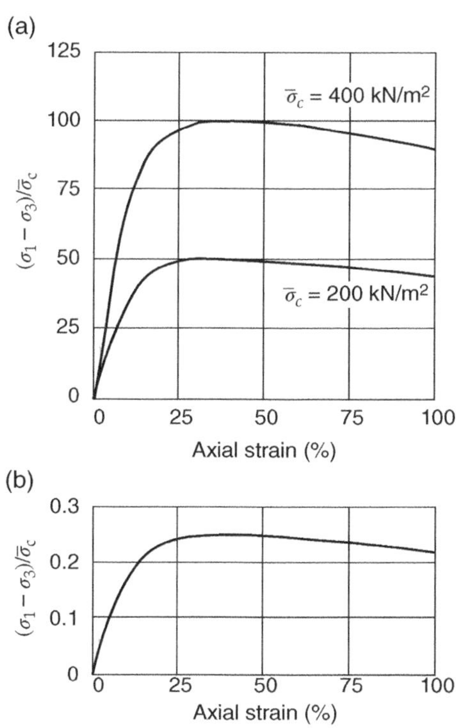

Figure 7.11 Example of normalized soil behavior for normally consolidated clay: (a) triaxial compression test data; and (b) normalized plot of triaxial test data. Reproduced from Ladd and Foott 1974 by permission of ASCE.

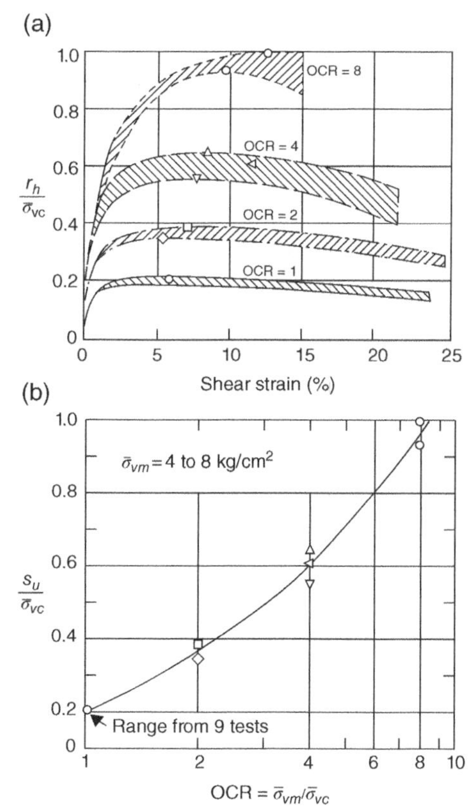

Figure 7.12 Example of normalized soil behavior for overconsolidated clay: (a) normalized stress versus strain; and (b) normalized undrained shear strength versus OCR. Reproduced from Ladd and Foott 1974 by permission of ASCE.

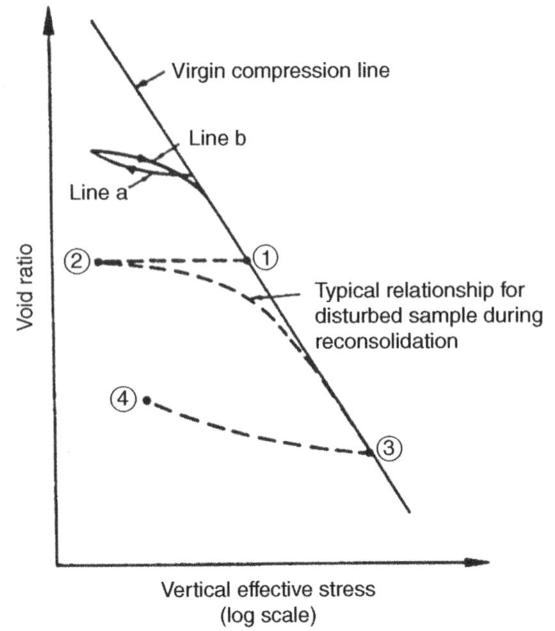

Figure 7.13 Schematic diagram showing effect of sample disturbance (1–2), reconsolidation (2–3) and unloading to create a specimen with known overconsolidation ratio (3–4). Reproduced from Ladd and Foott 1974 by permission of ASCE.

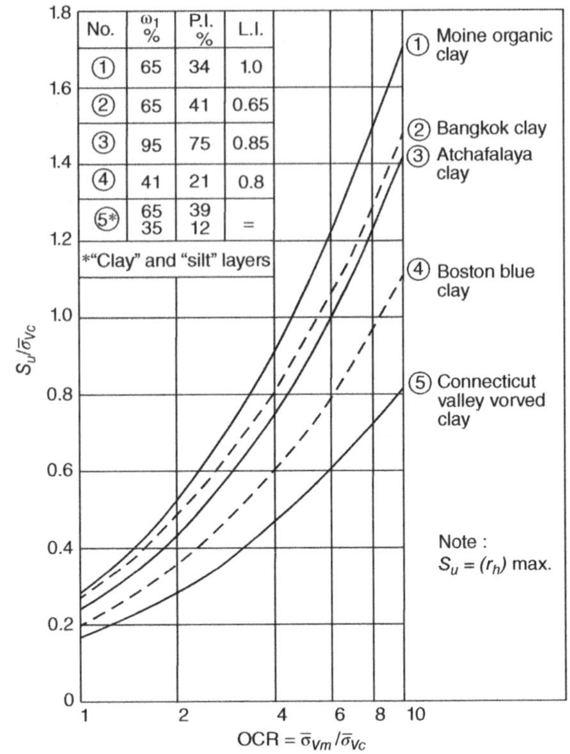

Figure 7.14 Variation of normalized undrained shear strength with overconsolidation ratio for five clays. Reproduced from Ladd and Foott 1974 by permission of ASCE.

7.2.5 Very sensitive clay

The SHANSEP method is not recommended for sensitive clays in which the fabric may be disturbed by increasing the effective confining pressure to magnitudes significantly above that in the ground.

7.3 Coefficient of consolidation

The coefficient of consolidation, c_v, contains information relating to the rate of consolidation:

$$c_v = \frac{k}{\gamma_w \cdot C_d} \tag{7.5}$$

in which k is the hydraulic conductivity, γ_w is the unit weight of water and C_d is the volume compressibility of the soil skeleton. The value of c_v may be used to determine the strain rates to be employed during shearing in drained as well as in undrained tests to ensure that a required degree of consolidation or a required degree of pore pressure equalization is achieved in these tests.

7.3.1 Effects of boundary drainage conditions

The value of c_v may be determined from one of the expressions given in Table 7.1. Each expression relates to the type of drainage conditions used in the triaxial test. End drainage is usually provided by filterstones and radial boundary drainage may be provided by filter paper side drains.

7.3.2 Determination of time for 100% consolidation

End drainage alone

The time for 100% consolidation, t_{100}, may be determined from the consolidation time curve plotted as volume change versus log of time, as shown in Fig. 7.15.

Alternatively, the $\sqrt{t}$ -method may be employed. The $\sqrt{t}$-method for end drainage, that is the

Table 7.1 Expressions for determination of the coefficient of consolidation, c_v, from triaxial tests (after Bishop and Henkel 1962) for isotropic clay, that is c_v(vertical) = c_v(horizontal)

Drainage conditions	t_{100}	t_{100} (for $h = 2R$)
Drainage from one end only	$\frac{\pi \cdot h^2}{c_v}$	
both ends	$\frac{\pi \cdot h^2}{4 \cdot c_v}$	
radial boundary only	$\frac{\pi \cdot R^2}{16 \cdot c_v}$	$\frac{\pi \cdot h^2}{64 \cdot c_v}$
both ends and radial boundary	$\frac{\pi \cdot h^2}{4 \cdot c_v}\left[\frac{1}{(1+2h/R)^2}\right]$	$\frac{\pi \cdot h^2}{100 \cdot c_v}$

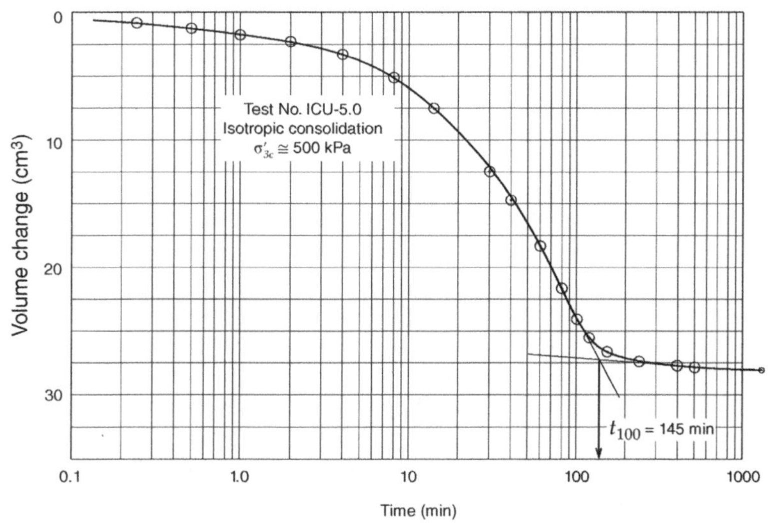

Figure 7.15 The time for 100% consolidation, t_{100}, may be determined from the consolidation–time curve plotted as volume change versus log of time, as shown for isotropic consolidation of an Edgar Plastic Kaolinite clay specimen consolidated at an effective cell pressure of 500 kPa.

pore water percolates vertically to the end drains, was presented by Taylor (1948). This method is based on a linear relation between the degree of consolidation and the square root of time to about 50% of consolidation, after which it curves and becomes asymptotic at 100% consolidation. When plotting the volume change on the vertical axis against the square root of time on the horizontal axis, and drawing a straight line from degree of consolidation, $U = 0\%$ to the point on the curve corresponding to $U = 90\%$, then this point has an abscissa of $x \cdot (1+a) = 1.15$ times the abscissa of the straight portion of the initial curve. The value of x is the horizontal distance from the vertical axis to the straight-line consolidation curve. This is demonstrated in Fig. 7.16. Once the point of 90% consolidation has been determined, then 1/9 of the distance from U_0 to U_{90} is added on the U-axis to reach the point of 100% consolidation, as shown in Fig. 7.16. This method of analyzing the time curve is advantageous if the point is required at or near the time of 100% consolidation. In comparison, analyzing the

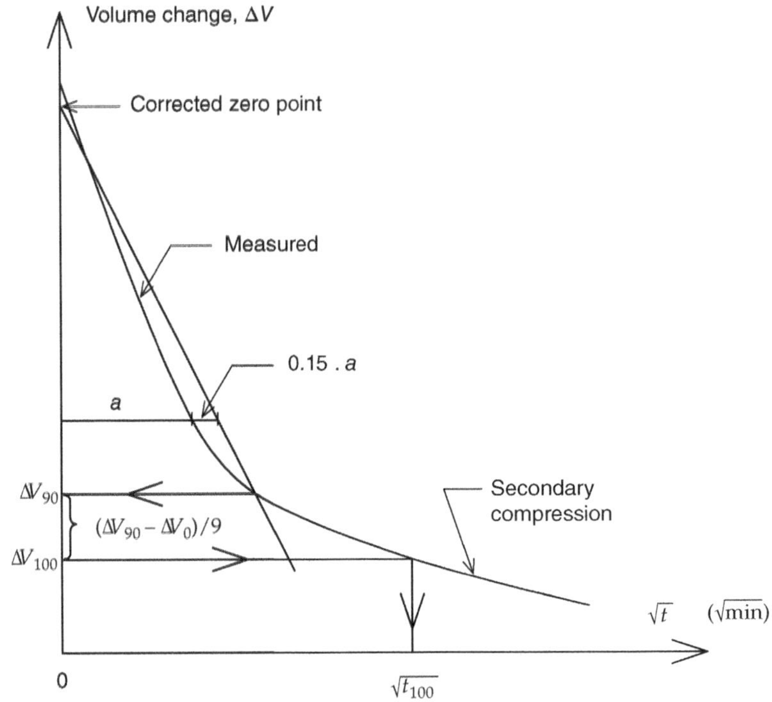

Figure 7.16 The $\sqrt{t}$ -method in which the volume change is plotted versus the square root of time, and analyzed as indicated to determine t_{100}.

time effects on a *U*–log(time) plot does not provide this information, because the straight line obtained from the secondary compression is not available until later after a fair amount of creep has occurred.

All-around drainage

Speeding up the consolidation of cylindrical specimens of clay for triaxial testing by combined end and radial drainage and subsequently determining the time for end-of-consolidation may be important for studying time effects such as creep and stress relaxation, which clearly occurs after 100% consolidation has been achieved. Radial drains are more effective than end drains in triaxial testing, because the clay may be cross-anisotropic and more permeable in the horizontal direction and because the distance to the radial drainage boundary is smaller than the height of the specimen.

Many authors have presented research on the results of radial drainage (e.g., Carillo 1942; Barron 1948; Gibson and Lumb 1953) and they have produced equations for analysis of the time for end (Terzaghi 1925) and radial consolidation (Gibson and Lumb 1953; Silveira 1953; McKinlay 1961). The combined action of end and radial consolidation has also been presented and the following expression for the average degree of consolidation, U_{vr}, has been determined (Carillo 1942):

$$U_{vr} = 1 - (1 - U_v)(1 - U_r) \tag{7.6}$$

in which U_v is the degree of consolidation by end drainage and U_r is the degree of consolidation by radial drainage.

The time to reach a certain degree of consolidation for end drainage alone can be expressed as:

$$t_v = \frac{T_v \cdot (H/2)^2}{c_v} \tag{7.7}$$

in which T_v is the theoretical time factor for one-dimensional consolidation, H is the height of

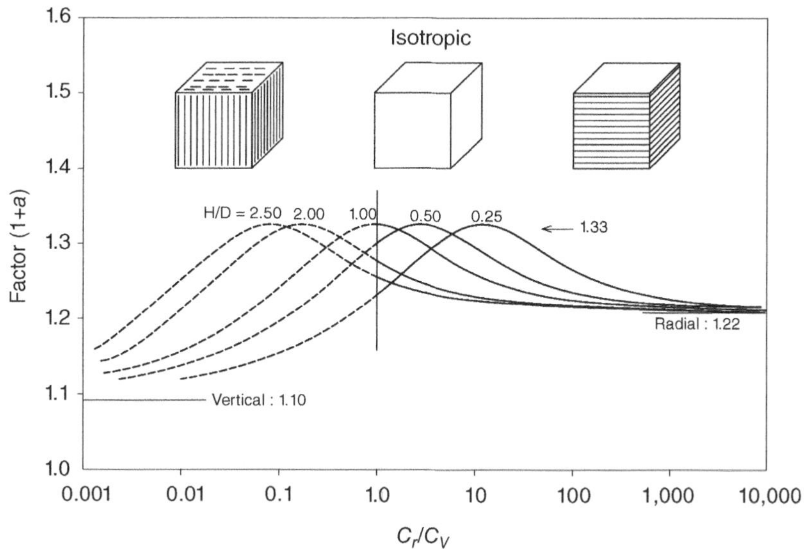

Figure 7.17 Factor (1+*a*) for determination of U = 90% used for the determination of t_{100} for all-around drainage consisting of end and radial drains by plotting volume change versus (time)$^{0.465}$ and analyzing the time curve as indicated.

the cylindrical specimen, $H/2$ is the maximum distance to a free surface of the specimen and c_v is the coefficient of one-dimensional consolidation. Similarly, the time to reach a certain degree of consolidation for radial drainage alone can be written as:

$$t_r = \frac{T_r \cdot (D/2)^2}{c_r} \quad (7.8)$$

in which T_r is the theoretical time factor for radial consolidation, D is the diameter of the cylindrical specimen, $D/2$ is the maximum distance to a free surface of the specimen and c_r is the coefficient of radial consolidation.

For the same time, $t_v = t_r$, the expressions in Eqs (7.7) and (7.8) can be set equal to produce:

$$T_v = T_r \cdot \left(\frac{c_v}{c_r}\right)\left(\frac{D}{H}\right)^2 \quad (7.9)$$

For drainage boundaries other than end drains, the value of $a \neq 0.15$. For radial drainage alone, McKinlay (1961) pointed out that the relation between degree of consolidation U and (time)$^{0.465}$ (not $\sqrt{\text{time}}$) is straight up to about U = 50%. The point on the curve corresponding to U = 90% has an abscissa of 1.22 of the abscissa of the straight portion of the consolidation curve. For combinations of end and radial drainage, the values of (1+*a*) are determined as shown below.

Equation (7.9) shows that the coefficients of consolidation and the dimensions of the specimen with all-around drainage controls the time for consolidation. The factors (1+*a*) corresponding to 90% consolidation have been determined for radial drainage (Carillo 1942; Barron 1948) and plotted in Fig. 7.17 for various degrees of cross-anisotropy and for various geometries of the cylindrical specimens. Since the cross-anisotropy expressed through c_r/c_v is greater than or equal to 1.0 (isotropic behavior), the factors for $c_r/c_v \geq 1.0$ are given by solid lines, while those for $c_r/c_v < 1.0$ are shown by dashed lines, because they are not likely to be employed.

Figure 7.17 shows that the specimen geometry plays an important role in determination of the factor for 90% consolidation. The curves for different geometries are similarly shaped, but shifted according to the ratio of *H/D*. If the degree of cross-anisotropy expressed by the value of c_r/c_v is known in advance, then the factor (1+a) can be determined from Fig. 7.17. For isotropic clay, that is remolded clay specimens, the values may be selected on the basis of H/D. However, for cases where the degree of cross-anisotropy is not known in advance of testing, the value of (1+a) is between 1.22 and 1.33. For a given value of H/D, the value of (1+a) may be evaluated from the diagram in Fig. 7.17. The value will be higher than 1.15 used for the $\sqrt{t}$ -method.

It is possible to get a feel for the extent to which pore pressures have been dissipated by monitoring them after the drainage valve is turned off and before the shearing proceeds. If full consolidation has been achieved, the pore pressure should settle down to the value previously set as the back pressure. Generally, it will be a bit above that value, because full consolidation has not been achieved.

The coefficient of consolidation, c_v, may be determined by substitution of t_{100} and the dimensions of the specimen into the appropriate expression from Table 7.1.

Note that the calculations presented above are for ideal cross-anisotropic soils with no effects of the possible radial paper drain inefficiency or smear at the specimen surfaces. The value of c_v determined from the triaxial test may not be applicable to the field deposit due to anisotropic soil behavior, inefficiency of side drains, and so on. However, it is applicable for determination of appropriate strain rates for drained or undrained shearing of the same specimen. This will be discussed in Chapter 8.

8 Testing Stage II: Shearing

8.1 Introduction

It is during the shearing stage of the triaxial test that the stress–strain, volume change or pore pressure relations, and the drained or undrained strengths are determined. In addition to the behavior itself, the most important factors that control the measured soil behavior are the strain rate, the effects of lubricated ends, and the specimen shape.

8.2 Selection of vertical strain rate

8.2.1 UU-tests on clay soils

The strain rates chosen for unconsolidated-undrained (UU) tests are often determined by how fast the readings can be recorded manually, and strain rates for such tests may be up to 1%/min. These tests typically produce too low shear strength due to disturbance, and effects of strain rate due to the viscous behavior of clays cannot make up for this loss.

8.2.2 CD- and CU-tests on granular materials

Freely draining granular materials, such as gravel, sand, and non-plastic silts, whether tested under consolidated-drained (CD) or consolidated-undrained (CU) conditions, may be sheared at reasonably high rates. The only consideration to be made for such soils is the influence of the strain rate on the soil behavior due to time effects. However, granular materials do not exhibit significant time rate effects, that is essentially the same stress–strain, volume change or pore pressure and strength behavior is obtained for any strain rate.

8.2.3 CD- and CU-tests on clayey soils

Clayey soils, in which the clay fraction dominates the soil behavior and in which the particles consist of physico-chemically active minerals, exhibit classic viscous behavior, and they show greater stiffness and higher shear strength with increasing strain rate, as indicated on Fig. 8.1. For triaxial compression tests on soft clay, each log cycle increase in strain rate is typically accompanied by 10 ± 5% increase in undrained shear strength (Ladd and Foott 1974), the actual variation being a function of the plasticity and the susceptibility to time effects of the soil.

However, soils whose hydraulic conductivities are relatively low require shearing at low strain rates to assure that pore pressures of only negligible magnitudes develop under drained

Triaxial Testing of Soils, First Edition. Poul V. Lade.

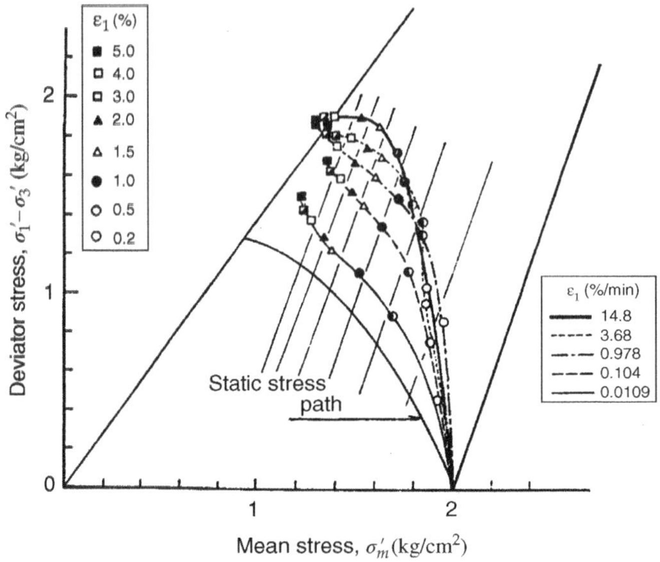

Figure 8.1 Effect of strain rate in a CU-test on effective stress path, pore pressure generation, shear strength and stiffness of clay. Reproduced from Akai *et al.* 1975 by permission of Elsevier.

conditions, and that the pore pressures are practically equalized in undrained tests with end restraint. Each of these two cases is treated below.

Drained tests

From the theory of consolidation it is possible to express the time, *t*, to reach a certain degree of consolidation or dissipation of pore pressure generated during shear as:

$$t = \frac{T \cdot H^2}{c_v} \tag{8.1}$$

in which T is the time factor, H is half of the specimen height, and c_v is the coefficient of consolidation. The time factor is a function of the degree of consolidation, U:

$$T = f(U) \tag{8.2}$$

It is possible to develop theoretical expressions for this relation which depends on the drainage conditions in the triaxial test (Gibson and Henkel 1954; Bishop and Henkel 1962). The relation between T and U may also be obtained from experiments performed in triaxial tests with appropriate drainage conditions. Blight (1963) expressed the volume change occurring in drained triaxial tests in relation to the stresses that caused the volume change:

$$A_v = \frac{\Delta V / V_0}{\Delta(\sigma_1 - \sigma_3)} \cdot \sigma_{3c}' \tag{8.3}$$

And the degree of drainage was then expressed as:

$$U = \frac{A_v}{A_{v100}} \tag{8.4}$$

in which A_{v100} is the ultimate value of A_v reached when full drainage has occurred in a very slow test.

Figure 8.2(a) shows the relations obtained between the degree of drainage and the time factor determined from Eq. (8.1). It is seen that the experimental results are displaced relative to the theoretical predictions for both types of drainage applied in the tests.

A degree of drainage and therefore a degree of dissipation of pore pressures in the drained test of 95% is sufficient to ensure negligible error in the measured soil properties. The required time to reach 95% dissipation may be calculated from Eq. (8.1) with a time factor

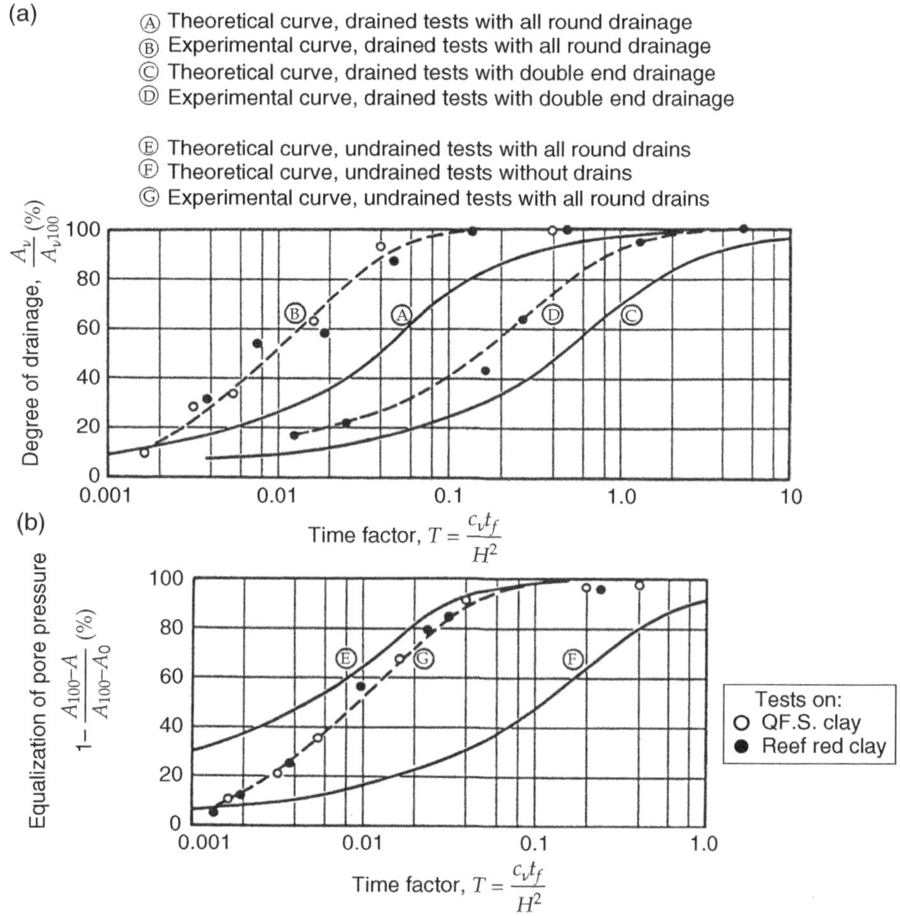

Figure 8.2 (a) Relations between degree of drainage and the time factor in drained tests and (b) relations between equalization of pore pressure and the time factor in undrained tests. Reproduced from Blight 1963 by permission of ASTM International.

obtained from Fig. 8.2(a). For tests with all around drains, that is filter stones at the ends and side drains of slotted filter paper, the time for 95% dissipation is (Blight 1963):

$$t = 0.07 \cdot \frac{H^2}{c_v} \tag{8.5}$$

For tests with double end drainage, that is with filter stones at both ends, the time for 95% dissipation is (Blight 1963):

$$t = 1.6 \cdot \frac{H^2}{c_v} \tag{8.6}$$

The time factors used in these expressions are those corresponding to the experimental results in Fig. 8.2(a). Note that H is half of the specimen height and c_v is the coefficient of consolidation determined during the preceding consolidation stage.

The time t required for 95% dissipation is the time for the first significant measurement in the drained test. The first significant measurement may be the strength, in which case t is the time to failure. However, if it is desirable to obtain the correct stress–strain relation, then the first significant stress measurement may occur much earlier in the test, say at 25 or 50% of the strength.

Since the volume change development is nonlinear and becomes slower with increasing axial strain, choosing the first significant measurement to occur at, say, 25% of the strength results in better than 95% equalization for the remainder of the stress–strain relation. Thus, for cases where the stress–strain and volume change behavior is important, such a condition may be employed when determining the test duration and the strain rate.

In either case, it is necessary to estimate the strain, ε_{sig}, to reach the first significant measurement. The axial strain rate for the drained test may then be calculated as:

$$\frac{d\varepsilon_a}{dt} = \frac{\varepsilon_{sig}}{t} \tag{8.7}$$

The strain rate may be converted to a vertical deformation rate to be set on the loading machine.

It can be rather time consuming to perform drained tests on soils with low permeability. Unfortunately, there is no effective method of speeding up the drainage of pore water from the soil other than using small specimens with both end and side drains.

Determination from drained experiments

Using the Cam Clay model for normally consolidated clay, Newson *et al.* (1997) predicted the deviation from the required effective stress path for a drained test. They then made corrections to the experimental findings to reach a deformation rate that would yield essentially drained conditions. Thus, a trial-and-error method was employed to obtain a suitable strain rate for the drained test.

Similarly, the trial-and-error procedure to determine the proper axial strain rate for the true triaxial drained tests on normally consolidated kaolin clay performed by Anantanasakul *et al.* (2012) can be summarized as follows: the most critical testing condition (in terms of drainage) in which the axial strain-to-failure was smallest and the specimen hydraulic conductivity was lowest was first determined. Three isotropic consolidation tests at an effective confining stress of 250 kPa were performed on cross-anisotropic specimens with all three orientations using the same installation procedure in the true triaxial apparatus. The results indicated that the hydraulic conductivity of the specimen with the z-axis parallel to the σ_3-direction was the lowest of the three. Judging from the stress–strain results presented by Wang and Lade (2001), the strains-to-failure of tests with $b = (\sigma_2 - \sigma_3)/(\sigma_1 - \sigma_3) = 0.67$ tests were assumed to be the smallest of the four b-values considered. Four true triaxial tests with $b = 0.67$ on horizontal specimens whose axes of material symmetry coincided with the σ_3-direction ($\theta = 160^{\circ}$) were performed at axial strain rates of 0.01, 0.005, 0.001, and 0.0005 %/min. The mean stress was kept constant at 250 kPa and drainage from the specimens was allowed during shear. In each test, shearing was halted and the stresses were held constant when the vertical strain was 0.25%. The drain line to the volume change device was then closed, creating undrained conditions. The pore pressures increased and stabilized within about 2 h. The drain line was then opened and the test continued. The same procedure was repeated at vertical strains of 2.5 and 6%. The stabilized pore pressures were found to decrease with decreasing strain rate and increasing vertical strain. The strain rates of 0.001%/min resulted in very slightly larger pore pressures than those of 0.0005%/min. The pore pressure changes at the rate of 0.001%/min were 7, 4, and 2% of the cell pressures when the vertical strains were 0.25, 2.5, and 6%, respectively.

An axial strain rate of 0.001%/min (corresponding to a deformation rate of 0.003 in./min) was adopted for the true triaxial testing program. Although the axial strain rate resulted in partially drained conditions at the beginning of shear, it was still chosen because practically drained conditions were obtained at higher strains, and lower strain rates did not significantly decrease the excess pore pressures. This strain rate also allowed true triaxial tests to finish in a reasonable amount of time in which the apparatus functioned reliably. During the production testing, pore pressure changes were also monitored in selected true triaxial tests with other b-values using the same procedure discussed above. The monitored pore pressure changes were practically equal to or smaller than those obtained in the trial tests. In the subsequent development, fully drained conditions were assumed when the specimens were sheared under vertical strain control with a strain rate of 0.001%/min.

Alternatively, the volume changes in drained experiments may be measured and plotted to observe the required loading rate required to obtain practically drained conditions. When the volume change–axial strain relations do not change anymore as the loading rate decreases, then the loading rate may be sufficiently slow to produce drained conditions.

For cases where only the effective strength parameters are to be determined, drained tests may be replaced by CU-tests with pore pressure measurements, which are then interpreted in terms of effective stresses to yield the same strength envelope.

Undrained tests

In undrained tests with end restraint, the nonuniform strain distribution produces nonuniform pore pressure distribution. The extent to which equalization of pore pressure occurs within the specimen depends on the hydraulic conductivity of the soil, the dimensions of the specimen, the drainage conditions, and the strain rate employed in the test. Using Skempton's pore pressure parameter A, Blight (1963) expressed the equalization of pore pressures in a triaxial test as:

$$U = 1 - \frac{A_{100} - A}{A_{100} - A_0} \tag{8.8}$$

in which A_{100} corresponds to a degree of equalization of 100%, A_0 is the value that would have been measured in a test performed so rapidly that no equalization could occur, and A corresponds to the average pore pressure measured over the surface of the specimen using end filter stones and filter paper on the sides. Because the pore pressure in a soft clay is higher in the middle of the specimen, the measured value of A increases as the pore pressure equalizes with time.

Fig. 8.2(b) shows the relations obtained between degree of equalization and the time factor determined from Eq. (8.1). Both theoretical and experimental relations for undrained tests are shown in this diagram. The experimental results, which correspond to all around drains, fit the theoretical prediction only at very high degrees of equalization. The reason for the lack of fit at the lower degrees of equalization is the inefficiency of the drains (Blight 1963).

Final recommendation for CD- and CU-tests

Comparison of the two diagrams in Fig. 8.2 indicates that the results for drained tests with double end drainage practically coincides with those from the undrained tests without drains, and both will reach a particular degree of equalization at the same time factor. This also applies to the drained and undrained tests with all around drains. Therefore, the time required for 95% equalization of pore pressures and the first significant measurements in undrained tests are the same as those for the drained tests.

Thus, for specimens with *all around drains*, that is with filter stones at the ends and side drains of slotted filter paper, the time for 95% pore pressure equalization may be calculated from Eq. (8.5). For tests with *double end drainage*, that is with filter stones at both ends, the time for 95% dissipation may be calculated from Eq. (8.6). The strain rates are also calculated as indicated before.

Figure 8.3 shows a chart for finding test durations for a degree of drainage of 95% in drained tests and a degree of pore pressure equalization of 95% in undrained tests, both at failure, here considered being the first significant measurement. The appropriate strain rate is subsequently determined from Eq. (8.7).

However, if the first significant measurement occurs at, say, 25% of the strength, then the strain rate determined as explained in the previous paragraph is divided by the ratio between the axial strain-to-failure and the axial strain to 25% strength resulting in a lower strain rate than that corresponding to 95% drainage or 95% pore pressure equalization at failure.

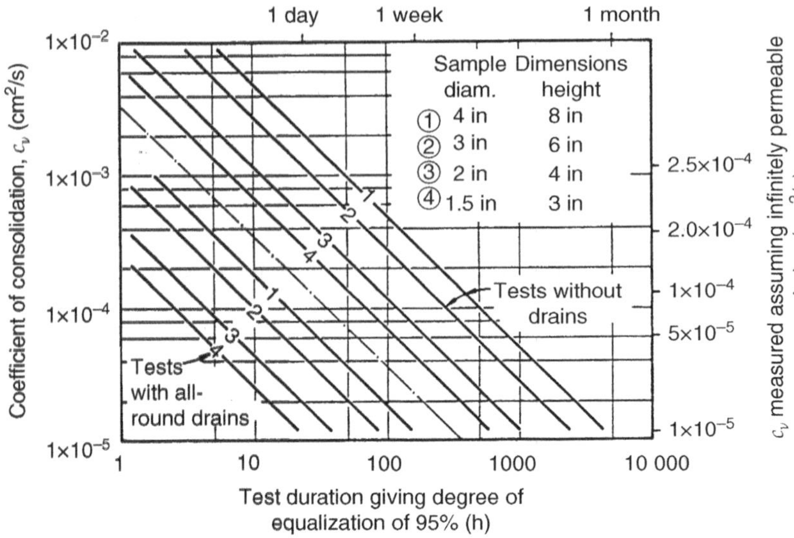

Figure 8.3 Chart for finding durations of drained and undrained tests for 95% dissipation at failure. Reproduced from Blight 1963 by permission of ASTM International.

8.2.4 Effects of lubricated ends in undrained tests

Lubricated ends may be employed to reduce the nonuniform distribution of strains in the triaxial specimen. This produces a reduction of nonuniformity of the pore pressure distribution in undrained tests, and these may therefore be performed with higher strain rates than indicated above. Calculations to ascertain the strain rate required for pore pressure equalization are not applicable since the stress and strain distributions are assumed to be uniform. Barden and McDermott (1965) showed that failure may be reached within 1 or 2 h in undrained tests with lubricated ends without any adverse effects due to nonuniformities in pore pressure distribution.

Because drained tests still depend on drainage of water out of the specimens, their strain rates cannot be reduced by application of lubricated ends.

8.3 Effects of lubricated ends and specimen shape

Ideally, the stresses and strains in laboratory triaxial specimens should be completely uniform. Although this ideal may seldom if ever be achieved, it is generally considered that adequately uniform conditions will be achieved by using tall specimens with heights greater than or equal to two diameters ($H/D = 2.0$) or by using lubricated ends, or both. Even under conditions where end restraint is negligibly small, however, severe nonuniformities in strain distributions can develop in triaxial tests, and these nonuniformities can have significant effects on the stress–strain and strength behavior of soils. The best boundaries for enforcing uniform strains are stiff, lubricated flat surfaces. Such boundaries reduce friction greatly with the result that the stresses are essentially uniform.

8.3.1 Strain uniformity and stability of test configuration

Figure 8.4 displays a typical axisymmetric triaxial compression test for any type of material (Yamamuro and Lade 1995). The specimen is divided into sections, with section 3 initially having a slightly smaller diameter, as shown in the diagram on the left-hand side of the figure. This represents a geometric defect or a material defect such as lower density or strength. Compressive load is then applied as shown on the right-hand side of the figure. The initial stresses in section 3 are larger than in the

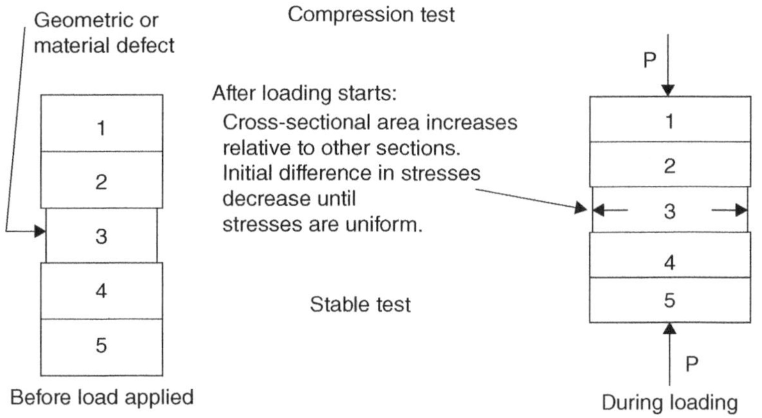

Figure 8.4 Axisymmetric triaxial compression specimen divided into sections, with section 3 initially having a slightly smaller diameter, representing a geometric defect or a material defect such as lower density or strength. Reproduced from Yamamuro and Lade 1995 by permission of ASCE.

surrounding sections, because the cross-sectional area is smaller. Therefore, the strains in both the vertical and radial directions in section 3 are relatively larger than in the surrounding sections. However, as loading progresses, the difference in the initial stress states of section 3 and surrounding sections actually decreases, even though the total stresses are increasing. Eventually, the outward directed radial strains in section 3 are large enough that the cross-sectional areas become equal to those in the surrounding sections. At this point the stresses in all sections of the specimen are equal. Thus, in a triaxial compression test the initial concentration of stresses at a geometric or material defect tend to distribute as the loading progresses and the resulting stress and strain distributions tend toward uniform conditions. Therefore, the triaxial compression test can be considered to be a stable test, excluding instability by a buckling condition.

The stability of the triaxial compression test is further enhanced by using short specimens. If a defect in the specimen is located to one side, as indicated in the tall and the short specimens in Fig. 8.5, then the tall specimen would be more likely than the short specimen to buckle. Further stability is provided by a stiff connection between the piston and the cap that minimizes rotation of the cap and forces the two end plates to remain essentially parallel. The lubricated ends

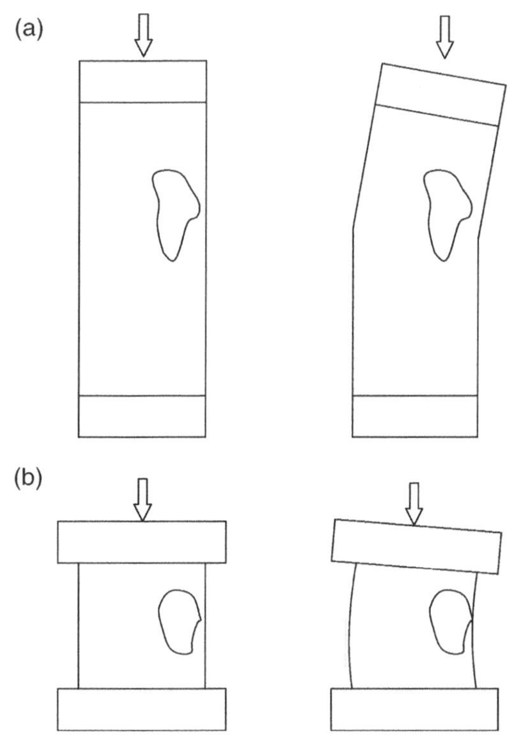

Figure 8.5 A defect in the specimen located to one side, as indicated in (a) the tall and (b) the short specimens. The tall specimen would be more likely to buckle than the short specimen.

employed with the shorter specimens will enhance to strain and stress uniformity in the specimen, which will tend to produce density uniformity within the specimen by a mechanism similar to that explained for the configuration in Fig. 8.4. Thus, stresses will initially be higher in the denser than in the looser portion of the specimen, but the average deformation will allow the looser portion to become denser and therefore take more load, thus tending to catch up with the initially dense portion. The average stress–strain response is reasonably representative of the response corresponding to the average density of the specimen.

8.3.2 Modes of instability in soils

Three basically different modes of instability can occur in laboratory specimens. The most common type is smooth peak failure occurring at constant effective confining pressure, which may be observed in all types of soils. Another type of instability is initiated by localization of plastic strains and subsequent development of shear bands, which is followed by a decrease in load bearing capacity. This mode most often occurs in the softening regime in triaxial compression tests on normally consolidated soils but may occur in the hardening regime for overconsolidated soils. While these two modes may occur under either drained or undrained conditions, the third mode of instability occurs under undrained conditions in granular materials, mostly loose, fine silty sand, inside the effective stress failure surface, and it may lead to liquefaction. This mode of instability is not seen in ordinary clays, but it may be observed in quick clays in which failure is very much like liquefaction failure.

8.3.3 Triaxial tests on sand

All three modes of instability mentioned above are observed in sands, and their occurrences are described in detail by Lade (2002).

Smooth peak failure

Smooth peak failure characterizes the sand behavior under drained conditions and constant confining pressure. The internal structure of the granular material is loaded and deformed sufficiently to cause collapse of load bearing chains of particles. The resulting reduction in bearing capacity is characterized as material softening, and it involves macroscopically uniform deformations and stresses such that the granular material can be characterized as a continuum.

While specimens with uniform density tend to deform uniformly, the boundary conditions in the form of frictional or lubricated ends as well as the height-to-diameter (H/D) ratio may affect the measured soil behavior. Stress–strain and volume change in the form of void ratio change curves for triaxial compression tests on dense and loose Santa Monica Beach sand are shown in Fig. 8.6 for two types of triaxial compression tests. The first is a conventional triaxial test performed on a specimen 7.1 cm in diameter and 19.0 cm high for H/D = 2.7, and with normal (un-lubricated) cap and base. The second test was performed on a cylindrical specimen with both height and diameter of 9.7 cm (i.e., H/D = 1.0) and lubricated cap and base. The initial average void ratios of the specimens were 0.61 and 0.81 corresponding to relative densities of 90 and 20%, respectively. All tests were performed with σ_3' = 196 kPa (2.00 kg/cm^2).

Dense sand

Figure 8.6(a) shows that the stress–strain curve for the conventional test specimen is steeper at small strains, as a result of the end restraint imposed by the rough cap and base. These have the effect of imposing higher confining pressures at the ends and this result in higher initial modulus. It may be noted that the stress–strain curve of this specimen breaks over more sharply and that the strain-to-failure is considerably smaller than observed for the specimen with lubricated ends. These effects are due to the fact that more of the deformation occurs within the mid-section of the tall specimen than near the ends. The decrease in strength following smooth peak failure (softening) occurs at a faster rate than in the specimen with lubricated ends. Figure 8.6(c) shows that a shear band occurs well into the softening regime as visibly observed and as indicated by the abrupt cessation of the

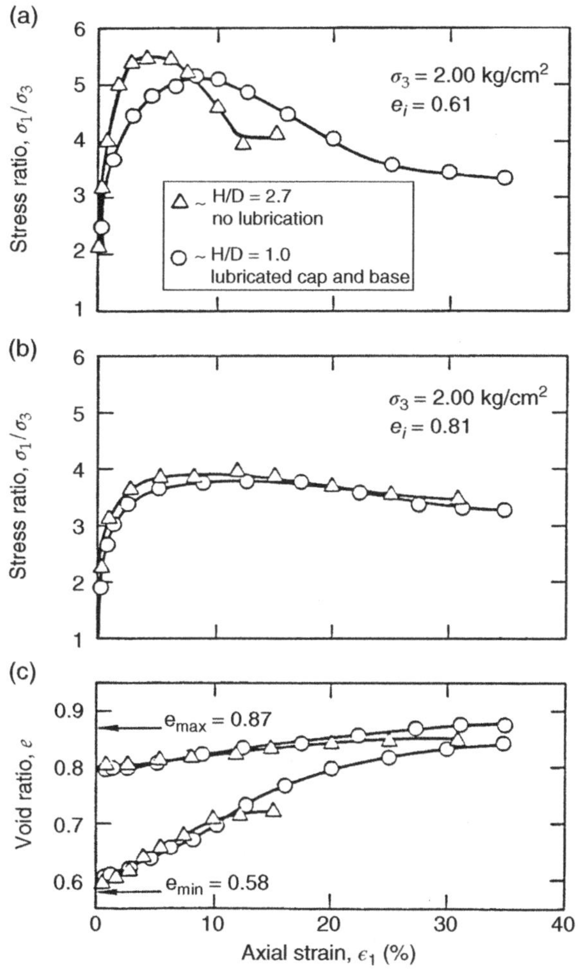

Figure 8.6 Comparison of (a, b) stress–strain and (c) void ratio changes in triaxial compression tests on specimens with *H/D* = 1.0 and 2.7 for (a) dense and (b) loose Santa Monica Beach sand (after Lade 1982a).

volumetric dilation at around 12% axial strain. The nonuniform strains in this test most likely resulted in premature initiation and development of shear banding.

In comparison, the stress–strain curve for the short specimen with lubricated ends breaks over more gradually, and the strain-to-failure is considerably larger, owing to the fact that the entire specimen is undergoing essentially uniform strains. The decrease in strength after the smoother peak failure (softening) is more gradual than in the specimen without lubricated ends, and significant development of shear bands was likely impeded due to the presence of the stiff end plates. While they may have been present, the measured stress–strain and void ratio change curves do not indicate any effects of development of shear bands in this test.

In the particular tests shown in Fig. 8.6 the strength of the tall specimen without lubricated ends is slightly higher than that from the short specimen with lubricated ends. Inspection of a larger collection of experiments does not indicate any particular trend with regard to which types of boundary conditions produce the higher strength.

Loose sand

The results for loose specimens are shown in Fig. 8.6(b) and (c). Loose sand contracts more or dilates less than dense sand at the same confining pressure. The horizontal expansion is therefore smaller in the loose specimens, and the initial parts of the stress–strain curves for the loose specimens are not much affected by end restraint. The initial portions of the stress–strain curves for the tall specimen with lubricated ends consequently have the same shape. As for the dense sand the effect of boundary conditions on strength of loose sand is negligible.

Void ratio change

The results in Fig. 8.6(c) show that the volume change or the change in void ratio of the dense sand is affected by the development of shear band(s) after which the rate of dilation decreases abruptly to zero. This occurs because strains are now localized to within the relatively thin shear band, whose thickness may vary in the range from 10 to 20 grain diameters. The volumetric expansion ceases almost immediately after their initiation, because they quickly exhaust their capacity to expand. Consequently, the void ratio change or the volumetric strain at the end of the test on the tall specimen is only half of that for the short specimen.

The maximum rate of dilation occurs near peak failure for both dense and loose sand, and they are about the same for the two types of specimen, independent of geometry and end lubrication.

Nevertheless, the average void ratio at failure is not the same for the two types of specimen. The short specimens of both dense and loose sand have experienced larger changes in void ratios at failure because of the larger axial strain-to-failure. Volume changes continue to occur uniformly in the short specimens, even at 35% axial strain, at which the void ratio approaches the critical void ratio attained at large strains. In comparison, the void ratios in the tall specimens are far apart and do not approach a common value at the end of the tests.

Note that the void ratio curve for the dense sand in the short specimen approaches but does not quite reach the corresponding curve for the loose specimen. Desrues *et al.* (1996) showed that a very similar void ratio is actually attained inside the shear bands as reached in the loose specimen.

Shear banding

Shear bands develop in sands that dilate. To avoid impeding their initiation and development, it is necessary to provide boundary conditions that allow their development by using specimens with high H/D ratios, so the shear band(s) can protrude through the soft membrane. The orientation of shear bands in triaxial compression tests may be evaluated from simple analyses involving stress considerations or strain considerations, as reviewed in Section 2.3. The two calculations do not produce the same result, but the stress consideration yields the tallest specimen. Accordingly, the shear band(s) are oriented at $\pm(45° + \varphi/2)$ to the planes of the cap and base (referred to as the Coulomb direction), and they transcend a length of $D{\cdot}\tan\,(45° + \varphi/2)$, as shown in the upper left in Fig. 8.7. If the length of the specimen is equal to or

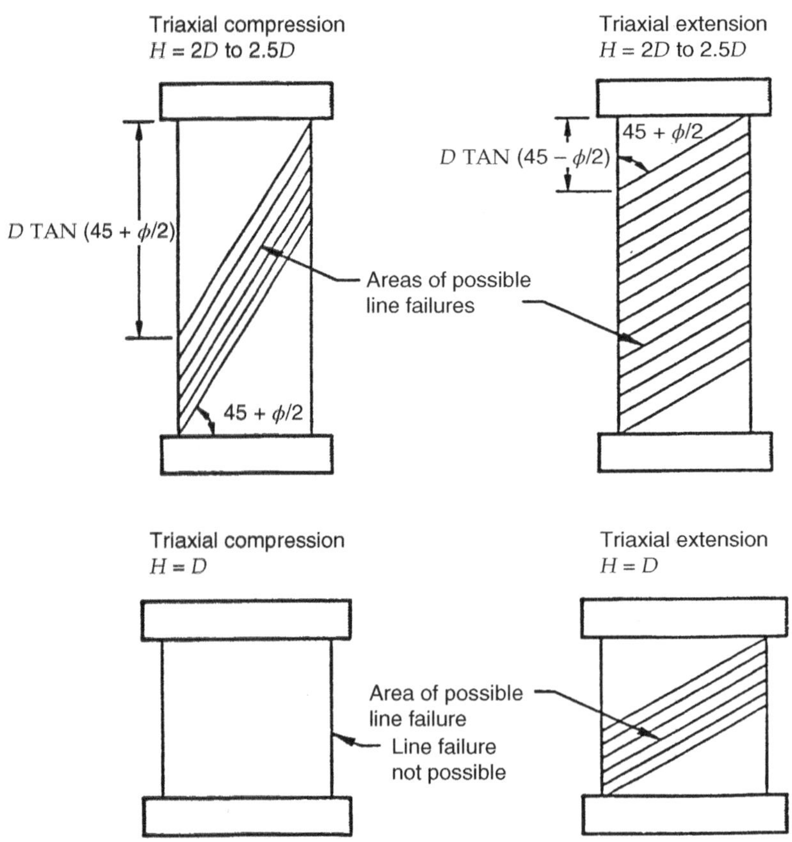

Figure 8.7 Boundary condition effects on failure type: shear bands are free to develop and they transcend a vertical height of $D{\cdot}\tan\,(45° + \phi/2)$.

greater than the length transcended by the shear band, then such a shear band is free to occur, because the soft latex rubber membrane does not impede the development. For friction angles greater than 37°, the required H/D ratio is greater than 2.0 for shear bands to clear the end plates. However, based on strain considerations, the shear band(s) are oriented at $\pm(45°+\psi/2)$ (referred to as the Roscoe direction), in which ψ is the angle of dilation, whose value is approximately $\psi = \varphi - 30°$. Thus, for a very high friction angle of 60° in triaxial compression, the H/D ratio becomes only 1.73. Observed shear bands in triaxial compression tests may be closer to the inclinations determined from the strain considerations (Arthur *et al.* 1977b), thus indicating that H/D ratios near 2.0 are sufficient to allow their free development in the specimens.

On the other hand, to avoid the premature initiation and development of shear bands, uniform strains should be maintained in the specimen, and this requires use of lubricated ends. Thus, to study shear banding it is necessary to employ specimens with end lubrication and with H/D ratio of 2.0–2.5 or higher. Note that specimens with H/D ratios higher than approximately 3.0 may become unstable due to structural buckling.

The above considerations are based on assumed isotropic behavior of the specimen. However, sand deposits are rarely isotropic and triaxial compression specimens are most often cross-anisotropic with directional variations in friction angles and rates of dilation. It is therefore not so easy to express the direction of the shear bands relative to the plane of the major principal stress. An extreme example of this was experienced with a specimen of dense Cambria sand deposited with a cross-anisotropic structure and essentially horizontally oriented grain directions (Lade 2004). A triaxial compression test performed on a specimen of this sand with $H/D = 2.68$ and lubricated ends exhibited a shear band oriented at 45.0° to the σ_1-plane. With a measured friction angle of 38.8° and an angle of dilation of 13.3°, the Coulomb inclination is 64.4° and the Roscoe inclination is 51.6°. It is clear that none of the theoretical inclinations support the measured inclination of 45.0°. This clearly shows that comparisons of experimental and theoretical values of shear band inclinations may be affected by the anisotropic nature of most sand deposits, whether created in nature or in the laboratory.

The triaxial compression test shown in Fig. 8.8 was performed on a dense specimen with lubricated ends and with diameter of 7.1 cm and $H/D = 2.66$. The specimen was carefully observed during the test for early detection of developing shear band(s). The location on the stress–strain curve at which the first observation of a shear plane was made is indicated on the diagram. Considerable straining beyond the peak failure point may be required before the shear band(s) occur in triaxial compression tests, as also supported by theoretical calculations (Lade 2003). The stress–strain curve clearly shows a

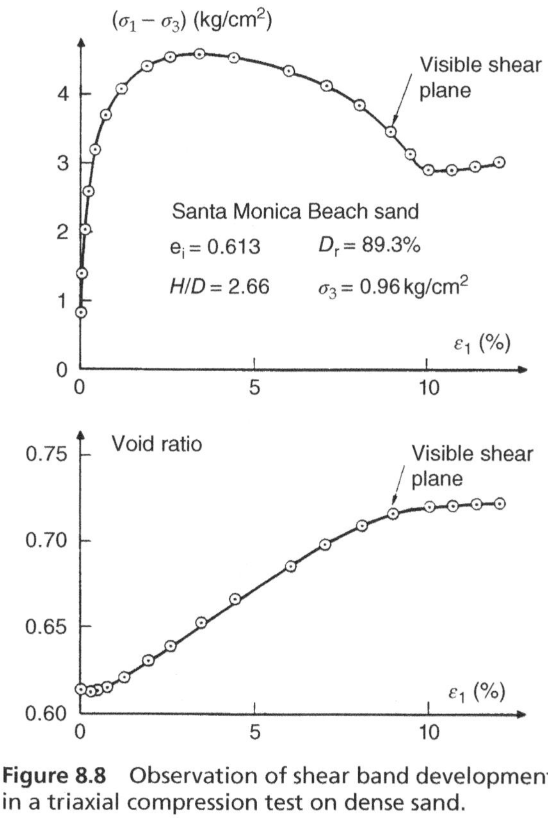

Figure 8.8 Observation of shear band development in a triaxial compression test on dense sand. Reproduced from Lade 1988a by permission of John Wiley & Sons.

drop in strength and the rate of dilation diminishes substantially immediately before the shear band becomes visible. Once the shear band has developed fully, the stresses and the volume changes level off, the specimen outside the developing shear band unloads elastically, and the material inside the shear band loosens up to the critical void ratio (Desrues *et al.* 1996) and rapidly becomes weaker than the remaining major parts of the specimen.

Instability inside the failure surface

Experimental evidence from tests on several types of soils have clearly indicated that the use of conventional associated flow rules in formulation of elasto-plastic constitutive models results in prediction of too large volumetric expansion. To characterize the volume change correctly, it is necessary to employ a nonassociated flow rule. The plastic potential surfaces do therefore not coincide with the yield surfaces, but the two families of surfaces cross each other. A summary of experimental findings was given by Lade (2002). Questions regarding the stability of materials with nonassociated plastic flow therefore emerged and consequently triaxial tests on fully saturated and partly saturated specimens of sand were performed under drained and undrained conditions to study the regions of stable and unstable behavior. It was found that granular materials may become unstable *inside* the failure surface if the state of stress is located between the instability line and the failure surface for the material. Thus, instability is not synonymous with failure, although both may lead to catastrophic events such as gross collapse of earth structures. Conventional slope stability methods do not capture the mechanics of instability and subsequent liquefaction that may occur in gently inclined submarine slopes, in tailings dams, and that develop in earth and snow avalanches.

It is a fact that loading of a contracting soil (resulting in large plastic strains) can occur under decreasing stresses that leads to unstable behavior under undrained conditions. Loose, fine sands and silts have sufficiently low hydraulic conductivities that small disturbances in load or even small amounts of volumetric creep may temporarily produce undrained conditions in such soils, and instability of the soil mass follows. As long as the soil remains drained, it will remain stable in the region of potential instability.

Region of potential instability

When the condition of instability is reached, the sand may not be able to sustain the current stress state. This stress state corresponds to the top of the current yield surface, as shown schematically on the p'–q diagram in Fig. 8.9. Following this top point the sand can deform plastically under decreasing stresses. The top of the undrained effective stress path, corresponding to $(\sigma_1 - \sigma_3)_{max}$, occurs slightly after but close to the top of the yield surface.

For a granular material to become unstable, the state of stress must be located on or above the instability line. Figure 8.10 shows a schematic p'–q diagram in which the line connecting the tops of a series of effective stress paths from undrained tests on loose sand provides the lower limit of the region of potential instability. In the region above this instability line the soil can deform plastically under decreasing stresses. Experiments show that this line is straight. Since it goes through the top points of the yield surfaces which evolve from the origin, the instability line also intersects the stress origin. Figure 8.10 also shows a region of

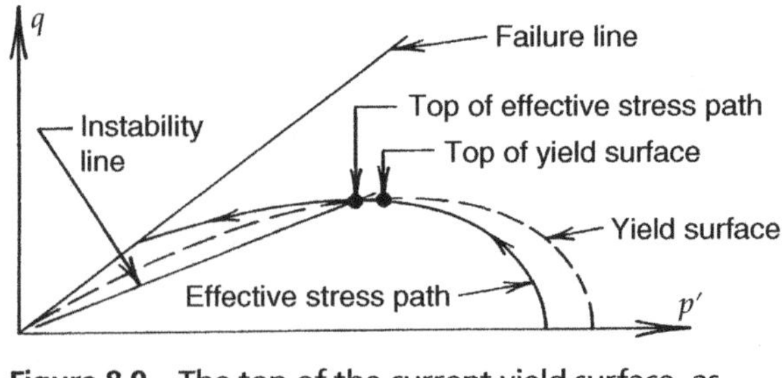

Figure 8.9 The top of the current yield surface, as shown schematically on the p'–q diagram, at which the sand may not be able to sustain the current stress state and instability occurs. Reproduced after Lade (1993a) by permission of Canadian Science Publishing.

temporary instability, which is located in the upper part of the dilating zone. It is a region where instability may initially occur, but conditions allow the sand to dilate after the initial instability, thus causing the sand to become stable again. For very loose sands the region of potential instability reaches down to the origin of the stress diagram.

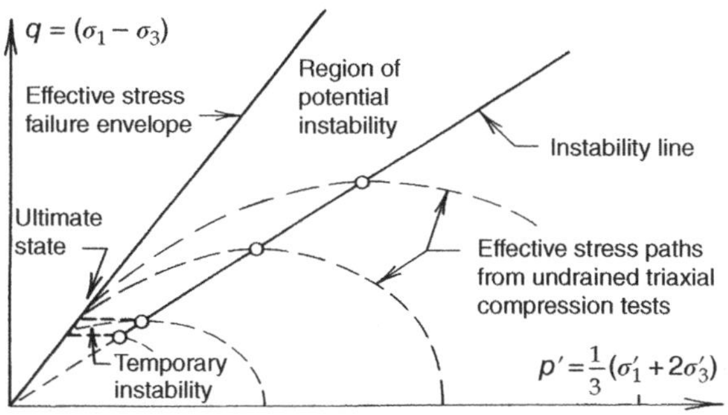

Figure 8.10 Schematic p'–q diagram in which the line connecting the tops of a series of effective stress paths from undrained tests on loose sand provides the lower limit of the region of potential instability. Reproduced after Lade (1993a) by permission of Canadian Science Publishing.

Occurrence of instabilities

The different types if instabilities discussed above, smooth peak failure, shear banding, and instability inside the failure surface, occur for different conditions of sand density and confining pressure. Figure 8.11 shows a critical state diagram for triaxial compression tests on granular materials in which the conditions for stable and unstable behavior are summarized. Several investigators (e.g., Hettler and Vardoulakis 1984; Molenkamp 1985; Peters *et al.* 1988; Lade 2003) have shown that the plastic hardening modulus (= a normalized expression for the slope of the stress–strain curve), is required to reach a critical value for the development of shear planes. For triaxial compression conditions, the critical hardening modulus has to reach a certain negative value before shear planes can theoretically develop.

The type of instability resulting in shear planes is not the same as that producing instability inside the failure surface. Based on experimental observations, the two types of instability appear to be mutually exclusive, that is shear bands are observed after peak failure in granular materials

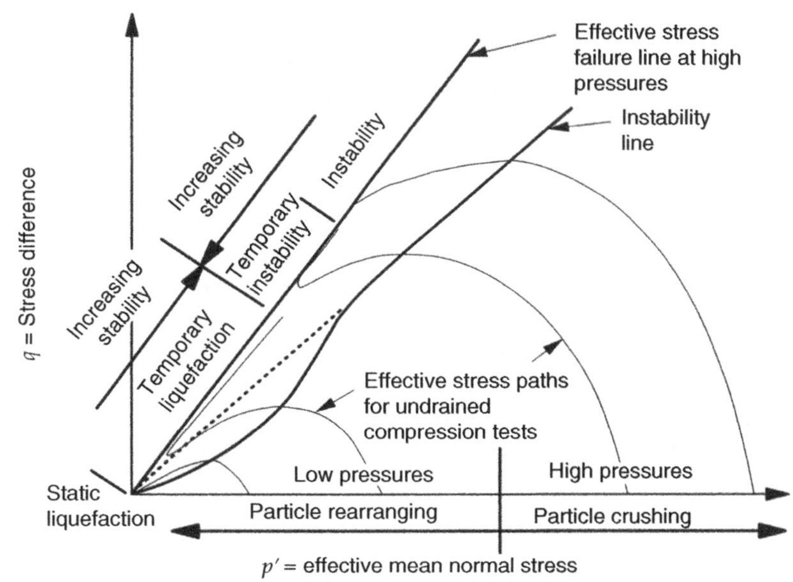

Figure 8.11 Schematic p'–q diagram for triaxial compression tests on granular materials in which the conditions for stable and unstable behavior are summarized. Reproduced from Yamamuro and Lade 1997 by permission of Canadian Science Publishing.

that dilate, and these materials are perfectly stable for all possible stress path directions before the failure surface has been reached. On the other hand, shear bands are not observed in granular materials that contract during shear, but they can become unstable for certain stress paths directions *inside* the failure surface.

Conclusions for triaxial compression tests on sand

Based on the uniformity of strains discussed and the modes of instability presented above, it is clear that the best definition of the actual properties of sands is obtained from tests on specimens with uniform strains. These are best achieved by employing lubricated ends on specimens with $H/D = 1.0$, because this specimen shape tends to overcome effects of nonuniform density within the sand specimen and promote uniform strains. This is true for both dense and loose sand at any confining pressure and under drained and undrained conditions. The modes of instability developing for this test configuration are smooth peak failure or instability inside the failure surface for contracting sand. Development of shear band(s) occurs in dilating sand after peak failure in triaxial compression tests. Their developments therefore do not interfere with peak failure. However, if shear banding is the object of study, then it is advantageous to use specimens with lubricated ends and with H/D>2.0–2.5 to avoid impeding their development.

8.3.4 Triaxial tests on clay

The results from CU-tests on remolded kaolin clay are used to demonstrate effects of end lubrication and specimen shape. The experiments presented here were part of a larger study of overconsolidation of remolded clays under three-dimensional stress conditions (Tsai 1985). The CU-tests with pore pressure measurements were performed on saturated specimens of Edgar Plastic Kaolinite clay with the following characteristics: liquid limit = 60, plastic limit = 30, clay fraction = 60%, and activity = 0.50. The specimens were prepared from a clay slurry mixed at a water content of two times the liquid limit and consolidated in a double draining consolidometer at a vertical pressure of 196 kPa (2.0 kg/cm^2). After thoroughly remolding the clay, specimens were trimmed and consolidated isotropically at 294 kPa (3.00 kg/cm^2). The overconsolidated clay specimens were subsequently allowed to swell at isotropic effective confining pressures of 147, 59, and 20 kPa (1.50, 0.60, and 0.20 kg/cm^2) corresponding to overconsolidation ratios of 2, 5, and 15. Tests were also performed on normally consolidated specimens, that is corresponding to overconsolidation ratio (OCR) = 1.

Typical normalized stress–strain curves and effective stress ratio–strain curves for three undrained triaxial compression tests on Edgar Plastic Kaolinite clay with OCR = 5 are shown in Fig. 8.12. The first test was performed on a specimen with diameter of 7.1 cm, $H/D = 2.3$, and normal (unlubricated) cap and base. The consequent end restraint caused nonuniform stress and strain distributions and therefore nonuniform pore pressure distribution within the specimen. Filter paper drains were employed around the specimen to help equalize the pore pressures and the test was performed with sufficiently slow strain rate to allow a high degree of pore pressure equalization.

Two tests were performed on short specimens with $H/D = 1.0$ and with lubricated cap and base. A cubical specimen with side length of 7.6 cm was used in one test, and a cylindrical specimen with diameter of 7.1 cm was employed in the other test. Uniform strains were observed in these two short specimens. The strain rate used in tests on specimens with lubricated end plates is not subject to calculation, because the stresses and strains and therefore also the pore pressures are presumed to be uniform.

While the tall specimen without lubricated ends was sheared at an axial strain rate of 0.0025%/min, the two specimens with lubricated ends were sheared at rates that were approximately 16 times faster, determined according to a procedure recommended by Barden and McDermott (1965).

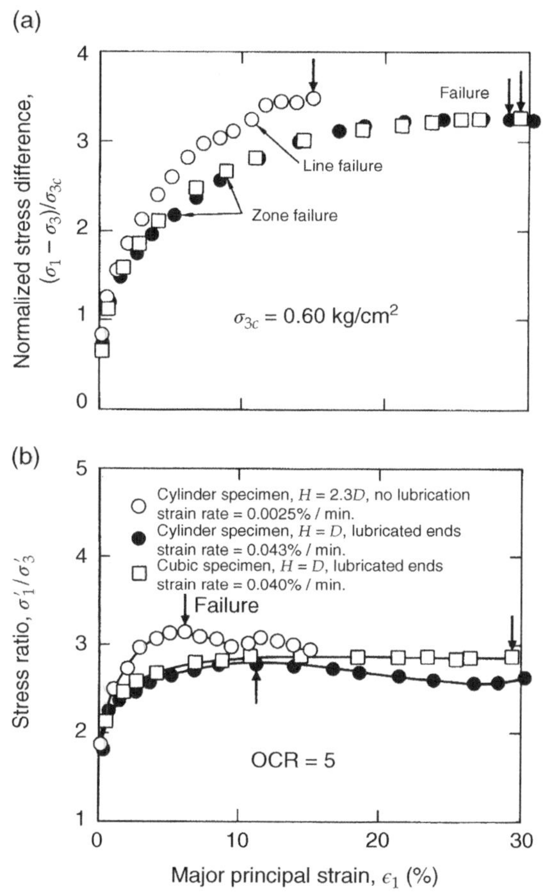

Figure 8.12 Typical (a) normalized stress–strain curves and (b) effective stress ratio–strain curves for three undrained triaxial compression tests on Edgar Plastic Kaolinite clay with OCR = 5. The specimens have different shapes and end conditions, as indicated (after Lade and Tsai 1985).

Figure 8.12(a) shows that the normalized stress–strain curve for the conventional specimen is steeper at small strains as a result of the end restraint imposed by the rough cap and base. The effective stress ratio–strain curve shown in Fig. 8.12(b) for the tall specimen breaks over more sharply and the strain-to-failure is considerably smaller than those obtained from the short specimens with lubricated ends. In comparison, the stress–strain curves for the short specimens break over much more gradually and the strains- to-failure are considerably larger, owing to the fact that the specimens undergo essentially uniform strains.

Results similar to those shown in Fig. 8.12 were also obtained from the other triaxial compression tests performed with different OCRs. The strengths of the tall specimens with end restraint were slightly higher than those of the short specimens with lubricated ends. The average effective friction angle for the tall specimens with end restraint was 31.0°, whereas the average effective friction angle for all the tests on short specimens with lubricated ends was 29.3°.

The results in Fig. 8.12 indicate that the behavior measured for the short cylindrical specimens and for the cubical specimens were very similar. The average friction angle obtained from the cylindrical specimens was 29.0° and that obtained from the cubical specimens was 29.7°. Thus, the influence of specimen shape is not pronounced.

Tests on very short specimens

As part of a larger study of the behavior of San Francisco Bay Mud, series of triaxial tests were performed on specimens with different H/D ratios and with and without lubricated ends. Cylindrical block samples with both diameter and height of 1 ft (30.5 cm) were recovered from a depth of 6.5 m in an excavation at a site located about 1 mile (1.6 km) south of the San Francisco International Airport, CA, USA (Kirkgard and Lade 1991, 1993). The following index properties were obtained for this clay: liquid limit = 85, plastic limit = 48, clay fraction = 45%, and activity = 0.82.

CU-tests were performed in triaxial compression on specimens trimmed from the blocks with the following dimensions: (1) D = 35.6 mm (1.4 in.), H = 88.9 mm (3.5 in.), H/D = 2.50; (2) D = 71.1 mm (2.8 in.), H = 71.1 mm (2.8 in.), H/D = 1.00; and (3) D = 71.1 mm (2.8 in.), H = 25.4 mm (1.0 in.), H/D = 0.36. The tall specimens were tested as conventional tests without lubricated ends, while the two shorter specimens were tested with lubricated ends. The lubricated ends consisted of one or two thin (0.002 in. = 0.05 mm) rubber sheets cut from Trojan® condoms with

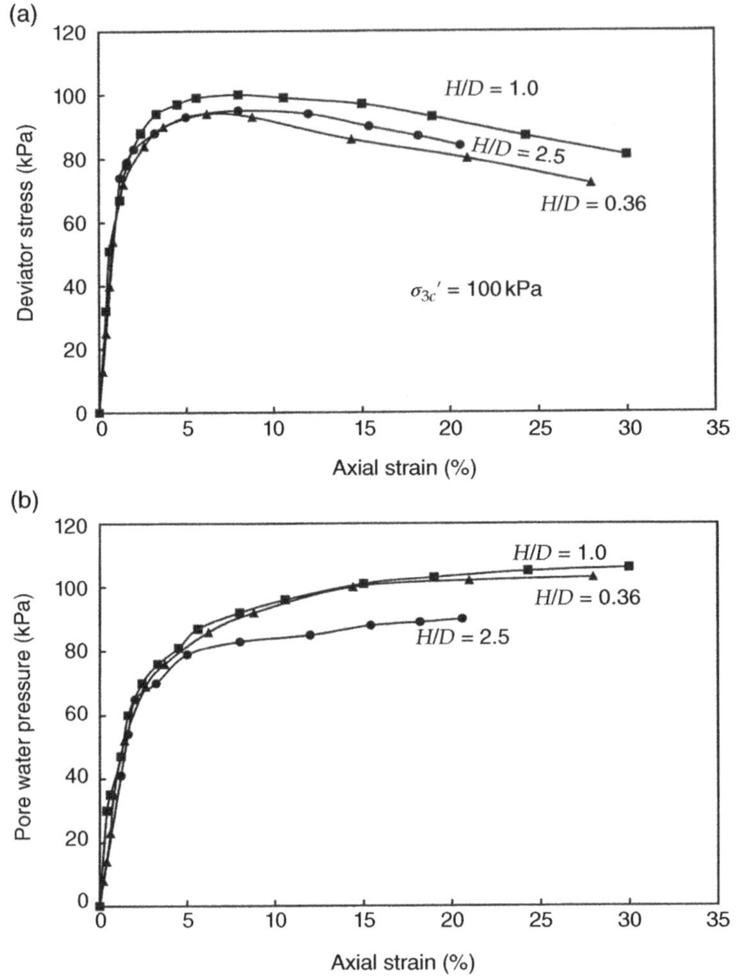

Figure 8.13 (a) Stress–strain curves and (b) pore pressure relations from CU triaxial compression tests on specimens of intact San Francisco Bay Mud with different shapes and end conditions. They indicate that short specimens may be tested with results similar to those from taller specimens.

smears of silicone grease between. Slotted filter paper was installed around all specimens to help pore pressure equalization.

The results of the CU-tests are shown in Fig. 8.13. The diagram shows that the stress–strain curves and the pore water relations are very similar, and it is consequently possible to test very short specimens with results that are acceptable relative to those from taller specimens.

8.4 Selection of specimen size

The specimen diameter relative to the maximum grain size in a granular material plays a role in the results from the triaxial tests. Thus, if the maximum grain size becomes too large relative to the specimen diameter, then the results are affected. Experiments have shown that a diameter of at least $D_{spec} = 6 \cdot d_{max}$ is the minimum specimen diameter for nonuniformly graded granular materials, while $D_{spec} = 10 \cdot d_{max}$ is reasonable for uniformly graded granular materials.

8.5 Effects of membrane penetration

8.5.1 Drained tests

In drained tests, membrane penetration due to changes in confining pressure causes erroneous volume changes. However, the errors from membrane penetration may be corrected as indicated in Chapter 9.

8.5.2 Undrained tests

In undrained tests, the pore pressures are affected by membrane penetration, which responds to changes in effective stresses. The resulting pore pressures are in error, but it is difficult to impossible to correct for these erroneous pore pressures. A constitutive model is required for prediction of pore pressures in which the additional volume change due to changes in effective confining pressure are included in the calculations.

It may be possible to eliminate membrane penetration effects in undrained tests, as explained by Tokimatsu and Nakamura (1986) and Nicholson *et al.* (1993a, b). Their studies explain how the amount of membrane penetration is affected by primarily the grain size at the specimen surface and by the effective confining pressure.

8.6 Post test inspection of specimen

Following the end of a triaxial test, the specimen may be inspected and measured for comparison with actual calculated deformations, and to see if any abnormalities are present. It is recommended to make a sketch of the specimen indicating the failure planes or to photograph it. Then the whole specimen should be weighed and the water content determined from a representative portion of the specimen. The specimen may also be broken apart to describe the soil. Thus, for example, it may be observed that a large pebble was present and may have caused unusual pore pressure developments, or unusual volume changes. Similarly, a specimen may consist of different layers, that is layers with different consistency, soft and hard, resulting in pore pressures or volume changes that are not representative of either of the two consistencies.

9 Corrections to Measurements

9.1 Principles of measurements

Because physical processes may be measured by many different types of instruments, it is necessary to determine which ones are appropriate for measuring a particular quantity, and this may be determined from the principle of *maximum signal for minimum interference* in the physical process. Thus, it may be necessary to take energy from the process which is to be measured to activate the instrument. This should be done according to the above principle, that is the physical process should be disturbed as little as possible by the measurement method.

While there are measurement techniques that require no corrections, the measured quantities with conventional methods may include errors that are too large to ignore. Thus, the measured vertical load may include loads to compress the rubber membrane, the filter paper used for lateral drainage, and to overcome the piston friction from the bushing in the triaxial cell top plate. The measured vertical deformation may be affected by compression of interfaces in the triaxial apparatus and compression of lubricated ends, as well as penetration of sand grains into the lubricating ends. The measured pore pressures may be in error due to membrane penetration effects in granular materials and sand penetration into the lubricating ends. The measured volume change may be in error due to membrane penetration occurring due to changing effective confining pressure, sand grain penetration into the lubricating ends, and water or air penetrating through the membrane.

Methods of correcting for these errors or circumventing the errors by alternate measurement methods are presented below.

9.2 Types of corrections

It may be necessary to apply corrections to all measured quantities, unless measures are taken to avoid the experimental problems that require corrections. Thus, corrections may be required to the measured values of vertical load, vertical deformation, volume change, and cell and/or pore pressure.

9.3 Importance of corrections – strong and weak specimens

The importance of corrections depends on the stress and the strain to which the soil specimen is exposed. If the specimen to be tested is strong and stiff, the corrections to the measured vertical

Triaxial Testing of Soils, First Edition. Poul V. Lade.

load due to for example the load taken by the membrane and piston friction may be small compared with the load required for deformation of the specimen. Corrections to the measured vertical deformation due to for example compression of lubricating rubber sheets and compressing interfaces between parts of the equipment may be important if they are large relative to the deformation of the specimen. But on the other hand, if the specimen is weak and flexible, then corrections to the vertical load due to for example the load taken by the membrane and the side drains may be very important, while the corrections to the vertical deformation due to for example compression of lubricating rubber sheets may not be of serious consequence relative to the vertical compression of the specimen. Thus, the corrections need to be evaluated relative to their importance for the final outcome of the experiments.

9.4 Tests on very short specimens

Triaxial compression and extension tests may be performed on very short specimens. Such specimens may have H/D ratios as low as 0.36 and still produce acceptable results. These experiments require use of lubricated ends to minimize the end restraints, and if performed with great care they may produce high quality results. It is particularly important to apply reliable corrections to such experiments. Because the heights of such specimens are relatively low, the vertical strains are sensitive to compression of the lubricating rubber sheets. On the other hand, the vertical load is not especially sensitive to corrections of the types suggested above.

9.5 Vertical load

9.5.1 Piston uplift

The vertical load measured on the load cell outside the triaxial cell includes the piston uplift force created by the cell pressure that tends to push the piston out. As shown in Fig. 9.1 , it is determined from

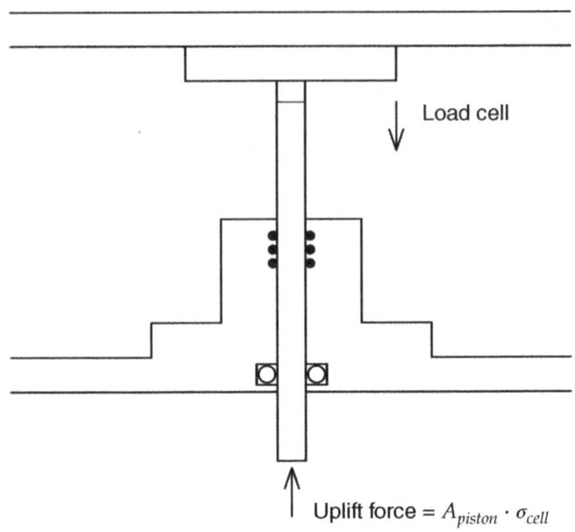

Figure 9.1 Piston uplift force. This is calculated as the cross-sectional area times the cell pressure.

$$F_{piston} = A_{piston} \cdot \sigma_{cell} \tag{9.1}$$

in which A_{piston} is the cross-sectional area of the piston and σ_{cell} is the total cell pressure. This force is subtracted from the measured vertical force before this is divided by the cross-sectional area of the specimen.

If the piston is not attached to the cap (not screwed into the cap or otherwise attached), as explained in Section 3.3, it is possible to eliminate the effect of piston uplift by zeroing the reading while the piston is pushed up against the load cell by the cell pressure.

9.5.2 Piston friction

The triaxial specimen is loaded in the vertical direction through a piston that passes through a bushing in the top plate of the triaxial cell. Many designs of this bushing have been made with the intent of reducing the friction between the piston and the bushing while minimizing or eliminating leakage of cell fluid, as reviewed in Section 3.3.1. Measures to avoid this friction are reviewed in Section 3.3.2.

For the conventional setup, the vertical load is applied through a stainless steel piston guided by ball bushings and is sealed by an O-ring near the lower end of the bushing, as shown in Fig. 9.2 The piston friction force is affected by the fit of the O-ring around the piston and the cell pressure and by any side force that tends to apply a moment to the piston. The latter may occur because the specimen deforms in a non-symmetric manner or a shear band develops so that the upper part of the specimen slides off such as to create a force in a direction perpendicular to the piston. Note that shear banding in triaxial compression tests occurs after smooth peak failure and therefore does not affect the shear strength of the soil. Note that O-rings have a tendency to shrink with time and therefore require periodical replacement to maintain a predictable friction force. Assuming that the specimen is simply compressed in the vertical direction and no side force develops, the frictional component may be

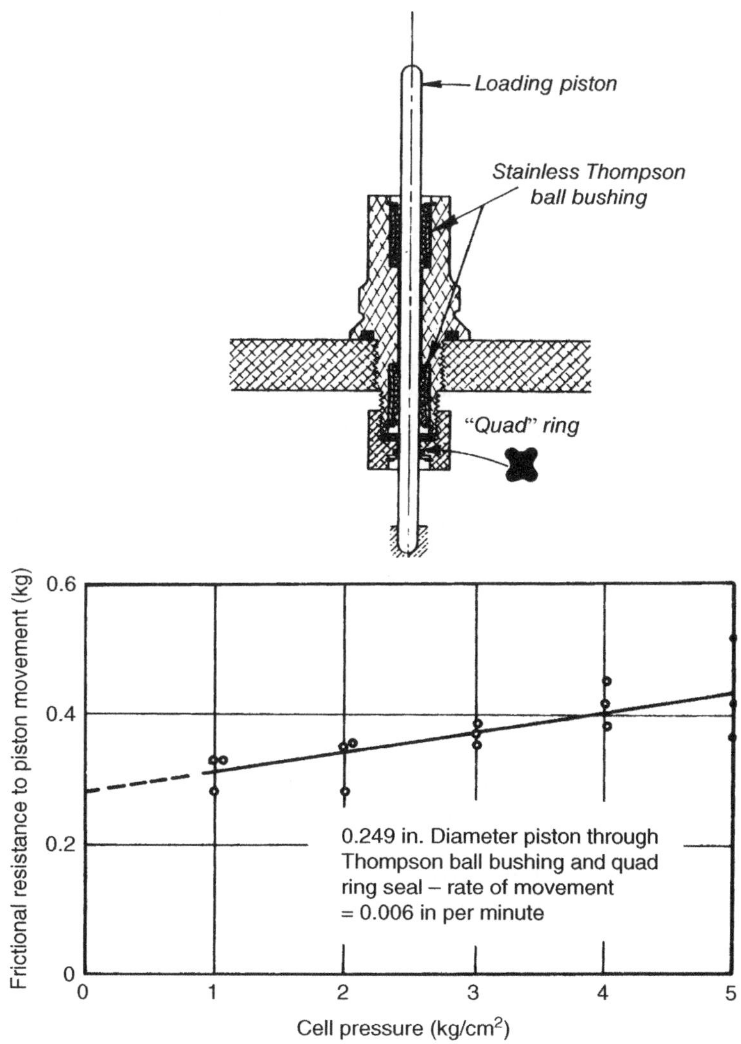

Figure 9.2 Example of typical increasing piston friction with confining pressure. Reproduced from Duncan and Seed 1967 by permission of ASCE.

measured using a small load cell and loaded in deformation control, as shown in Fig. 9.2.

Since the piston uplift is predictable from the confining pressure and the piston cross-sectional area, as indicated in Eq. (9.1), the difference between the measured load and the piston uplift constitutes the piston friction for the case of no specimen present. The piston friction typically increases with confining pressure. An example of piston friction increasing with cell pressure is given in Fig. 9.2.

9.5.3 Side drains

The vertical load correction for side drains (see Section 3.1.4) consisting of filter paper or other radial drainage materials has been investigated in several studies (Henkel and Gilbert 1952; Ladd and Lambe 1963; Olson and Kiefer 1964; Duncan and Seed 1967; Ramanatha Iyer 1973; Balasubramaniam and Waheed-Uddin 1978; Berre 1982; Leroueil *et al.* 1988; Mitachi *et al.* 1988; Sivakumar *et al.* 2010; Yamamuro *et al.* 2012). From the studies performed on filter paper drains the recommendations range from no correction up to 0.8 N/cm of width of filter strip drains (Leroueil *et al.* 1988). The corrections may be greater for non-woven geotextiles.

Yamamuro *et al.* (2012) conducted an experimental study of vertical load corrections for radial drainage materials consisting of hygroscopic filter paper and non-woven geotextiles. Effective confining pressures from 100 to 400 kPa were used for the Whatman No. 1 filter paper, which is effective as a side drain up to about 500 kPa. Confining pressures up to approximately 3000 kPa were employed for Reemay 2214 non-woven geotextile. Both types of side drains were tested with configurations consisting of the classical vertical strips (eight 10 mm strips were positioned around a 70 mm diameter rubber specimen) and frames with inclined slotted material, as shown in Fig. 9.3. Both configurations were tested in compression and extension.

The testing setup consisted of a rubber dummy specimen with diameter of 70 mm loaded with and without filter material at the same confining pressure. The filter material was contained between two membranes with and without lubrication between the rubber specimen and the inner membrane, as shown in Fig. 9.4. Tests were also performed on a more realistic setup with filter material applied directly to specimens consisting of kaolinite clay. These allowed additional conclusions regarding the vertical load corrections.

Filter paper

For Whatman No. 1, most often used at lower effective confining pressures, two configurations were employed: Fig.9.5 (a) shows the measured vertical load corrections for compression and extension tests with the pre-wetted, inclined slotted configuration, and Fig. 9.5(b) shows compression and extension results for the pre-wetted, eight-vertical-strip configuration. Results with lubrication show slightly smaller corrections for the inclined slotted configuration, indicating slippage, while the eight-vertical-strip shows little difference between the two setups. Based on the contact friction between the filter paper and the kaolinite clay specimens, the load correction was estimated and indicated with the long dashed lines in Fig. 9.5. The correction is nearly constant at 20 N for both types of configurations in compression. This limit on the vertical load correction is due to buckling of the filter paper. For extension the vertical load correction increases with effective confining pressure from zero at 100 kPa for the vertical strip configuration and −20 N for the inclined slotted configuration. At a confining pressure of 400 kPa both configurations require a correction of −50 N.

For compression, the load correction for the vertical strips corresponds to 2.5 N/cm of filter paper. For variations in the number and width of the strips of Whatman No. 1, the measured load corrections can be scaled accordingly. For the inclined slotted filter paper the load correction in compression corresponds to 0.9 N/cm of specimen circumference.

For extension, the load correction for vertical strips varies with confining pressure from zero

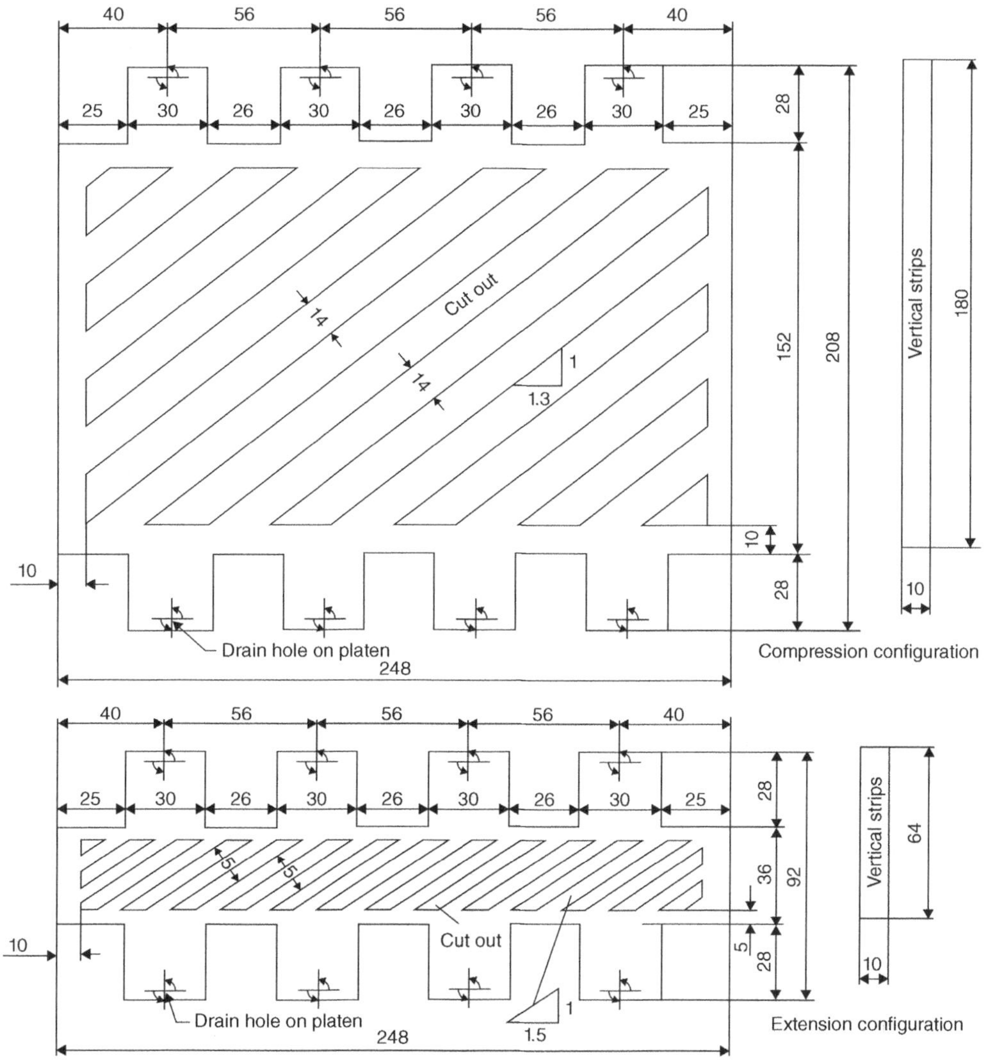

Figure 9.3 Dimensions of inclined slotted and vertical filter strip configurations of radial drainage material used for compression tests on a tall specimen with *H/D* = 152 mm/70 mm = 2.17 and for an extension test specimen with *H/D* = 36 mm/70 mm = 0.51. The dimensions of the vertical strips are shown on the right-hand side. Reproduced from Yamamuro *et al.* 2012 by permission of ASTM International.

at a confining pressure of 100 kPa to about 6 N/cm of Whatman No. 1 filter paper at 400 kPa. For inclined slotted filter paper, the vertical load correction varies from −0.9 N/cm of specimen circumference at 100 kPa to −2.3 N/cm at 400 kPa. On real soils the filter paper was never observed to tear. This supports the observation that filter paper has more significant strength than previously measured.

In addition to these observations and measurements, Saada *et al.* (1994) pointed out that shear band propagation in clays can be greatly affected by the strength of the filter paper.

Non-woven geotextile

The results for Reemay 2214 showed that the vertical load correction in compression increased

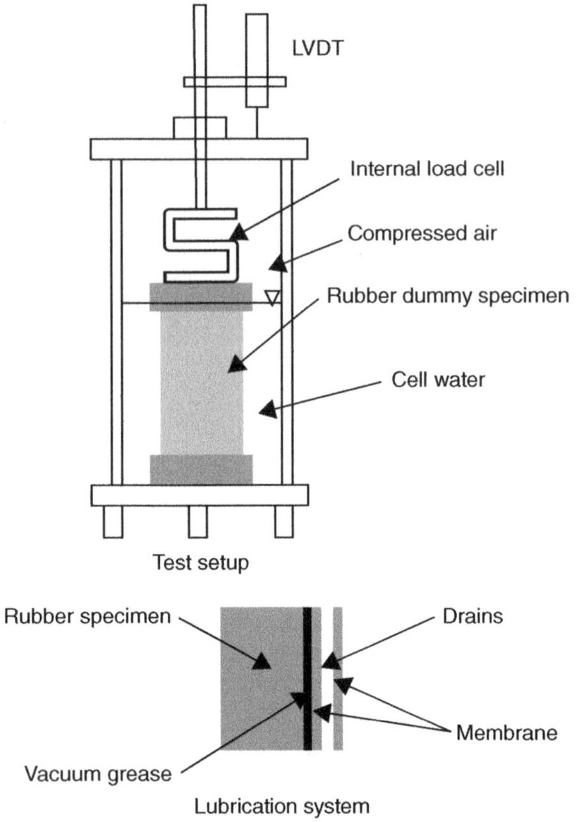

Figure 9.4 Test setup to determine the vertical load correction of different drainage materials and the lubrication system. Reproduced from Yamamuro *et al.* 2012 by permission of ASTM International.

from zero until about 6% axial strain was reached, after which it became relatively constant. Figure 9.6(a) shows the measured vertical load corrections for compression and extension tests with Reemay 2214 in the inclined slotted configuration, and Fig. 9.6(b) shows compression and extension results for the eight-vertical-strip configuration. Results with lubrication produced smaller corrections for the inclined slotted configuration, indicating slippage, while the eight-vertical-strip showed little difference between the two setups. The correction is greater for extension than compression. Buckling of the filter material may limit the vertical load correction in compression. Based on the contact friction between the geotextile drains and the kaolinite clay specimens, the load correction was estimated and indicated with the long dashed lines in Fig. 9.6.

Independent of the side drain configuration and confining pressure, the maximum vertical load correction for Reemay 2214 is approximately 50 N in compression and −100 N in extension.

For compression, the load correction for the vertical strips corresponds to 6 N/cm of geotextile. For variations in the number and width of the strips of Reemay 2214, the measured load corrections can be scaled accordingly. For the inclined slotted filter paper the load correction in compression corresponds to 2.3 N/cm of specimen circumference.

For extension, the load correction for vertical strips corresponds to −12 N/cm of geotextile. The load correction can be scaled according to the number and width of the strips of Reemay 2214. For inclined slotted geotextile, the vertical load correction in extension corresponds to −4.5 N/cm of specimen circumference.

Effects of vertical load corrections

The influence of vertical load corrections and residual pore pressures on the effective stress friction angles of kaolinite clay in compression and extension were determined by Yamamuro *et al.* (2012), as indicated in Fig. 9.7. Corrections for filter paper are employed at lower confining pressures, while corrections for geotextile filter material are applied at higher confining pressures. Friction angles in compression are not greatly affected, especially at high pressures, because the failure loads are much larger. In extension the friction angles are significantly affected, especially at lower pressures, because the failure loads are much smaller.

The measured vertical load corrections observed in the study by Yamamuro *et al.* (2012) were much greater than previous studies have shown. The effects on extension test friction angles were very significant.

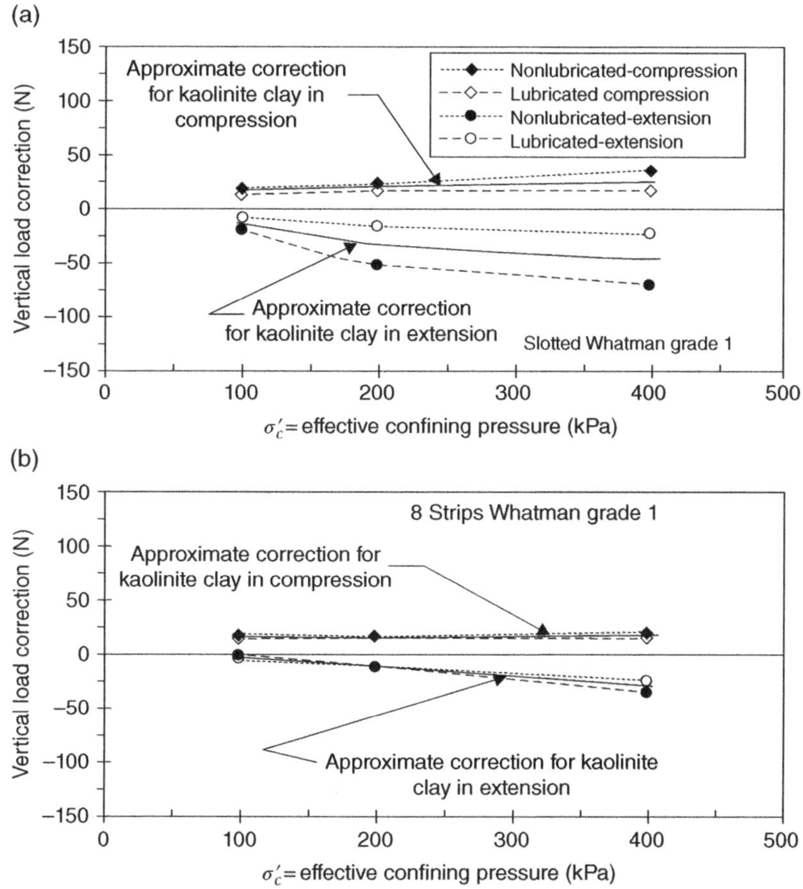

Figure 9.5 Vertical load correction for Whatman No. 1 filter paper in compression and extension loading at lower confining pressures with and without lubrication for (a) the inclined slotted configuration and (b) the eight-vertical-strip configuration. Reproduced from Yamamuro *et al.* 2012 by permission of ASTM International.

9.5.4 Membrane

Expressions for corrections to the stresses acting on a triaxial specimen due to the rubber membrane may be derived from elasticity theory on the basis of the following assumptions summarized by DeGroff *et al.* (1988):

1. The specimen deforms as a right cylinder, that is bulging does not occur and shear planes do not develop.
2. The membrane and the specimen deforms as a unit with no wrinkles and no slip between membrane and specimen, that is the membrane can sustain compression and $\varepsilon_{am} = \varepsilon_1$ and $\varepsilon_{\theta m} = \varepsilon_3$.
3. The modulus of rubber is the same in compression and tension (it is easiest to measure the modulus of the rubber membrane, E_m, in tension).
4. The rubber is incompressible such that Poisson's ratio for the membrane, $\nu_m = 0.5$.

The stresses and strains in a cylindrical rubber membrane are indicated in Fig. 9.8.

In the simplest method of calculating the correction to the deviator stress due to the load taken by the membrane, the following expression

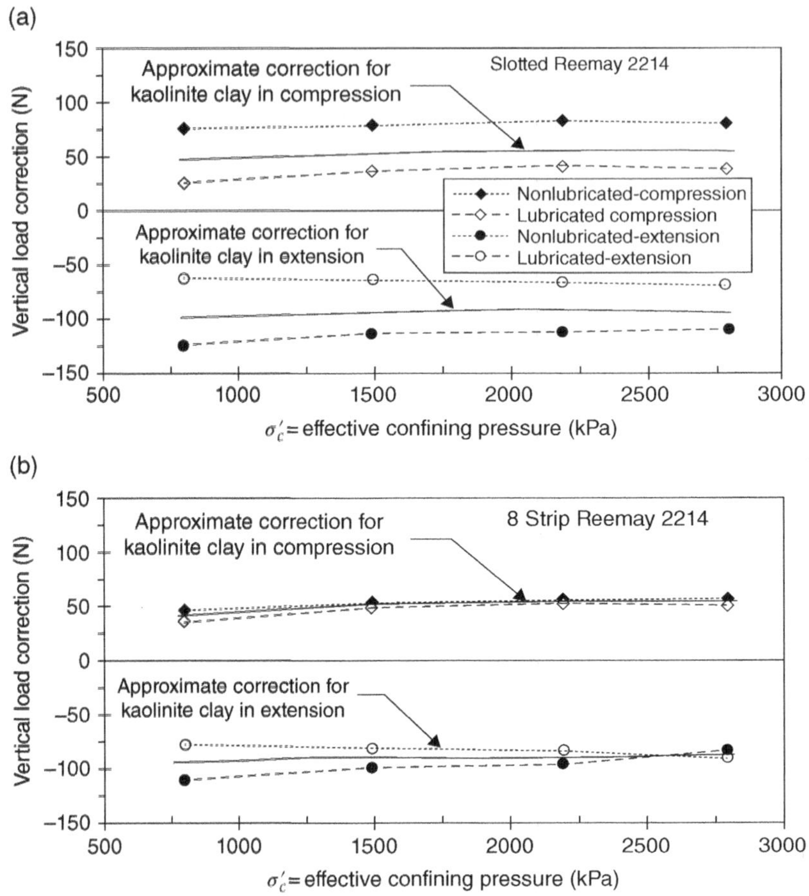

Figure 9.6 Vertical load correction for non-woven geotextiles in compression and extension loading at higher confining pressures with and without lubrication for (a) the inclined slotted configuration and (b) the eight-vertical-strip configuration. Reproduced from Yamamuro *et al.* 2012 by permission of ASTM International.

was given by Henkel and Gilbert (1952) for undrained tests:

$$\Delta(\sigma_1-\sigma_3)_{corr} = -\frac{P_m \cdot A_m}{A_s} = -\frac{\pi \cdot D \cdot t \cdot \sigma_m}{A_0 \cdot \dfrac{1}{1-\varepsilon_a}}$$

$$= -\frac{\pi \cdot D \cdot t \cdot E_m \cdot \varepsilon_a \cdot (1-\varepsilon_a)}{A_0} \qquad (9.2)$$

in which

D = diameter of specimen

t_m = thickness of rubber membrane

E_m = modulus of elasticity of the rubber membrane

ε_a = axial strain of the specimen

A_0 = initial cross-sectional area of specimen

Henkel and Gilbert (1952) argued that in the undrained test the volumetric strain is zero corresponding to a Poisson's ratio for the soil of 0.5. The deformations of the soil and the membrane (which has $\nu_m = 0.5$) are therefore compatible, and hoop stresses in the membrane are consequently zero.

In another simple method of calculating the stresses taken by the membrane it is assumed that the axial deformation of the membrane occurs independently of the radial and circumferential deformations. The equations for stress

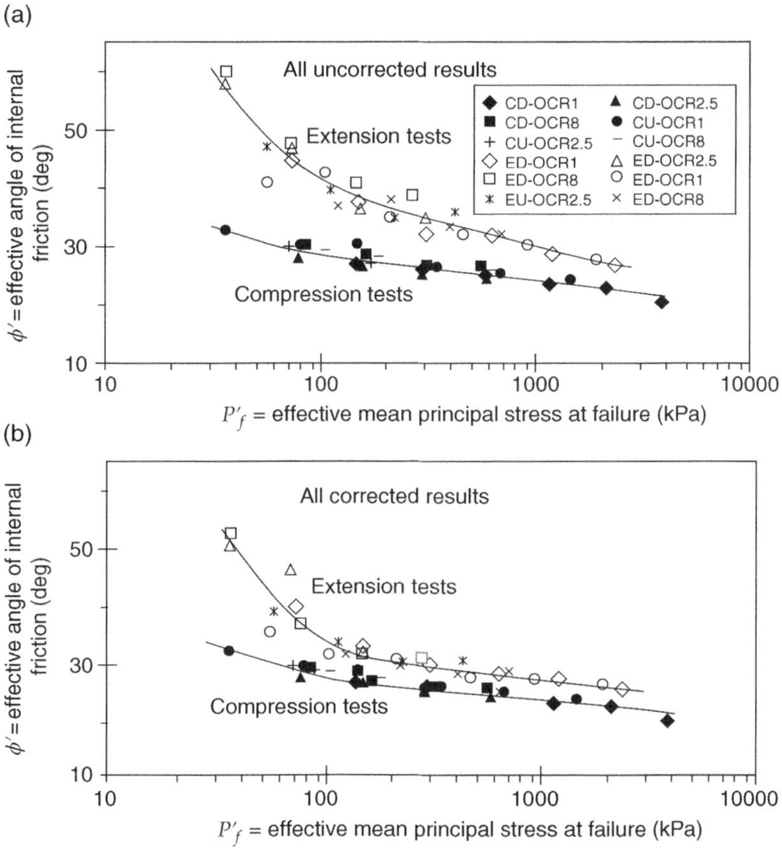

Figure 9.7 Effective drained and undrained friction angles of kaolinite clay: (a) without correction for strength contribution of drainage material and residual pore pressure in drained tests; and (b) with all corrections. Reproduced from Yamamuro *et al.* 2012 by permission of ASTM International.

corrections are then derived from simple elasticity theory without Poisson effects:

$$\Delta\sigma_{acorr} = -\frac{P_m}{A_{spec}} = -\frac{\Delta\sigma_{am}\cdot A_m}{A_{spec}} = -\frac{E_m\cdot\varepsilon_a\cdot\pi\cdot D_{spec}\cdot t_m}{\frac{\pi}{4}\cdot D_{spec}^2}$$
$$= -4\cdot E_m\cdot\varepsilon_a\cdot\frac{t_m}{D_{spec}} \tag{9.3}$$

and from horizontal equilibrium in a vertical section:

$$\Delta\sigma_{rcorr} = -\frac{P_{\theta m}}{H_{spec}\cdot D_{spec}} = -\frac{2\cdot\sigma_{rm}\cdot H_{spec}\cdot t_m}{H_{spec}\cdot D_{spec}}$$
$$= -\frac{2\cdot E_m\cdot\varepsilon_r\cdot H_{spec}\cdot t_m}{H_{spec}\cdot D_{spec}} = -2\cdot E_m\cdot\varepsilon_r\cdot\frac{t_m}{D_{spec}} \tag{9.4}$$

The axial and radial stresses in the triaxial test are corrected as follows:

$$\sigma_{acorr} = \sigma_a + \Delta\sigma_{acorr} \tag{9.5}$$

$$\sigma_{rcorr} = \sigma_r + \Delta\sigma_{rcorr} \tag{9.6}$$

For triaxial compression $\sigma_1 = \sigma_{acorr}$ and $\sigma_3 = \sigma_{rcorr}$, and for triaxial extension $\sigma_1 = \sigma_{rcorr}$ and $\sigma_3 = \sigma_{acorr}$. Equations (9.3) and (9.4) were indicated by Fukushima and Tatsuoka (1984), but they were found to be unrealistic in view of elasticity theory. The corrections reviewed above do not take into account the Poisson effect in the membrane and the change in membrane thickness, and they are only developed for small strains.

The following developments for membrane corrections are based on the same assumptions

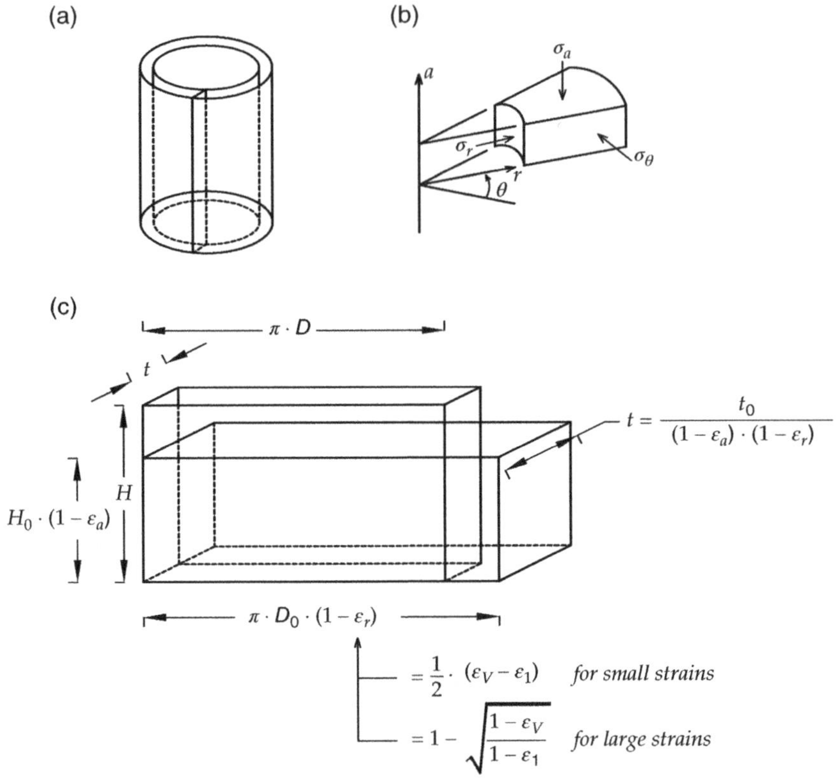

Figure 9.8 Stresses and strains in a cylindrical rubber membrane: (a) cylindrical membrane; (b) stresses in membrane element; and (c) deformation of membrane viewed as a rectangular prism.

as listed above, but they include Poisson effects in the membrane as well as the change in membrane thickness, and they are developed for both small and large strains. These developments also include the remaining corrections presented in the literature as special cases. Thus, the correction formulas presented in previous studies are in basic agreement, with the exceptions reviewed above.

Hooke's law in polar coordinates (subscript m for membrane has been dropped, except in the final expressions):

$$\varepsilon_a = \frac{1}{E} \cdot (\sigma_a - \nu \cdot \sigma_\theta - \nu \cdot \sigma_r) \qquad (9.7)$$

$$\varepsilon_\theta = \frac{1}{E} \cdot (-\nu \cdot \sigma_a + \sigma_\theta - \nu \cdot \sigma_r) \qquad (9.8)$$

$$\varepsilon_r = \frac{1}{E} \cdot (-\nu \cdot \sigma_a - \nu \cdot \sigma_\theta + \sigma_r) \qquad (9.9)$$

Solving Eqs (9.7) and (9.8) for σ_a and σ_θ gives:

$$\sigma_a = \frac{E}{1-\nu^2} \cdot (\varepsilon_a + \nu \cdot \varepsilon_\theta) + \frac{\nu}{1-\nu} \cdot \sigma_r \qquad (9.10)$$

$$\sigma_\theta = \frac{E}{1-\nu^2} \cdot (\nu \cdot \varepsilon_a + \varepsilon_\theta) + \frac{\nu}{1-\nu} \cdot \sigma_r \qquad (9.11)$$

Setting Poisson's ratio for the membrane, $\nu = 0.5$ produces:

$$\sigma_{am} = \frac{2}{3} \cdot E_m \cdot (2 \cdot \varepsilon_a + \varepsilon_\theta) + \sigma_r \qquad (9.12)$$

$$\sigma_{\theta m} = \frac{2}{3} \cdot E_m \cdot (\varepsilon_a + 2 \cdot \varepsilon_\theta) + \sigma_r \qquad (9.13)$$

These expressions for σ_{am} and $\sigma_{\theta m}$ indicate the axial and circumferential stresses *in the membrane*. The circumferential strain in the expressions equals the radial strain in the specimen, $\varepsilon_\theta = \varepsilon_r$.

Corrections to the axial stress *in the specimen* are calculated as follows:

$$\Delta\sigma_{acorr} = -\frac{P_{membrane}}{A_{specimen}} = -\frac{\sigma_{am}\cdot A_m}{A_s} \quad (9.14)$$

in which (for $\nu_m = 0.5$, i.e., $\Delta V_{membrane} = 0$):

$$A_m = A_{0m}\cdot\frac{1}{1-\varepsilon_a} \quad (9.15)$$

and

$$A_s = A_{0s}\cdot\frac{1-\varepsilon_v}{1-\varepsilon_a} \quad (9.16)$$

Then

$$\Delta\sigma_{acorr} = -\frac{4\cdot t_0}{D_0\cdot(1-\varepsilon_v)}\cdot\sigma_{am} \quad (9.17)$$

Substitution of Eq. (9.12) into Eq. (9.17) produces:

$$\Delta\sigma_{acorr} = -\left[\frac{2}{3}\cdot E_m\cdot(2\cdot\varepsilon_a+\varepsilon_r)+\sigma_r\right]\cdot\frac{4\cdot t_0}{D_0\cdot(1-\varepsilon_v)} \quad (9.18)$$

The corrected axial stress is obtained from Eq. (9.5).

Correction for the radial stress *in the specimen* is calculated from horizontal equilibrium in a vertical section:

$$\Delta\sigma_{rcorr} = -\frac{P_{\theta m}}{H\cdot D} = -\frac{2\cdot\sigma_{\theta m}\cdot H\cdot t}{H\cdot D} = -\frac{2\cdot t}{D}\cdot\sigma_{\theta m} \quad (9.19)$$

in which the thickness of the membrane is given by:

$$t = \frac{t_0}{(1-\varepsilon_a)(1-\varepsilon_r)} \quad (9.20)$$

and the diameter of the specimen is given by:

$$D = D_0\cdot(1-\varepsilon_r) \quad (9.21)$$

Then

$$\Delta\sigma_{rcorr} = -\frac{2\cdot t_0}{(1-\varepsilon_a)(1-\varepsilon_r)\cdot D_0\cdot(1-\varepsilon_r)}\cdot\sigma_{\theta m} = -\frac{2\cdot t_0}{D_0\cdot(1-\varepsilon_v)}\cdot\sigma_{\theta m} \quad (9.22)$$

Substitution of Eq. (9.13) into Eq. (9.22) gives:

$$\Delta\sigma_{rcorr} = -\left[\frac{2}{3}\cdot E_m\cdot(\varepsilon_a+2\cdot\varepsilon_r)+\sigma_r\right]\cdot\frac{2\cdot t_0}{D_0\cdot(1-\varepsilon_v)} \quad (9.23)$$

The corrected radial stress is obtained from Eq. (9.6).

Equations (9.18) and (9.23) indicate the general formulas for correction of the axial and radial stresses in triaxial compression and extension tests. These formulas are accurate for small as well as for large engineering strains (i.e., $\varepsilon_a = \Delta H/H$, etc.).

For *large strains* the radial strain to be substituted in these equations is obtained from the solution to the quadratic strain equation:

$$(1-\varepsilon_v) = (1-\varepsilon_a)(1-\varepsilon_\theta)(1-\varepsilon_r) \quad (9.24)$$

For triaxial tests: $\varepsilon_\theta = \varepsilon_r$, and solving for ε_r:

$$\varepsilon_r = 1-\sqrt{\frac{1-\varepsilon_v}{1-\varepsilon_a}} \quad (9.25)$$

Substitution of Eq. (9.25) into Eqs (9.18) and (9.23) produces:

$$\Delta\sigma_{acorr} = -\left[\frac{2}{3}\cdot E_m\cdot\left(2\cdot\varepsilon_a+1-\sqrt{\frac{1-\varepsilon_v}{1-\varepsilon_a}}\right)+\sigma_r\right]\cdot\frac{4\cdot t_0}{D_0\cdot(1-\varepsilon_v)} \quad (9.26)$$

and

$$\Delta\sigma_{rcorr} = -\left[\frac{2}{3}\cdot E_m\cdot\left(\varepsilon_a+2-2\cdot\sqrt{\frac{1-\varepsilon_v}{1-\varepsilon_a}}\right)+\sigma_r\right]\cdot\frac{2\cdot t_0}{D_0\cdot(1-\varepsilon_v)} \quad (9.27)$$

in which

E_m = modulus of elasticity of the rubber membrane
ε_a = axial strain in the specimen
ε_v = volumetric strain of specimen
σ_r = radial stress applied to specimen
t_0 = initial thickness of the rubber membrane
D_0 = initial diameter of the specimen

The expressions in Eqs (9.26) and (9.27) were given by Ponce and Bell (1971). Duncan and Seed (1965) argued that the radial stress, σ_r, has no overall effect on the deformation and the corrections, because Poisson's ratio of the membrane, ν_m = 0.5 (i.e., rubber is incompressible). Thus, any change in radial stress would cause equal changes in the axial and circumferential stresses and therefore no resulting deformation of the membrane. Duncan and Seed (1965, 1967) presented the following expressions in which σ_r has been set equal to zero:

$$\Delta\sigma_{acorr} = -\frac{2}{3}\cdot E_m \cdot \left[2\cdot\varepsilon_a + 1 - \sqrt{\frac{1-\varepsilon_v}{1-\varepsilon_a}}\right]\cdot\frac{4\cdot t_0}{D_0\cdot(1-\varepsilon_v)} \tag{9.28}$$

and

$$\Delta\sigma_{rcorr} = -\frac{2}{3}\cdot E_m \cdot \left[\varepsilon_a + 2 - 2\cdot\sqrt{\frac{1-\varepsilon_v}{1-\varepsilon_a}}\right]\cdot\frac{2\cdot t_0}{D_0\cdot(1-\varepsilon_v)} \tag{9.29}$$

These equations are the same as those given in Eqs (9.26) and (9.27) for $\sigma_r = 0$.

Duncan and Seed (1965, 1967) demonstrated the effects of large strains in the corrections by writing the equations as follows:

$$\Delta\sigma_{acorr} = \Delta\sigma_{am} = -C_{am}\cdot\left(\frac{2}{3}\right)\cdot E_m\cdot\frac{4\cdot t_{0m}}{D_{0m}} \tag{9.30}$$

and

$$\Delta\sigma_{rcorr} = \Delta\sigma_{lm} = -C_{lm}\cdot\left(\frac{2}{3}\right)\cdot E_m\cdot\frac{2\cdot t_{0m}}{D_0} \tag{9.31}$$

in which

$$C_{am} = \frac{1+2\varepsilon_a - \sqrt{\frac{1-\varepsilon_v}{1-\varepsilon_a}}}{1-\varepsilon_v} \tag{9.32}$$

$$C_{lm} = \frac{2+\varepsilon_a - 2\cdot\sqrt{\frac{1-\varepsilon_v}{1-\varepsilon_a}}}{1-\varepsilon_v} \tag{9.33}$$

The variations of C_{am} and C_{lm} with axial and volumetric strains, ε_a and ε_v, are shown in the diagrams in Fig. 9.9. These diagrams are particularly useful for triaxial compressions tests with zero or contractive volumetric strains.

For *small strains* Eqs (9.19) and (9.23) with $\sigma_r = 0$ reduce as follows. For stress correction:

$$\Delta\sigma_{acorr} = -\frac{2}{3}\cdot E_m\cdot(2\varepsilon_a+\varepsilon_r)\cdot\frac{4\cdot t_0}{D_0\cdot(1-\varepsilon_v)}$$

$$= -E_m\cdot\left(\varepsilon_a+\frac{\varepsilon_v}{3}\right)\cdot\frac{4\cdot t_0}{D_0\cdot(1-\varepsilon_v)} \tag{9.34}$$

If the correction for volumetric strain in the denominator, $(1-\varepsilon_v)$, is neglected, Eq. (9.34) reduces to:

$$\Delta\sigma_{acorr} = -4\cdot E_m\cdot\frac{t_0}{D_0}\cdot\left(\varepsilon_a+\frac{\varepsilon_v}{3}\right) \tag{9.35}$$

For radial stress correction, Eq. (9.23) reduces to:

$$\Delta\sigma_{rcorr} = -\frac{2}{3}\cdot E_m\cdot(\varepsilon_a+2\varepsilon_r)\cdot\frac{2\cdot t_0}{D_0\cdot(1-\varepsilon_v)}$$

$$= -4\cdot E_m\cdot\frac{t_0}{D_0}\cdot\frac{\varepsilon_v}{3} \tag{9.36}$$

in which $(1-\varepsilon_v)$ in the denominator has been neglected.

Equations (9.35) and (9.36) were presented by Molenkamp and Luger (1981) and Berre (1982).

Unconsolidated-undrained tests on saturated soil

For unconsolidated-undrained tests on saturated soil the volumetric strain is zero. Using the small strain expressions in Eqs (9.35) and (9.36), the following membrane corrections are obtained:

$$\Delta\sigma_{acorr} = -4\cdot E_m\cdot\frac{t_0}{D_0}\cdot\varepsilon_a \tag{9.37}$$

$$\Delta\sigma_{rcorr} = 0 \tag{9.38}$$

Note that Eq. (9.37) is the same as Eq. (9.3) and very similar to Eq. (9.2), which were both derived from simplified considerations. However, Eq. (9.4) for the radial correction does not produce zero correction as indicated in Eq. (9.38). This is

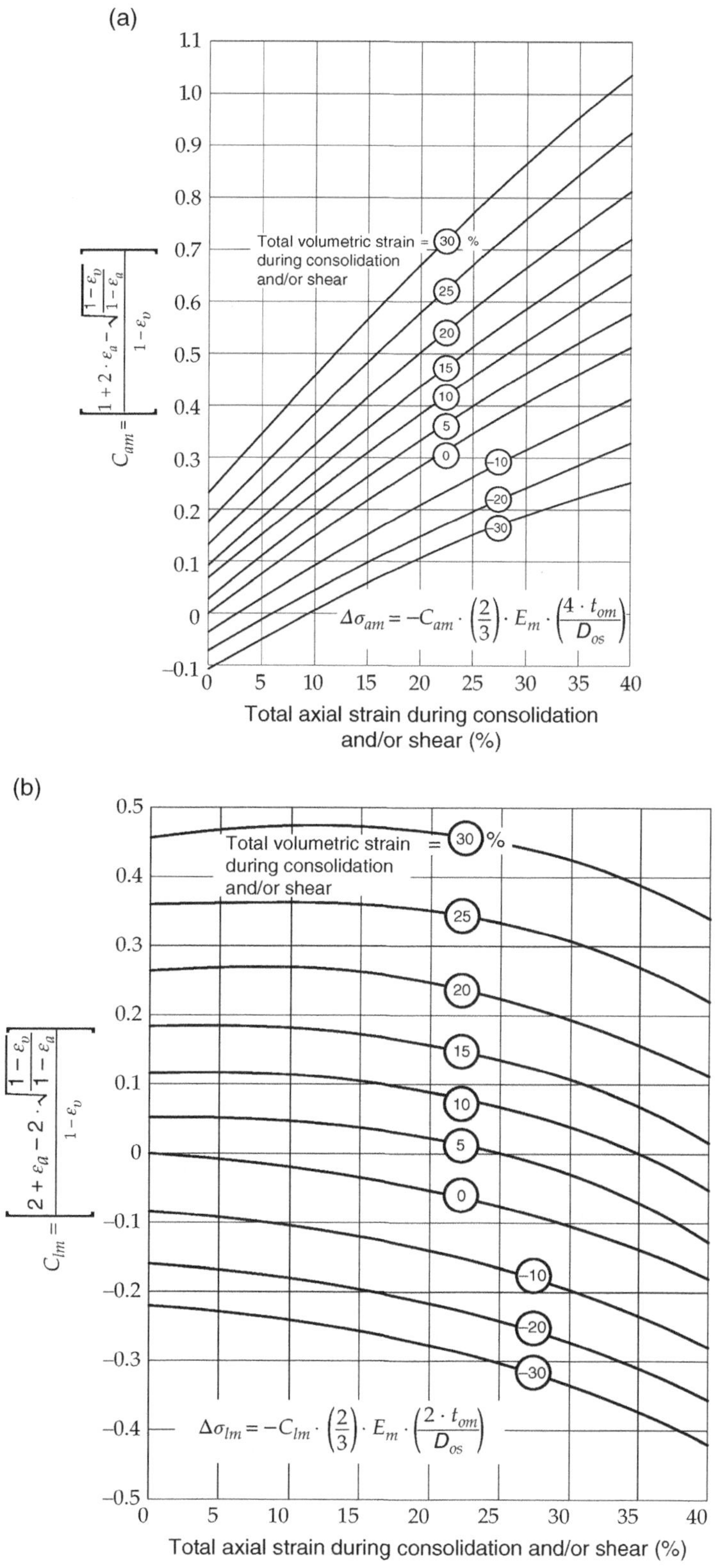

Figure 9.9 The variations of (a) C_{am} and (b) C_{lm} with axial and volumetric strains, ε_a and ε_v, respectively (modified after Duncan and Seed 1967 by permission of ASCE).

because Eq. (9.4) was derived without consideration of the Poisson effect in the membrane.

Equations (9.28) and (9.29) for large strains may also be used for membrane corrections. For small strains they produce the same corrections as given by Eqs (9.37) and (9.38). However, the corrections at large strains are different and probably more accurately determined from Eqs (9.28) and (9.29).

Consolidated-undrained tests on saturated soil

For consolidated-undrained tests the corrections may be performed using either Eqs (9.28) and (9.29) for large strains or Eqs (9.37) and (9.38) for small strains. The strains to be substituted in these equations are the total strains sustained by the specimen during consolidation and shearing. Since the volumetric strain is zero for the undrained shearing of a saturated specimen, the value of ε_v is that obtained at the end of the consolidation stage.

For K_0-consolidation, in which $\varepsilon_v = \varepsilon_a$, the small strain equations yield:

$$\Delta\sigma_{acorr} = -\frac{16}{3} \cdot E_m \cdot \frac{t_0}{D_0} \cdot \varepsilon_a \quad (9.39)$$

$$\Delta\sigma_{rcorr} = -\frac{4}{3} \cdot E_m \cdot \frac{t_0}{D_0} \cdot \varepsilon_a \quad (9.40)$$

Additional membrane correction to the radial stress is not required during the following undrained shearing stage, but the axial stress should be corrected according to Eq. (9.37) for small strains or Eq. (9.28) for large strains.

Drained tests

For consolidated-drained tests the corrections may be performed according to either Eqs (9.28) and (9.29) for large strains or Eqs (9.37) and (9.38) for small strains. The strains to be substituted in these equations are the total strains sustained by the specimen during consolidation and shearing.

Equation (9.37) for small strains shows that the correction to the axial stress is larger for a specimen with volumetric contraction than for a specimen with volumetric expansion (dilation). Equation (9.38) indicates that the radial stress decreases for a specimen with volumetric compression, whereas the radial stresses increase for a specimen with volumetric expansion (for a test with constant cell pressure). Similar, but more accurate membrane corrections are obtained from Eqs (9.28) and (9.29) for large strains.

Other types of membrane behavior

The case for large strains reviewed above is appropriate for soft soils that undergo large strains to reach failure, and for which the membrane corrections play an important role in the final strength evaluation. In addition to the relatively simple case reviewed above, in which the specimen and the membrane deform in unison, three types of behavior of the membrane relative to the specimen can be identified:

1. The specimen with the membrane bulges.
2. The membrane develops uniformly spaced horizontal wrinkles as the triaxial compression test progresses. In this case there is no correction to the axial stress, but the radial stress is corrected for hoop tension in the membrane. This case has been addressed by Henkel and Gilbert (1952) and Fukushima and Tatsuoka (1984). The latter authors recommend employing Eq. (9.4) for the radial stress correction.
3. The specimen develops a shear plane. This case has been treated by Chandler (1966), Blight (1967), La Rochelle (1967), Symons (1967), Pachakis (1976), and La Rochelle *et al.* (1988).

9.5.5 Buoyancy effects

The stresses vary along the specimen height due to gravitational effects on soil and water. To include the effects of self-weight in the calculation, the vertical stresses are calculated at mid-height of the specimen, and the effect of buoyancy in the saturated specimen therefore plays a role in calculation of the vertical stresses.

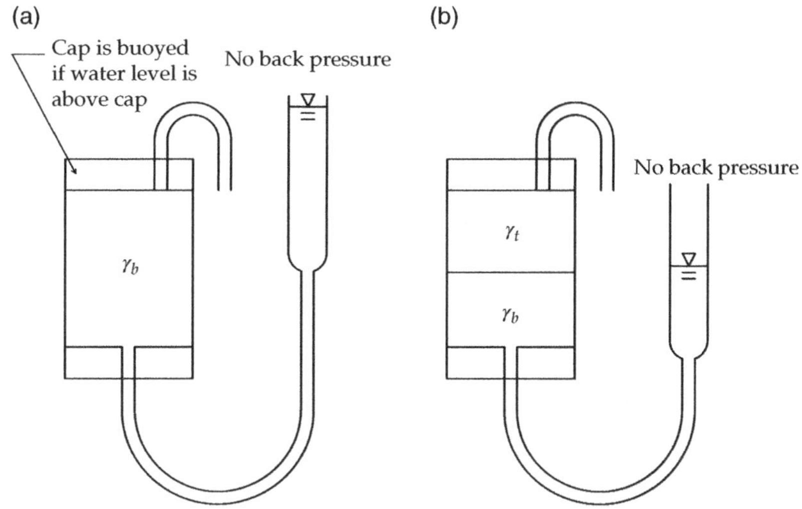

Figure 9.10 Effects of buoyancy on the vertical stress in the triaxial test for (a) full submergence and (b) partial submergence. γ_b, buoyant unit weight; γ_t, total unit weight.

If the water level in the volume change device is maintained at a certain level, as seen in Fig. 9.10, then for the purpose of calculating the vertical stress, the unit weight above this level is the total unit weight and it is the buoyant unit weight below this level. Similarly, if the water level in the volume change device is above the cap, then it is also buoyed, as seen in Fig. 9.10(a). If a back pressure is applied, then the buoyant unit weight is applied for calculation of the vertical stress due to self-weight.

9.5.6 Techniques to avoid corrections to vertical load

It is possible to avoid the corrections due to piston friction and piston uplift by placing the vertical load cell inside the triaxial cell, as explained in Section 3.3.1. All other corrections to the vertical load cannot be avoided.

9.6 Vertical deformation

9.6.1 Compression of interfaces

Interfaces external to the specimen are reviewed in Section 3.1.8. They will compress during vertical loading and cause errors in the vertical deformation. They may be determined by a calibration and employed for correction. However, they are easily and most often avoided by measuring the vertical deformation closer to or directly on the specimen.

9.6.2 Bedding errors

The externally measured vertical deformation is also influenced by the bedding errors associated with lubricated ends on the triaxial specimen (Sarsby *et al.* 1980, 1982; Molenkamp and Tatsuoka 1983; Lo *et al.* 1989). These bedding errors are shown in Fig. 9.11, and they consist of (1) sand grains penetrating into the lubricating ends and rearrangement of the grains at the surface, $\Delta\alpha$, (2) compression of the lubricating sheets due to Poisson's effect, Δt, and (3) change in the average thickness of the lubricating grease layer due to lateral squeezing and adjustment of lubricating sheet/grease due to lack of initial fit between specimen and lubricating sheet, Δh. They are additive and comprise the total bedding error, which affects the vertical deformation.

If more than one lubricating rubber sheet is employed, as explained in Section 3.1.8, then the rubber sheet next to the end plate is lubricated

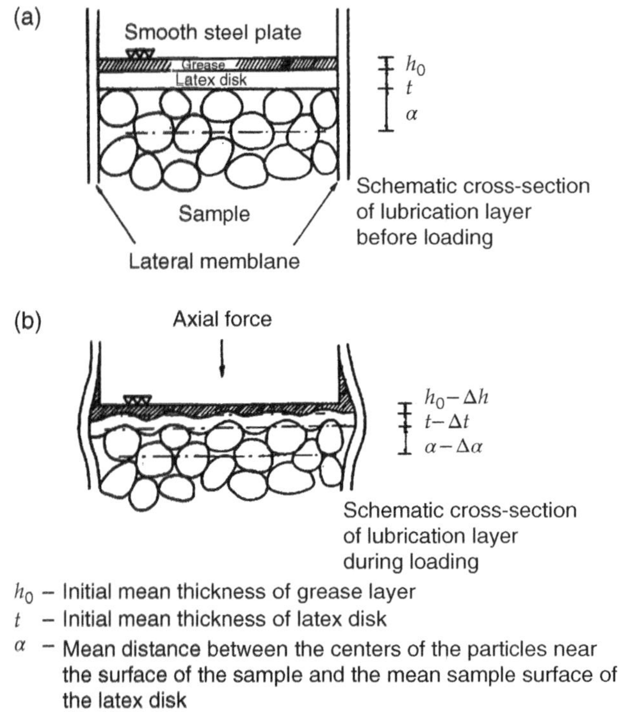

Figure 9.11 (a) Uncompressed and (b) compressed specimen ends in which the bedding errors consist of (1) sand grains penetrating into the lubricating ends and rearrangement of the grains at the surface, $\Delta\alpha$, (2) compression of the lubricating sheets due to Poisson's effect, Δt, and (3) change in the average thickness of the lubricating grease layer due to lateral squeezing and adjustment of lubricating sheet/grease due to lack of initial fit between specimen and lubricating sheet, Δh. Reproduced from Molenkamp and Tatsuoka 1983 by permission of ASCE.

on both sides and it may be compressed and squeezed out between the end plate and the sheet next to the specimen, which follows the lateral expansion of the specimen. Whether this happens will have to be determined by observation, and additional correction to the vertical deformation due to compression of the rubber sheet may be made by calculation from Hooke's law due to the Poisson effect.

The component $\Delta\alpha$ consisting of sand grain penetrating into the lubricating ends and rearrangement of grains at the surface of the lubricating sheet also cause a volume change, which is equal to $\Delta\alpha$ times the cross-sectional area of the specimen. Since the grease and the lubricating sheet consisting of a rubber membrane have relatively large bulk moduli, these components do not contribute to the error in volume change.

Based on the data presented by Sarsby *et al.* (1982), Molenkamp and Tatsuoka (1983) analyzed the bedding errors in terms of dimensionless quantities, as follows:

(t/d_{50}) = ratio of thickness of latex rubber disk, t, and mean grain size, d_{50}
(Δ_{BE}/d_{50}) = ratio of bedding error, Δ_{BE}, and mean grain size, d_{50}
(σ/E) = ratio of normal stress, σ, and Young's modulus of latex rubber, E

The estimates of the bedding errors, given below, are valid in the ranges 10 kPa $< \sigma <$ 2000 kPa and 0.135 mm $< d_{50} <$ 2 mm corresponding to the experimental ranges $0.009 < \sigma/E < 1.8$ and $0.17 < t/d_{50} < 2.6$. The bedding error is divided into two components: the reversible; and the irreversible.

Reversible bedding error

At very low stresses ($\sigma/E < 0.1$, where E is approximately 1100 kPa), where the grease has not penetrated yet, the reversible bedding error, Δ_{BE}, may be approximated by an expression for the conventional membrane penetration ($\Delta\alpha$):

$$\frac{\Delta\alpha}{d_{50}} \approx \left\{0.68 + 0.51 \cdot \exp\left(-2\frac{t}{d_{50}}\right)\right\}\left(\frac{\sigma}{E}\right)^{2/3} \quad (9.41)$$

The bedding error then transitions to the following expression at higher stresses:

$$\left(\frac{\Delta_{BE}}{d_{50}}\right) = 0.126 \cdot \left(\frac{t}{d_{50}}\right)^{0.25}\left\{0.651 + \log\frac{\sigma / E}{(\sigma / E)_t}\right\} \quad (9.42)$$

in which $(\sigma/E)_t$ is the normalized transitional stress between the two expressions, given by:

$$\left(\frac{\sigma}{E}\right)_t = \left[\frac{0.082 \cdot (t / d_{50})^{0.25}}{0.68 + 0.51 \cdot \exp(-2t / d_{50})}\right]^{3/2} \quad (9.43)$$

Figure 9.12 shows the reversible bedding error according to these equations for the ranges $0.8 < t/d_{50} < 2.5$ and $0 < \sigma/E < 2.0$. The transitional stresses according to Eq. (9.43) are also indicated in this diagram.

Irreversible bedding error

The irreversible bedding error may be estimated from:

$$\left(\frac{\Delta_{BE}}{d_{50}}\right) \approx \left\{0.221 \cdot \left(\frac{t}{d_{50}}\right)^{0.3} - 0.126 \cdot \left(\frac{t}{d_{50}}\right)^{0.25}\right\} \cdot \log\left(\frac{\sigma}{E}\right) + function\left(\frac{t}{d_{50}}\right) \quad (9.44)$$

Figure 9.13 shows the irreversible bedding error according to Eq. (9.44) for the range of normalized stress of $0.023 < \sigma/E < 1.8$, corresponding to σ = 25–2000 kPa. It increases with (t/d_{50}) and (σ/E). This bedding error is not caused by lateral squeezing of grease or by the Poisson effect of the latex rubber disk. Therefore, Δ_{BE} corresponds to $\Delta\alpha$ alone.

9.6.3 Techniques to avoid corrections to vertical deformations

The vertical deformation may be measured error free by attachments on the triaxial specimen, as reviewed in Section 4.7.

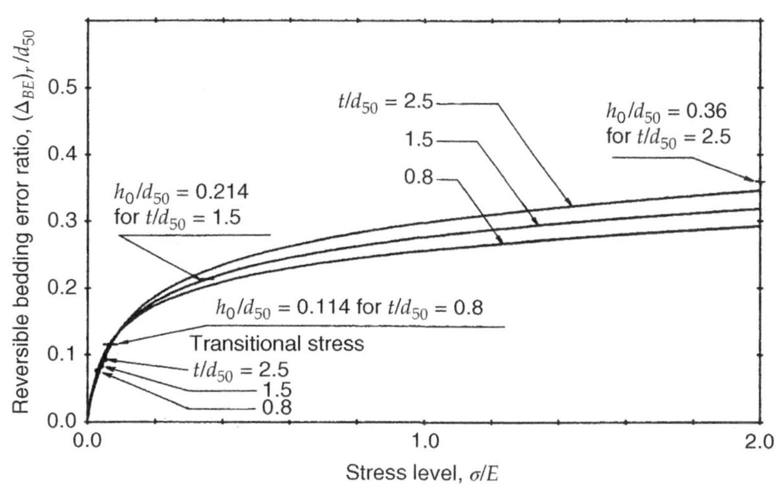

Figure 9.12 Reversible bedding error according to Eqs (9.41)–(9.43) for the ranges $0.8 < t/d_{50} < 2.5$ and $0 < \sigma/E < 2.0$. Reproduced from Molenkamp and Tatsuoka 1983 by permission of ASCE.

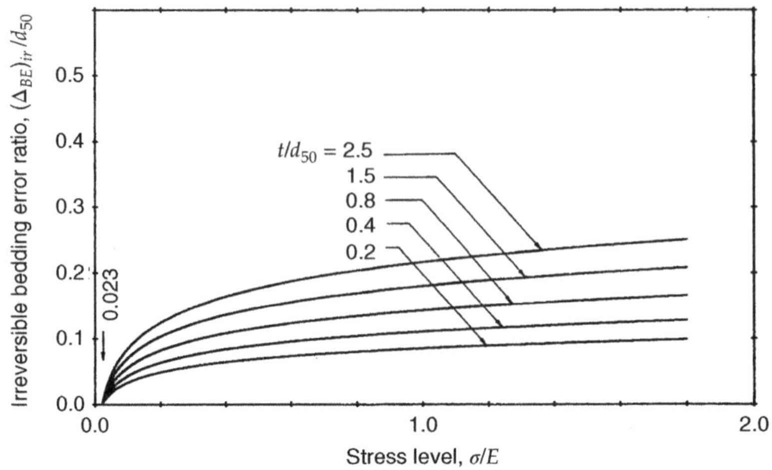

Figure 9.13 Irreversible bedding error according to Eq. (9.44) for the range of normalized stress of $0.023 < \sigma/E < 1.8$, corresponding to σ = 25–2000 kPa. Reproduced from Molenkamp and Tatsuoka 1983 by permission of ASCE.

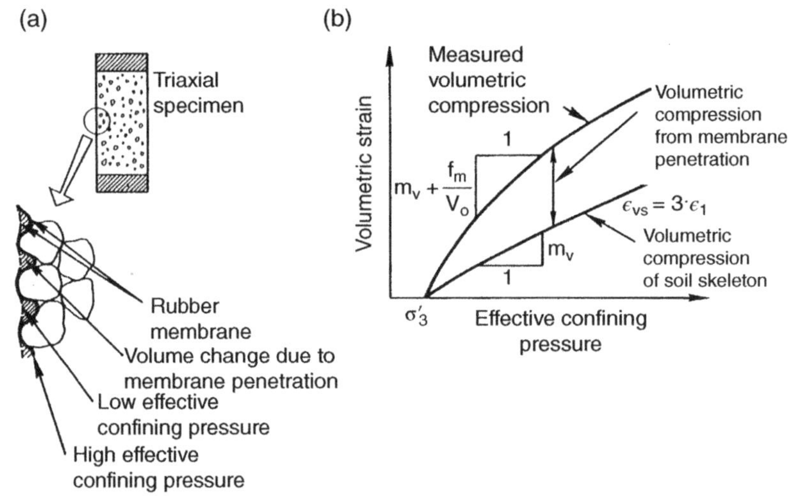

Figure 9.14 (a) Schematic diagram showing penetration of the rubber membrane enclosing the triaxial specimen into the voids between the particles in granular soils and (b) the resulting volume changes in tests with changing effective confining pressures (after Lade and Hernandez 1977).

9.7 Volume change

9.7.1 Membrane penetration

Penetration of the rubber membrane enclosing the triaxial specimen into the voids between the particles in granular soils causes volume changes in tests with changing effective confining pressures, as shown in Fig. 9.14(a). This creates an experimental error since the measured volume change is not solely representative of soil skeleton compression, but also includes that volume of water forced out by the penetrating membrane. The magnitude of this volume change primarily depends on the average particle size, the magnitude and change in the effective confining pressure, the modulus and the thickness of the rubber membrane, and the surface area covered by the membrane (the change is proportional to the surface area). The particle size distribution, the particle shape, and the initial void ratio have only minor effects within conventional effective confining pressures (Frydman *et al.* 1973). Membrane penetration is negligible for soils with average particle sizes below 0.1–0.2 mm (Frydman *et al.* 1973). Continuous increases in effective cell pressure result in further penetration until a maximum is reached and no further penetration is possible. However, for a given sand above a certain effective cell pressure, the volume change, due merely to membrane penetration, decreases with further increase in effective cell pressure. This is due to particle crushing. Over a large pressure range the initial void ratio and the amount of particle crushing play significant roles in membrane penetration (Bopp and Lade 1997a).

Since Newland and Allely (1959) first recognized membrane penetration as a source of measured volume change, both experimental and theoretical methods have been employed to compensate for membrane penetration. Most of the methods attempted to either: (1) approximate the experimentally determined volume change due to membrane penetration and make appropriate corrections in the experimental results (Newland and Allely 1959; Lade and Hernandez 1977); (2) devise analytical solutions to predict volume change from membrane penetration and correct the measured results (Molenkamp and Luger 1981; Baldi and Nova 1984; Kramer and Sivaneswaran 1989; Kramer *et al.* 1990); (3) modify the flexural characteristics of the membrane to limit the degree of penetration (Kiekbusch and Schuppener 1977; Lo *et al.* 1989); or (4) adjust the water volume within the specimen to counterbalance the volume change due to the penetrating membrane (Raju and Venkastaramana 1980; Nicholson *et al.* 1993a, b).

Experimental determination

Various techniques have been employed for experimental determination of the volume change due to membrane penetration (Newland and Allely 1959; Roscoe *et al.* 1963; Lee 1966; Raju and Sadasivan 1974; Ramana and Raju 1982).

Newland and Allely (1959) and Lade and Hernandez (1977) assumed the specimen to be isotropic and an increase in cell pressure therefore resulted in a volume change due to membrane penetration of

$$\varepsilon_{vmemb} = \varepsilon_{vmeas} - 3 \cdot \varepsilon_1 \tag{9.45}$$

In which $3 \cdot \varepsilon_1$ represents the isotropic compression of the specimen. Since granular specimens are typically stiffer in the vertical than the horizontal direction, the volume change due to membrane penetration will be overestimated. However, granular specimens are rarely isotropic and this method is not considered to correctly indicate the volume change due to membrane penetration.

The method employed by Roscoe *et al.* (1963) involved isotropic compression of three (or more) specimens with enclosure of centrally located metal dummy-rods with different diameters in two (or more) specimens, as shown in Fig. 9.15. The concept of the Roscoe dummy-rod method is to reduce the volume change due to soil skeleton compression while keeping constant the volume change from membrane penetration. This is accomplished by isotropically consolidating a series of soil specimens with the same outside diameter and height, prepared around steel rods with increasing diameter. The soil is deposited around the dummy-rod by the preparation method and the void ratio to be used in the triaxial tests. The measured volume change at any particular effective confining pressure reflects the combined result of soil skeleton compression and membrane penetration. By plotting the measured volume change versus the volume of soil in each specimen at discrete cell pressures for the three (or more) specimens, a diagram such as that shown in Fig. 9.16 is obtained. Extrapolation to zero soil volume indicates the volume change due to

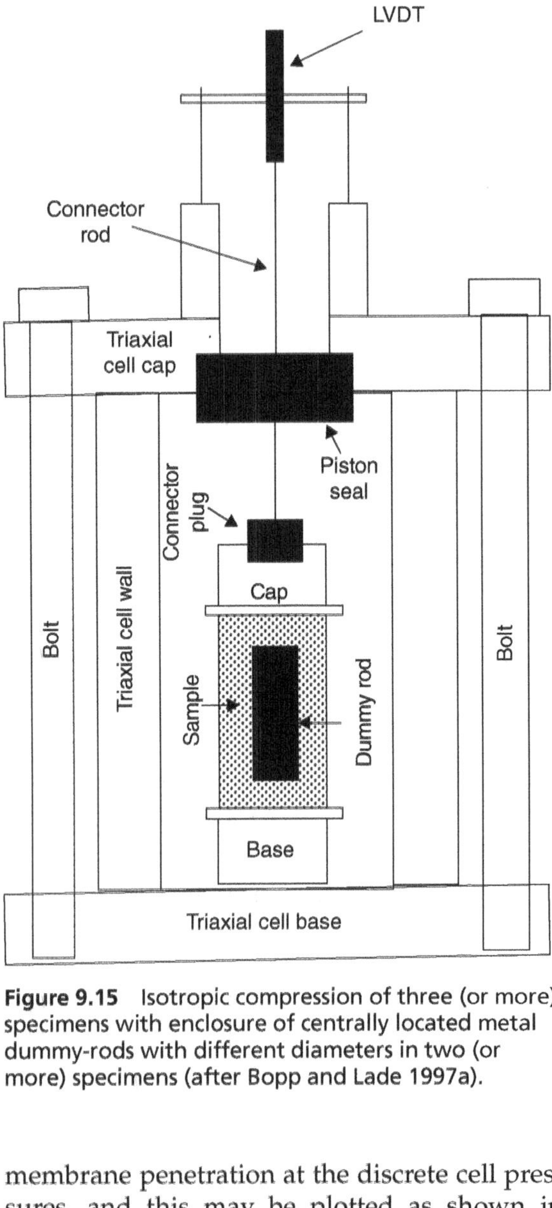

Figure 9.15 Isotropic compression of three (or more) specimens with enclosure of centrally located metal dummy-rods with different diameters in two (or more) specimens (after Bopp and Lade 1997a).

membrane penetration at the discrete cell pressures, and this may be plotted as shown in Fig. 9.17. This method was employed by Bopp and Lade (1997a) to determine the membrane penetration at high confining pressures where particle crushing was important and it was determined that the membrane penetration reduces beyond a certain pressure due to particle crushing.

Small problems with the Roscoe dummy-rod method have been pointed out by Raju and Sadisivan (1974), Wu and Chang (1982), and Vaid and Negussey (1984a). This method requires testing of separate, additional specimens, but it is considered to be reasonably reliable.

Rather than using dummy-rods, Ali *et al.* (1995) used specimens that were cemented. In their experiments the cell pressure was increased from 35 to 235 kPa and decreased again to 35 kPa. The specimen was sufficiently rigid due to the cementation that the specimen compression was assumed to be zero. Thus, all measured volume change was due to membrane penetration and this method requires only one specimen.

Theoretical characterization

Theoretical relationships for the volume change due to membrane penetration have been derived by Molenkamp and Luger (1981), Baldi and Nova (1984), and Kramer *et al.* (1990). These theoretical relationships all have the same general form and may be used for estimates of the volume change due to membrane penetration, V_m, based on the specimen volume, V_0, the specimen diameter, D, the average particle size, d_{50}, the magnitude of the effective cell pressure, σ'_{cell}, the thickness of the membrane, t_m, and the elastic modulus of the membrane, E_m, as follows:

$$V_m = A \cdot V_0 \frac{d_{50}}{D} \cdot \left[\frac{\sigma'_{cell} \cdot d_{50}}{E_m \cdot t_m} \right]^{1/3} \tag{9.46}$$

in which the constant A has been determined in several studies:

A=0.64 [Molenkamp and Luger 1981 (see Ali *et al.* 1995)]
A=0.50 (Baldi and Nova 1984)
A=0.924 (simplified solution by Kramer *et al.* 1990)

An improved solution by Kramer *et al.* (1990) employs:

$$A = 1.58 \cdot \left(\frac{1-\alpha}{5+64\alpha^2+80\alpha^4} \right)^{1/3}$$

$$\text{in which } \alpha = 0.15 \left(\frac{d_{50} \cdot \sigma'_{cell}}{E_m t_m} \right)^{0.34} \tag{9.47}$$

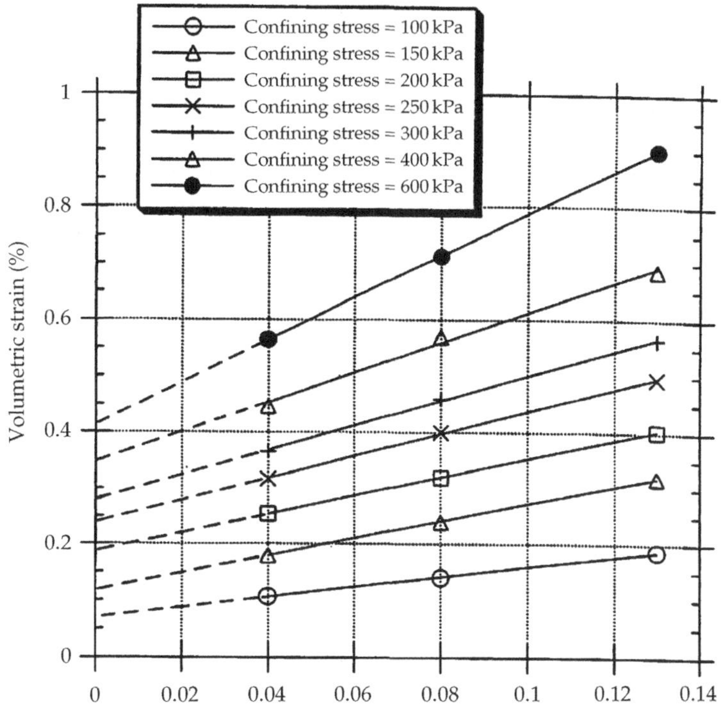

Figure 9.16 Measured volume change plotted versus the volume of soil in each specimen at discrete cell pressures for the three (or more) specimens. Extrapolation to zero soil volume indicates the volume change due to membrane penetration at the discrete cell pressures. Reproduced from Nicholson *et al.* 1993a by permission of Canadian Science Publishing.

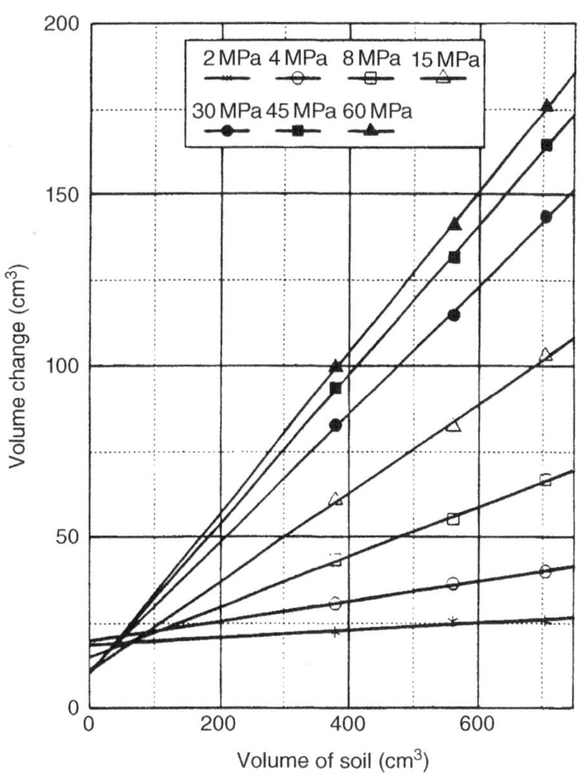

Figure 9.17 Diagram of membrane penetration at discrete cell pressures. Reproduced from Bopp and Lade 1997a by permission of ASTM International.

Figure 9.18 shows a comparison of the membrane penetration from the theoretical expressions with experimental results for increasing effective confining pressure from 35 to 235 kPa. The simple expression proposed by Molenkamp and Luger (1981) appears to best fit the experimental membrane penetration.

The effect of particle size on membrane penetration has been compared with experimental results by Baldi and Nova (1984). For the common range of particle sizes and cell pressures, the difference between their predictions and the experimental relation proposed by Frydman *et al.* (1973) is small when plotting S versus $\log(d_{50})$, in which S is the membrane penetration per unit surface area of the specimen divided by $\Delta\log(\sigma_{cell}')$. The two relations are compared with experimental results in Fig. 9.19.

Nicholson *et al.* (1993a) found that the fines in the granular materials played a role in filling out the voids near the membrane and that d_{20} represented a better grain diameter for correlation

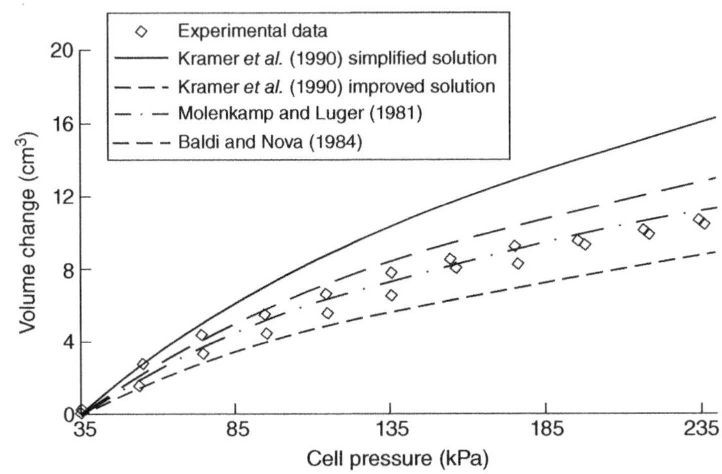

Figure 9.18 Comparison of the membrane penetration from theoretical expressions with experimental results for increasing effective confining pressure from 35 to 235 kPa. Reproduced from Ali *et al.* 1995 by permission of Geotechnique.

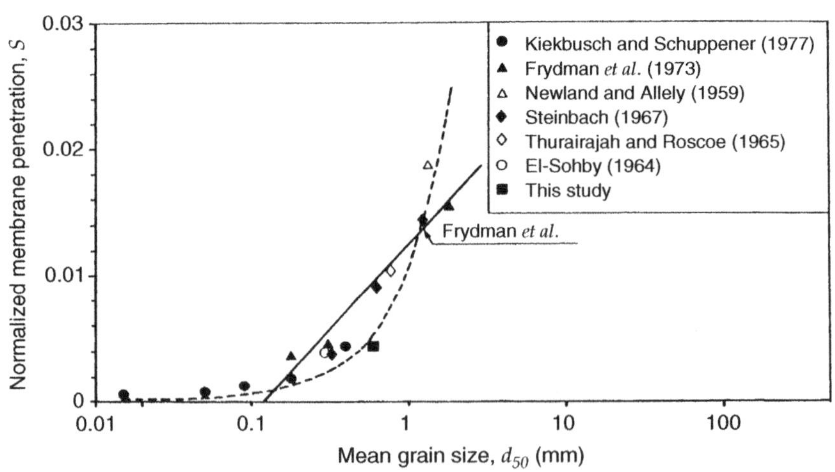

Figure 9.19 Effect of particle size on membrane penetration has been compared with experimental results by Baldi and Nova (1984) and the experimental relation proposed by Frydman *et al.* (1973). Reproduced after Baldi and Nova 1984 by permission of ASCE.

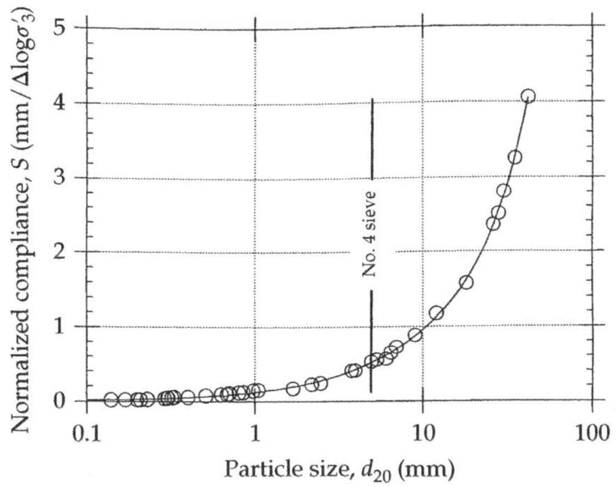

Figure 9.20 Diagram of normalized compliance S ($mm/\Delta\log\sigma_3' = ml/cm^2$) plotted versus d_{20} (mm) for all available membrane penetration data. Relationship given by Eq. (9.48). Reproduced from Nicholson *et al.* 1993a by permission of Canadian Science Publishing.

with membrane penetration. Figure 9.20 shows a diagram of normalized compliance S ($mm/\Delta\log\sigma_3' = ml/cm^2$) plotted versus d_{20} (mm) for all available membrane penetration data. The best-fit curve through the data points is given by a simple polynomial equation:

$$S = 0.0019 + 0.0095 \cdot d_{20} + 0.0000157 \cdot d_{20}^2 \quad (9.48)$$

This equation provides a better empirical relationship for estimation of the unit membrane penetration as seen in Fig. 9.20.

Minimization of membrane penetration

Minimization of membrane penetration effects may be achieved by increasing the specimen diameter, increasing the membrane thickness, and using a stiffer membrane.

Kiekbusch and Schuppener (1977) and Lo *et al.* (1989) modified the flexural characteristics of the membrane to limit the degree of penetration. In both studies the inside of the membrane was coated with a layer of liquid latex rubber before deposition of the soil. A confining pressure was then applied and the liquid rubber would stiffen in the outer layer of grains and consequently reduce the membrane penetration. However, the membrane would in effect become thicker and enclose a layer of particles thereby increasing its stiffness and its load carrying ability. This would result in much larger corrections to the measured axial load, and these corrections would be difficult to measure or predict.

Elimination of membrane penetration

It may be possible to adjust the water volume within the specimen to counterbalance the volume change due to the penetrating membrane. This was proposed by Raju and Venkastaramana (1980), Tokimatsu and Nakamuro (1986), and Nicholson *et al.* (1993a, b). Using the relation between effective confining pressure and volume change due to membrane penetration for the particular specimen being tested, water is injected into the specimen to circumvent the effects of membrane penetration. While manual adjustments are possible, closed-loop computer-controlled injection may be employed to continuously and automatically adjust for the membrane penetration. This method of circumventing the problem with membrane penetration may be used in both drained and undrained tests.

9.7.2 Volume change due to bedding errors

Sand penetration into the lubricated ends causes volume changes due to bedding errors, as reviewed in Section 9.6.2. This volume change may be calculated as Δa (= Δ_{BE}) times the cross-sectional area of the specimen for one end of the specimen.

9.7.3 Leaking membrane

Rubber membranes are permeable to water and to gases of various types. Rubber membranes may also develop leaks due to holes created by sharp grains that cut through the membrane as the effective confining pressure is increased in the course of an experiment. These problems are more typically encountered in long-term experiments at high confining pressures. Both types of leaks may be counteracted by applying several membranes with silicone grease between them.

Leaks due to diffusion

Since the membrane is much more permeable to gas than to fluid, gas dissolved in the cell fluid may penetrate through the membrane and come out of solution inside the specimen. This will show up as a volume change in a drained test and it will increase the pore pressure in an undrained test. Figure 9.21 shows an example (Karimpour 2012) in which a four-layer thick assembly of 0.3 mm thick membranes sealed with two O-rings at each end showed leakage in a long-term triaxial test conducted at an effective confining pressure of 8000 kPa. In test A, the specimen was sheared to 22% axial strain in 33 min, while the specimen in test B reached the same axial strain in 528 min. No obvious leakage was observed in these two tests. However, after the specimen in test C reached an axial strain of 1.20% in about 460 min, the volume change began to deviate seriously from the other two volume change curves due to leakage. The rate of volume change increased continuously and reached a value at which it was not possible to continue the test.

The nature of this leak indicated that it was due to diffusion of gas. The nitrogen from the bottle that supplied the high cell pressure dissolved in the initially de-aired water in the cell, traveled by diffusion towards the specimen and penetrated through the four-layer membrane. Once the nitrogen was inside the membrane, it came out of solution and formed bubbles which increased with time in the much lower back pressure (200 kPa). This resulted in a false volume change and the observation of leakage.

Rather than applying the nitrogen pressure directly to the top of the triaxial cell, it was

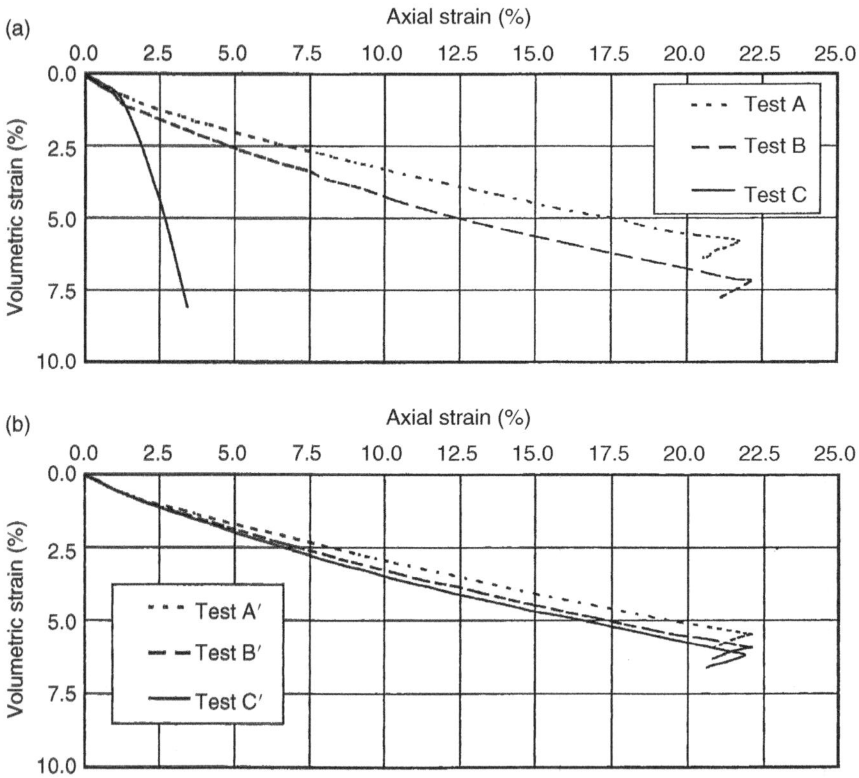

Figure 9.21 (a) Example in which a four-layer thick assembly of 0.3 mm thick membranes sealed with two O-rings at each end showed leakage in a long-term triaxial test conducted at an effective confining pressure of 8000 kPa and (b) similar tests performed without membrane leakage (after Karimpour 2012).

applied at the end of a long spiral of stainless steel tubing full of de-aired water. Since the gas travels by diffusion, a sufficiently long tube would prevent it from reaching the specimen within the time of the test. For experiments of the type shown in Fig. 9.21(a), a 3.6 m long spiral tube was used, and this length was sufficient to avoid diffusion of nitrogen into a specimen exposed to an effective confining pressure of 8000 kPa over a period of 2 months.

Figure 9.21(b) shows results of repeated experiments similar to those in Fig. 9.21(a) in which the long spiral tube had been installed. Test A′, which lasted 33 min, exhibited 0.30% less volume change than test A in Fig. 9.21(a). Similarly, test B′ showed 1.30% less volume change than the corresponding test B. The greater difference in tests B′ and B than between tests A′ and A is due to the longer test time. Thus, some nitrogen diffusion may have occurred even for the two shorter term tests. It is clear that removal of the nitrogen pressure source from the proximity of the triaxial specimen is beneficial and removes the possibility of gas diffusion into the specimen.

Leaks due to puncture

In most of the experiments performed at high confining pressure, the grains will puncture the innermost membrane(s), and these need to be exchanged as the series of experiments proceeds. The outer membranes are protected from puncture by the inner membranes and they maintain the functions of the membrane in the triaxial tests. Trial-and-error is required to determine how many membranes are necessary for a given sand and a given confining pressure.

9.7.4 Techniques to avoid corrections to volume change

It is possible to determine the volume change from the linear deformations measured directly on the specimen, as explained in Section 4.8, thus avoiding the corrections discussed above.

9.8 Cell and pore pressures

9.8.1 Membrane tension

The effective confining pressure is calculated as the difference between the cell and pore pressures, but corrections to the radial stress may be required for tension in the membrane. This correction is indicated in Eq. (9.6) and the expressions following this equation.

9.8.2 Fluid self-weight pressures

If the specimen is saturated and the confining pressure is applied through a fluid (e.g., water), then the fluid self-weight inside and outside the membrane are compensating and there is no effect of fluid self-weight pressures. If the specimen is dry, then the external water pressure increases downwards along its height and the effective confining pressure increases accordingly. If the specimen is saturated and the confining pressure is applied by compressed air, then the effective confining pressure decreases downwards along its height. The effects of fluid self-weight on the vertical stress are reviewed in Section 9.5.5.

9.8.3 Sand penetration into lubricated ends

Sand penetration into the lubricated ends has an effect on the measured pore pressure in undrained tests. The correction will be similar to that of membrane penetration, reviewed below.

9.8.4 Membrane penetration

Membrane penetration effects on the pore pressure in undrained tests were observed by Lade and Hernandez (1977), Kiekbusch and Schuppener (1977), and Martin *et al.* (1978). It is possible to include effects of volume changes due to membrane penetration and bedding errors in simple calculations of pore pressures such as those for the value of Skempton's pore pressure parameters *A* and *B* (Lade and Hernandez 1977; Baldi and Nova 1984; Yamamuro and Lade

1993a). It is not possible to perform simple, analytical corrections for membrane penetration to the stress–strain and pore water pressure in undrained triaxial compression tests. It will require an advanced analysis involving a constitutive model for the soil to include this effect in corrections of stress–strain, strength and pore pressure measurements. In such an advanced analysis, Molenkamp and Luger (1981) included the effects of membrane penetration in the calculation of pore pressures and effective stress paths in undrained triaxial compression tests using a double hardening elasto-plastic constitutive model.

9.8.5 Techniques to avoid corrections to cell and pore pressures

It is also possible to experimentally avoid the effects of membrane penetration by volume compensation as explained in Section 9.7.1. Following that procedure the pore water pressure is measured as though no effects from membrane penetration are present.

10 Special Tests and Test Considerations

10.1 Introduction

A number of issues emerge as various types of soils are tested under different conditions, and this may require special attention to some of the details. Some of these issues are simple and may be handled by a few remarks, while others require some discussion. The simple issues are discussed in this section, while the issues requiring discussion are dealt with in subsequent sections.

10.1.1 Low confining pressure tests on clays

Clay specimens may be tested using the paraffin method developed at the Norwegian Geotechnical Institute and described by Ramanatha Iyer (1973, 1975) and by Iversen and Moum. (1974). In this method a rubber membrane is not employed and membrane strength corrections are therefore not required. The liquid paraffin (kerosene) method depends on the difference in surface tension between the paraffin (0.023–0.032 N/m), used as the cell fluid, and the water (0.076 N/m) inside the specimen, and it depends on the menisci created at the surface pores between the small clay particles. Non-fissured clay is required and larger pores cannot be present, because they will break the surface tension. Only small centrally located drains in the cap and base can be present, that is no outside filter paper drains can be employed. The clay specimens have to be securely attached to the cap and base so that the fluid paraffin cannot intrude into the junction between end plates and specimen ends to reach the central drains. Small lengths of neoprene rubber membranes may be attached to the cap and base to keep the paraffin from entering the drains between the specimen and the end plates. The surface tension acts as the membrane and it is possible to develop and maintain a pore water pressure in the undrained specimen, and it is also possible to measure volume changes from the expelled water. Experiments were performed with effective confining pressures up to 20 kPa, but most tests were performed with somewhat lower pressures.

10.1.2 Conventional low pressure tests on any soil

In experiments at very low confining pressures, a water column may be employed to provide low and accurate cell pressures. The fact that the rubber membrane may be stretching and therefore applying some confining pressure to the specimen has to be taken into account to get an

Triaxial Testing of Soils, First Edition. Poul V. Lade.

accurate value of the confining pressure. Either a perfectly fitting membrane with no initial stretch has to be used, or the actual stretch of the membrane has to be measured to account for the additional confining pressure. Based on the strain in the membrane the additional confining pressure may be determined on the basis of the modulus and Poisson's ratio.

10.1.3 High pressure tests

The most significant problem with high pressure triaxial tests is puncturing of the latex rubber membranes in tests on granular materials with large or sharp grains. The sharp grains may cut the membrane, and the large grains will allow holes to develop as the membrane is pushed into the voids in the surface. To overcome this problem, several membranes are used around the soil specimen. Layers of grease are used between the membranes to disconnect punctures and not allow water to travel between consecutive membranes.

One thick membrane around the specimen is not as advantageous as several membranes with grease between them. The reason is that a hole generated by cutting cannot travel to the next membrane as easily as in one thick membrane in which the cut will simply continue through to the specimen surface.

10.1.4 Peats and organic soils

Peat specimens may be trimmed by an electric knife. For highly decomposed peats and organic soils the specimen may not be able to support its own weight, but the results may not be significantly affected by disturbance. Peats contain at least 10% gas. Peats behave elastically, but they exhibit high rates of creep. The deformation may be essentially vertical due to the strength of fibers. More information is given by Landva and Pheeney (1980) and Landva *et al.* (1983).

10.2 K_0-tests

To perform K_0-tests, the radial deformation has to be held at zero. However, this is not easy to accomplish, because the cell pressure has to be increased in response to the application of axial stress so as to maintain the value of K_0 constant. Thus, the cell pressure is first increased to cause a radial compressive strain of no more than 0.005%. This corresponds to a change in diameter of 0.0035 mm for a specimen with a diameter of 70 mm. The axial stress is then increased to counter the radial strain, that is the radial strain has to be reduced by the same amount (or 0.005%) to result in zero strain in the radial direction. This sequence of stress changes is required to correctly increase the stresses along the K_0-line.

Note that if the sequence of changes in radial and axial stresses is inverted, that is the axial stress is applied first to increase the diameter, and then the radial stress is applied to decrease the diameter, then a higher and incorrect value of K_0 is obtained. This is because of the sensitivity of the yield surface location to the changes in stresses.

K_0-experiments may also be performed by using a stiff triaxial cell, that is with a steel cell wall, and with a piston that has the same diameter as the specimen. As the piston is advanced into the fully saturated cell, the cell fluid will maintain the volume and prevent the specimen from expanding, while the horizontal pressure is registered by a pressure transducer. The vertical pressure on the specimen is not affected by the cell pressure, and $K_0 = \sigma_{cell}/\sigma_{vertical}$, in which σ_{cell} is the cell pressure and $\sigma_{vertical}$ is the vertical pressure.

10.3 Extension tests

Analyses of most laboratory tests to determine the shear strength of soils assume that the soil deforms such that uniform strain conditions exist throughout the specimen. Cylindrical compression and extension test specimens are assumed to deform as right cylinders, so that all parts of the specimen participate equally in the overall strains. If the specimen does not behave in this manner, the computations of axial strain, volumetric strain, and stresses are not correct.

Extension tests may be performed by using pistons with diameters that are adjusted relative to the specimen diameter.

10.3.1 Problems with the conventional triaxial extension test

Strain localization occurs when a majority of the overall deformations in a test specimen are contained within a small portion of the specimen. Shear planes and severe "necking" are common examples of strain localization occurring in cylindrical compression and extension tests. Roscoe *et al.* (1963) convincingly demonstrated that the conventional triaxial extension test has substantially more problems associated with strain localization than the triaxial compression test, and they concluded that the results could not be relied upon to evaluate failure criteria. This conclusion resulted in many investigators developing true triaxial testing equipment, in which the full range of the intermediate principal stress could be investigated. Some of the apparatus could enforce uniform strains by employing rigid boundary conditions around the specimen (Lade and Duncan 1973; Reades and Green 1976; Nakai and Matsuoka 1983; Lam and Tatsuoka 1988).

Yamamuro and Lade (1995) explained why there are problems maintaining uniform strains in an extension test. Figure 10.1 shows an extension test with a specimen that has an assumed geometric or material defect in section 3, which would have the effect of triggering necking in the specimen. When the upward load is applied, the initial vertical stress in section 3 is larger than in the surrounding sections, owing to its smaller cross-sectional area. Correspondingly, the vertical and radial strains are also larger. However, these larger lateral strains are directed inward, thus decreasing the cross-sectional area. Simple analysis clearly shows that as upward loading progresses, the inward directed radial strains continue to decrease the cross-sectional area in section 3 at a faster rate than in the surrounding sections. Therefore, the initial difference in stresses between section 3 and the surrounding ones increases very rapidly. Eventually, the material fails in the region of section 3, well in advance of the material in the other sections of the specimen. This is accentuated in soils by the highly nonlinear stress–strain behavior of the material, in which increasing stress increments produce larger amounts of strain as shearing progresses. Therefore, in an extension test the highest stresses and strains are produced and concentrated in the weakest part of the

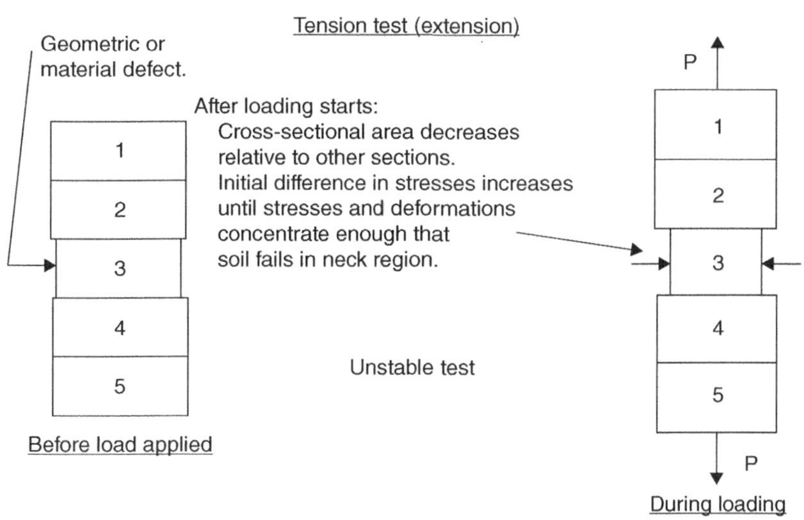

Figure 10.1 Extension test with a specimen that has an assumed geometric or material defect in Section 3, which could trigger necking in the specimen. Reproduced from Yamamuro and Lade 1995 by permission of ASCE.

specimen. This results in the formation of strain localization and leads to eventual premature failure.

The conclusion from this analysis agrees with the observed experimental results in that the conventional triaxial extension test is inherently unstable. Necking is always observed in conventional extension tests. Necking occurred in various locations, but generally within the middle half of the specimens, indicating that the weaker zones in the specimen at which stress concentrations developed were not necessarily at the same exact place. Apparently, the only possible way to acquire uniform strains in a conventional triaxial extension test would be to create a perfectly isotropic, uniform specimen with no geometric or material defects, and then apply loads to the specimen in a perfectly uniform fashion. Even for this idealized condition, such a test would only be conditionally stable, because any small variation in stress or strain distribution developed within the specimen during the test would create an opportunity for strain localization to initiate. This unstable nature of the conventional extension test itself partially explains the wide scatter in reported experimental results.

Another factor in the scatter of results from conventional extension tests comes from the fact that it possesses much higher sensitivity to errors in measured quantities than does the comparable compression test as shown by, for example, Proctor and Barden (1969), Lade (1972), and Wu and Kolymbas (1991). These investigators have analytically demonstrated that errors in load, cross-sectional area, and cell pressure significantly affect the friction angle in the conventional extension test depending on the assumed amount of error in the measurement and the magnitude of the stress ratio. Assuming reasonable experimental errors in extension can easily result in scatter of the friction angle of 5–7 times that in compression.

In view of all these problems just discussed it appears that the conventional extension test does not provide a reliable means to determine soil properties in extension.

10.3.2 Enforcing uniform strains in extension tests

Because the conventional triaxial extension test rarely produces uniform stresses and strains due to the effects of strain localization, it becomes necessary to modify the test such that uniform strains can be achieved. The conventional extension test allows necking or shear planes to develop, because the flexible boundary conditions provided by the soft latex membranes surrounding the specimen do not provide any means to redistribute the concentrated stresses and strains at the neck location to the rest of the soil in the specimen. Therefore, the approach undertaken to enforce uniform strains was to provide external support to the soil in the radial direction that would be flexurally rigid, but provide free movement for the specimen during consolidation and shearing. This method would redistribute the stress and strain concentrations throughout the specimen, allowing the development of strain localization at any point in the specimen. However, the system would also have to move with the specimen during both the isotropic consolidation phase (decreasing specimen diameter and height) and the shearing phase (increasing height and decreasing diameter). In addition, no friction between the soil and the uniform-strain enforcement system could be allowed, because the vertical forces on the soil were measured externally by a load cell, and the magnitude of any frictional resistance could not be measured, leading to an incorrect value of deviator stress.

Various methods were developed and tested with varying degrees of success (Yamamuro and Lade 1995). During these early attempts, it was observed that by strengthening one area of the specimen against localization, it was possible to move the neck location to a different, less protected area. The location of strain localization in the extension test appeared to always initiate in the weakest portion of the specimen. This observation supports the contention that the conventional extension test is inherently unstable, because of its tendency to form stress concentrations.

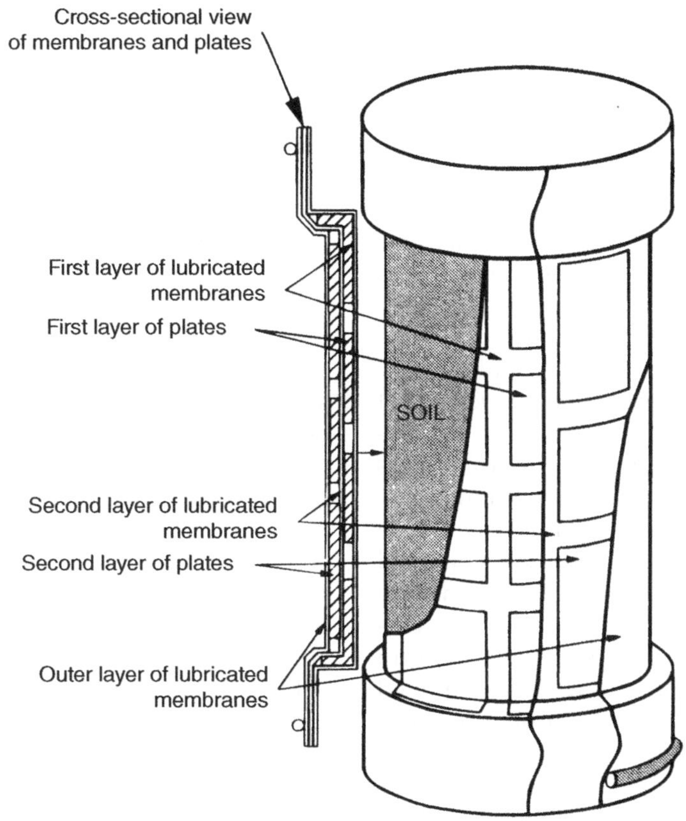

Figure 10.2 Method developed to enforce uniform strains utilized small stainless steel or tempered steel plates fabricated from shim stock (0.25, 0.64 and 0.81 mm thickness) and bent to curve and fit the contours of the cylindrical specimen. Reproduced from Yamamuro and Lade 1995 by permission of ASCE.

The method developed to enforce uniform strains used small stainless steel or tempered steel plates fabricated from shim stock (0.25, 0.64 and 0.81 mm thickness), and bent to curve such that they fit the contours of the cylindrical specimen, as shown in Fig. 10.2. Pieces located adjacent to the cap and base have one end bent so that a small section is perpendicular to the rest of the plate, such that the cell pressure can press onto this area and keep these plates moving with the cap and base. The first layer of plates was placed over the first layer of membranes. The plates were spaced apart to allow the specimen to decrease in diameter during isotropic consolidation and shearing. All membranes and plates were lubricated with vacuum grease to minimize friction. A second greased membrane and then a second layer of greased plates were placed over the first layer of plates. This layer was also placed in a spaced pattern, but it overlapped the plates in the first layer as shown in Fig. 10.2. Additional layers of lubricated membranes and plates were used to increase the resistance to necking at higher pressures. Lastly, external membranes were placed over the last layer of plates to seal the specimen against the cell fluid.

The conceptual basis behind this method was that as the confining pressure is applied to the specimen outside the plates and the outer membranes, the overlapped layers of plates will interlock under that pressure to form a relatively rigid boundary around the specimen to prevent formation of a neck. Deformation from isotropic consolidation and shearing is accommodated through the spacing afforded between the plates. Because the greased plates are separated by greased membranes and are not actually in physical contact with each other, there was little or no friction induced. The major concern was to ensure that the soil grains do not penetrate the first layer of membranes, and directly induce friction between soil grains and the first layer of plates. Therefore, adequate thickness of the first layer of membranes was essential. The specimen cap and base were enlarged to provide a horizontal seat on which the bent end of the adjacent plates could rest.

While this method appears to work well at lower confining pressures, the inherent instability in the extension test makes uniform strains more difficult to achieve at high confining pressures, because the strain magnitudes are very high. However, the use of the thicker tempered steel plates appeared to help achieve uniform strains at higher confining pressures. This method was successful in achieving uniform strain conditions in drained as well as undrained extension tests with consolidation pressures up to 52.0 MPa (Yamamuro and Lade 1995).

Another method of impeding the shear bands was presented by Lade *et al.* (1996), in which very short specimens were tested in extension, and they provided higher friction angles than those on taller specimens.

Further investigations were presented by Lade and Wang (2012a,b) in which results of tests on cross-anisotropic sand deposits were compared for various boundary conditions in triaxial compression and extension. These results also indicated the importance of using stiff boundaries to produce uniform strains.

10.4 Tests on unsaturated soils

The soil near the ground surface is very often not water saturated such as in pavements, shallow foundations, retaining walls, slopes, embankments, and so on. The stress–strain and strength behavior is therefore not as determined in drained or undrained tests on fully saturated specimens. Assuming the soil to be fully saturated for undrained conditions is often too conservative and may result in over-dimensioning of soil structures, whereas a total stress approach for design does not reveal the details of the behavior in terms of the effective stresses and the pore pressures.

The assumed states for design and for back-analysis of soil structures for property determination are different. For design, a sufficient factor of safety is desirable without excessive conservatism. In back-analysis, evaluation of the performance of a structure is desirable with a view toward improving the design of future structures of similar type. For example, it may be useful to know how much capillary stresses contribute to the stability of a slope, even if in a design context one may not want to depend on maintaining the dry state contributing to these stresses. Similarly, one may be hesitant to assume any tensile strength for design purposes, but it is difficult to deny the presence of some tensile strength in a vertically standing slope.

Bishop and Blight (1963) proposed expressing the pore pressure in unsaturated soils for various degrees of saturation as follows:

$$\sigma' = (\sigma - u_a) + \chi \cdot (u_a - u_w) \tag{10.1}$$

in which $(\sigma - u_a)$ is referred to as the net normal stress, $(u_a - u_w)$ is the matric suction, and the parameter χ is related to the degree of saturation S and it varies between zero and unity. For a fully saturated soil the value of $\chi = 1$ and the pore pressure is entirely represented by the pore water pressure, u_w. Many articles have been produced in which the determination of χ has been discussed.

To determine the salient behavior of unsaturated soils, experiments are done in which the pore pressure is separately determined as the pore water pressure (u_w) and the soil matric suction ($\psi = u_a - u_w$) in which u_a is the air pressure. Therefore, drying an initially saturated soil produces an increase in suction, and wetting an initially dry soil leads to a decrease in suction.

10.4.1 Soil water retention curve

The pressures in the water and the air depend on the volumetric water content (indicated by θ in unsaturated soil mechanics) or the degree of

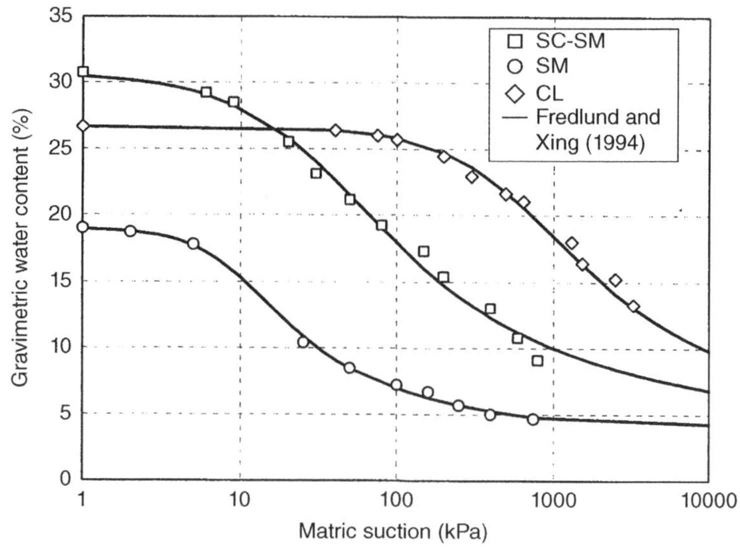

Figure 10.3 Soil water retention curve describing the relation between the degree of saturation and the soil suction. In the key SC-SM is clayey, silty sand, SM is silty sand, and CL is clay with low plasticity (after Fredlund and Xing 1994).

saturation (S) of the soil. Thus, the soil water retention curve describes the relation between the degree of saturation and the soil suction, as exemplified by the three relations in Fig. 10.3. Several mathematical fitting expressions have been proposed for description of this relation (e.g., van Genuchten 1980; Fredlund and Xing 1994). The van Genuchten equation is highly flexible and most often used:

$$\chi = S_e = \left[\frac{1}{1 + \left[\alpha \left(\psi \right) \right]^n} \right]^{\left(1 - \frac{1}{n}\right)} \tag{10.2}$$

in which α and n are fitting parameters, and S_e is the effective degree of saturation, which is given by:

$$S_e = \frac{S - S_r}{1 - S_r} \tag{10.3}$$

in which S_r is the residual saturation.

10.4.2 Hydraulic conductivity function

The other information for water flow analysis in unsaturated soil (in addition to the soil water characteristic curve) is the hydraulic conductivity function. This is a relation between the suction and the hydraulic conductivity, an example of which is shown in Fig. 10.4.

To measure both the pore water pressure and the pore air pressure or to control one of these pressures, a high or low air entry ceramic disk is required at the bottom of the triaxial specimen and a regular filter stone is used at the top of the specimen, as shown in Fig. 10.5. Grooves may be machined into the base plate for drainage and as a channel for continuous circulation adjacent to the ceramic disk, as shown in Fig. 3.9. This allows flushing before testing begins to remove air bubbles that may accumulate by diffusion through the ceramic disk. Flushing may be done by venting the effluent port and injecting de-aired water into the influent port using a manual syringe.

10.4.3 Low matric suction

Matric suction from zero (saturated soil) to about 50 kPa requires a low air entry disk and may be controlled using a hanging column setup, as explained in ASTM Standard D6836 (2014), and shown in Fig. 10.6. This method is used for coarse granular materials that drain

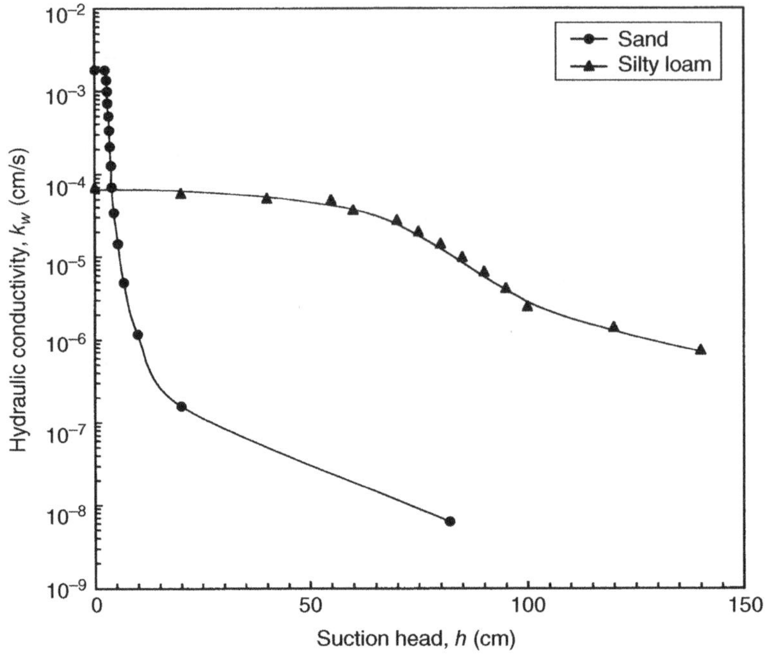

Figure 10.4 Examples of relations between suction and hydraulic conductivity for different soils. Reproduced from Lu and Likos 2004 by permission of John Wiley & Sons.

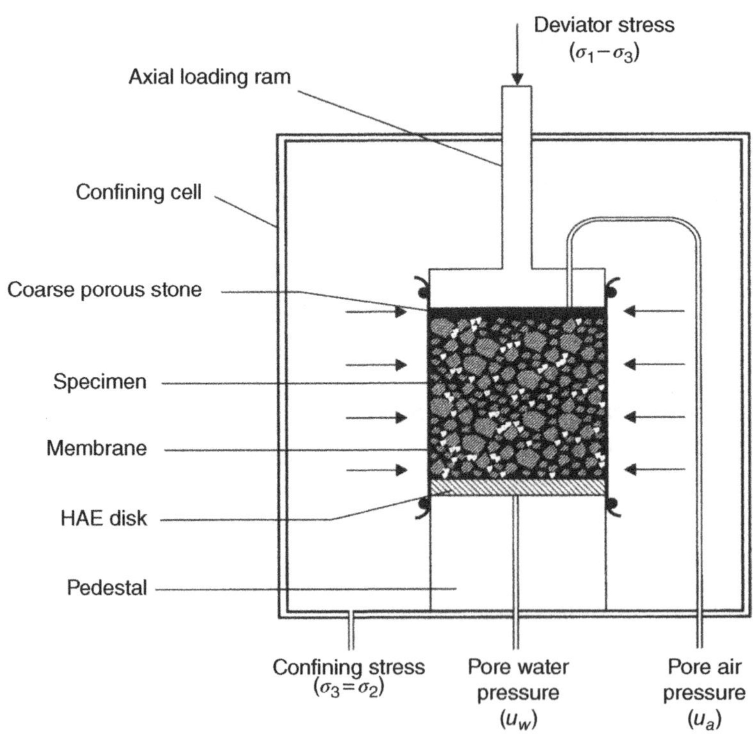

Figure 10.5 Setup with a high air entry (HAE) porous stone at the specimen base with an HAE value of 100 kPa. Air flow through the stone is prevented up to this air entry value. Water pressure is measured at the base, and the air pressure is measured at the regular filter stone at the top of the specimen. Reproduced from Lu and Likos 2004 by permission of John Wiley & Sons.

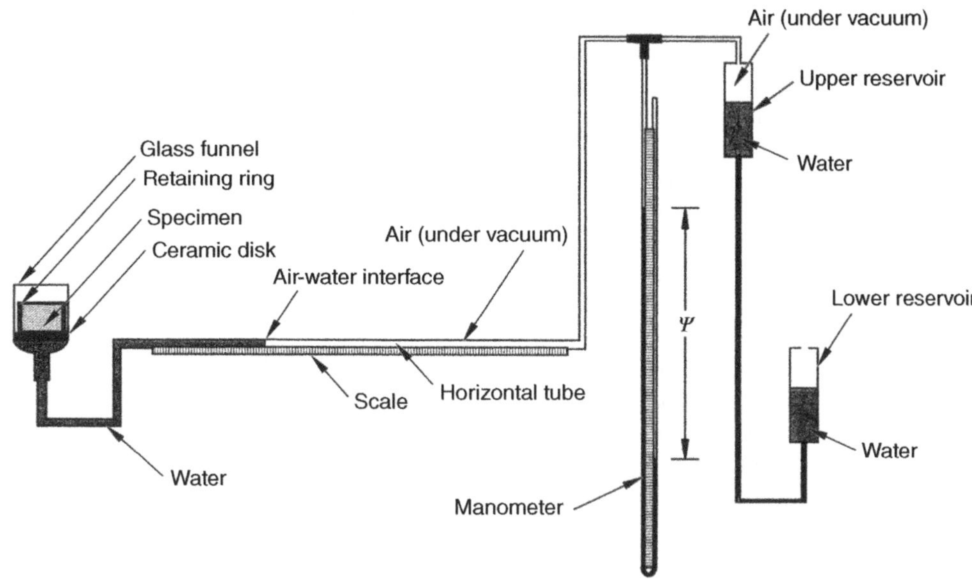

Figure 10.6 Matric suction from zero (saturated soil) to about 50 kPa requires a low air entry disk and may be controlled using a hanging column setup. Reprinted from ASTM Standard D6836 2014 by permission of ASTM International.

readily. Control of matric suction and net normal stress at very low magnitudes and over a very small range is required.

The hanging column setup applies negative pore water pressure to the specimen. The equipment consists of an upper and a lower reservoir with water and connected with flexible tubes, as seen in Fig. 10.6. The top of the lower reservoir is vented to the atmosphere, and suction is induced in the upper reservoir by the hanging water column such that the difference in height of the surfaces of the two water reservoirs times the unit weight of water represents the negative pressure. The air–water interface in the horizontal tube transfers the vacuum in the air phase to negative water pressure, which is directly connected to the water in the lower saturated ceramic disk in the triaxial setup. In this technique the pore air pressure remains at atmospheric pressure ($u_a = 0$) by connection with the upper filter stone, and the matric suction is determined from the magnitude of the negative water pressure.

The change in volumetric water content is determined from the movement of the air–water interface on the horizontal tube in Fig. 10.6. This tube has a rather small cross-sectional area, A_c, to produce high accuracy. The change in volumetric water content, $\Delta\theta$, corresponding to a given change in suction is then calculated from

$$\Delta\theta = \Delta L \cdot A_c / V \tag{10.4}$$

in which ΔL is the change in length of the horizontal string of water and V is the volume of the specimen.

10.4.4 High matric suction

For matric suction ranging from 1 to 1500 kPa, an axis translation technique (e.g., Ho and Fredlund 1982; Escario and Saez 1986; Fredlund and Rahardjo 1993) is most applicable to sandy and silty soils. The axis translation technique avoids dealing with negative pore water pressure, which can be especially difficult if it becomes less than zero absolute. The procedure involves a translation of the reference pressure, which is the pore air pressure. The pore water pressure is then referenced to a positive pore air pressure. If unsaturated specimens are exposed to an externally controlled air pressure at the cap, the pore air pressure becomes equal to that applied air pressure, and the pore water pressure increases by the same amount. The matric suction therefore remains constant regardless of the magnitude of the translation of both pressures. Using this axis translation technique the pore water pressure can be increased to a positive value that can be measured without problems relating to cavitation of the water. For soils that contain significant amounts of occluded pore air, the measurements can be incorrect. In this case the actual soil suction will be overestimated. Nevertheless, the axis translation technique is just as important for unsaturated soils as the back pressure technique is for saturated soils.

The upper limit of 1500 kPa, mentioned above, is controlled by the high air entry pressure ceramics, and is also influenced by the time required to reach equilibrium, and the uncertainty regarding the continuity of the water phase near the lower end of saturation. The lower end is controlled by the accuracy with which the air pressure can be regulated.

The two pressures, in water and in air, are measured separately, and this is done at the two ends of the triaxial specimen. Since water tends to flow to the base of the specimen due to gravity, the water pressure is measured at the base, and the air pressure is measured at the cap. Figure 10.5 shows a setup in which the lower porous stone can be a high air entry stone. If the high air entry value of the porous stone is 100 kPa, then air flow through the stone is prevented up to this air entry value.

The ceramic stone is seated directly on the base of the triaxial setup to minimize leakage and this also allows soil specimens to be compacted or vibrated directly on the disk. Drainage grooves can be machined into the base to form a water reservoir and a continuous channel below the ceramic disk. Ports for flushing and access to this reservoir are used to saturate the channel during initialization of the test. The channel may be flushed between steps in multi-stage testing to remove air bubbles that may accumulate by diffusion through the ceramic disk. This may be achieved by injecting de-aired water through the influent port using a manual syringe and venting the effluent port. The valve on the influent port is closed during shearing while the effluent port is connected to the hanging column setup.

10.4.5 Modeling

The hydraulic conductivity function may be modeled by the van Genuchten–Mualem model (Mualem 1976):

$$k = k_s \cdot \frac{\left\{1-(\alpha\psi)^{n-1}\left[1+(\alpha\psi)^{n-1}\right]^{(1-1/n)}\right\}}{\left[1+(\alpha\psi)^{n-1}\right]^{p(1-1/n)}} \tag{10.5}$$

in which k_s is the hydraulic conductivity of the saturated soil and p is a pore interaction term. The value of p is typically 0.5 for clean coarse grained soils and it ranges between −1 and −3 for fine grained soils. Figure 10.7 shows the hysteresis obtained when wetting and drying the soil for which the expression in Eq. (10.5) may be used to fit the results of experiments performed to determine the variation of hydraulic conductivity with suction.

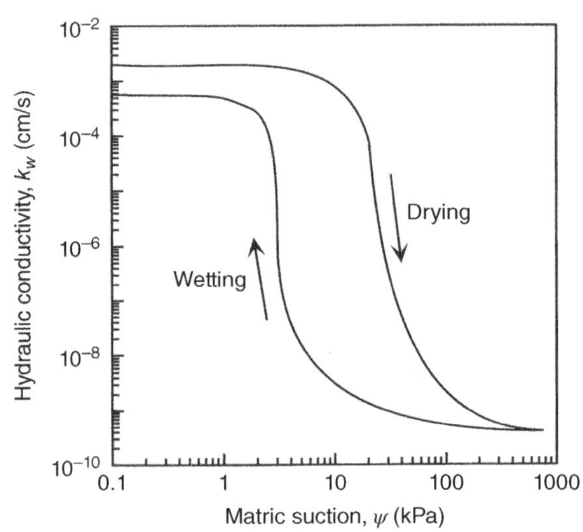

Figure 10.7 The hydraulic conductivity function may be modeled by the van Genuchten–Mualem model for the drying and wetting portions of the relation. Reproduced from Lu and Likos 2004 by permission of John Wiley & Sons.

10.4.6 Triaxial testing

The stress–strain behavior and the soil water characteristic curve that describes the relation between matric suction and degree of saturation may be obtained concurrently by shearing the specimen at select increments along a drainage or sorption path. The mechanical response of unsaturated soils is usually determined from a series of constant suction tests. The difference in air and water pressures is maintained constant and the tests are, in fact, drained tests. They represent the response of the soil subjected to additional effective confining pressure. Figure 10.8 shows the results of tests performed with increasing amounts of suction. For each constant amount of suction the shear strength increases for a given net normal stress (total stress minus air pressure). If, however, the shear strength results are interpreted in terms of effective stress [with χ from e.g., Eq. (10.2) inserted into Eq. (10.1)], then one curve is obtained for the shear strength. This is very similar to the effective stress interpretation of triaxial tests performed on saturated soils with various amounts of back pressure, that is in terms of effective stresses the same strength envelope is obtained.

Thus, the shear strength of unsaturated soils may be determined using an apparatus with a high air entry stone at the base for suction control and a conventional filter stone at the cap. The axis translation technique is then employed under drained conditions, and this requires relatively slow loading rates to maintain constant suction during the test. Thus, the loading rate must be lower than for tests on saturated soils, because the hydraulic conductivity for the unsaturated soil can be much lower.

The shear strength may also be determined from undrained tests, but this requires a differential pressure transducer to measure the excess pore pressures produced due to shearing. An increase in suction produces an increase in shear strength. The effect of suction and net normal stress (difference between total stress and air pressure) will affect the failure envelope, as shown in Fig. 10.8. Both diagrams may be employed to indicate the shear strength of unsaturated soils.

10.5 Frozen soils

Because saturated, clean sands cannot be sampled without disturbance or disintegration of the soil structure by conventional sampling techniques, the sand may be frozen in situ, cored and brought to the laboratory for testing. The special procedure of freezing prior to sampling and handling may provide a reliable means for preserving the in situ characteristics of the sand fabric. Adachi (1988) reviewed the

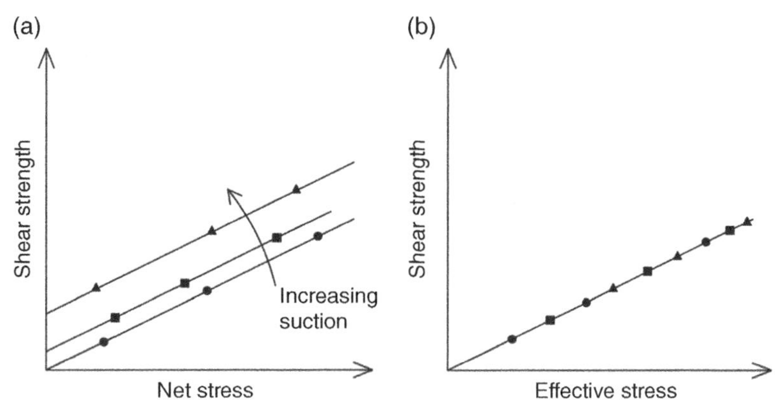

Figure 10.8 (a) Drained failure envelopes for different suction failures and (b) drained failure envelope with single-value effective stress.

state-of-the-art on using freezing techniques for intact sampling of granular soils. Yoshimi *et al.* (1978) reported a procedure in which a mixture of ethanol and crushed dry ice was circulated through a thin-walled steel tube and inserted vertically into the ground. The steel tube together with the surrounding frozen column of sand was then pulled out of the ground and a column of sand sample was successfully obtained. They concluded that high quality undisturbed sand samples can be obtained using this method. Studies have shown that the most important process in the freezing of soils is the redistribution of water, which accompanies the freezing process (Tystovich 1975). Freezing may create conditions that cause the soil volume to increase due to migration of water towards the freezing front with associated formation of ice lenses. Yoshimi *et al.* (1978) showed that the best way to freeze sands while maintaining their in situ conditions is by unidirectional freezing, and not impeding the drainage, while maintaining the confining pressure.

Yoshimi *et al.* (1978) studied the variables that could affect the soil properties, including time of freezing (up to 35 min), surcharge while freezing (0.3–33.6 kPa), the coolant temperature (−20 to −70 °C), the relative density (40–90%), the soil type, and the fines content (0–6%). None of these physical processes affected the sand behavior once it had been thawed and tested. Triaxial specimens were also prepared and one set of specimens was frozen and thawed and another set was tested without freezing. Both methods resulted in the same stress–strain, strength and volume change behavior, thus showing that the freezing of the specimens had no effect on the resulting behavior.

Singh *et al.* (1982) and Seed *et al.* (1982) compared results from push tube sampling and block sampling. They found that push tube samples at 60% relative density were changing their relative densities and this process would therefore involve relative movement of the sand grains, that is disturbance of the soil fabric. In comparison, block sampling by advance trimming and sampling caused minimal density change in the Monterey #0 sand deposit placed at a relative density of 60%.

Table 10.1 shows a comparison of characteristics of clean sand behavior following block sampling, tube sampling and controlled freezing, followed by testing. It is clear from this comparison that freezing of saturated, clean sand followed by coring, thawing, and testing produces the least disturbance. It causes no change in density, no change in effects of long-term loading, and no change in effects of seismic loading.

The effect of fixing the sand particles and maintaining their position may also be achieved by injecting water soluble high density polymer solution into the voids in the ground. Then after coring the specimen, it is brought into the triaxial cell, surrounded by a confining pressure

Table 10.1 Comparison of characteristics of clean sand behavior following block sampling, tube sampling and controlled freezing, followed by testing. Reproduced from Seed *et al.* 1982 by permission of ASCE

Factor	Block sampling	Tube sampling	Controlled freezing
Density	Slight loosening (strength reduced about 5%)	Some densification (strength increased = 15%)	No change
Long-term loading	Some loss of strength (estimated = 5%)	Some loss of strength (estimated = 25%)	Probably no change
Structure and fabric	Little change	Little change	No change
Seismic history	Slight loss of strength (about 5%)	Slight loss of strength (about 5%)	No change
Net effect	Some loss of strength (about 15%)	Some loss of strength (about 15%)	Probably no change

and the high density polymer solution is dissolved to produce an intact cylindrical specimen in the triaxial cell.

10.6 Time effects tests

Time effects may be divided into two categories: viscous effects; and aging effects. While viscous effects may result in aging, aging may also be caused by positive [digenesis process (cementation)] and negative (weathering) time effects as well as by physical and chemical changes.

10.6.1 Creep tests

In creep tests in which the initial portions of the tests are performed under deformation control, great care has to be exercised in switching from deformation control to load control, under which the creep is determined under constant vertical stress. Depending on the setup, the vertical load in the load control cylinder has to match the vertical load reached under deformation control before switching occurs.

10.6.2 Stress relaxation tests

There is a fundamental experimental difficulty in performing stress relaxation tests. While the axial displacement of the loading machine is completely stopped at a given deviator stress (and the triaxial setup is stiff to avoid significant compression of interfaces, etc.), it is necessary to measure the drop in axial load on the specimen. This requires a load cell in series with the triaxial specimen inside the load frame that holds the deformation across the setup at zero. To register a decrease in load due to stress relaxation, the load cell must expand, and this expansion is countered by the specimen, that is the specimen is compressed in the axial direction. The load applied to the specimen and measured on the load cell is consequently too high to correspond to true stress relaxation at zero axial strain.

To investigate the sensitivity of the measured stress relaxation to the small amounts of axial deformation imposed by expansion of the load cell during decreasing load, three experiments were performed on dense Virginia Beach sand at a confining pressure of 8000 kPa in which the specimens were initially loaded at the same three strain rates used in the experiments presented above, namely 0.00260, 0.0416 and 0.666%/min, that is a 256-fold change in axial loading strain rate (Lade and Karimpour 2015). The axial deformation of the specimen in these sensitivity check tests was continuously monitored by a digital dial gage. The dial gage reading was very accurately maintained constant by small adjustments in the displacement of the deformation control loading machine. The results of these three experiments are compared in Fig. 10.9 with the results of the stress relaxation experiments in which no special adjustment for load cell expansion was made. From the lowest to the highest shearing rate, the corresponding stress–strain points at the initiation of stress relaxation were $[(\sigma_1-\sigma_3), \varepsilon_1]$ = (10600 kPa, 2.86%), (11240 kPa, 2.81%), and (11880 kPa, 2.75%), respectively. The total energy input up to these points were equal to 390 kN·m/m^3 in each test. In the magnified diagram in Fig. 10.10, the slightly inclined lines indicate stress relaxation without correction, while the vertical lines include experimental correction so the results correspond to true stress relaxation at zero strain. Comparing individual pairs of tests from those initially loaded at the lowest to the highest strain rates, the difference in amount of stress relaxation in 1 day is (1800 – 1530) = 270 kPa, (2470 – 2340) = 130 kPa, and (3370 – 2970) = 400 kPa. These errors show that the true stress relaxation values are underestimated by 5–15% when no adjustments are made. The stiffness of the load cell employed in these experiments relates to the slopes of the slightly inclined relaxation curves and it is approximately $36\cdot10^3$ kN/m. It is possible that an even stiffer load cell would have reduced the errors in the amounts of stress relaxation measured, thus avoiding the continuous adjustment in displacement of the deformation control loading machine.

In stress relaxation tests, the axial load is measured while the axial strain is maintained

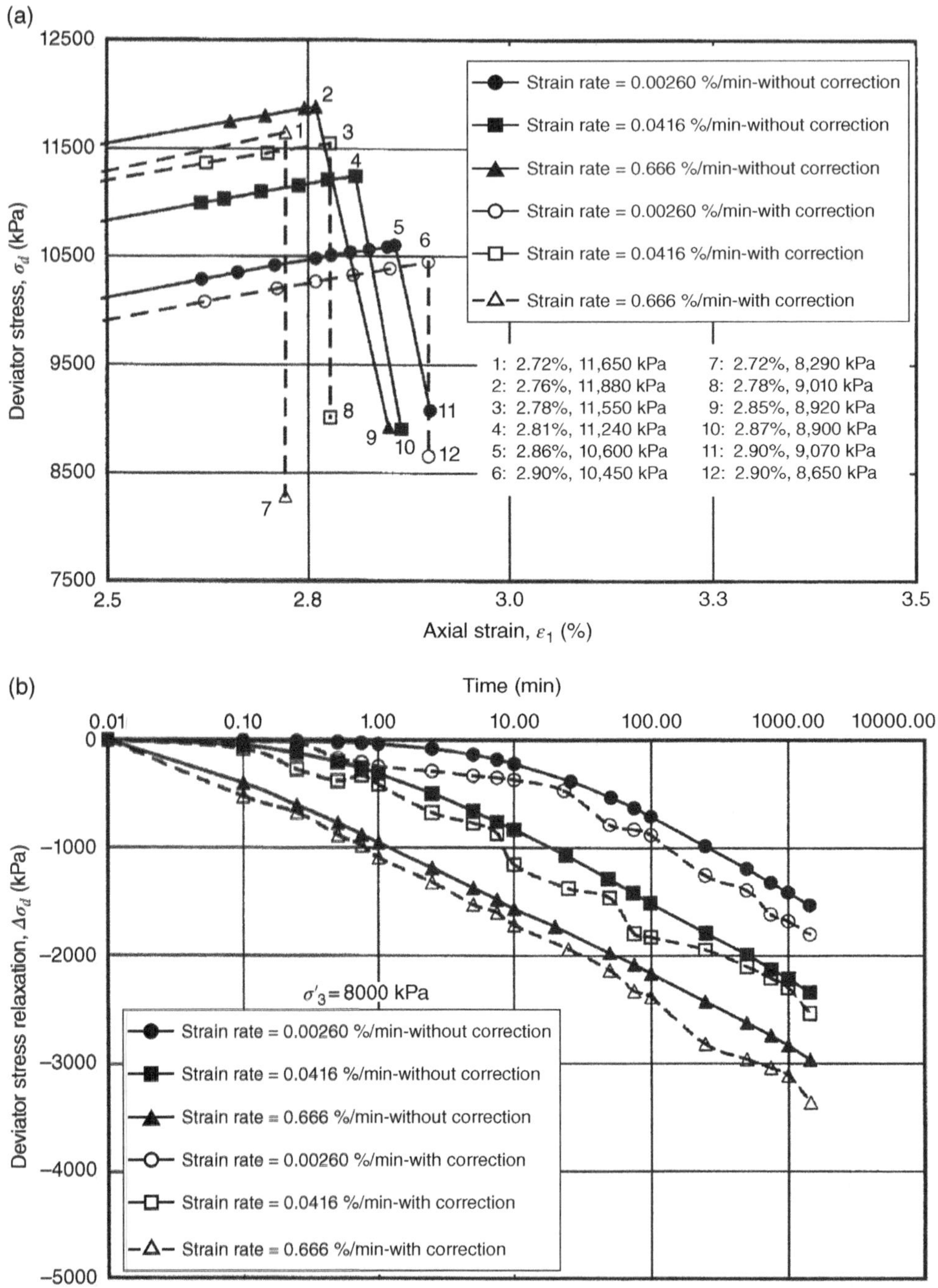

Figure 10.9 (a) Results of the three stress relaxation experiments with and without special adjustment for load cell expansion and (b) deviator stress relations with time for conditions with and without correction. Reproduced from Lade and Karimpour 2015 by permission of Canadian Science Publishing.

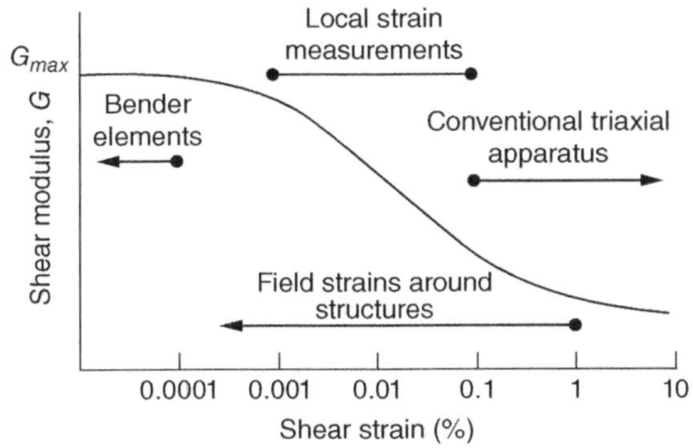

Figure 10.10 Diagram indicating the accuracy of different methods of measurements: bender elements are required for accurate determination of shear moduli.

at zero. In addition, the drainage condition may be maintained at zero drainage, that is undrained conditions may be imposed. In the experiments on dense Virginia Beach sand, the measured volume change followed the drained volume change curve, and so no special effect was noticeable when the experiments were performed as undrained tests (Lade and Karimpour 2015).

10.7 Determination of hydraulic conductivity

It is possible to perform tests to determine the hydraulic conductivity (or permeability) in the triaxial setup in which two filter stones form the ends of the specimen setup. These filter stones and the tubing (unless ¼ in. tubings and ¼ in. fittings are used) may provide some further resistance to the percolation of the soil, and corrections are therefore required to the measured values of hydraulic conductivity. For such tests, a falling head test is preferred, because the time response gives an indication of the nonlinear head loss in the fittings.

Considering the triaxial specimen as a sandwich with a filter stone at each end, that is a three-layer system with flow perpendicular to the stratified layers, the following expression may be derived for the hydraulic conductivity of the soil:

$$k_{soil} = \frac{H \cdot k_m}{H + 2t\left(1 - \frac{k_m}{k_f}\right)} \tag{10.6}$$

in which H is the height of the soil specimen, $2t$ is the the combined thickness of the two filter stones, k_m is the measured permeability, and k_f is the permeability of the filter stones. This may be determined separately by simply measuring across the two filter stones in the setup with no specimen between them.

10.8 Bender element tests

Bender element tests to determine shear wave (V_s) and compression wave (V_p) velocities in the soil may be part of the triaxial test setup. These velocities are determined from very small-strain responses measured by bender elements. Figure 10.10 shows comparisons of measurement methods and for which purposes they are required. Bender elements are embedded in the cap and base, as shown in Fig. 10.11, or through the membrane, as indicated in Fig. 10.12, so that both vertical and horizontal wave velocities may be determined. These velocities may be used to determine the degree to which the specimens are intact by comparison with similar velocities measured in the field, and they may be used as an indicator of the degree of saturation by comparison with V_p = 1450 m/s for water in a fully saturated specimen.

Compression waves are longitudinal, that is the soil particles move in the same direction as the wave propagation. Shear waves are transverse, that is the particles move perpendicular to the propagation. The velocity of a P-wave (pressure wave) is controlled by the bulk and shear moduli, and the P-waves are transmitted through the pore water and specimen saturation has an important effect on the measured P-wave velocity. The velocity of the shear wave is controlled by the shear modulus of the soil

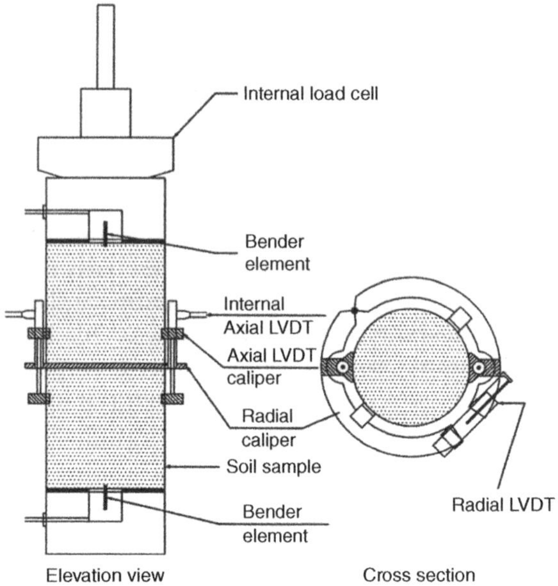

Figure 10.11 Location of bender elements in end plates of a triaxial specimen. LVDT, linear variable differential transformer. Reproduced from Finno and Kim 2012 by permission of ASCE.

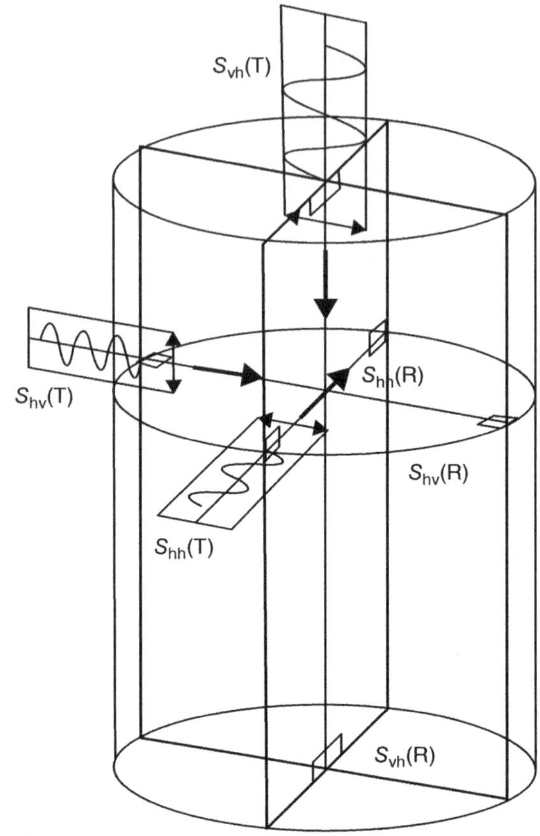

Figure 10.12 Location of bender elements at the ends and at opposite sides of the triaxial specimen for measurements of shear waves in the vertical and the horizontal directions. Reproduced from Jardine 2014 by permission of Geotechnique.

and is not affected by the specimen saturation, because water cannot transfer shear stresses. The compression wave velocity is greater than the shear wave velocity.

10.8.1 Fabrication of bender elements

A bender element is a piezoelectric transducer that converts electrical energy to mechanical energy or mechanical energy to electrical energy. These elements are composed of two conductive outer electrodes, that is two ceramic plates, and a conductive metal shim at the center, as shown in Fig. 10.13. Bender elements may vary in size, but they are typically 10 by 16 by 0.5 mm. They are fabricated from 0.5 mm thick piezoelectric bimorph strips that are cut into 16 mm segments using a diamond edge cutter. The ceramic plates are polarized so that either the dipole deforms with one plate elongating and the other shortening, or conversely when a mechanical deformation of the ceramics is applied the transducer generates a voltage. This blade is activated by alternating currents (AC), which make the blade move back and forth at the frequency of the AC. Electrical connections to the element are made with 1.8 mm diameter coaxial cable. These two components are then encapsulated in resin to provide waterproofing and protection to the ceramic and circuit. This component is further protected by an aluminum split potting mold that is approximately 40 by 30 by 20 mm. This mold is put in the oven at 105°C for 24 h and upon removal from the mold placed in a brass cup fitted to the dimensions. The protrusion length is approximately half the length of the element to ensure sufficient

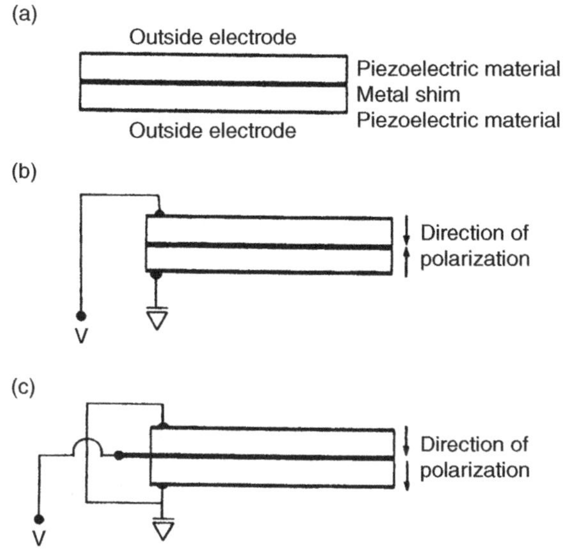

Figure 10.13 Bender elements: (a) schematic representation of a bender element; (b) series type; and (c) parallel type. Reproduced from Lee and Santamarina 2005 by permission of ASCE.

coupling with the soil specimen. A polyurethane coating and subsequent conductive paint is applied to the bender element itself. This conductive paint creates an electric shield that is grounded to avoid electromagnetic coupling and cross-talk between source and receiver (Cha and Cho 2007). This component is then placed in a housing unit and fixed with epoxy. These elements are then placed in the cap and the base prior to specimen preparation. One provides the signal and the other measures the signal. These elements, when coupled with a soil specimen, operate in the same frequency range and are relatively tuned to one another. The time it takes for this non-destructive signal to go from one element to the other is divided by the distance between the two elements to determine the shear wave velocity, V_s.

10.8.2 Shear modulus

The initial maximum shear modulus of a soil specimen, G_{max}, is calculated from:

$$G_{max} = \rho \cdot V_s^2 \quad (10.7)$$

in which ρ is the soil density. G_{max} is a key parameter in small strain dynamic analysis that provides significant information on the soil properties and factors that control the soil behavior. Its location is shown on the ordinate in Fig. 10.10.

Note that the wave speeds in discrete media (e.g., lattices, granular media) depend on wavelength, where smaller wavelengths may not propagate at all. This causes dispersion of waves and loss of definition under the best of conditions.

While Shirley and Hampton (1978) introduced the bender element method into soil testing practice using piezoelectric transducers, Dyvik and Madshus (1985) went into detail on the installation and use of the piezoelectric bender elements in testing equipment such as triaxial, simple shear and oedometer test apparatus. The bender element test has also been applied in cyclic triaxial apparatus, stress-path cells, resonant column experiments, centrifuges, calibration chambers, and true triaxial equipment.

The signal generator is connected to both the source bender element and a digital oscilloscope to both generate the voltage and give the oscilloscope the time zero of the source wave. The receiver element is connected with a signal conditioner that reads the arrival of waves and also a signal amplifier to improve the performance. This is also connected with the oscilloscope that will output the data of all waves through the soil specimen that generate a voltage at the receiver element.

Typically the soil specimen is 7 cm in diameter and 14 cm in height (e.g., Camacho-Tauto *et al.* 2012). The transmitter is located at the base and the receiver is located at the cap. A current amplifier stabilizes the input signal generated by the function generator and sends it to the transmitter. The received output signal is amplified and both the input and output signals are recorded by a digital oscilloscope.

Polarity is verified by using a single sine pulse through bender elements in direct contact with each other. Positive polarity is either confirmed or the wiring is adjusted to acquire positive polarity. To measure equipment time delay, aluminum rods of varying heights are placed between the two bender elements. For each rod length the travel distance is plotted against time of arrival of the shear wave. Through interpolation, a line is formed on this graph and the intersection of this line on the x-axis is determined to be the time delay to be incorporated into the computation of the shear modulus. The time delay determined using aluminum rods may be checked with specimens of completely dry sand at different effective confining pressures, and experiments show that such comparisons confirm the results from the aluminum rods (Camacho-Tauto *et al.* 2012).

To avoid cross-talk when two parallel type bender elements are used, they should be shielded and connected to the ground, and the first signal arrival will be the arrival of the S-wave (shear wave). It is therefore recommended to both shield and ground the bender elements, or to use two parallel bender elements as both the output element and receiver element.

10.8.3 Signal interpretation

To interpret the signals the time method of determining the arrival of the shear wave include visual picking, first major peak-to-peak, cross correlation and cross spectrum. Reliability in results relies heavily on employing the same method for a given study, coupled with user judgment on interpretation of results to determine the velocity of the shear wave and the resulting small-strain shear modulus.

10.8.4 First arrival time

Figure 10.14 shows the determination of the first arrival time using a single sine wave as the input function, while Fig. 10.15 indicates the variation in arrival times with the normal stress in the specimen during loading and unloading in a consolidation test on kaolinite clay.

Visual picking

The shear wave arrival time occurs at the peak of the first major deflection of the received signal. Depending on polarity, the first major

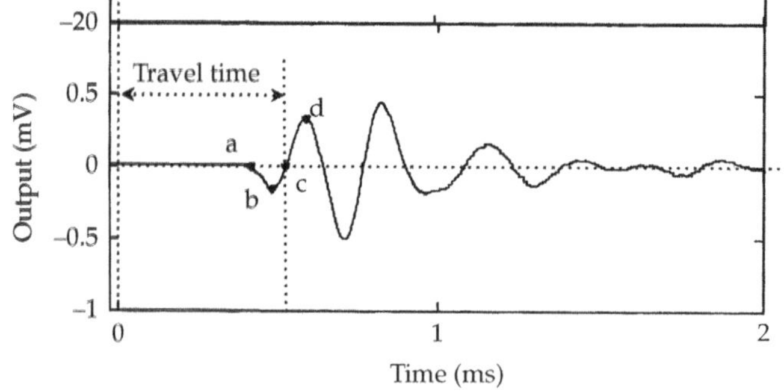

Figure 10.14 Determination of the first arrival time using a single cycle sine wave in the input function. Point c is taken as the time of first arrival. Reproduced from Kang *et al.* 2014 by permission of ASTM International.

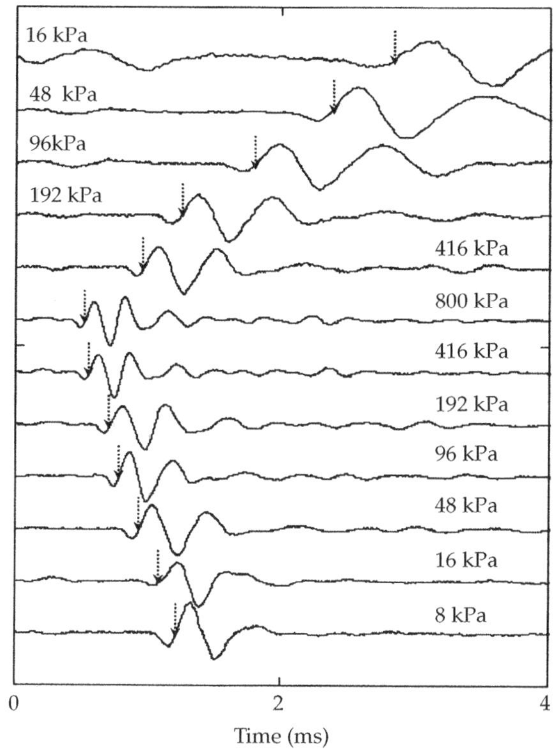

Figure 10.15 Typical received shear wave signals under different applied normal stresses in a consolidation test on kaolinite clay. Reproduced from Kang *et al.* 2014 by permission of ASTM International.

amplitude could be either positive or negative, either of which would be considered the arrival of the shear wave. Changing the polarity would reverse the entire waveform which would cause a negative signal to become positive and vice versa. A major disadvantage to this method is interpretation when a distinct and sharp deflection point cannot be located. The lack of a distinguishable peak is often due to the ambiguous nature of near field effects or other interference such as background noise.

First major peak-to-peak

In the first major peak-to-peak method of determining the shear wave velocity, the arrival time is based on the assumption that the transmitted wave and the received wave closely resemble one another. The shear wave velocity is obtained by dividing the distance traveled by the time between the peak of the transmitted signal and the first major peak of the received signal.

Distortion in the signal is typical when varying the specimen length due to the energy absorbing nature of soil, which is known as damping. Many localized peaks with slight differences in amplitude make defining the first major peak difficult. The advantage of this time method over the visual picking method is that distortion of the signal and the near field effects are minimized, but accuracy in determining the shear wave arrival time depends on the quality of the signal. A phenomenon known as dispersion, which is a significant difference in frequency between the transmitted signal and the received signal, would lead to lower confidence in the accuracy of the reading, but it is noted that this could be due to the damping properties of the soil or the coupling between the transducers and the soil.

Cross correlation

The cross correlation method measures the level of correspondence or interrelationship between the transmitted signal and the received signal, and it is examined in the frequency domain with the shear wave velocity being calculated by a phase shift. In converting the signal from a time domain to a frequency domain, the decomposition produces a group of harmonic waves with known amplitude and frequency. The fast Fourier transform puts the received signal in a linear spectrum that gives the magnitude and phase shift of the harmonic component in the signal. In a similar manner the transmitted signal is transformed and the cross power spectrum of the two signals is examined and a cross correlation coefficient is established by plotting the two transformed signals against the cross correlation function. The maximum cross correlation coefficient between the two values would be the travel time of the shear wave. Although this is a more consistent method, problems arise when the transmitted signal does not have a frequency similar to the received signal. Some other setbacks to this method include "the complex nature of the received signal, incompatible phase frequency manipulation, non-plane wave propagation characteristics and near field effects" (Chan 2010).

Cross spectrum

The cross spectrum method is in essence an extension of the cross correlation method. In this method the manipulation done in the cross correlation method is implemented with further manipulation of the signals to produce the absolute cross power spectrum. An algorithm is used on the cross power spectrum phase angle with results given in a plot of the absolute cross power spectrum phase diagram. A linear plot is fitted to the data points compiled over a range of frequencies and the slope of the best fitting line gives the group travel time. Nonlinearity in the plot would be considered dispersion in this method. This method displays a normal distribution of the results of all tests and therefore could arguably be considered reliable.

10.8.5 Specimen size and geometry

As mentioned above, bender element tests are typically performed on specimens with diameter of 7 cm and height of 14 cm. The wavelength ratio, which is defined as the tip-to-tip distance of the source and receiver elements divided by the wavelength of the shear wave, is a parameter which constrains the development and propagation of the shear wave. If the soil specimen is too short, and the distance between the two bender elements is relatively small, the resulting output would be a precursory signal which is a reading prior to the arrival of the S-wave. This is known as the near field effect. To eliminate this phenomenon, it is proposed that the wavelength ratio be no less than two. One parameter tied to the wavelength ratio is the input frequency. It was determined through experimental results that the highest amplitude from the receiver element corresponded to an input frequency equal to the resonant frequency of the bender element and any greater input frequency would result in a response matching the resonant frequency of the bender element, but with decreased amplitude. For this reason it was recommended that frequencies greater than the resonant frequency of the bender element should not be used.

10.8.6 Ray path analysis

With varying geometries of both the specimen and the bender element, the received signal will be entirely different, which is why there is a need to look into ray path analysis to properly interpret the output signal. When waves such as the ones generated by bender elements travel through a confined space, such as a triaxial specimen, the signal indicating the arrival of the S-wave can be affected due to what is known as directivity. In wave mechanics, the two main types of elastic body waves are the P-waves and the S-waves. The P-waves oscillate in the direction of propagation while the S-waves oscillate perpendicular to the direction of propagation. The P-waves travel much quicker than the S-waves, reflect off the specimen walls, and generate a response from the receiver element prior to the S-wave response. One solution is to manipulate the P-wave to arrive significantly sooner than the S-wave. Another directivity topic discussed is in-plane directivity. The receiver element could either be arranged perpendicular to the signal or the receiver element could be parallel to the receiver. The amplitude at 0° in the transverse or perpendicular orientation is proposed to be approximately 75% of that measured in the parallel configuration. The received signal will typically consist of multiple peaks, each one corresponding to a different wave that has traveled on its specific path through the specimen (Marjanovic and Germaine 2013). The optimal length of the bender element depends on the voltage used, the maximum stress level, and the material being tested.

10.8.7 Surface mounted elements

The bender elements may be mounted on the surface of the soil specimen with axes parallel to one another. With this parallel axes configuration, it has been determined that the amplitude of the output signal would be approximately 50% of that of the amplitude of the conventional tip-to-tip configuration. With the bender elements mounted on the soil surface the shear waves propagate horizontally with the soil particles also vibrating in the horizontal direction. The major issue is the determination of the travel distance of the shear wave. In conventional setups the distance is simply the tip-to-tip distance but with surface mounted bender elements the distance could either be the center-to-center distance or the edge-to-edge distance. The results of one study (Zhou *et al.* 2008) show that while varying the frequency and keeping the travel distance constant, the response is quite similar to that of the conventional method. Although the amplitude of the response in surface mounted bender elements is weaker than that of the conventional test, the versatility of the former method to further study attributes of the soil sample such as damping characteristics makes this method of interest for future studies.

10.8.8 Effects of specimen material

Early studies were conducted on fine grained sands and clays. To generalize the use of bender elements, recent studies have looked into the feasibility of measuring the shear modulus of coarse materials by comparing results of the test with those of resonant columns (e.g., Anderson and Stokoe 1978) as well as comparing the results with empirical relationships that have been previously established. The limitations of bender elements with regards to size and shape of the grains, ranges of densities, the state of stress, and the influences of these factors are also all issues of concern. Three different types of sand, a medium angular, a coarse angular, and a medium round, as well as glass beads, have been used in examining these differences (Nazarian and Baig 1995). The results were compared with empirical relationships proposed by Iwasaki and Tatsuoka (1977) and Hardin (1978). It was shown that varying grain size and angularity, as well as the density of the sample would dictate how well the test results agreed with each empirical relationship.

10.8.9 Effects of cross-anisotropy

Shear wave propagation may also be explored in the horizontal direction of the vertical specimen by mounting bender elements across from each other, as shown in Fig. 10.12. The horizontal shear modulus is then calculated from a similar formula to that for the vertical shear modulus:

$$G_{h,max} = \rho \cdot V_{h,s}^{\ 2} \tag{10.8}$$

in which ρ is the soil density and $V_{h,s}$ is the shear wave velocity in the horizontal direction. The ratio between the vertical and the horizontal values of the shear moduli will indicate the small strain anisotropy of the specimen fabric.

11 Tests with Three Unequal Principal Stresses

11.1 Introduction

To place the results of triaxial compression tests in perspective, it is necessary to (1) realize to which degree the stress conditions in these tests match the stress conditions in the field and (2) understand how the results of triaxial compression tests compare with results from other types of laboratory shear tests in which three different principal stresses are applied to the soil element under conditions of (a) no principal stress rotation and (b) rotating principal stress directions. While the results of such three-dimensional (3D) tests will be briefly reviewed, emphasis will be placed on the operational principles of the equipment and special procedures required in such equipment different from those employed in triaxial compression tests.

Field conditions almost always involve 3D stress conditions, and only rarely are axisymmetric stress conditions encountered in situ. For example, the stress and strain conditions in many geotechnical structures can be simulated with good accuracy by plane strain conditions. Figure 11.1 shows an embankment in which all soil particles move in parallel planes during shear and failure. Stress and strain conditions in such an embankment may be simulated in plane strain and in simple shear equipment.

Figure 11.2 illustrates some of the stress conditions under a centrally loaded square footing. Only along the axis below the footing may the stress conditions be axisymmetric as in a triaxial compression test. Outside this axis, the state of stress on each element is different and they are all general 3D stress states, as indicated in Fig. 11.2. Thus, there are three different principal stresses acting on each soil element, and these principal stresses are rotating as the footing is loaded.

Only a few types of laboratory equipment are available to apply 3D stress conditions to a soil specimen. The states of stress that can be generated in this equipment are relatively limited. It is, however, possible to get an impression of the effects of various 3D stress conditions on the behavior of soils through the results from tests in such equipment.

The 3D tests are divided into two groups: one in which only principal stresses are applied; and another in which normal as well as shear stresses are applied at the boundaries of the specimen. In the latter group of tests the principal stresses usually rotate during shear.

Triaxial Testing of Soils, First Edition. Poul V. Lade.

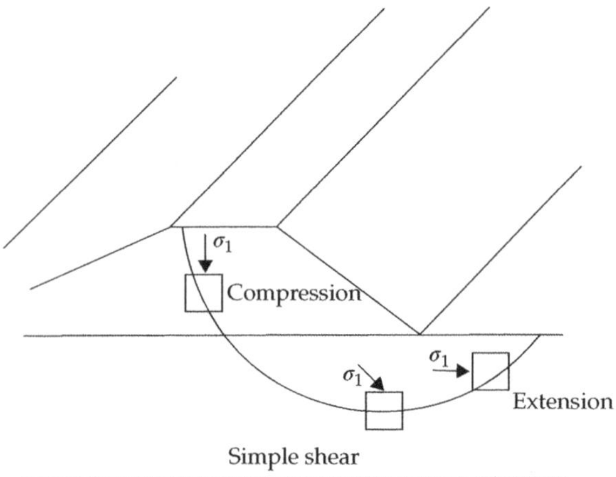

Figure 11.1 Embankment with all soil particles moving in parallel planes during shear and failure, which may be reproduced in a plane strain test and in a simple shear test.

11.2 Tests with constant principal stress directions

Basically two types of tests are available in the group in which three different principal stresses are applied without stress rotation: plane strain tests; and true triaxial tests. Both of these tests are used as tools in research studies of the 3D behavior of soils.

11.2.1 Plane strain equipment

In the plane strain test, the intermediate principal strain is maintained at zero:

$$\varepsilon_2 = 0 \tag{11.1}$$

This is usually accomplished by mounting two rigid plates on opposite sides of a prismatic specimen and connecting them rigidly to each other. Figure 11.3 shows the principle of the plane strain apparatus. To avoid development of significant shear stresses, the rigid plates are provided with lubrication just as employed at the end plates of triaxial tests. Strains different from zero occur only in the major and minor principal stress directions, and the value of the intermediate principal stress, σ_2, is such that the value of $b = (\sigma_2 - \sigma_3)/(\sigma_1 - \sigma_3)$ is in the order of 0.2–0.4 depending on soil density and confining pressure.

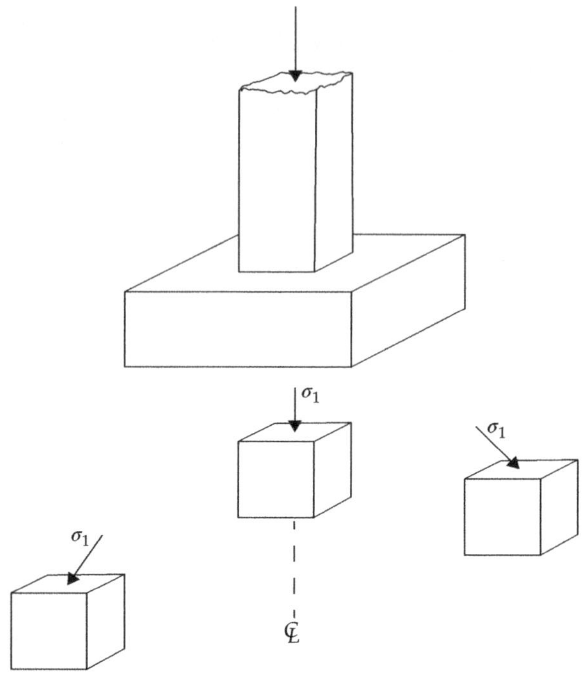

Figure 11.2 Schematic illustration of stress conditions under a centrally loaded square footing.

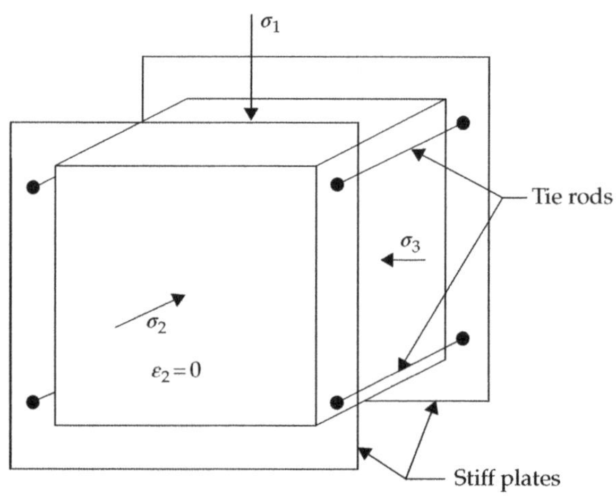

Figure 11.3 Principle of the plane strain apparatus.

Results of plane strain tests fit in the pattern of test results obtained from true triaxial tests, and these will be briefly reviewed below.

11.2.2 True triaxial equipment

Studies of the influence of the intermediate principal stress on the behavior of soils have been performed with the help of different types of true triaxial apparatus designed and constructed since the 1960s. They may be classified into a few categories according to the type of boundary conditions and the loading procedures, as illustrated in Fig. 11.4:

(I) Apparatus in which all six faces of a cubical specimen are loaded by flexible pressure bags or fluid cushions. Ko and Scott (1967) made the original design. Many series of tests have been performed in various studies employing a version of this apparatus. Tests in this type of apparatus are often performed with constant mean normal stress and constant value of $b = (\sigma_2 - \sigma_3)/(\sigma_1 - \sigma_3)$ proposed by Habib (1953).

Figure 11.5 shows the original version of this apparatus. Each pair of opposite membranes is interconnected and can be pressurized individually, thus providing for three different principal stresses. To avoid a tendency for the membranes to interfere with each other at the edges, a rigid frame is used to separate the pressurized membranes. This apparatus offers the advantage of easy application of normal stresses and ensures that no shear stresses are induced on the faces of the specimen. Uniformity of strains across each pressure bag, especially as the strains become of significant magnitudes, is not guaranteed.

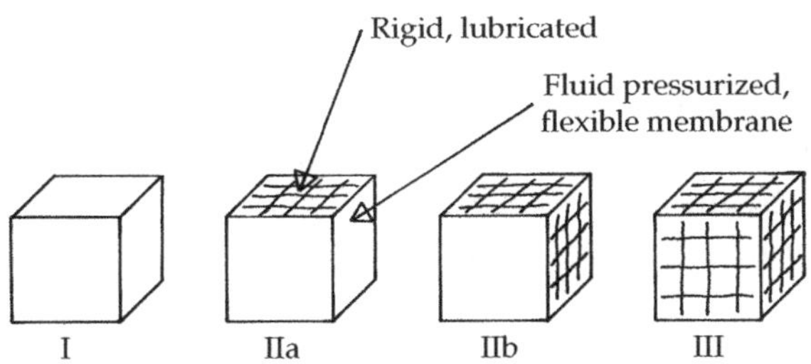

Figure 11.4 Types of boundary conditions employed in different true triaxial apparatus for soil testing. Reproduced from Lade 2006 by permission of John Wiley & Sons.

Cubical specimens are usually employed. Any other shape of specimen requires the construction of new apparatus.

(IIa) Apparatus in which the vertical stress is applied through rigid, lubricated cap and base, one horizontal stress is applied by the cell pressure on the membrane surrounding the specimen, and the other horizontal stress is provided by the cell pressure and an additional deviator stress supplied by two pressurized rubber bags sitting on opposite sides of the specimen. Shibata and Karube (1965) constructed the first apparatus of this type.

Figure 11.6 shows the original version of this apparatus. The two rubber bags are interconnected and provide the intermediate or major principal deviator stress. Application of shear

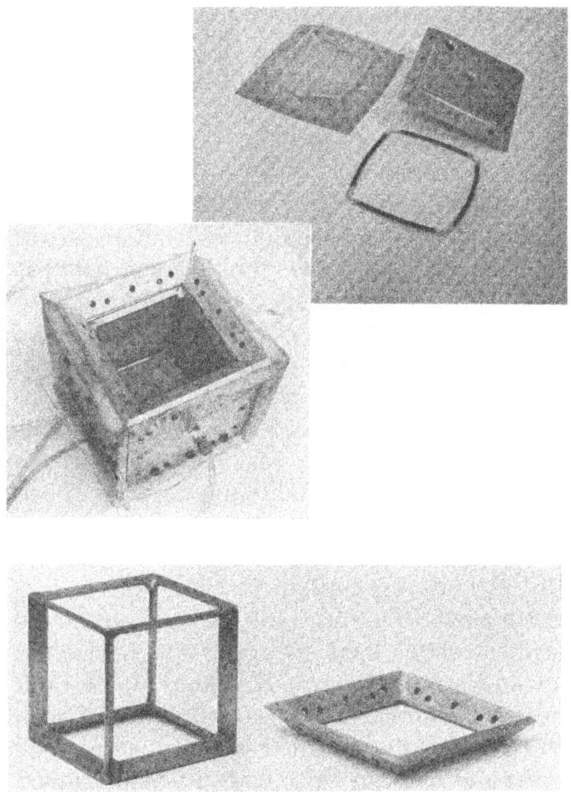

Figure 11.5 Fluid cushion cubical triaxial apparatus. Reproduced from Ko and Scott 1967 by permission of Geotechnique.

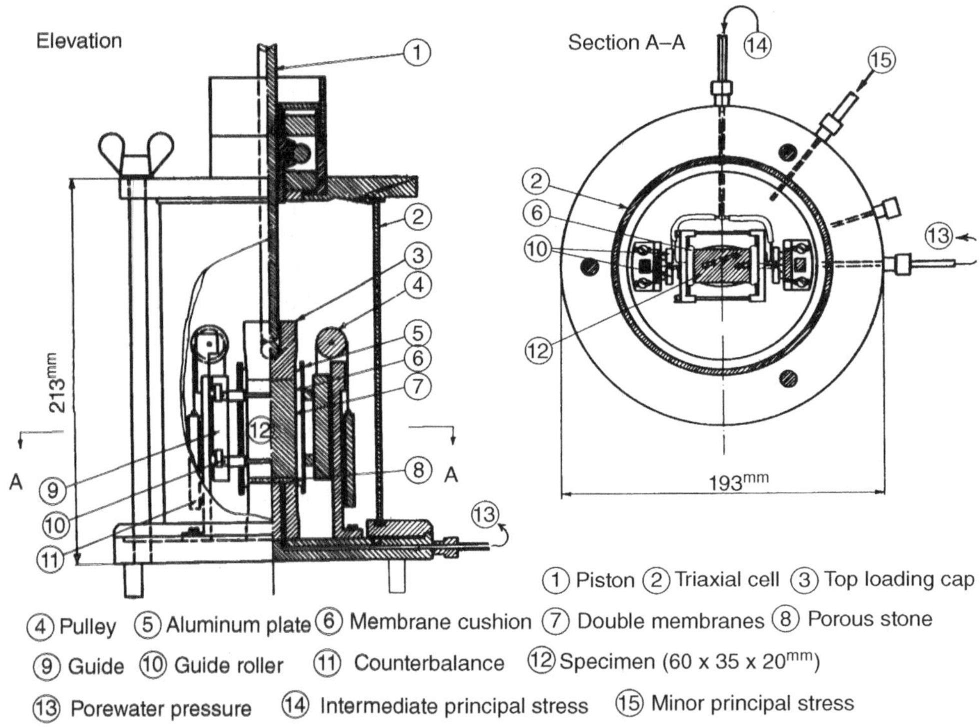

Figure 11.6 Shibata and Karube's true triaxial apparatus (after Shibata and Karube 1965).

stresses on the four vertical sides is avoided and lubricated cap and base reduce shear stresses at the ends. The specimen is simply constructed as a conventional triaxial specimen with rectangular cross-section. Uniformity of strains along the vertical rubber bags as well as containment of the rubber bags and interference with the vertical loading plates may be issues of concern. This type of apparatus can easily be modified to accept any other rectangular prismatic specimen shape.

(IIb) Apparatus similar to type IIa, but in which a pair of vertical, stiff platens replace the flexible rubber bags to apply the horizontal deviator stress. Green and Bishop (1969) built the first version of this type of apparatus (Green 1971). A slightly modified version of this type (Lade and Duncan 1973; Lade 1978) employed vertically compressible, stiff side plates, thus avoiding any interference between the two sets of stiff loading plates. Tests in types IIa and IIb apparatus are performed on cubical or rectangular prismatic specimens, and they are conducted easiest and most often with constant confining pressure (σ_3).

Figure 11.7 shows the apparatus designed and constructed by Lade (1972, 1978). The vertically compressible, stiff side plates consist of alternating stainless steel and balsa wood lamellas that compress symmetrically around the middle of the vertical sides, thus essentially following the compressing specimen. Lubricated ends on the vertical as well as the horizontal loading plates help reduce any shear stresses along the sides. The minor principal stress is applied by the cell pressure, while the major and intermediate principal stresses may be interchanged between the vertical and horizontal loading plates.

The specimen is simply constructed as a conventional triaxial specimen with square cross-section. This type of apparatus can easily be modified to accept any other rectangular prismatic specimen shape (Wang and Lade 2001).

(III) Apparatus in which all six faces of the cubical specimen are loaded by rigid stainless steel plates that are keyed together and can slide relative to each other to produce strains along all three perpendicular directions. The cubical specimen is enclosed in a membrane and surrounded by six interconnected rigid plates, one on each face. Hambly (1969) described the basic idea, and Pearce (1970, 1971) built the first apparatus of this type. As for type I, tests are often performed with constant mean normal stress and constant value of $b = (\sigma_2 - \sigma_3)/(\sigma_1 - \sigma_3)$.

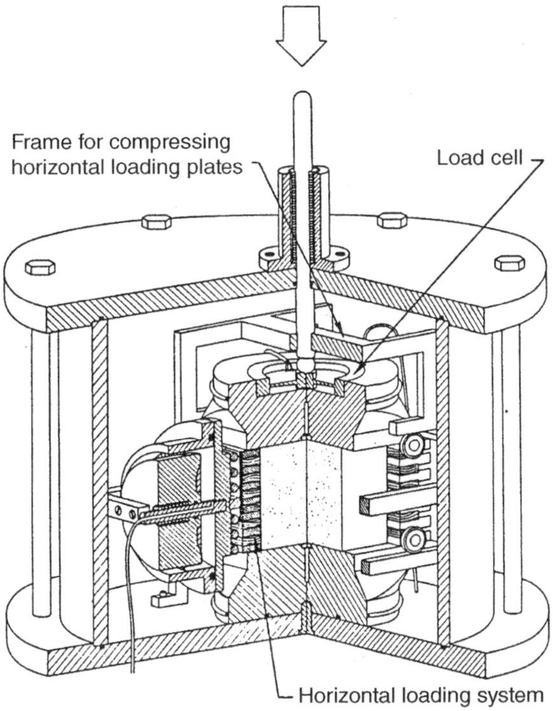

Figure 11.7 Lade's cubical triaxial apparatus. Reproduced from Lade and Duncan 1973 by permission of ASCE.

Figure 11.8 shows a recent version of this apparatus. Because of development of excessive friction between the sliding parts, reliable forces on the loading plates cannot be measured from outside load cells. Instead, load cells are mounted internally in the loading plates to measure the load on a given area in each of the three perpendicular directions. The best method of preparation of sand specimens for this apparatus involves depositing the specimens in a mold, saturating them with water, freezing them and inserting them in a prefabricated membrane with the shape of the cubical specimen, followed by thawing before testing. All loading plates are lubricated to avoid application of significant shear stresses. While cubical specimens are usually employed, any other shape of specimen requires the construction of new apparatus.

The apparatus described above represent the design principles that have been used for equipment with independent control of the three principal stresses. Some of these apparatus are easier to employ than others, but in reviewing the experimental results from the various types of equipment, it appears that they all produce similar and reasonable results when interpreted correctly (Lade 2006). Only a few sets of data presented in the literature appear to be questionable.

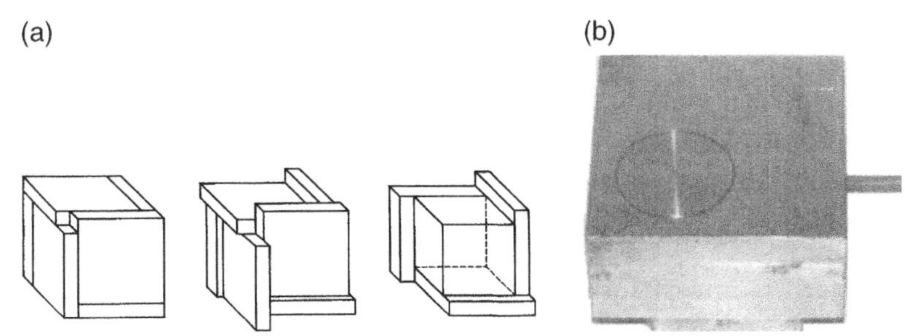

Figure 11.8 (a) Schematic diagrams of true triaxial apparatus with (b) load cells in three plates. Reproduced from Ibsen and Praastrup 2002 by permission of ASTM International.

11.2.3 Results from true triaxial tests

In presenting the results of true triaxial tests, the issue is the influence of the intermediate principal stress on the soil behavior. The relative magnitude of the intermediate principal stress may be indicated by the Lode angle (see Section 2.7.4), but is often indicated by the value of b:

$$b = \frac{\sigma_2 - \sigma_3}{\sigma_1 - \sigma_3} \tag{11.2}$$

The value of b is zero for triaxial compression in which $\sigma_2 = \sigma_3$, and it is unity for triaxial extension in which $\sigma_2 = \sigma_1$. For intermediate values of σ_2 the value of b is between zero and unity.

Behavior of sand

The behavior of sand under 3Dt stress conditions is illustrated below.

Stress–strain characteristics for sand

Examples of stress–strain curves obtained from true triaxial tests on dense and loose Monterey No. 0 sand are shown in Figs 11.9 and 11.10. It may be seen that the strength increases with increasing value of the intermediate principal stress, especially from $b = 0.0$ to 0.2. The strength increases further until the value of b reaches 0.75–0.90 and then it decreases slightly at $b = 1.00$.

The data in Figs 11.9 and 11.10 show that for a constant value of σ_3 the initial slope of the stress–strain curve increases continually with increasing value of the intermediate principal stress for both dense and loose sand. This behavior indicates that for small stress levels, that is for stress levels close to the hydrostatic axis, the influence of σ_2 on the stress–strain curves may be accounted for, at least qualitatively, by Hooke's law. The strain-to-failure is greatest and the strength is lowest for triaxial compression ($b = 0.0$). For loose sand the strain-to-failure decreases initially with increasing value of b and remains approximately constant for b-values greater than 0.6. It may also be seen that the major deviator stress, $(\sigma_1 - \sigma_3)$, drops off more rapidly after the peak when the value of b is large.

Volume change behavior

The volumetric strains measured in the true triaxial tests are also shown in Figs 11.9 and 11.10. The initial rate of contraction increases with increasing b-value for both dense and loose sand. This characteristic is indicative of elastic behavior at low stress levels. As the stress level increases, plastic dilation begins to dominate the volume change. The rate of dilation, expressed as $\Delta\varepsilon_v/\Delta\varepsilon_1$, increases with increasing stress level from a negligible value at low stress levels to such a magnitude at high stress levels as to completely dominate the elastic contraction. Whereas this behavior is observed for both dense and loose sand, the rate of dilation at failure was much higher for the dense sand than for the loose sand. The rate of dilation increases with increasing value of b.

Relations between principal strains

The intermediate and minor principal strains, ε_2 and ε_3, are plotted versus ε_1 in Fig. 11.11 for both dense and loose sand. The upper diagrams in Figure 11.11 show that the intermediate principal strains, ε_2, are expansive for b-values smaller than those corresponding to the plane strain condition and contractive for higher b-values. The minor principal strains, ε_3, are expansive in all cases and decrease with increasing b-values, as shown in the lower diagrams of Fig. 11.11. A given increment in b has a greater effect on the relation between the principal strains at small b-values than at high b-values.

Lines are drawn through the points corresponding to failure in Fig. 11.11. The major principal strain-to-failure decreases with increasing b-value for dense sand. For loose sand the major principal strain-to-failure first decreases with increasing b-value and then remains approximately constant for b-values greater than about 0.6.

In the extension tests performed in the true triaxial apparatus, the major principal stress, σ_1 ($= \sigma_2$), was applied in the vertical direction and the intermediate principal stress, σ_2 ($= \sigma_1$), was

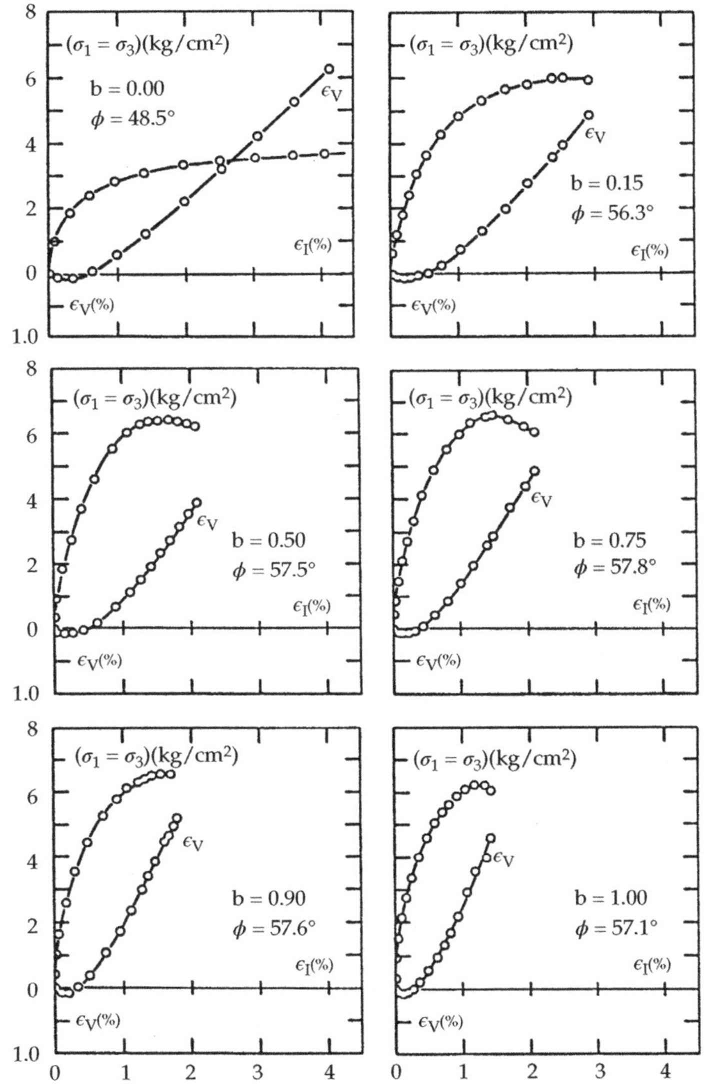

Figure 11.9 Stress–strain and volume change characteristics obtained in cubical triaxial tests on dense Monterey No. 0 sand (e = 0.57). All tests performed with σ_3 = 58.8 kN/m². Reproduced from Lade and Duncan 1973 by permission of ASCE.

applied in one of the horizontal directions. The specimen was consequently loaded symmetrically around a horizontal axis, but since the specimen was deposited in the vertical direction, the principal strains, ε_1 and ε_2, would not be expected to be equal unless the sand were isotropic.

The upper diagrams in Fig. 11.11 show that the values of ε_1 for practical purposes are equal to the values of ε_2 for extension tests on both dense and loose sand. It was also observed that in some of the triaxial extension tests failure occurred in the horizontal direction, in others failure occurred in the vertical direction, and in others failure occurred in both directions simultaneously. The strengths measured in these tests were approximately the same. Both of these observations indicate that the sand specimens were essentially isotropic.

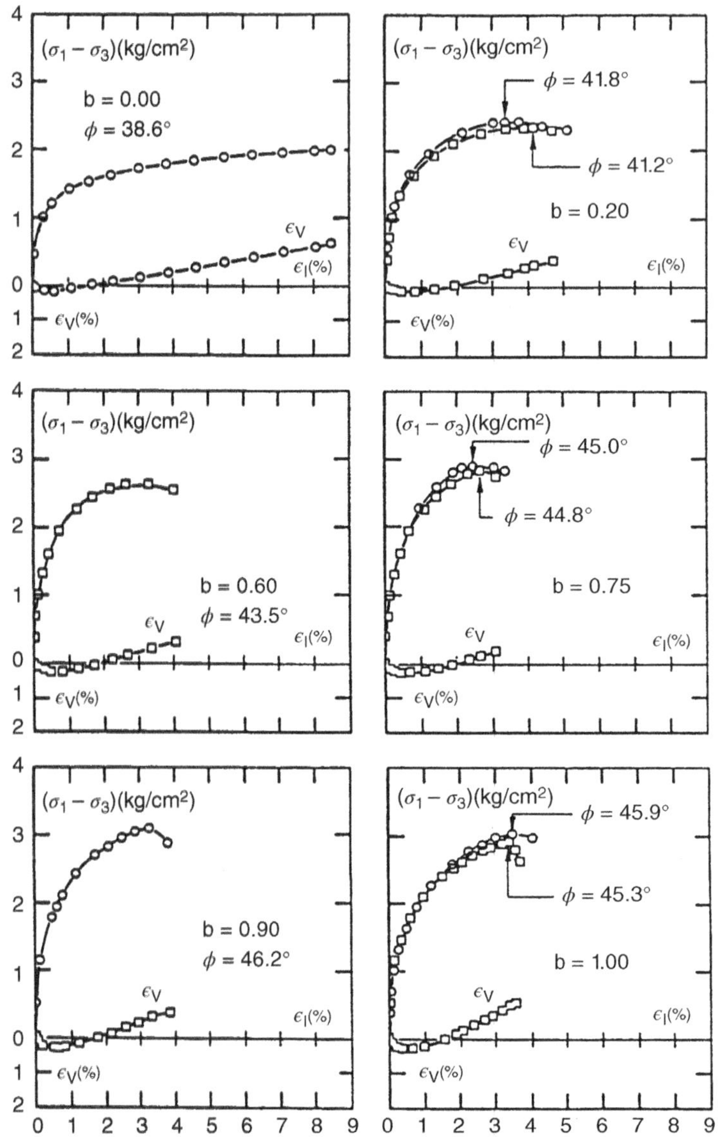

Figure 11.10 Stress–strain and volume change characteristics obtained in cubical triaxial tests on loose Monterey No. 0 sand ($e = 0.78$). All test performed with $\sigma_3 = 58.8$ kN/m². (○) One lubricating sheet on each of four interfaces; (□) one lubricating sheet on bottom, two on each of three other interfaces. Reproduced from Lade and Duncan 1973 by permission of ASCE.

While the results reviewed above indicated isotropic sand specimens, experiments on other sand specimens clearly show cross-anisotropic behavior, as will be illustrated in Section 11.2.4.

Behavior of clay

The behavior of clay under 3D stress conditions is illustrated below.

Stress–strain and pore pressure characteristics

Examples of stress–strain and pore pressure behavior for normally consolidated, remolded Grundite clay under undrained conditions are shown in Fig. 11.12 for specimens consolidated at 147 kPa (1.50 kg/cm²). The relationships obtained from triaxial compression tests with

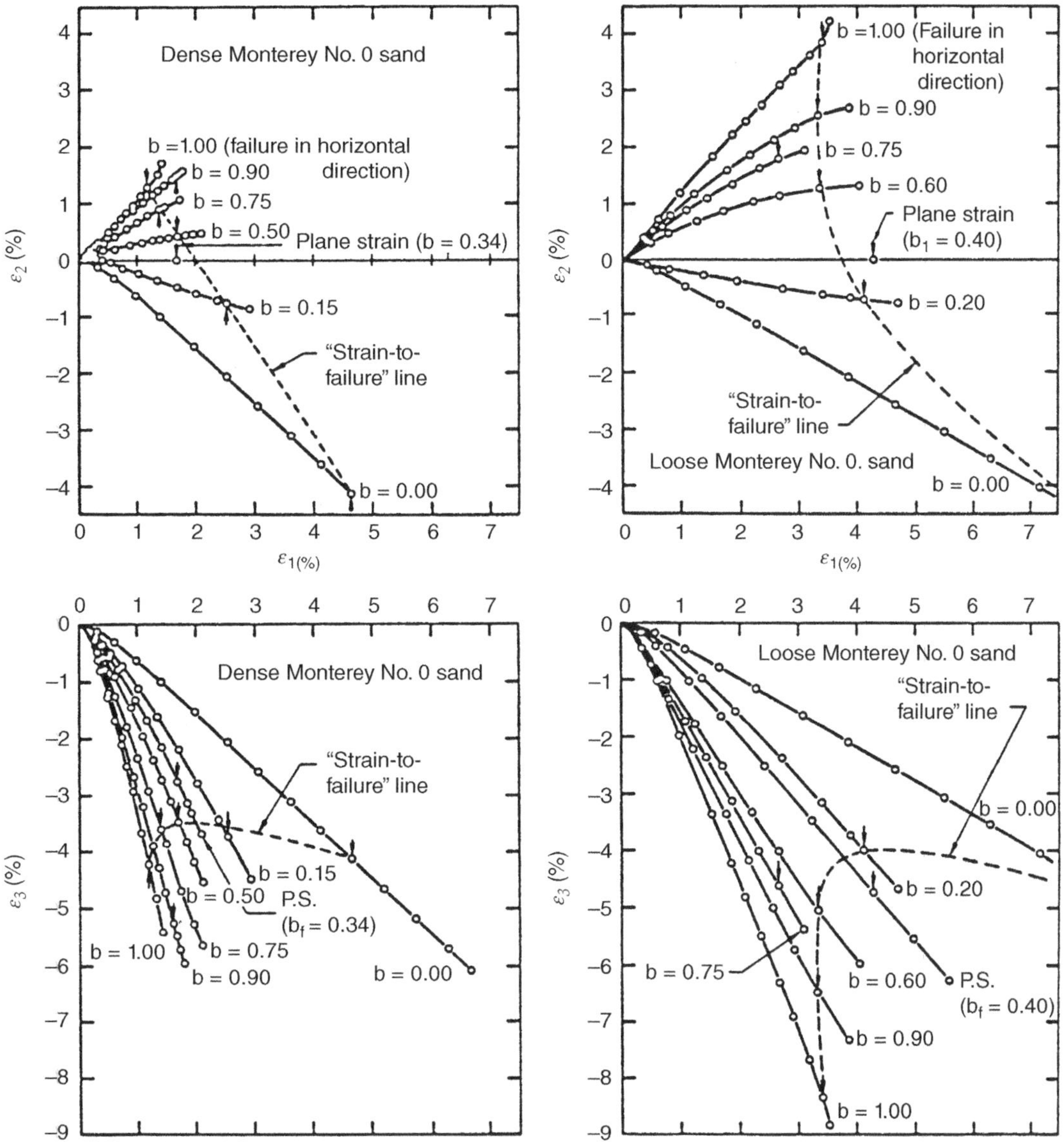

Figure 11.11 Relations between principal strains obtained from cubical triaxial tests on dense and loose Monterey No. 0 sand. Reproduced from Lade and Duncan 1973 by permission of ASCE.

consolidation pressures of 98 and 196 kPa (1.00 and 2.00 kg/cm^2) are also shown in Fig. 11.12(a). The normalized stress differences, $(\sigma_1 - \sigma_3)/\sigma_c'$, the effective stress ratio, σ_1'/σ_3', and the normalized pore pressure changes, $\Delta u/\sigma_c'$, are plotted versus the major principal strain, ε_1, in these diagrams, and the relative magnitudes of the intermediate principal stresses are indicated by the values of b.

Considering that the stress–strain relations in Fig. 11.12(a) are normalized, it may be seen that the initial undrained modulus increases and the initial slope from the effective stress ratio diagram increases with increasing consolidation pressure. The pore pressures shown in Fig. 11.12(a) increase to values at failure that are almost proportional to the initial consolidation pressure. Thus, the ratio of pore pressure change to consolidation pressure

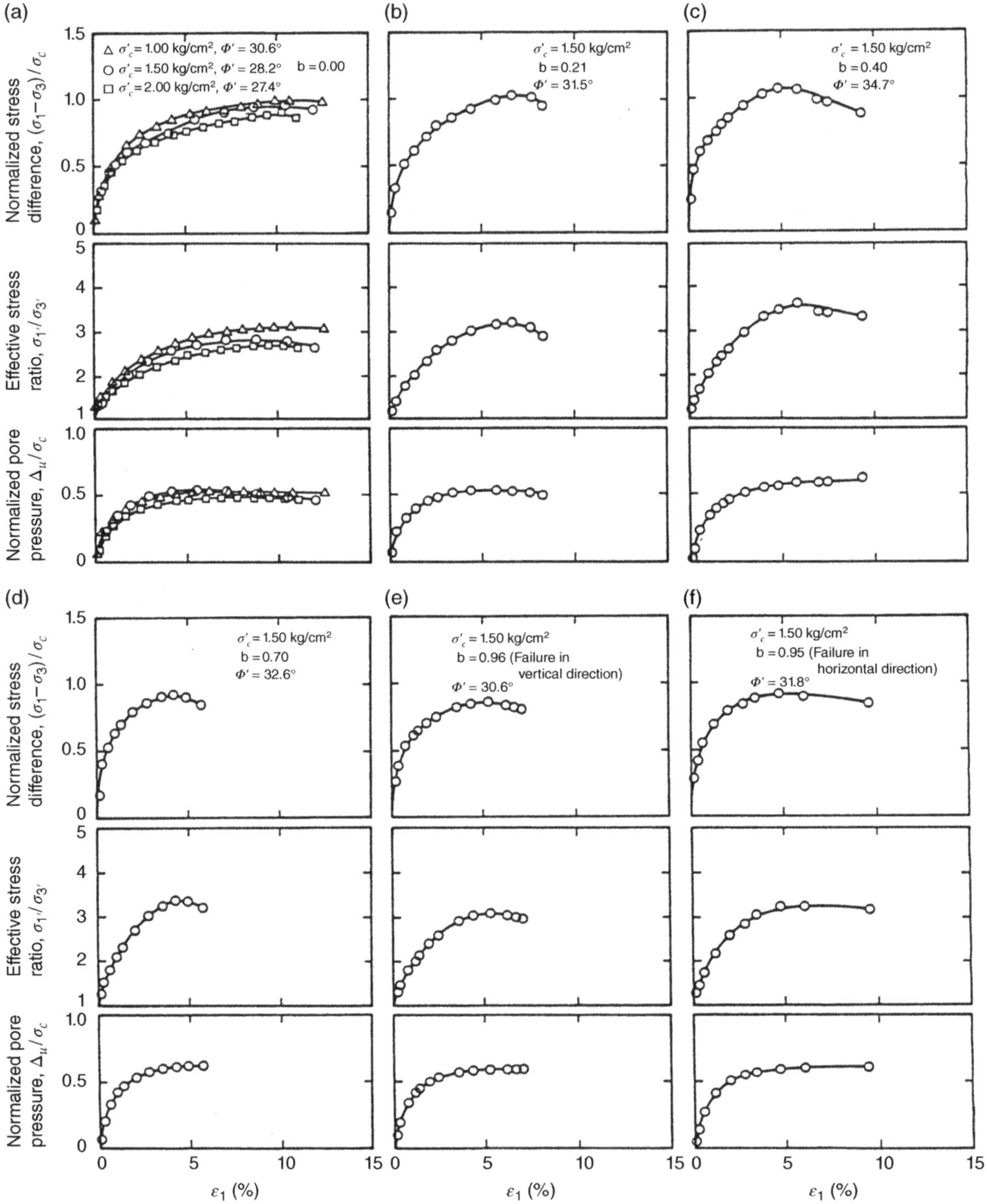

Figure 11.12 Stress–strain and pore pressure characteristics obtained from cubical triaxial tests on normally consolidated, remolded Grundite clay. Tests performed with $\sigma_3' = 147$ kN/m². Reproduced from Lade and Musante 1978 by permission of ASCE.

decreases slightly with increasing consolidation pressure. This pattern corresponds to the pattern of decreasing effective stress ratio with increasing consolidation pressure.

The results of the true triaxial tests shown in Fig. 11.12 indicate that for a constant consolidation pressure the initial slope of the stress difference–strain and the effective stress ratio–strain relations increase continually with increasing value of the intermediate principal stress. The strain-to-failure is greatest, the pore pressure developed at failure is lowest, the effective strength is lowest, and the undrained strength is highest for triaxial compression ($b = 0.0$). The strain-to-failure decreases and the pore pressure change increases initially with increasing b-value and both remain approximately constant for b-values greater than about 0.6. Note that the tests performed with initial consolidation pressures of 98 and 196 kPa (1.00 and 2.00 kg/cm^2) showed similar patterns of behavior as those in Fig. 11.12.

Relations between principal strains

The intermediate and minor principal strains, ε_2 and ε_3, respectively, are plotted versus ε_1 in Fig. 11.13 for specimens consolidated at 147 kPa (1.50 kg/cm^2). Figure 11.13(a) shows that the intermediate principal strains, ε_2, are expansive for b-values smaller than about 0.4 and contractive for higher values of b. The minor principal strains, ε_3, are expansive in all cases and decrease with increasing b-values, as shown in Fig. 11.13(b). Since the tests were performed under undrained conditions, the sum of the three principal strains is always equal to zero. Figure 11.13 indicates that a given increment in b has a greater effect on the relations between the principal strains at small b-values than at high b-values.

The points corresponding to failure according to the maximum effective stress ratio are indicated on each curve in Fig. 11.13. The major principal strain-to-failure decreases initially with increasing b-value and remains approximately constant for b-values greater than about 0.6.

11.2.4 Strength characteristics

Strength of sand

The variation of measured friction angles with b is shown in Fig. 11.14 for dense and loose Monterey No. 0 sand. The friction angles were calculated from the expression for the Mohr–Coulomb failure criterion for non-cohesive soils, which does not depend on the intermediate principal stress, σ_2:

$$\sin\varphi = \frac{\sigma_1 - \sigma_3}{\sigma_1 + \sigma_3} \tag{11.3}$$

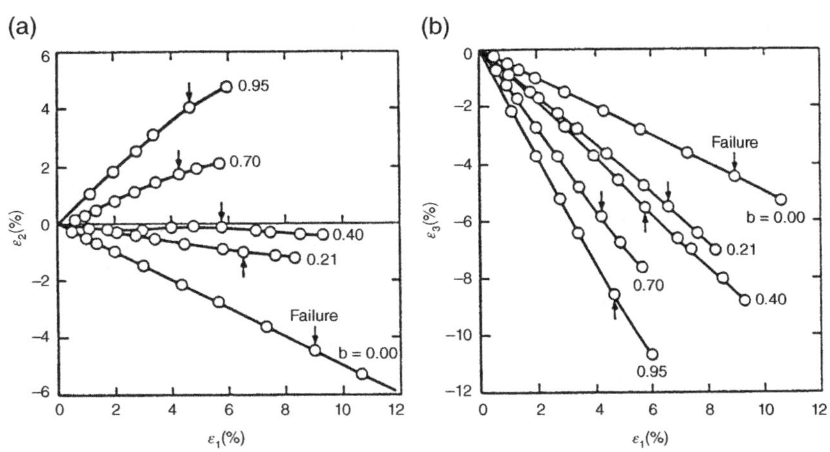

Figure 11.13 Relations between principal strains obtained from cubical triaxial tests on normally consolidated, remolded Grundite clay. Tests performed with $\sigma_3' = 147$ kN/m^2. Reproduced from Lade and Musante 1978 by permission of ASCE.

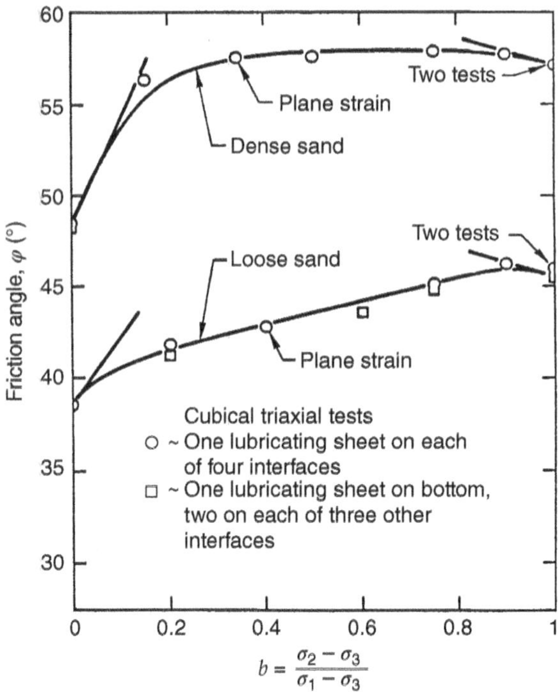

Figure 11.14 Failure surfaces for dense and loose Monterey No. 0 sand shown in φ–b diagram for tests in cubical triaxial apparatus. Reproduced from Lade and Duncan 1973 by permission of ASCE.

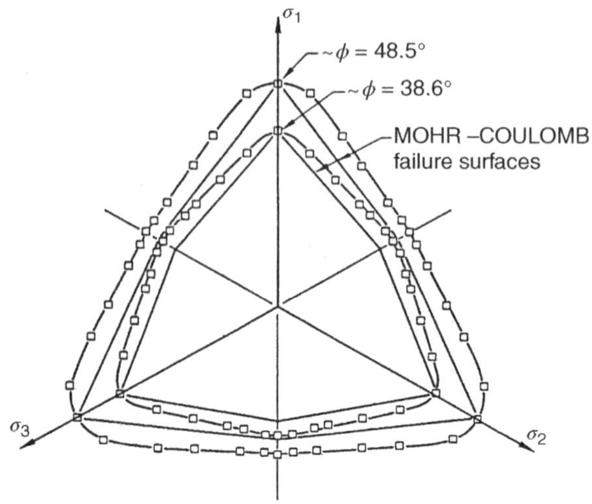

Figure 11.15 Failure surfaces for dense and loose Monterey No. 0 sand shown on an octahedral plane for tests in cubical triaxial apparatus. The Mohr–Coulomb failure surfaces are shown for comparison. Reproduced from Lade and Duncan 1973 by permission of ASCE.

They are smallest in triaxial compression for both the dense and the loose sand, and the use of values of φ measured in triaxial compression may be seen to be quite conservative. As the magnitude of b increases, the friction angle increases to a maximum before decreasing slightly close to the extension condition.

The data in Fig. 11.14 show that the amount of lubrication has little influence on the strength as long as the specimens deform uniformly and the predominant part of the end restraint has been removed.

Figure 11.15 shows the test results plotted on an octahedral plane. It has been assumed in plotting this diagram that the confining pressure (= σ_3) has no influence on the friction angle, so that the failure surfaces in the principal stress space are cones for which the cross-sectional shapes are shown in Fig. 11.15. This is only approximately true, but it is believed that the assumption involves little error.

The cross-sections of the Mohr–Coulomb failure surfaces corresponding to the strengths obtained in triaxial compression for dense and loose specimens are also shown in Fig. 11.15. These cross-sections have shapes of irregular hexagons, with acute and obtuse angles at the points corresponding to the states of stress in triaxial compression and extension, respectively. In contrast, the traces of the experimental failure surfaces in the octahedral plane are smooth throughout their lengths.

It was pointed out in Section 11.2.3 that the specimens were essentially isotropic, and it was also noticed that the strengths of the extension test specimens were the same whether they failed horizontally or vertically. Interchanging the principal stress directions will therefore not have any effect on the strength of this sand. The traces of the failure surfaces in the octahedral plane are consequently symmetric around the

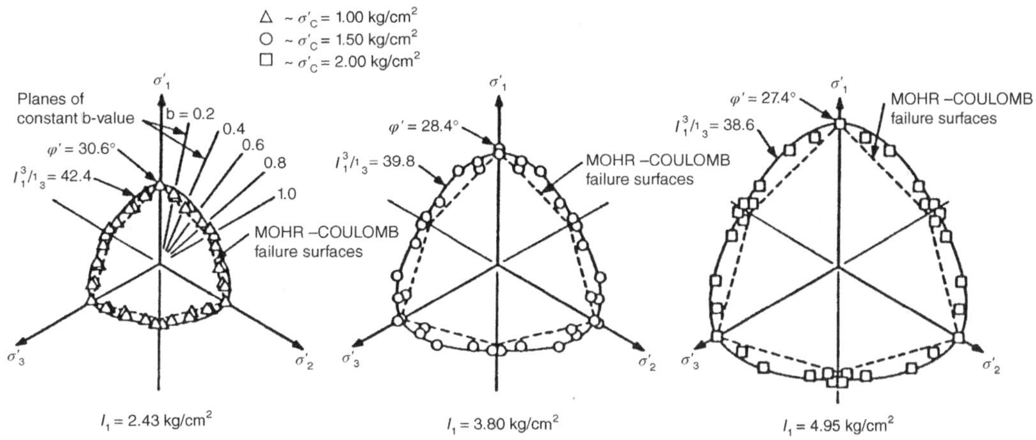

Figure 11.16 Traces of failure surfaces on an octahedral plane for normally consolidated, remolded Grundite clay. Reproduced from Lade and Musante 1978 by permission of ASCE.

projections of the three principal stress axes and intersect these at right angles.

Effective strength of clay

The traces of the failure surface in terms of effective stresses are shown in Fig. 11.16 for tests on normally consolidated, remolded Grundite clay. The locations of the octahedral planes are indicated by the values of the first stress invariant, I_1. The failure points shown in Fig. 11.16 have been projected on the octahedral planes along curved failure surfaces observed in planes that contain the hydrostatic axis and have constant b-values.

Undrained shear strength of clay

The results of true triaxial tests on normally consolidated Edgar Plastic Kaolinite are shown in terms of normalized, undrained shear strength, s_u/σ_c', plotted versus b in Fig. 11.17(a). The effect of the intermediate principal stress is to reduce the normalized, undrained shear strength with increasing b-value. It is clear from this diagram that assuming s_u/σ_c' = constant, as implied in the Tresca failure criterion, is not conservative if the value has been determined from triaxial compression tests. The decrease in this ratio from $s_u/\sigma_c' = 0.54$ in triaxial compression ($b = 0.0$) to $s_u/\sigma_c' = 0.38$ in triaxial extension ($b = 1.0$) corresponds to a change of 30%. This substantial drop in normalized, undrained shear strength is not accounted for by the Tresca failure criterion, which is most often assumed to hold for total stress stability of soil structures. The decrease in s_u/σ_c' from triaxial compression to plane strain conditions, $b = 0.40$–0.45, for which most analyses procedures have been developed, is in the order of 20% for the remolded Edgar Plastic Kaolinite. The solid line drawn through the experimental data is predicted by a constitutive model (Lade 1990).

The experimental results are also shown in terms of total stresses on the octahedral plane in Fig. 11.17(b). Again, it is clear that the Tresca failure criterion is not conservative when applied in stability analyses of geotechnical structures.

11.2.5 Failure criteria for soils

Many frictional materials such as gravels, sands, silts, clays, rockfill, mine tailings, coal, feed grain, and so on do not have effective cohesion. According to Mitchell (1976) "tests over large ranges of effective stress show that the actual effective stress failure envelope is curved," and

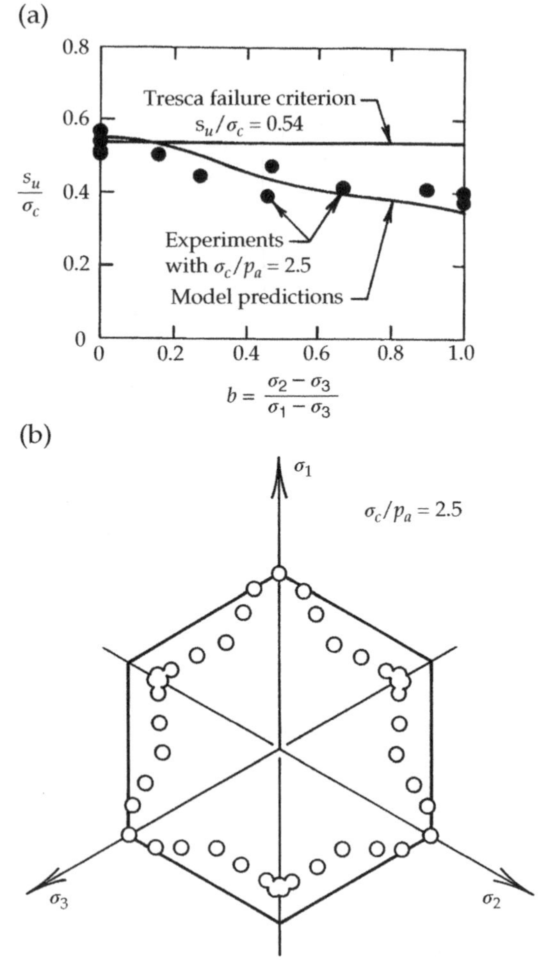

Figure 11.17 (a) Results of true triaxial tests on normally consolidated Edgar Plastic Kaolinite shown in terms of (a) normalized, undrained shear strength, s_u/σ_c', plotted versus b and (b) compared with the Tresca failure criterion. Note that the solid line drawn through the experimental data is predicted by a constitutive model. Reproduced from Lade 1990 by permission of ASCE.

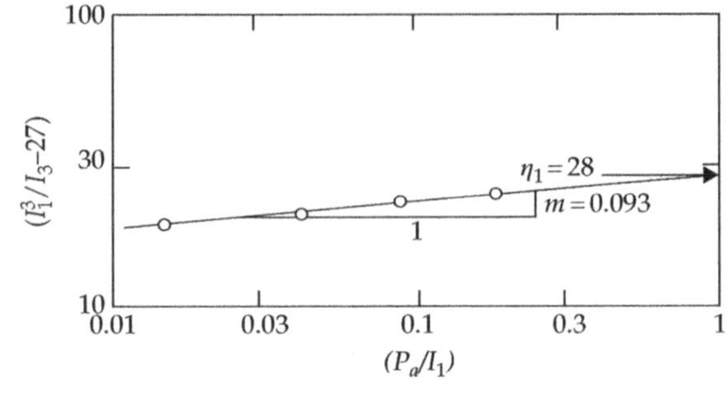

Figure 11.18 Parameters η_1 and m in the 3D failure criterion can be determined by plotting $(I_1^3/I_3 - 27)$ versus (p_a/I_1) at failure in a log–log diagram and locate the best fitting straight line. The intercept of this line with $(p_a/I_1) = 1$ is the value of η_1 and m is the geometric slope of the line.

"that cohesion is either zero or very small, even for heavily overconsolidated clays. Thus, a significant true cohesion, if defined as strength present at zero effective stress, does not exist in the absence of chemical bonding (cementation)." Several studies of materials without effective cohesion under two-dimensional (2D) and 3D stress conditions show that these materials have many characteristics in common.

Three-dimensional, isotropic failure criterion

The 3D failure criterion for frictional materials without effective cohesion presented here was developed for soils with curved failure envelopes (Lade 1977). This criterion is expressed as follows:

$$\left(\frac{I_1^3}{I_3} - 27\right) \cdot \left(\frac{I_1}{p_a}\right)^m = \eta_1 \qquad (11.4)$$

in which I_1 and I_3 are the first and the third stress invariants of the stress tensor (see Section 2.62) and p_a is atmospheric pressure expressed in the same units as the stresses. The value of I_1^3/I_3 is 27 at the hydrostatic axis where $\sigma_1 = \sigma_2 = \sigma_3$. The parameters η_1 and m in Eq. (11.4) can be determined by plotting $(I_1^3/I_3 - 27)$ versus (p_a/I_1) at failure in a log–log diagram and locate the best fitting straight line, as shown in Fig. 11.18. The intercept of this line with $(p_a/I_1) = 1$ is the value of η_1 and m is the geometric slope of the line.

In principal stress space the failure surface defined by Eq. (11.4) is shaped like an asymmetric bullet with the pointed apex at the origin of the stress axes, as shown in Fig. 11.19(a). The apex angle increases with the value of η_1. The failure surface is concave towards the hydrostatic axis, and its curvature increases with the value of m. For $m = 0$ the failure surface is straight. Figure 11.19(b) shows typical cross-

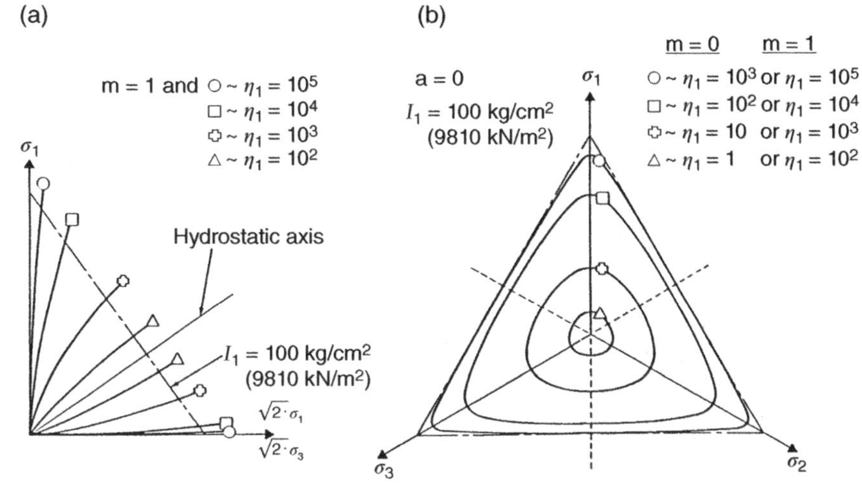

Figure 11.19 Characteristics of failure surfaces shown in principal stress space. Traces of failure surfaces shown on (a) a triaxial plane and (b) an octahedral plane. Reproduced from Lade 1984 by permission of John Wiley & Sons.

sections in the octahedral plane (I_1 = constant) for $m = 0$ and $\eta_1 = 1, 10, 10^2$, and 10^3. As the value of η_1 increases, the cross-sectional shape changes from circular to triangular with smoothly rounded edges in a fashion that conforms to experimental evidence. The shape of these cross-sections does not change with the value of I_1 when $m = 0$. For $m > 0$, the cross-sectional shape of the failure surface changes from triangular to become more circular with increasing value of I_1. Similar changes in cross-sectional shape are observed from experimental studies on frictional materials. The cross-sections in Fig. 11.19(b) also correspond to $m = 1$ and $\eta_1 = 10^2, 10^3, 10^4$, and 10^5.

Comparison of failure criterion and test data

The failure criterion described above has been shown to model the experimentally determined 3D strengths of many frictional materials (Lade 1982b, 1993b; Kim and Lade 1984) including sands and clays with good accuracy in the range of stresses where the failure envelopes are concave towards the hydrostatic axis (Lade 1977, 1978; Lade and Musante 1978).

Figure 11.20 shows examples of comparisons between failure criterion and test data in terms of effective stresses for dense Monterey No. 0 sand and normally consolidated, remolded Edgar Plastic Kaolinite. The values of η_1 and m suitable for description of failure in the two soils are given in Fig. 11.20. The data points were projected on the common octahedral planes along curved meridians (see Section 2.7.5) to provide a correct comparison between failure criterion and experimental data. It may be seen that the failure criterion models the experimentally obtained 3D failure surfaces with good accuracy for both sand and clay.

Effects of shear banding on 3D failure

True triaxial experiments were performed on tall prismatic specimens of Santa Monica Beach sand at three different relative densities to study the occurrence of shear banding under 3D stress conditions (Lade and Wang 2001; Wang and Lade 2001). For this purpose the true triaxial apparatus was modified to accept tall specimens with $H/D = 2.4$, which allowed free development of shear bands. The results showed that shear bands occurred in the hardening regime in the mid-range of b-values for all three relative densities tested. Because shear bands occur in dilating sands and are weaker than the surrounding sand, their development

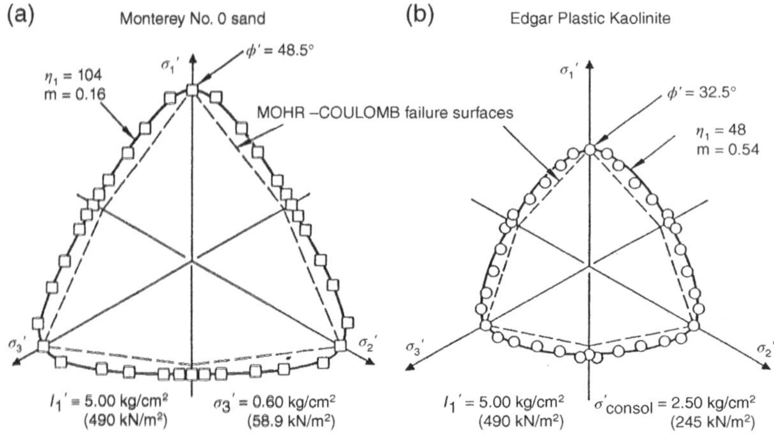

Figure 11.20 Comparisons between failure criterion and test data in terms of effective stresses for (a) dense Monterey No. 0 sand and (b) normally consolidated, remolded Edgar Plastic Kaolinite. Reproduced from Lade 1984 by permission of John Wiley & Sons.

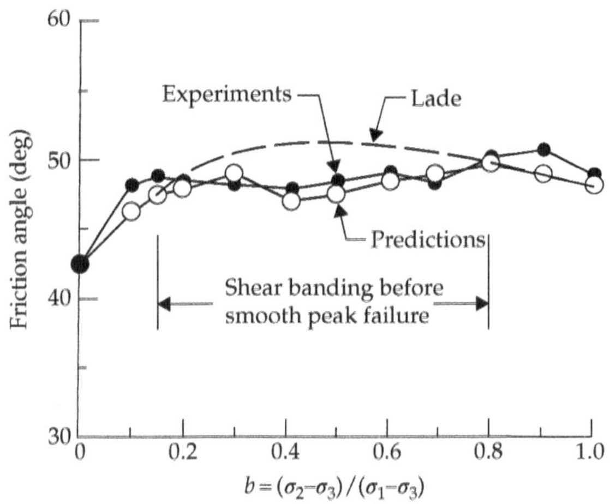

Figure 11.21 Comparison between experiments and predictions of strength including shear banding predicted by a constitutive model that involves the 3D failure criterion. Reproduced from Lade 2003 by permission of Elsevier.

in the hardening regime implied that smooth peak failure was not achieved in these tests and that the shear strength was lower than it would have been if shear bands would not occur. The shear banding was predicted with good accuracy by a constitutive model that involved the 3D failure criterion presented above and the comparison between experiments and predictions are shown in Fig. 11.21 (Lade 2003). Thus, shear banding plays an important role in the 3D strength of granular materials. Both experiments and predictions show that peak failure is caused by shear banding in the hardening regime in the approximate range of b from 0.18 to 0.85, while it occurs in the softening regime outside this range of b-values. A smooth, continuous failure surface is therefore not generally obtained for granular materials. Nevertheless, the smooth, continuous failure criterion, which seems to match experimental results for materials with homogeneous deformations, was part of the model used to predict the points of shear banding.

Three-dimensional cross-anisotropic failure criterion

Most soils are cross-anisotropic due to their mode of deposition in the field. True triaxial experiments on cross-anisotropic sand deposits in all three sectors of the octahedral plane show clearly that the experimental failure surface is symmetric around the vertical axis for the vertically pluviated sand, as shown in Fig. 11.22 (Lade 2008). The sand is stronger in the vertical direction than in the horizontal directions, and this strength variation is captured in a modified version of the failure

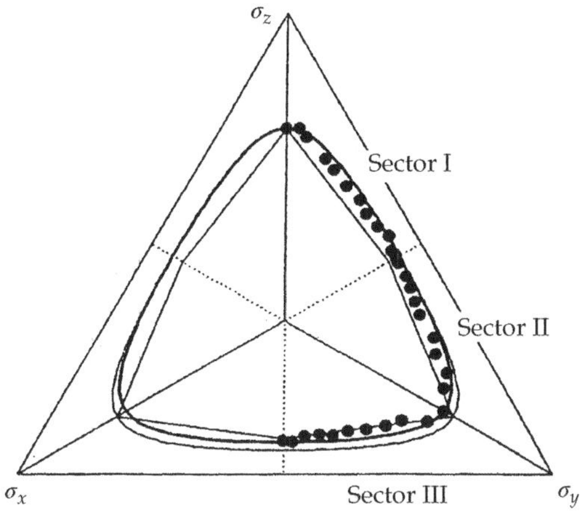

Figure 11.22 True triaxial experiments on cross-anisotropic sand deposits in all three sectors of the octahedral plane show the experimental failure surface is symmetric around the vertical axis for the vertically pluviated sand. Reproduced from Lade 2008 by permission of ASCE.

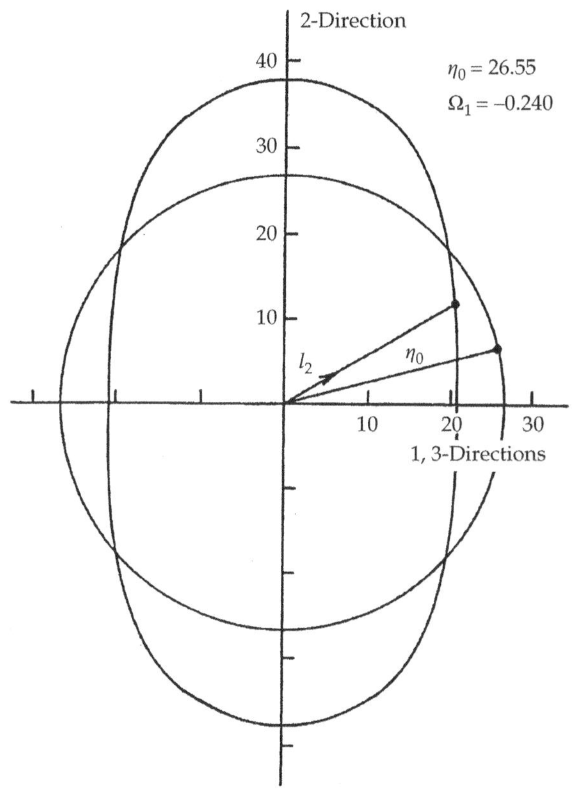

Figure 11.23 Spatial variation of factor $\eta_0 \cdot \left[1+\Omega_1\left(1-3\ell_2^2\right)\right]$ indicated by rotationally symmetric shape (around 2-direction) for medium dense Santa Monica Beach sand tested in torsion shear. Reproduced from Lade 2008 by permission of ASCE.

criterion given in Eq. (11.4) for isotropic frictional materials (Lade 2007, 2008):

$$\left(\frac{I_1^3}{I_3}-27\right)\left(\frac{I_1}{p_a}\right)^m = \eta_0 \cdot \left[1+\Omega_1\left(1-3\ell_2^2\right)\right] \quad (11.5)$$

in which η_0 is the radius of a sphere, as shown in Fig. 11.23 and represents an average value of η_1 from the isotropic criterion. The factor $\left[1+\Omega_1\left(1-3\ell_2^2\right)\right]$, which is controlled by the scalar material parameter Ω_1 and the loading direction ℓ_2, describes the deviation in three dimensions from the sphere. Thus, the right-hand side of Eq. (11.5) describes a rotationally symmetric shape, as indicated in Fig. 11.23. For cross-anisotropic materials tested in common laboratory experiments in which up to three different, orthogonal normal stresses and one shear stress can be applied the expression for ℓ_2 becomes:

$$\ell_2 = \sqrt{\frac{\sigma_x^2 \sin^2\beta + \sigma_y^2 \cos^2\beta}{\sigma_x^2+\sigma_y^2+\sigma_z^2}} \quad (11.6)$$

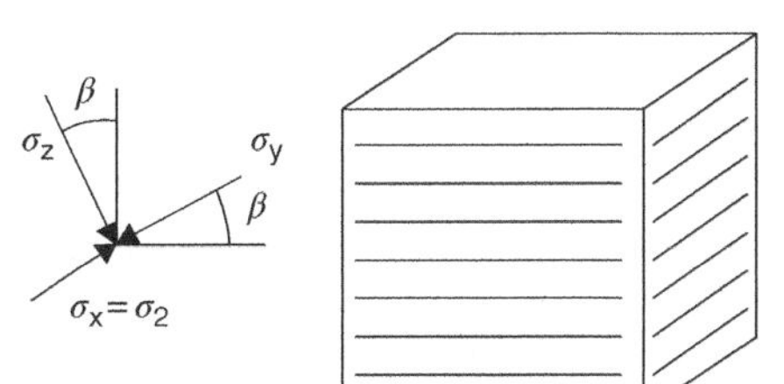

Figure 11.24 Principal stress conditions in true triaxial and torsion shear tests on cross-anisotropic soil.

in which $(\sigma_x, \sigma_y, \sigma_z)$ are principal stresses as indicated in Fig. 11.24.

The three parameters, m, η_0, and Ω_1 for the failure criterion in Eq. (11.5) may be determined from (1) conventional triaxial compression tests on vertical specimens and (2) either triaxial compression tests on horizontal specimens or conventional triaxial extension tests on vertical specimens. Other experiments may be employed, but the shear strengths from experiments in the mid-range of *b*-values may be affected by shear banding and these strengths are therefore not representative of the homogeneous deformation required for the failure criterion for cross-anisotropic soils. The test data employed should therefore be produced near $b = 0$ and/or $b = 1$. Actual parameter determination is described in detail by Lade (2007, 2008).

The cross-anisotropic failure criterion is compared with the experimental results for dense Santa Monica Beach sand in Fig. 11.22, where it essentially traces the same surface on the octahedral plane. Figure 11.22 also shows that shear banding, resulting in lower strength, occurred in the mid-range of *b*-values in all three sectors. Thus, 3D failure of soils is affected by (1) the intermediate principal stress, (2) shear banding in the mid-range of *b*-values, and (3) cross-anisotropy.

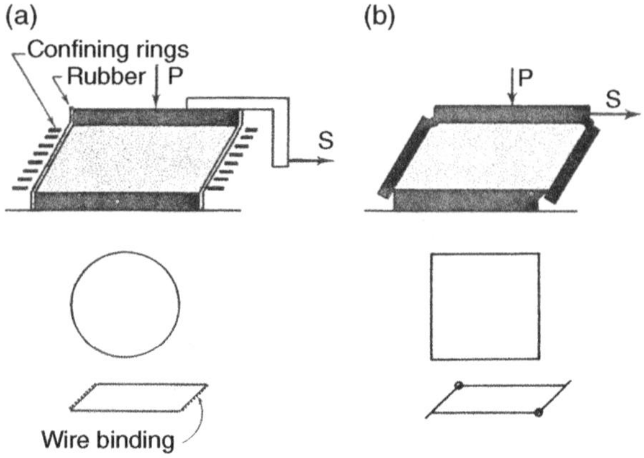

Figure 11.25 (a) SGI simple shear equipment. The hockey puck-shaped specimen is restrained from lateral expansion by a stack of circular flat rings fitting tightly around the specimen outside the membrane. Reproduced from Kjellman 1951 by permission of Geotechnique. In the NGI apparatus the rubber membrane is reinforced by a spiral of metal wire. Reproduced from Bjerrum and Landva 1966 by permission of Geotechnique. (b) The Cambridge simple shear apparatus employs a square specimen surrounded by stiff boundaries. Reproduced from Roscoe 1953 by permission of Geotechnique.

11.3 Tests with rotating principal stress directions

Equipment in which normal as well as shear stresses can be applied include: simple shear, directional shear, and torsion shear apparatus. The simple shear apparatus is sometimes used to obtain soil parameters for design, whereas the other two apparatus are used for research purposes. The states of stress that can be created in these pieces of equipment in many respects simulate those applied under field conditions where the direction of the major principal stress rotates as shear stresses are applied. This was illustrated in Figs 11.1 and 11.2.

11.3.1 Simple shear equipment

Two versions of the simple shear apparatus are available. In the Swedish Geotechnical Institute/Norwegian Geotechnical Institute (SGI/NGI) equipment a circular disk of soil is contained between end plates and a rubber membrane, as shown in Fig. 11.25(a). In the SGI equipment (Kjellman, 1951) the specimen is restrained from lateral expansion by a stack of circular flat rings fitting tightly around the specimen outside the membrane. In the apparatus developed at NGI (Bjerrum and Landva 1966), the rubber membrane is reinforced by a spiral of metal wire, as shown in Fig. 11.25(a). The Cambridge simple shear apparatus (Roscoe 1953) employs a square specimen surrounded by stiff boundaries, as indicated in Fig. 11.25(b). The simplicity of operation and the circular cross-section of the SGI/NGI apparatus, which makes it suitable for testing tube samples, account for its wide use for practical purposes.

Interpretation of simple shear tests

In both pieces of equipment, the specimen is first consolidated under a K_0-stress state and then sheared by application of shear load or displacement to the top or bottom of the specimen.

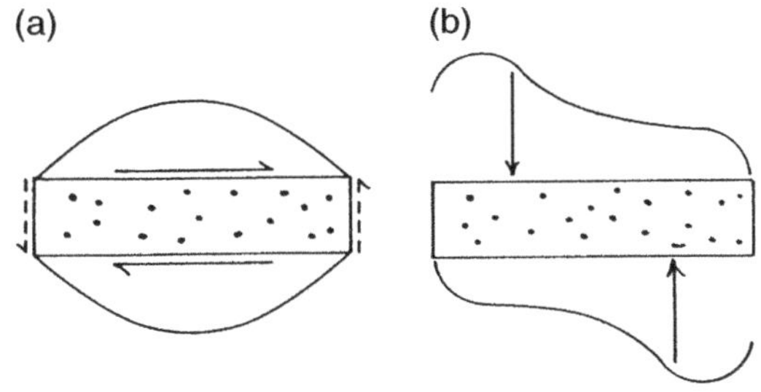

Figure 11.26 (a) Nonuniform distribution of shear stress due to absence of complementary shear stresses at the ends of the specimen and (b) nonuniform distribution of normal stresses to preserve moment equilibrium of specimen. Reproduced from Wood *et al.* 1979 by permission of ASTM International.

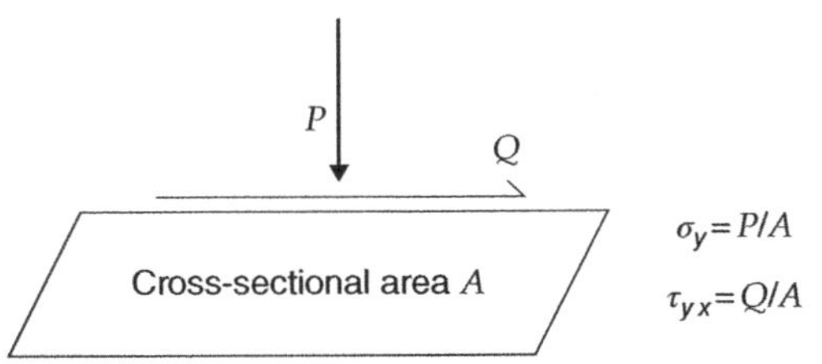

Figure 11.27 Load measurements in simple shear apparatus in which σ_y and τ_{yx} are the only components determined.

Deformation occurs in a plane and the test is therefore a plane strain test. Complementary shear stresses along the vertical surfaces of any significant magnitude cannot be sustained in any of the two types of equipment, and the state of stress acting on the specimen is consequently nonuniform as indicated in Fig. 11.26. Furthermore, the normal stresses acting on the sides of the specimen are unknown. Thus, the measured loads are limited to the vertical normal load and the applied shear load, as indicated in Fig. 11.27. In addition, the vertical and shear deformations of the specimen are measured. The interpretation of the results of simple shear tests is therefore difficult, and assumptions are required regarding the behavior of the soil.

Drained test on sand

The assumption most often employed for drained simple shear tests is that near failure the soil behaves as a plastic material, in which the principal axes of strain increment coincide with the principal axes of stress. Experimental evidence to this effect has been presented by Roscoe *et al.* (1967), Lade (1975, 1976), Wood *et al.* (1979), and Hong and Lade (1989a, b), as shown in Fig. 11.28. On this assumption Davis (1968) showed that:

$$\tan\varphi_d = \frac{\tau_{yx}}{\sigma_y} = \frac{\sin\varphi_p \cdot \cos\psi}{1-\sin\varphi_p \cdot \sin\psi} \tag{11.7}$$

in which φ_d and φ_p are the friction angles in direct shear or simple shear and in plane strain tests, and ψ is the angle of dilation defined as shown in Fig. 11.29:

$$\sin\psi = -\frac{\dot{\varepsilon}_v}{\dot{\gamma}_{max}} = -\frac{\dot{\varepsilon}_1+\dot{\varepsilon}_3}{\dot{\varepsilon}_1-\dot{\varepsilon}_3} \tag{11.8}$$

Typical values of the angle of dilation, ψ, are up to 30° lower than the friction angle, φ. Interpretation of actual test data according to Eq. (11.7) was presented by Frydman (1974), as given in Table 11.1. They show that this equation represent the real sand behavior with good accuracy.

For ideal rigid-plastic Mohr–Coulomb material with associated plastic flow, the angle of dilation equals the friction angle ($\psi = \varphi$). This condition leads to the classical interpretation of a direct shear or simple shear test:

$$\tan\varphi_d = \tan\varphi_p = \left(\frac{\tau_{yx}}{\sigma_y}\right)_{max} \tag{11.9}$$

With this solution follows that the horizontal normal stress (σ_x) is greater than the applied vertical normal stress (σ_y) at failure ($\sigma_x > \sigma_y$).

On the other hand, at the critical state at which no further volume change occurs, the angle of dilation is zero, and Eq. (11.7) reduces to:

$$\tan\varphi_d = \left(\frac{\tau_{yx}}{\sigma_y}\right)_{cv} = \sin\varphi_p \tag{11.10}$$

In this case the horizontal normal stress equals the applied vertical normal stress at failure ($\sigma_x = \sigma_y$).

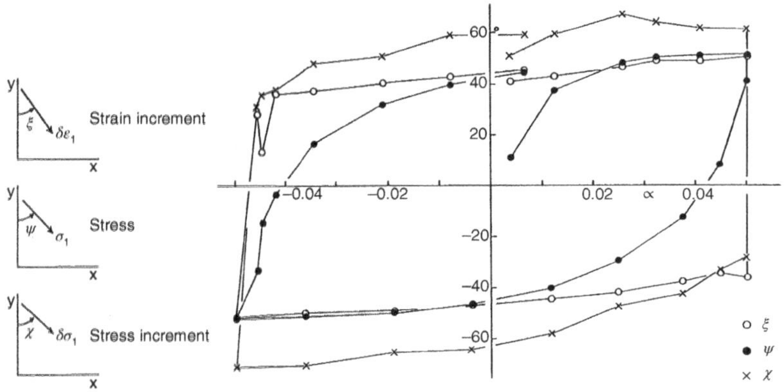

Figure 11.28 Variation of directions of strain increment, stress and strain increment in simple shear test on dense sand (after Wood *et al.* 1979).

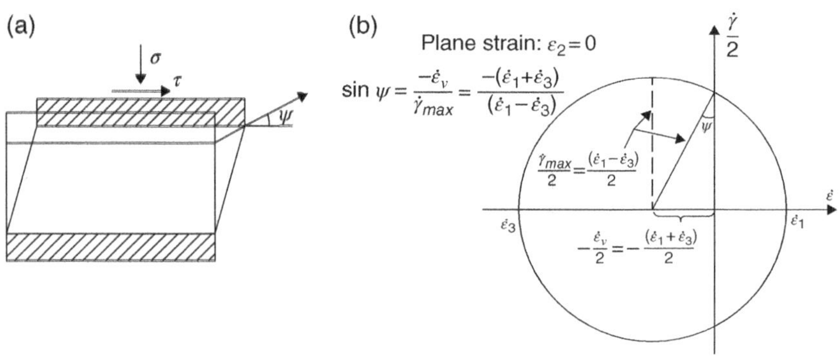

Figure 11.29 (a) Deformation of dilating sand in simple shear and (b) Mohr's circle for strain increments under plane strain conditions with the definition for the angle of dilation.

These two cases ($\psi < \varphi$ and $\psi = \varphi$ at failure) represent the extreme cases of interpretation and they show that $\sigma_x > \sigma_y$ at failure, while $\sigma_x = \sigma_y$ at the critical state reached at large shear strains.

Undrained tests on clay

In undrained tests on clay, the shear strength is most often presented in relation to the vertical consolidation stress such that:

$$\frac{s_u}{\sigma_{vc}'} = \left(\frac{\tau_{yx}}{\sigma_y}\right)_{max} \tag{11.11}$$

This expression is then used directly with the vertical consolidation pressure in the field to calculate the undrained shear strength for use in design procedures.

11.3.2 Directional shear cell

The directional shear cell is a piece of apparatus devised by Arthur *et al.* (1977a, b, 1981), but relatively few copies of this apparatus have been made (e.g., Arthur *et al.* 1981; Sture *et al.* 1985). Figure 11.30 shows the working principle of this device. It overcomes the problems of the simple shear device in that all stresses are uniformly distributed over the specimen surfaces and they are measured directly. It also produces uniform strain conditions inside the specimen. Only limited test data have been presented in the literature from the directional shear cell. However, this device is quite complex and useful only for research studies.

A variation on this piece of apparatus was presented by Joer *et al.* (1992, 1998) in which scissor jacks are used to distribute and produce

Table 11.1 Comparison of experimental and predicted values of φ_d for different granular materials. Reproduced from Frydman 1974 by permission of ASCE

Material	Data	φ_d, experimental	Predicted values	
			ψ	φ_d
Feldspar (dense)	$\varphi_u = 35.5$ $\varphi_{cv} = 41$ $\varphi_p = 53$	44	15.6	44.4
Feldspar (loose)	$\varphi_u = 35.5$ $\varphi_{cv} = 41$ $\varphi_p = 41.5$	32.5	6.3	35
Mersey sand (dense)	$\varphi_u = 26$ $\varphi_{cv} = 32$ $\varphi_p = 46$	42	18.3	41.3
Mersey sand (loose)	$\varphi_u = 26$ $\varphi_{cv} = 32$ $\varphi_p = 32$	28	5.7	29
Ballotini (dense)	$\varphi_u = 17$ $\varphi_{cv} = 24$ $\varphi_p = 39$	36.5	20.6	37
Ballotini (loose)	$\varphi_u = 17$ $\varphi_{cv} = 24$ $\varphi_p = 27$	25	9.1	25.8

φ_d, friction angle in direct shear or simple shear; φ_u, undrained friction angle; φ_{cv}, friction angle at constant volume; φ_p, friction angle in plane strain test.

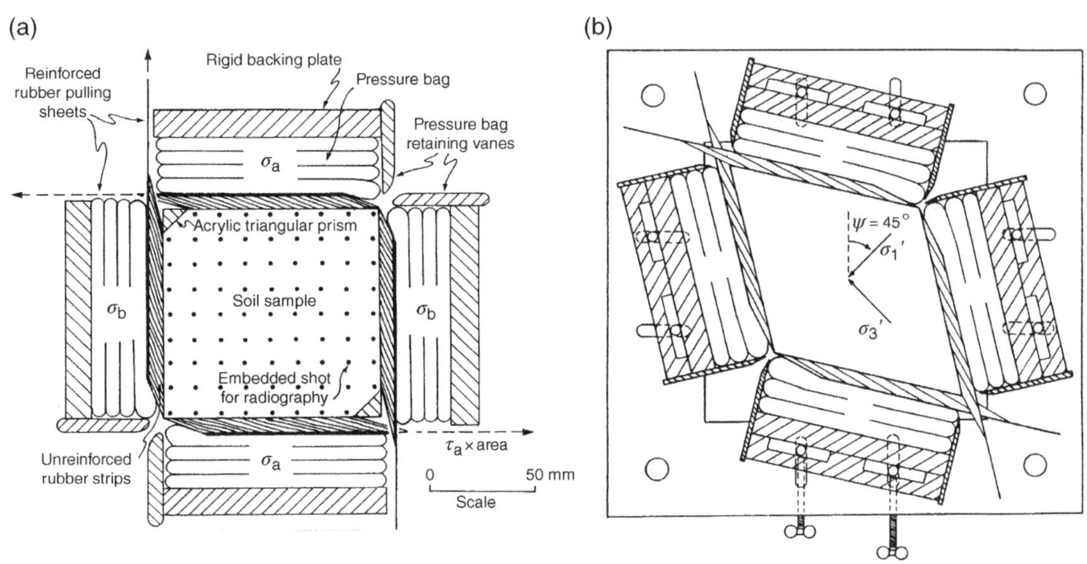

Figure 11.30 Directional shear cell in (a) undeformed and (b) deformed condition (after Arthur *et al.* 1981).

uniform deformation on the boundaries of a rectangular specimen with dimensions of 55 cm by 41 cm. These sides may be extended to 69 cm and 55 cm, respectively. The materials tested in this apparatus are rods with different cross-sections, which simulate the soil particles.

11.3.3 Torsion shear apparatus

Torsion shear equipment in which normal and shear stresses are applied to a hollow cylinder specimen is the preferable and most reliable apparatus in which to study effects of principal stress rotation. Figure 11.31 shows the stresses applied to the hollow cylinder specimen. The problem of applying correct complementary shear stresses, known from the simple shear apparatus, does not exist in the torsion shear apparatus, in which they are automatically generated to maintain equilibrium. Existing equipment differs mainly in the way the axial, normal stresses and shear stresses are generated outside the cell and applied to the specimen. Hollow cylinder specimens are employed in all cases, but their dimensions differ. Although complex in construction and operation, many pieces of apparatus have been designed and built for studying soil behavior under static and cyclic loading conditions (e.g., Broms and Casbarian 1965; Lade 1981; Sayao and Vaid 1991; Saada *et al.* 1994).

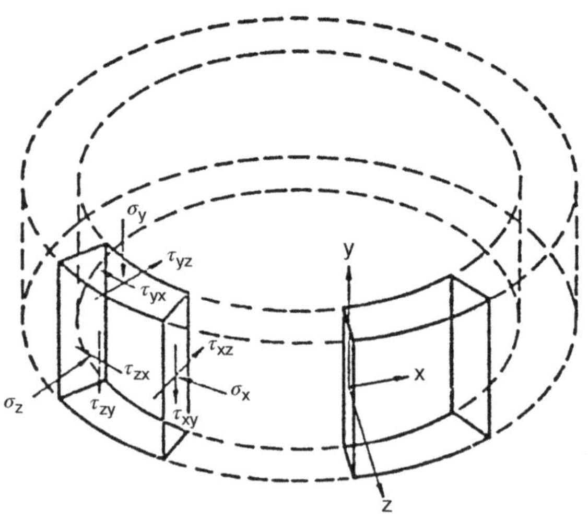

Figure 11.31 Stress components in a Cartesian coordinate system for a hollow cylindrical specimen.

Effects of stress rotation that may be studied in the torsion shear apparatus include: (a) the directions of strain increments relative to directions of stress increment and directions of stress in physical space; (b) plastic yielding and directions of plastic strain increments in various stress spaces suitable for depicting results of torsion shear tests to determine yield and plastic potential surfaces; (c) stress–strain and strength variation with β and with b on initially isotropic and cross-anisotropic soils; (d) soil behavior during cyclic loading; and (e) soil behavior during large 3D stress reversals.

Figure 11.32 shows a diagram of a torsion shear apparatus designed and built by Lade (1981). The hollow cylinder specimen has an average diameter of 20 cm, a wall thickness of 2 cm, and heights from 5 to 40 cm can be accommodated. The specimen is contained between outside and inside rubber membranes and between cap and base rings. The entire setup is contained in a pressure cell, and either (1) the same confining pressure is applied to the inside and the outside surfaces of the specimen or (2) higher or lower pressures are applied on the inside surface than to the outside surface.

If the same pressure is applied inside and outside the hollow cylinder specimen, then the inclination, β, of the major principal stress, σ_1, relative to vertical is related to the value of b as follows (Lade *et al.* 2008):

$$b = \sin^2 \beta \qquad (11.12)$$

To study the effects on soil behavior of the intermediate principal stress, indicated by b, and the effects of major principal stress inclination, β, separately, it is necessary to apply different pressures inside and outside the hollow cylinder specimen. This requires separation of the volume inside the hollow cylinder specimen, so it must be pressurized separately from the outside cell.

Although the difference in design is small, the apparatus shown in Fig. 11.32 applies the same pressure inside and outside the hollow cylinder. Shear stresses and vertical deviator stresses can be applied to the top and bottom of the hollow cylinder. The vertical load is transferred to the specimen by a cap plate, which is

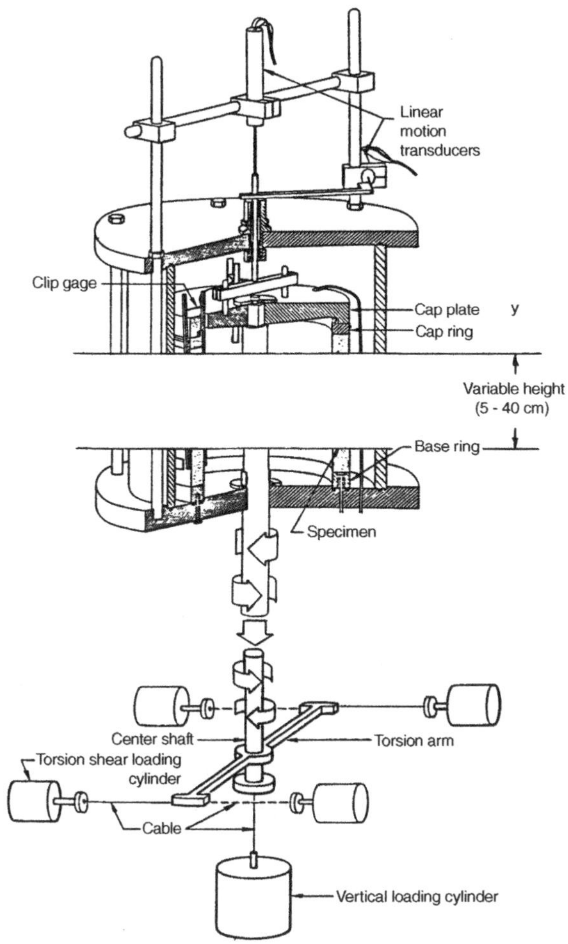

Figure 11.32 Torsion shear apparatus (after Lade 1981).

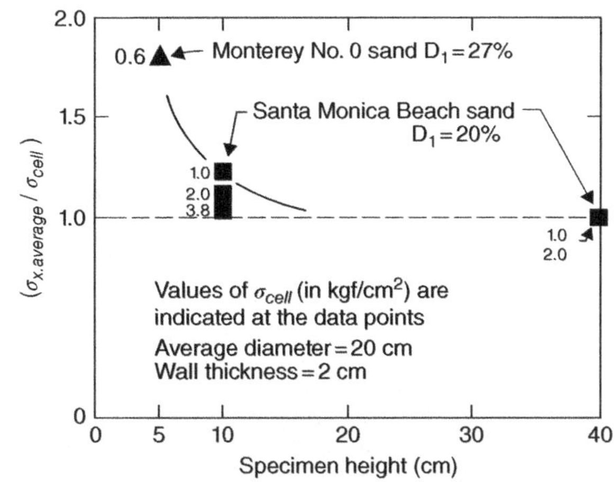

Figure 11.33 Effects of end restraint in torsion shear tests on a hollow cylinder specimen. Reproduced from Lade 1981 by permission of Elsevier.

connected to a shaft through the bottom plate of the cell, as seen in Fig. 11.32. The resulting deviator stress together with the cell pressure provides for a vertical normal stress larger or smaller than the confining pressure. The torque is transferred to the specimen through the center shaft and the cap plate. The shear stresses due to the torque cause rotation of the principal stress directions (when the vertical deviator stress is different from zero) and they generate a stress state with three unequal principal stresses. The cell pressure is always the intermediate principal stress, σ_2. In the limits of (1) compressive deviator stresses and no shear stresses, $\sigma_2 = \sigma_3$ and (2) extension deviator stresses and no shear stresses, $\sigma_2 = \sigma_1$.

During a torsion shear test, the vertical load, the torque, the cell pressure (or the inside and the outside cell pressures), the vertical normal deformation, the volume change of the inside cell, the volume change of the specimen, and the shear deformation are measured. This allows calculation of all stresses and strains in the hollow cylinder specimen.

Stress concentrations can occur in hollow cylinder specimens, especially if the height of the specimen is too small or if the ratio of inside and outside radii of the specimen is too low. The tangential, horizontal normal stress, σ_θ, in the cylinder wall (Fig. 11.31) is not measured, but assumed to be equal to the average of the inside and outside cell pressures. This is correct only if the height of the specimen is sufficient to reduce the influence of restraints at the cap and base rings, which are supplied with frictional ends to transfer the applied shear stresses, but also restrains the specimen from moving out horizontally. The restraint is most pronounced for dilating sands. Analysis of sufficient height was performed based on experiments on sand (Lade 1981) for the horizontal dimensions of the hollow cylinder specimen shown in Fig. 11.32. The results are shown in Fig. 11.33, and they indicate that a 20–25 cm tall specimen is sufficient to reduce the restraints at the cap and base rings.

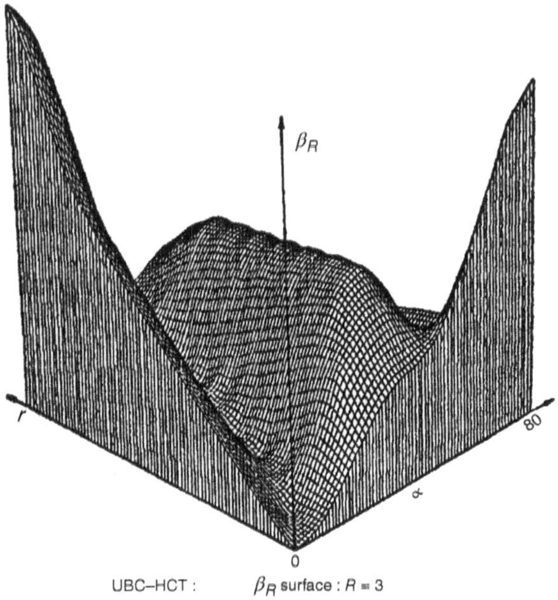

Figure 11.34 Diagram showing nonuniformity of stresses. $b = (\sigma_2 - \sigma_3)/(\sigma_1 - \sigma_3)$; $\beta°$, inclination of the major principal stress to the vertical; β_R, nondimensional expression for stress nonuniformity across the specimen wall; and R, principal stress ratio. Reproduced from Sayao and Vaid 1991 by permission of Elsevier.

Sayao and Vaid (1991) performed linear elastic analyses of the stress nonuniformities across the cylinder wall in specimens with different inside and outside pressures, and the results of these analyses are shown in Fig. 11.34. The most severe stress nonuniformities occur near the corners where (1) $b = 0.0$ and $\beta = 90°$ and (2) $b = 1.0$ and $\beta = 0°$, while the smallest stress nonuniformities occur for inclinations, β, and $b = 0.5$ and for all b-values and $\beta = 45°$.

Based on the results of these analyses Sayao and Vaid (1991) reviewed the dimensions of existing torsion shear apparatus and produced the diagram in Fig. 11.35, indicating the optimal dimensions of the hollow cylinder specimen for torsion shear testing.

Drained tests on sand

Examples of results from drained experiments on sand are indicated below. Results of such tests may be depicted and analyzed in other types of diagrams depending on the goal of the analyses.

Figure 11.36 shows the directions of the major principal stress and strain increments in physical space at failure from experiments on medium

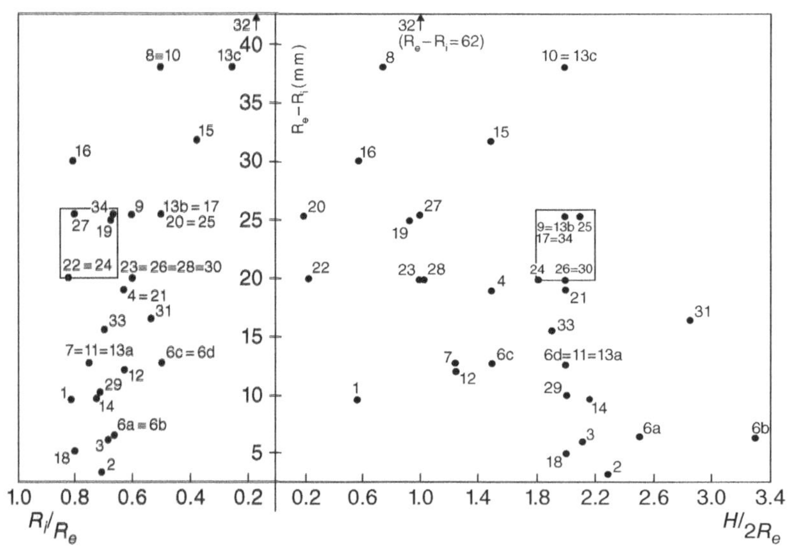

Figure 11.35 Diagram showing dimensions of various torsion shear apparatus. Sayao and Vaid (1991) reviewed the dimensions of existing torsion shear apparatus and produced this diagram indicating the optimal dimensions of the hollow cylinder specimen for torsion shear testing. R_i, internal specimen radius; R_e, external specimen radius; and H, height of the hollow cylinder specimen. Reproduced from Sayao and Vaid 1991 by permission of Elsevier.

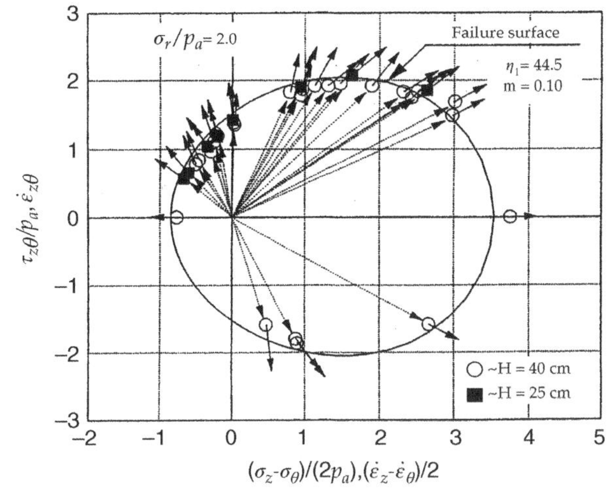

Figure 11.36 Comparison of directions of principal stress with directions of principal plastic strain increments at failure in physical space during rotation of principal stresses in torsion shear tests on Santa Monica Beach sand. Reproduced from Lade *et al.* 2009 by permission of Elsevier.

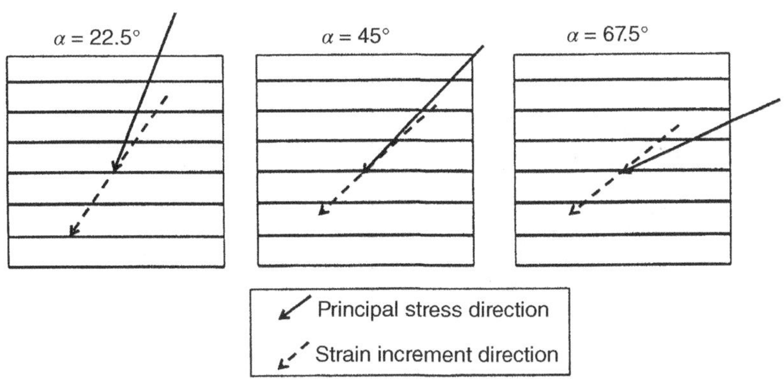

Figure 11.37 Schematic patterns of principal strain increment directions and principal stress directions observed in torsion shear tests on dense, cross-anisotropic Nevada sand. Reproduced from Rodriguez and Lade 2014 by permission of Elsevier.

dense Santa Monica Beach sand (Lade *et al.* 2009). Data presented by Rodriguez and Lade (2014) showed similar results for initially cross-anisotropic, dense, fine Nevada sand, and the directions were not coinciding, even at failure, as indicated in Fig. 11.37. Modeling of this behavior may require a cross-anisotropic elasto-plastic constitutive model framework.

Figure 11.38 shows the friction angle plotted versus the b-value at failure for the torsion shear tests on medium dense Santa Monica Beach sand. The failure surface described by the cross-anisotropic failure criterion given in Eq. (11.5) is also shown.

Drained tests on cross-anisotropic sand

Torsion shear tests on dense, fine Nevada sand deposited by dry pluviation formed a highly cross-anisotropic deposit. In these tests different inside and outside pressures were applied to perform tests with all combinations of the independent parameters of b and β over the entire range of these two variables, that is $0 \le b \le 1$ and $0° \le \beta \le 90°$. The value of b was changed by increments of 0.25 and the value of β was changed in increments of 22.5°. The consequent 3D variation in friction angle is seen in Fig. 11.39.

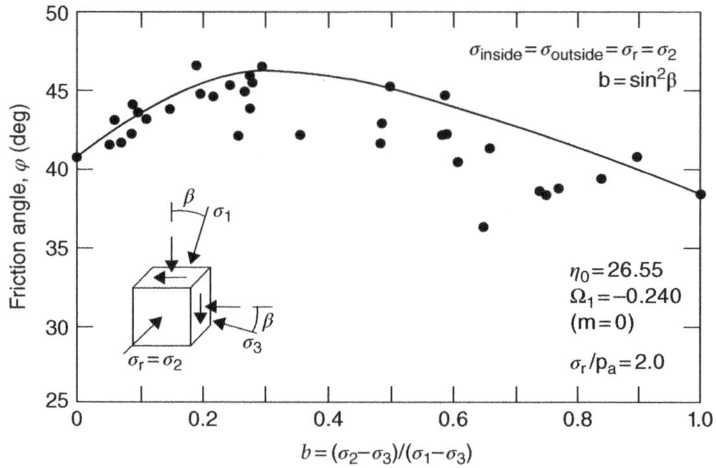

Figure 11.38 Comparison of failure criterion with torsion shear test data for medium dense, cross-anisotropic Santa Monica Beach sand. Shear banding in the hardening regime reduces friction angles in mid-ranges of *b*-values. Reproduced from Lade 2008 by permission of Canadian Science Publishing.

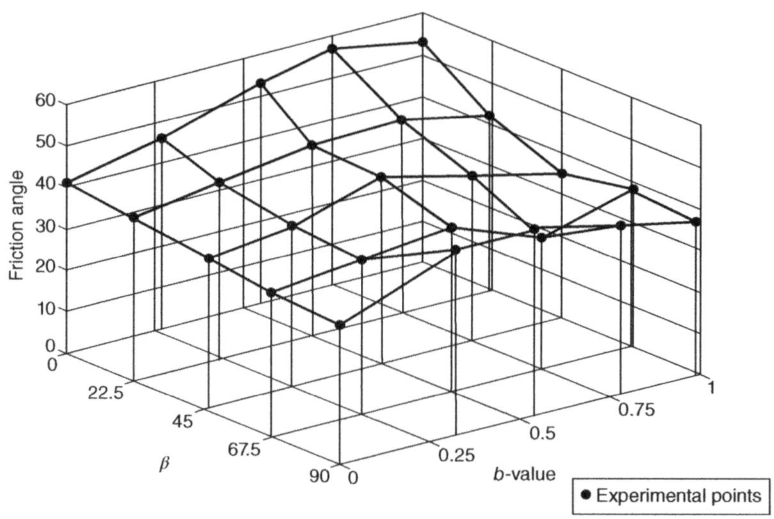

Figure 11.39 Three-dimensional variation in friction angle with *b* and principal stress inclination β for torsion shear tests on dense Nevada sand. Reproduced from Lade *et al.* 2014 by permission of ASCE.

There is a very significant variation in friction angles from these experiments. The highest friction angle of 57°, which occurs at $b = 0.75$ and $\beta = 0°$, is 25° higher than the lowest friction angle of 32°, which was obtained at $b = 0.75$ and $\beta = 67.5°$. Thus, there is a drastic drop in strength at $b = 0.75$ as the σ_1-direction changes from $\beta = 0°$ (vertical) to $\beta = 67.5°$. This is followed by a small increase from $\beta = 67.5°$ to $\beta = 90°$. Figure 11.39 also indicates a dip or a valley in the failure surface at $\beta = 67.5°$. This is likely due to alignment of the direction of shear banding with the horizontal plane, which is the weakest due to the bedding in this direction.

The triaxial compression test on a vertical specimen, corresponding to $b = 0.00$ and $\beta = 0°$, produced a friction angle of 41°, as seen in Fig. 11.39. Such a constant friction angle is often applied in a given geotechnical engineering project, but it could not possibly substitute

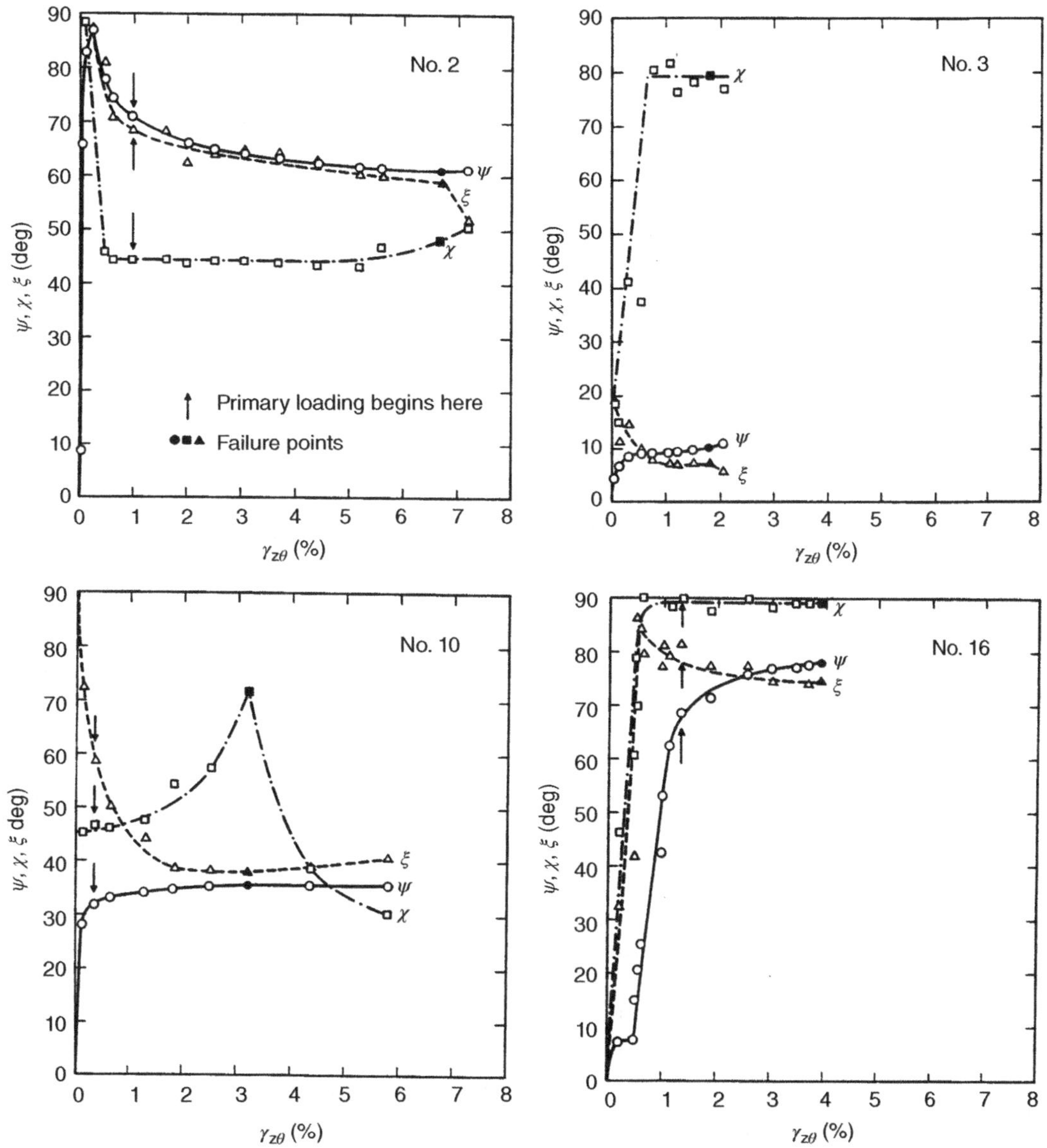

Figure 11.40 Directions of the major principal stress and strain increments in physical space at failure from experiments on reconstituted, normally K_0-consolidated Edgar Plastic Kaolinite. Reproduced from Hong and Lade 1989b by permission of ASCE.

for the strength variation with principal stress direction as observed in the experiments presented here.

Undrained tests on clay

Interpretations of results from undrained torsion shear tests on reconstituted, normally K_0-consolidated Edgar Plastic Kaolinite and on undisturbed, normally K_0-consolidated San Francisco Bay Mud are shown as examples of results from such tests.

Figure 11.40 shows the directions of the major principal stress and strain increments in physical space at failure from experiments on reconstituted, normally K_0-consolidated Edgar Plastic Kaolinite (Hong and Lade 1989b). Similar to the

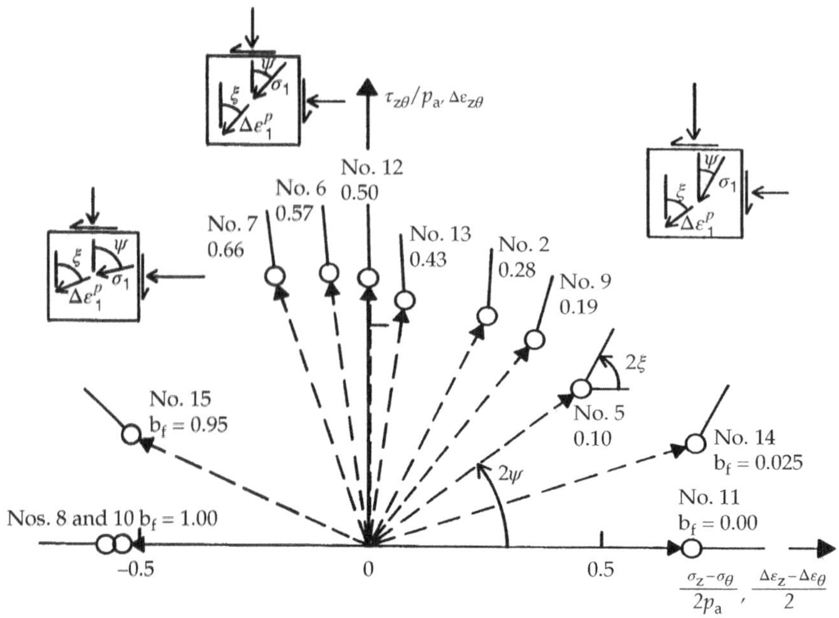

Figure 11.41 Data from torsion shear tests on undisturbed, normally K_0-consolidated San Francisco Bay Mud showing that directions of strain increments at failure are clearly not aligned with the directions of stress, thus indicating a significant influence of the cross-anisotropic fabric on the results. Reproduced from Lade and Kirkgard 2000 by permission of Elsevier.

results from experiments on sand, the directions are essentially coinciding at failure, thus indicating behavior that may be characterized by an isotropic elasto-plastic constitutive model.

For comparison with the results in Fig. 11.40, the data from torsion shear tests on undisturbed, normally K_0-consolidated San Francisco Bay Mud are shown in Fig. 11.41 (Lade and Kirkgard 2000). The directions of strain increments at failure are clearly not aligned with the directions of stress, thus indicating a significant influence of the cross-anisotropic fabric on the results.

The variation of the normalized, undrained shear strength with major principal stress inclination of the undisturbed, K_0-consolidated San Francisco Bay Mud is shown in Fig. 11.41. This variation of normalized, undrained shear strength is similar to that shown in Fig. 11.17(a) for Edgar Plastic Kaolinite. The highest value of normalized shear strength is obtained for vertical specimens, that is the usual and most convenient orientation of specimens tested in triaxial compression. The strength decreases rapidly as the stress is inclined relative to vertical, and at an inclination of 30° and beyond the normalized undrained shear strength is approximately constant at about 80% of the value for the vertical specimen. However, the undrained shear strength reduces slightly with increasing b-value, and it is lowest at the extension condition, where $b = 1.0$.

11.3.4 Summary and conclusion

It is clear from the observed variation in stress–strain, volume change, pore pressure, and effective strength behavior in 3D tests that the direction of the major principal stress is very important when testing cross-anisotropic soils. Experimentation is still required to determine the behavior variation with void ratio, effective confining pressure, relative value of σ_2, and inclination of the major principal stress relative to the bedding planes.

The above review has dealt with monotonic loading of soils under various 3D loading conditions. Unloading and reloading and stress reversals involving large stress changes under 3D conditions as well as pure stress rotation have not been discussed here. These subjects have been studied experimentally and the results indicate interesting soil behavior that requires modeling by more advanced mathematical models such as kinematic hardening. This is beyond the scope of the present book.

Appendix A: Manufacturing of Latex Rubber Membranes

Most soil specimens to be tested in triaxial compression are cylindrical, but some specimens for three-dimensional testing may have shapes with rectangular cross-sections. However, a cylindrical membrane can be coerced into just about any shape to surround the specimen using a forming mold with distributed vacuum on the inner surfaces, so it may not be necessary to produce membranes other than those with cylindrical shapes. Thus, corners and edges may not be required.

The fabrication process for latex rubber membranes has been described by Vaid and Campanella (1973) and comments and improvements were made by Raymond and Soh (1974). Additional observations were made by the author and his students and these are included in the following description.

A.1 The process

The process of making a rubber latex membrane consists of immersing a smooth mold made to the shape of the desired membrane into the liquid rubber latex and withdrawing it, after which the adhering film is allowed to dry in air at room temperature followed by curing at elevated temperature. The membrane is dusted with talcum powder and stripped off the mold while more talcum powder is applied to the inner surface to avoid the membrane sticking to itself.

A.2 Products for membrane fabrication

1. Ethyl alcohol (this is the same as ethanol).
2. Calcium nitrate: use "refined grade" (anything else is too expensive).
   The best product is 75% calcium nitrate trihydrate.
3. Latex: natural rubber latex compound.
4. Dipping tanks: one heavy-duty tank with lid to be fitted with a slow agitation device, often a slow moving paddle, to keep the latex from forming a thick skin on the surface, that is skimming on top. In addition, one tank to hold the calcium nitrate, in which the mold can be dipped for creation of a coagulation layer for the rubber membrane.
5. Aluminum form on which to create the rubber membrane.

Triaxial Testing of Soils, First Edition. Poul V. Lade.

A.3 Create an aluminum mold

The mold on which the membrane is created does not have to be circular in shape, but most shapes may be created out of circular membranes. The mold may be formed from a flat piece of aluminum plate with rounded edges, say with a radius of 0.5 in. (1.27 cm). Other metals like brass may also be used, but plastic and wood are not recommended. Sharp edges should be rounded to prevent thin areas during dipping. Latex will shrink as it dries and cures, and when the membrane is peeled off the mold it will be smaller than the mold. The amount of shrinkage depends on the type of latex used. Shrinkage from 4 to 7% may occur. Thus, the mold should be designed to be larger by the amount of shrinkage.

After the mold has been fabricated it may be outfitted with two eye-bolts at the top to affix a cord or rope for dipping and handling. This is to avoid touching the mold with the membrane during the process. Ensure the bolts are positioned so that the mold hangs in a relatively plumb condition.

The mold should be sanded using progressively finer sand paper, starting with 100 grit and working to a 400 grit wet sand paper. Emory cloth may not work as well as the wet/dry sand paper. After the mold is completely smooth, use metal polish to finish smoothing the surface.

A.4 Two tanks

One tank (stainless steel or PVC) should be filled with ethanol and then thoroughly mixed with 30% calcium nitrate, which acts as a coagulant for the rubber latex. The solution should be relatively clear when mixing is complete. The concentration of salt in the alcohol controls the thickness of the latex film deposit. Vaid and Campanella (1973) report that a 20% concentration by weight would result in membranes with thicknesses of 0.010–0.012 in. (0.25 mm) from a single dip. Higher or lower concentrations would produce thicker or thinner membranes.

The other tank should be filled with liquid latex. This tank should have sufficient capacity to allow submerging the aluminum mold completely, or at least until the membrane section is covered. A rotating paddle should keep the liquid latex in a state of gentle agitation to avoid skimming on the top surface. Care should be exercised to avoid any air bubbles from entering the liquid, because they may result in small pinholes in the membrane.

Both tanks should have tight fitting lids to keep the latex from forming a skim coat on the surface and curing. With a tight fitting lid, the latex may be stored for 2–3 years.

A.5 Mold preparation

The entire mold should be degreased and cleaned using soap and then acetone to make sure all oils and other impurities are removed from the surface of the mold. The mold should only be touched using sterile gloves from this point forward.

The mold should be preheated by placing it in an oven at about 100°C for 6 h or overnight.

A.6 Dipping processes

Make sure to work in a well-ventilated area as the ammonium that comprises much of the liquid latex compound is highly volatile and will form a considerable amount of vapor during the dipping process.

Also, a hook to hang the mold during drying should be present. It is advisable to place a mat beneath the hook as a little latex may drip during the curing.

Remove the mold from the oven and dip it directly in the ethanol mixture. Hang the mold on the hook and observe that the ethanol should "flash" off the mold leaving a uniform powder coating of calcium nitrate. If the mold has been sufficiently heated, this should take less than 5 min.

Depending on how well the latex dipping tank is agitated, the mold may be dipped directly into the liquid latex, or it may be allowed to cool, say for about 15 min, before dipping it in the liquid latex. If dipped directly, the heat will make the latex cure very quickly.

Dip the mold smoothly and evenly. Try to let the mold enter the latex evenly and as quickly as possibly without drawing air bubbles down into the latex.

Let the mold sit in the liquid latex for 20–60 s for a membrane thickness of between 0.012 in. (at the top) and 0.025 in. (at the bottom of the mold).

Pull the mold out of the latex at a rate of 1 cm/s, while any excess latex flows back into the tank.

Hang the mold with the newly formed membrane on the hook to dry for a minimum of 12 h. The curing time will vary based on the temperature of the mold, the ambient temperature, and the ventilation.

Make sure to dust the outside of the membrane with talcum powder (baby powder) prior to stripping it from the mold as the new membrane will tend to stick to itself otherwise. Also dust the inside of the membrane to make it "non-stick".

The vulcanization process can be completed by placing the membrane on paper or cloth in the oven at 140°C for about 12 h. Alternatively, the membrane may be hung in the laboratory and allowed to air dry. This will not achieve the 100% vulcanization, but there is no risk of "over-cooking" the membrane. Insufficient curing tends to leave membranes with excessive leakage rates.

The membrane may then be immersed in a water bath to eliminate any excess coagulant from the surfaces.

A.7 Post production

Inspect the membrane for any areas of weakness. Sometimes such an area is located near the edge of the mold. Any area near the middle of the mold that consistently produces poor quality must be re-sanded and thoroughly cleaned. It is likely that the calcium nitrate solution is not being spread evenly in this area due to surface irregularities.

A.8 Storage

The finished membranes should ideally be stored in a refrigerator, where the temperature is low, but not freezing, and where it is dark. This is to prevent or decrease the rate of deterioration, which occurs on account of light and higher temperatures.

A.9 Membrane repair

Leaky membranes may be repaired by sealing the holes with liquid latex, which then dries on the surface. Holes may be discovered by stretching the membrane up against the light. The holes may be circled by a waterproof felt pen to be able to find them again.

Appendix B: Design of Diaphragm Load Cells

It is on occasion practical to be able to design a diaphragm load cell to install at certain locations in the triaxial equipment. The diaphragm load cell is convenient, because this relatively flat disk may fit into the cap of the triaxial specimen or it may fit under the base without taking too much space. Therefore, the design of such load cells is reviewed here (Timoshenko and Woinowski-Krieger 1959).

B.1 Load cells with uniform diaphragm

Figure B.1 shows the deflection of a diaphragm load cell with uniform thickness t, and loaded in the center with P while supported along the circular rim. The maximum normal stress is proportional to P and inversely proportional to the thickness squared:

$$\sigma_{max} = \frac{k \cdot P}{t^2} \tag{B.1}$$

in which k is a non-dimensional factor related to the ratio a/b, in which a and b are the inner and outer radii of the diaphragm, respectively. The thickness, t, of the diaphragm is found from Eq. (B.1):

$$t = \sqrt{\frac{k \cdot P}{\sigma_{max}}} \tag{B.2}$$

And the deflection of the center point, w_{max}, is:

$$w_{max} = k_1 \cdot \frac{P \cdot a^2}{E \cdot t^3} \tag{B.3}$$

in which E is Young's modulus of the material used for the diaphragm load cell, and k_1 is related to the ratio a/b as follows:

a/b	1.25	1.5	2.0	3.0	4.0	5.0
k	0.115	0.220	0.405	0.703	0.933	1.13
k_1	0.00129	0.0064	0.0237	0.062	0.092	0.114

The physical properties (Young's modulus, E and yield stress, σ_{yield}) of metals for load cells are as follows:

Aluminum: E = 70 GPa; σ_{yield} = 415 MPa
Stainless steel: E = 200 GPa; σ_{yield} =520 MPa
Beryllium copper: E = 125 GPa; σ_{yield} =345 MPa

It is suggested to use 80–90% of the yield stress as σ_{max}.

Triaxial Testing of Soils, First Edition. Poul V. Lade.

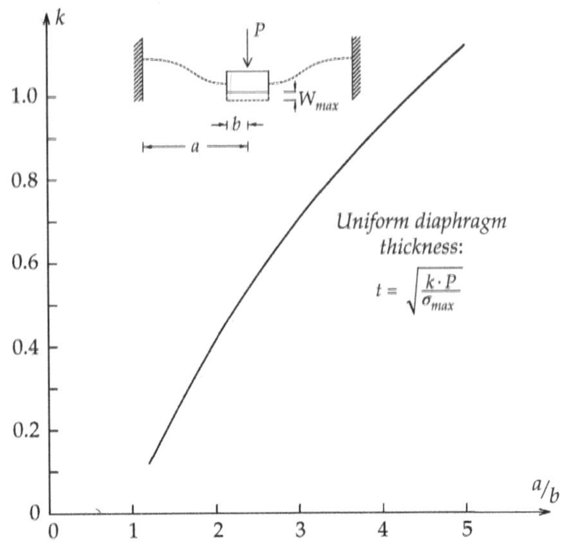

Figure B.1 Schematic deflection of a diaphragm load cell with uniform thickness loaded in the center with P and supported along the circular ring, and factor k used to calculate the uniform thickness.

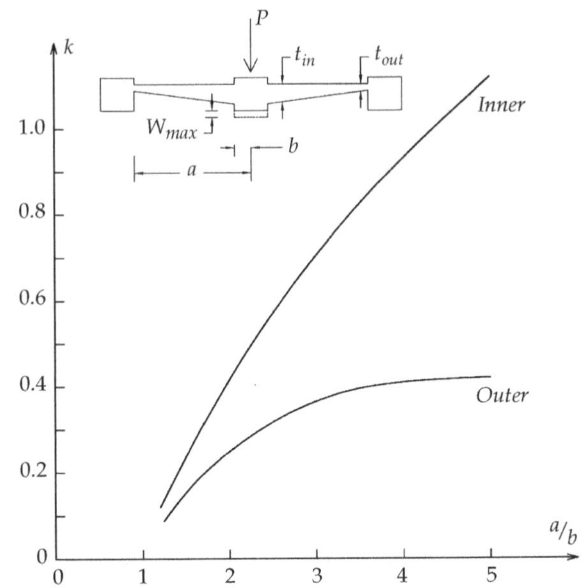

Figure B.2 Factors $k(in)$ and $k(out)$ for calculation of the diaphragm thicknesses at the inner and outer diameters.

B.2 Load cells with tapered diaphragm

The following pertain to a load cell with tapered diaphragm. The inner and outer thicknesses are calculated on the basis that the same moment per unit length of the circumference is achieved, while the signal from the strain gages is maximized.

a/b	1.25	1.5	2.0	3.0	4.0	5.0
$k(in)$	0.115	0.220	0.405	0.703	0.933	1.13
$k(out)$	0.0984	0.168	0.257	0.347	0.390	0.415
α	0.0013	0.0064	0.0237	0.062	0.092	0.114

The relations between a/b and $k(in)$ and $k(out)$ are shown in Fig. B.2.

The value of α is used to calculate the deflection, which allows design of overload protection:

$$w_{max} = \frac{\alpha \cdot P \cdot a^2}{E \cdot t^3} \tag{B.4}$$

in which $t = t_{avg} = (t_{in} + t_{out})/2$ is used.

B.3 Example: Design of 5 kN beryllium copper load cell

The outside diameter and the diameter of the inside knob are estimated on the basis of the space available for the load cell. Thus, the outside diameter is set to 10.00 cm and the knob (inner) diameter is 2.00 cm. The outside diameter of the diaphragm is $2a = 6.86$ cm and the inside diameter is $2b = 2.0$ cm for a ratio of $a/b = 3.43/1.00 = 3.43$.

For a *uniform thickness diaphragm* the value of $k = 0.81$ and $k_1 = 0.075$, which produces:

$$t = \sqrt{\frac{0.81 \cdot 5 \cdot 10^3 \mathrm{N}}{300 \cdot 10^6 \mathrm{N/m^2}}} = \sqrt{0.0000135}$$

$$= 0.00367\mathrm{m} = 3.67\mathrm{mm}$$

$$w_{max} = 0.075 \cdot \frac{5 \cdot 10^3 \mathrm{N} \cdot \left(3.43 \cdot 10^{-2} \mathrm{m}\right)^2}{125 \cdot 10^9 \mathrm{N/m^2} \cdot \left(0.00367\mathrm{m}\right)^3}$$

$$= 0.0000714\mathrm{m} = 0.0714\mathrm{mm}$$

For a *tapered diaphragm* the value of *k(in)* is the same as for the uniform diaphragm, while *k(out)* = 0.37 for the tapered diaphragm. Therefore,

$$t_{outer} = \sqrt{\frac{0.37 \cdot 5 \cdot 10^3 \mathrm{N}}{300 \cdot 10^6 \mathrm{N/m^2}}} = \sqrt{0.0000062}$$

$$= 0.00248\mathrm{m} = 2.48\mathrm{mm}$$

$$\mathrm{t}_{inner} = 3.67\mathrm{mm}\,(\text{as for the uniform diaphragm})$$

$$t_{avg} = \tfrac{1}{2}(2.48 + 3.67) = 3.08\mathrm{mm}$$

$$w_{max} = 0.075 \cdot \frac{5 \cdot 10^3 \mathrm{N} \cdot \left(3.43 \cdot 10^{-2}\mathrm{m}\right)^2}{125 \cdot 10^9 \mathrm{N/m^2} \cdot (0.00308\mathrm{m})^3}$$

$$= 0.000121\mathrm{m} = 0.121\mathrm{mm}$$

A gap between the knob and the cover plate for the load cell is therefore designed to be 0.121 mm, so the diaphragm lid stops at that deflection.

B.3.1 Punching failure

The load may produce punching failure at the *inner surface*, where the area is

$$A = 2\pi \cdot b \cdot t_{inner} = 2\pi \cdot 0.01 \cdot 0 \cdot 00367$$

$$= 0.000231\mathrm{m}^2$$

$$\text{Shear stress} = \frac{5 \cdot 10^3 \mathrm{N}}{0.000231\mathrm{m}^2}$$

$$= 21.6 \cdot 10^6 \mathrm{Pa} < 300 \cdot 10^6 \mathrm{Pa}$$

At the *outer surface*:

$$A = 2\pi \cdot a \cdot t_{outer} = 2\pi \cdot 0.0343 \cdot 0.00248$$

$$= 0.000535\mathrm{m}^2$$

$$\text{Shear stress} = \frac{5 \cdot 10^3 \mathrm{N}}{0.000535\mathrm{m}^2}$$

$$= 0.935 \cdot 10^6 \mathrm{Pa} < 3 \cdot 10^6 \mathrm{Pa}$$

⇨ No punching failure.

The design of a 5 kN load cell with tapered diaphragm made of beryllium copper is shown in Fig. B.3.

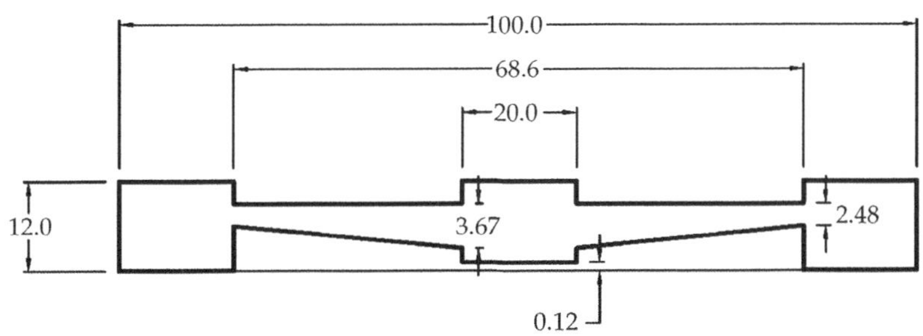

Figure B.3 Design of a 5 kN load cell with tapered diaphragm made of beryllium copper. Taper angle = 2.8°. All measurements are given in millimeters.

References

Abrantes, A.E. and Yamamuro, J.A. (2002) Experimental and data analysis techniques used for high strain rate tests on cohesionless soil. *Geotechnical Testing Journal*, **25**(2), 128–141.

Ackerly, S.K., Hellings, J.E., and Jardine, R.J. (1987) Discussion on a new device for measuring local strains on triaxial specimen. *Geotechnique*, **37**(3), 413–417.

Adachi, K. (1988) Sampling of cohesionless and gravelly soils. In: *The Art and Science of Geotechnical Engineering at the Dawn of the Twenty-First Century* (eds E.T. Cording, W.J. Hall, J.D. Haltiwanger, A.J. Hendron, and G. Mesri), pp. 206–220. Prentice Hall, Upper Saddle River, NJ.

Adams, J.I. and Radhakrishna, H.S. (1971) Loss of strength due to sampling in a glacial lake deposit. In: *Sampling of Soil and Rock*, ASTM STP 483 (eds B.B. Gordon and C.B. Crawford), pp. 109–120. ASTM, Philadelphia, PA.

Akai, K., Adachi, T., and Ando, N. (1975) Existence of a unique stress-strain-time relation for clays. *Soils and Foundations*, **15**(1), 1–16.

Ali, S.R., Pyrah, I.C., and Anderson, W.F. (1995) A novel technique for evaluation of membrane penetration. *Geotechnique*, **45**(3), 545–548.

Allam, M.M. and Sridharan, A. (1980) Influence of the back pressure technique on the shear strength of soils. *Geotechnical Testing Journal*, **3**(1), 35–40.

Altschaeffl, A.G. and Mishu, L.P. (1970) Capacitance techniques for radial deformations. *Journal of the Soil Mechanics and Foundations Division*, **96**(SM4), 1487–1491.

Alva-Hurtado, J.E. and Selig, E.T. (1981) Survey of laboratory devices for measuring soil volume change. *Geotechnical Testing Journal*, **4**(1), 11–18.

Ampuda, S. and Tatsuoka, F. (1989) An automated stress-path control triaxial system. *Geotechnical Testing Journal*, **12**(3), 238–243.

Anantanasakul, P., Yamamuro, J.A., and Lade, P.V. (2012) Three-dimensional drained behavior of normally consolidated anisotropic kaolin behavior. *Soils and Foundations*, **52**(1), 146–159.

Anderson, D.G. and Stokoe, K.H., II (1978) Shear modulus: a time-dependent soil property. In: *Dynamic Geotechnical Testing, ASTM STP 654* (eds M.L. Silver and D. Tiedemann), pp. 66–90. ASTM, Philadelphia, PA.

Andreasen, A. and Simons, N.E. (1960) Norwegian triaxial equipment and technique. *Proceedings of the ASCE Research Conference on the Shear Strength of Cohesive Soils*, Boulder, CO, USA, pp. 695–709.

Arthur, J.R.F., Bekenstein, S., Germaine, J.T., and Ladd, C.C. (1981) Stress path tests with controlled rotation of principal stress directions. In: *Laboratory Shear Strength of Soils, ASTM STP 740* (eds R.N. Young and F.C. Townsend), pp. 516–540. ASTM, Philadelphia, PA.

Arthur, J.R.F., Chua, K.S., and Dunstan, T. (1977a) Induced anisotropy in a sand. *Geotechnique*, **27**(2), 13–30.

Arthur, J.R.F., Dunstan, T., Al-Ani, Q.A.J.L., and Assadi, A. (1977b) Plastic deformation and failure in granular media. *Geotechnique*, **27**(1): 53–74.

Arthur, J.R.F., James, R.G., and Roscoe, K.H. (1964) The determination of stress fields during plane strain of a sand mass. *Geotechnique*, **14**, 283–308.

ASTM D698 (2014) Standard Test Methods for Laboratory Compaction Characteristics of Soil Using Standard Effort (12,400 ft-lb/ft^3 (600 kN-m/m^3)). ASTM International, West Conshohocken, PA.

ASTM D1557 (2014) Standard Test Methods for Laboratory Compaction Characteristics of Soil Using Modified Effort (56,000 ft-lb/ft^3 (2700 kN-m/m^3)). ASTM International, West Conshohocken, PA.

ASTM D2850 (2014) Standard Test Method for Unconsolidated-Undrained Triaxial Compression Test on Cohesive Soils. ASTM International, West Conshohocken, PA.

ASTM D4767 (2014) Standard Test Method for Consolidated Undrained Triaxial Compression

Triaxial Testing of Soils, First Edition. Poul V. Lade.

Test for Cohesive Soils. ASTM International, West Conshohocken, PA.

ASTM D5311 (2014) Standard Test Method for Load Controlled Cyclic Triaxial Strength of Soil. ASTM International, West Conshohocken, PA.

ASTM D6836 (2014) Standard Test Methods for Determination of the Soil Water Characteristic Curve for Desorption Using Hanging Column, Pressure Extractor, Chilled Mirror Hygrometer and/or Centrifuge. ASTM International, West Conshohocken, PA.

ASTM D7181 (2014) Standard Test Method for Consolidated Drained Triaxial Compression Test for Soils. ASTM International, West Conshohocken, PA.

Atkinson, J.H., Evans, J.S., and Ho, E.W.L. (1985) Non-uniformity of triaxial samples due to consolidation with radial drainage. *Geotechnique*, **35**(3), 353–355.

Aydilek, A.H. and Kutay, M.E. (2004) Development of an innovative computer-controlled water deairing system for hydraulic testing of geosynthetics. *Journal of Testing and Evaluation*, **32**(2), 161–166.

Balasubramaniam, A.S. (1976) Local strains and displacements patterns in triaxial specimens of a saturated clay. *Soils and Foundations*, **16**(1), 101–114.

Balasubramaniam, A.S. and Waheed-Uddin (1978) Reply to discussion on "Deformation characteristics of weathered Bangkok clay in triaxial extension," by Balasubramaniam, A.S. and Waheed-Uddin, Geotechnique, **27**(1):75-92. *Geotechnique*, **28**(3), 231–234.

Baldi, G., Hight, D.W., and Thomas, G.E. (1988) A reevaluation of conventional triaxial test methods. In: *Advanced Triaxial Testing of Soil and Rock, ASTM STP 977* (eds R.T. Donaghe, R.C. Chaney, and M.L. Silver), pp. 219–263. ASTM, Philadelphia, PA.

Baldi, G. and Nova, R. (1984) Membrane penetration effects in triaxial testing. *Journal of Geotechnical Engineering*, **110**(3): 403–420.

Baligh, M.M. (1976) Cavity expansion in sand with curved envelopes. *Journal of the Geotechnical Engineering Division*, **102**(GT11), 1131–1146.

Barden, L. (1972) The relation of soil structure to the engineering geology of clay soil. *Quarterly of the Journal of Engineering Geology*, **5**(1–2), 85–102.

Barden, L. and McDermott, R.J.W. (1965) Use of free ends in triaxial testing of clays. *Journal of the Soils Mechanics and Foundations Division*, **91**(SM6), 1–23.

Barron, R.A. (1948) Consolidation of fine-grained soils by drain wells. *Transactions, ASCE*, **113**, 718–754.

Becker, D.E. (2010) Testing in geotechnical design. *Geotechnical Engineering Journal of the SEAGS & AGSSEA*, **41**(1), 1–12.

Berre, T. (1982) Triaxial testing at the Norwegian Geotechnical Institute. *Geotechnical Testing Journal*, **5**(1/2), 3–17.

Bishop, A.W. (1958) The requirements for measuring the coefficient of earth pressure at rest. *Proceedings of the Brussels Conference on Earth Pressure Problems*, Brussels, Belgium, vol. 1, p. 214.

Bishop, A.W. (1973) The influence of an undrained change in stress on pore pressure in porous media of low compressibility. *Geotechnique*, **23**(3), 435–442.

Bishop, A.W. and Blight, G.E. (1963) Some aspects of effective stress in saturated and unsaturated soils. *Geotechnique*, **13**(3), 177–197.

Bishop, A.W. and Donald, I.B. (1961) The experimental study of partly saturated soil in the triaxial apparatus. *Proceedings of the 5th International Conference on Soil Mechanics and Foundation Engineering*, Paris, France, vol. I, pp. 13-21.

Bishop, A.W. and Eldin, A.K.G. (1950) Undrained triaxial tests on saturated sands and their significance in the general theory of shear strength. *Geotechnique*, **2**, 13–32.

Bishop, A.W. and Gibson, R.E. (1964) The influence of the provisions for boundary drainage on strength and consolidation characteristics of soils measured in the triaxial apparatus. In: *Laboratory Shear Testing of Soils, ASTM STP 361*, pp. 435–458. ASTM, Philadelphia, PA.

Bishop, A.W. and Green, G.E. (1965) Influence of end restraint on compression strength of a cohesionless soil," *Geotechnique*, **15**(3): 243–266.

Bishop, A.W. and Henkel, D.J. (1962) *The Measurement of Soil Properties in the Triaxial Test*, 2nd edn. St. Martin's Press, New York, NY.

Bishop, A.W. and Wesley, L.D. (1975) A hydraulic triaxial apparatus for controlled stress path testing. *Geotechnique*, **25**(4), 657–670.

Bjerrum, L. (1954) Geotechnical properties of Norwegian marine clays. *Geotechnique*, **4**(2), 49–69.

Bjerrum, L. and Landva, A. (1966) Direct simple shear tests on a Norwegian quick clay. *Geotechnique*, **16**(1), 1–20.

Black, D.K. and Lee, K.L. (1973) Saturating laboratory samples by back pressure. *Journal of the Soils Mechanics and Foundations Division*, **99**(SM1), 75–93.

Blight, G.E. (1963) The effect of nonuniform pore pressures on laboratory measurements of the shear strength of soils. In: *Laboratory Shear Testing of Soils, ASTM STP 361*, pp. 173–184. ASTM, Philadelphia, PA.

Blight, G.E. (1967) Observations on the shearing testing of indurated fissured clays. *Proceedings of the Geotechnical Conference*, Oslo, Norway, vol. 1, pp. 97–102.

Bopp, P.A. and Lade, P.V. (1997a) Membrane penetration in granular materials at high pressures. *Geotechnical Testing Journal*, **20**(3), 272–278.

Bopp, P.A. and Lade, P.V. (1997b) Effects of initial density on soil instability at high pressures. *Journal of Geotechnical and Geoenvironmental Engineering*, **123**(7), 671–677.

Broms, B. and Casbarian, A.O. (1965) Effects of rotation of principal stress axes and the intermediate principal stress on the shear strength. *Proceedings of the 6th International Conference on Soil Mechanics and Foundation Engineering*, Montreal, Canada, vol. I, pp. 179–183.

Brooker, E.W. and Ireland, H.O. (1965) Earth pressures at rest related to stress history. *Canadian Geotechnical Journal*, **2**(1), 1–15.

Brown, S.F., Austin, G., and Overy. R.F. (1980) An instrumented triaxial cell for cyclic loading of clays. *Geotechnical Testing Journal*, **3**(4), 145–152.

Brown, S.F. and Snaith, M.S. (1974) The measurement of recoverable and irrecoverable deformations in the repeated load triaxial test. *Geotechnique*, **24**(2), 255–259.

Burland, J.B. (1989) Ninth Laurits Bjerrum Memorial Lecture: "Small is beautiful, the stiffness of soils at small strains." *Canadian Geotechnical Journal*, **26**(4), 499–516.

Burland, J.B. and Symes, M. (1982) A simple axial displacement gauge for use in the triaxial apparatus. *Geotechnique*, **32**(1), 62–65.

Cabarkapa, Z., Cuccovillo, T., and Gunn, M. (1999) Some aspects of the pre-failure behavior of unsaturated soil. *Proceedings of the 2nd International Symposium on Pre-Failure Deformation Characteristics of Geomaterials*, Torino, Italy, vol. 1, pp. 159–165.

Camacho-Tauta, J.F., Jimenez Alvarez, J.D., and Reyes-Ortiz, O.J. (2012) A procedure to calibrate and perform the bender element test. *Dyna*, **176**, 10–18.

Carillo, N. (1942) Simple two- and three-dimensional cases in the theory of consolidation of soils. *Journal of Mathematics and Physics*, **21**(1), 1.

Carter, J.P. (1982) Predictions of non-homogeneous behaviour of clay in the triaxial test. *Geotechnique*, **32**(1), 55–58.

Casagrande, A. (1975) Liquefaction and cyclic deformation of sands – a critical review. *Proceedings of the 5th Pan-American Conference on Soil Mechanics and Foundation Engineering*, Buenos Aires, Argentina, vol. 5, pp. 79–133. A.A. Balkema, Rotterdam.

Cha, M. and Cho G. (2007) Shear strength estimation of sandy soils using shear wave velocity. *Geotechnical Testing Journal*, **30**(6), 484–495.

Chan, C.K. (1975) Low-friction seal system. *Journal of the Geotechnical Engineering Division*, **101**(GT9), 991–995.

Chan, C.K. and Duncan, J.M. (1967) A new device for measuring volume change and pressures in triaxial tests on soils. *Materials Research and Standards, ASTM*, **7**(7), 312–314.

Chan, C.M. (2010)Bender element test in soil specimens: identifying the shear wave arrival time. *Electronic Journal of Geotechnical Engineering*, **15**, 1263–1276.

Chandler, R.J. (1966) The measurement of residual strength in triaxial compression. *Geotechnique*, **16**(3), 181–186.

Clayton, C.R.I. and Khatrush, S.A. (1986) New device for measuring local axial strains on triaxial specimens. *Geotechnique*, **36**(4), 593–597.

Clayton, C.R.I., Khatrush, S.A., Bica, A.V.D., and Siddique, A. (1989) The use of Hall effect semiconductors in geotechnical instrumentation. *Geotechnical Testing Journal*, **12**(1), 69–76.

Cole, D.M. (1978) A technique for measuring radial deformation during repeated load triaxial testing. *Canadian Geotechnical Journal*, **15**(3), 426–429.

Colliat-Dangus, J.L., Desrues, J., and Foray, P. (1988) Triaxial testing of granular soil under elevated cell pressure. In: *Advanced Triaxial Testing of Soil and Rock, ASTM STP 977* (eds R.T. Donaghe, R.C. Chaney, and M.L. Silver), pp. 290–310. ASTM, Philadelphia, PA.

Costa-Filho, L. (1985) Measurement of axial strains in triaxial tests on London clay. *Geotechnical Testing Journal*, **8**(1), 3–13.

Coulomb, C.A. (1776) Sur Une Application des Regales de Maximis et Minimis a Quelques Problems es Statique, Relatifs a L'Architecture. *Memoirs Academic Royale des Sciences (pars Divers Savants)*, **7**, 343.

Cuccovillo, T. and Coop, M.P. (1997) The measurement of local strains in triaxial test using LVDTs. *Geotechnique*, **47**(1), 167–171.

Curry, J.R. (1956) The analysis of two-dimensional orientation data. *Journal of Geology*, **64**, 117–131.

Darley, P. (1973) Discussion on "Apparatus for automatic volume change suitable for automatic logging," by G.O Rowlands. *Geotechnique*, **23**(1), 140–141.

Davis, E.H. (1968) Theories of plasticity and the failure of soil masses. In: *Soil Mechanics, Selected Topics* (ed. I.K. Lee, pp. 341–380. Butterworths, London.

DeGroff, W., Donaghe, R., Lade, P.V., and La Rochelle, P. (1988) Correction of strength for membrane effects in the triaxial test. *Geotechnical Testing Journal*, **11**(1), 78–82.

Desrues, J., Chambon, R., Mokni, M., and Mazerolle, F. (1996) Void ratio evolution inside shear bands in triaxial sand specimens studied by computed tomography. *Geotechnique*, **46**(3), 529–546.

Donaghe, R.T., Chaney, R.C., and Silver, M.L. (eds) (1988) *Advanced Triaxial Testing of Soil and Rock, ASTM STP 977*. ASTM, Philadelphia, PA.

Duncan, J.M. and Seed, H.B. (1965) *Errors in strength tests and recommended corrections*. Report no. TE-65-4, University of California, Berkeley, CA.

Duncan, J.M. and Seed, H.B. (1967) Corrections for strength test data. *Journal of the Soil Mechanics and Foundations Division*, **93**(SM5), 121–137.

Duncan, J.M. and Wright, S. G. (2005) *Soil Strength and Slope Stability*. John Wiley & Sons, Inc., Hoboken, NJ.

Dunnicliff, J. (1988) *Geotechnical Instrumentation for Monitoring Field Performance*. John Wiley & Sons, Inc., New York, NY.

Dusseault, M.B. and Morgenstern, N.R. (1978) Shear strength of Athabasca Oil Sands. *Canadian Geotechnical Journal*, **15**(2), 216–238.

Dyvik, R. and Madshus, C. (1985) Lab measurements of Gmax using bender elements. *Proceedings of the ASCE Convention on Advances in the Art of Testing Soils under Cyclic Conditions*, Detroit, MI, pp. 186–197.

El-Ruwayih, A.A. (1976) Design manufacture and performance of a lateral strain device. *Geotechnique*, **26**(1), 215–216.

El-Sohby, M. (1964) *The behaviour of particulate materials under stress*. PhD thesis, University of Manchester.

Escario, V. and Saez, J. (1986) The shear strength of partly saturated soils. *Geotechnique*, **36**(3), 453–456.

Finn, W.D.L., Pickering, D.J., and Bransby, P.L. (1971) Sand liquefaction in triaxial and simple shear tests. *Journal of the Soil Mechanics and Foundations Division*, **97**(SM4), 639–659.

Finno, R.J. and Kim, T. (2012) Effects of stress path rotation angle on small strain responses. *Journal of Geotechnical and Geoenvironmental Engineering*, **138**(4), 526–534.

Fredlund, D.G. and Rahardjo, H. (1993) *Soil Mechanics for Unsaturated Soils*. John Wiley & Sons, Inc., New York, NY.

Fredlund, D.G. and Xing, A. (1994) Equations for the soil-water characteristic curve. *Canadian Geotechnical Journal*, **31**(4), 521–532.

Frydman, S. (1974) Yielding of sand in plane strain. *Journal of the Geotechnical Engineering Division*, **100**(GT5), 491–501.

Frydman, S., Zeitlen, J.G., and Alpan, I. (1973) The membrane effect in triaxial testing of granular soils. *Journal of Testing and Evaluation*, **1**(1): 37-41.

Fukushima, S. and Tatsuoka, F. (1984) Strength and deformation characteristics of saturated sand at extremely low pressures. *Soils and Foundations*, **24**(4), 30–48.

Fuller, W.B. and Thompson, S.E. (1907) The laws of proportioning concrete. *Transactions, ASCE*, **59**, 67–172.

Gachet, P., Geiser, F., Laluoi, L., and Vulliet (2007) Automated digital image processing for volume change measurement in triaxial cells. *Geotechnical Testing Journal*, **30**(2), 98–103.

van Genuchten, M.T. (1980) A closed-form equation for predicting the hydraulic conductivity of unsaturated soils. *Soil Science Society of America Journal*, **44**, 892–898.

Germaine, J.T. and Germaine, A.V. (2009) *Geotechnical Laboratory Measurements for Engineers*. John Wiley & Sons, Inc., Hoboken, NJ.

Germaine, J.T. and Ladd, C.C. (1988) Triaxial testing of saturated cohesive soils. In: *Advanced Triaxial Testing of Soil and Rock, ASTM STP 977* (eds R.T. Donaghe, R.C. Chaney, and M.L. Silver), pp. 421–459. ASTM, Philadelphia, PA.

Gibson, R.E. and Henkel, D.J. (1954) Influence of duration of tests at constant rate of strain on measured "drained" strength. *Geotechnique*, **4**(1), 6–15.

Gibson, R.E. and Lumb, P. (1953) Numerical solution of some problems in consolidation of clay. *Proceedings of the Institution of Civil Engineers*, **2**(2), 182–198.

Giroud, J.P. (1980) Introduction to geotextiles and their application. *Proceedings of the 1st Canadian Symposium on Geotextiles*, Calgary, AB, Canada, pp. 3–31. Canadian Geotechnical Society, Calgary, AB.

Goto, S., Tatsuoka, F., Shibuya, S., Kim, Y.-S., and Sato, T. (1991) A simple gage for local small strain measurements in the laboratory. *Soils and Foundations*, **31**(1), 169–180.

Green, G.E. (1971) Strength and deformation of sand measured in and independent stress control cell. *Proceedings of Stress-Strain Behaviour of Soils, Roscoe Memorial Symposium*, Cambridge University, Cambridge, UK, pp. 285–323. G.T. Foulis and Co., Ltd, Henley-on-Thames.

Green, G.E. and Bishop, A.W. (1969) A note on the drained strength of sand under generalized strain conditions. *Geotechnique*, **19**(1), 144–149.

Habib, P. (1953) Influence de la variation de la contrainte moyenne sur la resistance au cisaillement des sols. *Proceedings of the 3rd International Conference on Soil Mechanics and Foundation Engineering*, Zurich, Switzerland, vol. I, pp. 131–136.

Hambly, E.C. (1969) A new true triaxial apparatus. *Geotechnique*, **19**(2), 307–309.

Handy, R.L. (1981) Linearizing triaxial test failure envelopes. *Geotechnical Testing Journal*, **4**(4), 188–191.

Hansen, B. (1958) Line ruptures regarded as narrow rupture zones – basic equations based on kinematic considerations. *Proceedings of the Conference on Earth Pressure Problems*, Brussels, Belgium, vol.1, pp. 39–48.

Hardin, B.O. (1978) Nature of stress-strain behavior of soils. *Proceedings of the ASCE Geotechnical Engineering Division Specialty Conference*, Pasadena, CA, USA, vol. 1, pp. 3–90.

Hattab, M. and Hicher, P.-Y. (2004) Dilating behaviour of overconsolidated clay. *Soils and Foundations*, **44**(4): 27–40.

Henkel, D.J. (1960) The shear strength of saturated remoulded clays. *Proceedings of the ASCE Research Conference on Shear Strength of Cohesive Soils*, Boulder, CO, USA, pp. 533–554.

Henkel, D.J. and Gilbert, G.D. (1952) The effect of the rubber membrane on the measured triaxial compression strength of clay samples. *Geotechnique*, **3**(1), 20–29.

Hettler, A. and Vardoulakis, I. (1984) Behaviour of dry sand tested in large triaxial apparatus. *Geotechnique*, **34**(2), 183–198.

Heymann, G. (1998) *The stiffness of soils and weak rocks at very small strains*. PhD thesis, University of Surrey.

Hight, D.W., Gens, A., and Symes, M.J. (1983) The development of a new hollow cylinder apparatus for investigation of the effects of principal stress rotation in soils. *Geotechnique*, **33**(4), 355–383.

Hird, C.C. and Yung, C.Y. (1989) The use of proximity transducers for local strain measurements in triaxial tests. *Geotechnical Testing Journal*, **12**(4), 292–296.

Ho, D.Y.F. and Fredlund, D.G. (1982) A multi-stage triaxial test for unsaturated soils. *Geotechnical Testing Journal*, **5**(1–2), 18–25.

Hoek, E. and Brown, E.T. (1980) Empirical strength criterion for rock masses. *Journal of the Geotechnical Engineering Division*, **106**(GT9), 1013–1035.

Hoek, E. and Franklin, J.A. (1968) A simple triaxial cell for field and laboratory testing of rock. *Transactions of the Institution of Mining and Metallurgy*, **77**, A22–26.

Holubec, I. and Finn, P.J. (1969) A lateral deformation transducer for triaxial testing. *Canadian Geotechnical Journal*, **6**, 353–356.

Hong, W.P. and Lade, P.V. (1989a) Elasto-plastic behavior of K_0 - consolidated clay in torsion shear tests. *Soils and Foundations*, **29**(2), 127–140.

Hong, W.P. and Lade, P.V. (1989b) Strain increment and stress directions in torsion shear tests. *Journal of Geotechnical Engineering*, **115**(10), 1388–1401.

Hoque, E., Sato, T. and Tatsuoka, F. (1997) Performance evaluation of LDTs for use in triaxial tests. *Geotechnical Testing Journal*, **20**(2), 149–167.

Hryciw, R.D., Raschke, S.A., Ghalib, A.M., Horner, D.A., and Peters, J.F. (1997) Video tracking for experimental validation of discrete simulations of large discontinuous deformations. *Computers and Geotechnics*, **21**(3), 235–253.

Hudson, J.A., Crouch, S.L., and Fairhurst, C. (1972) Soft, stiff and servo-controlled testing machines: a review with reference to rock failure. *Engineering Geology*, **6**, 155–189.

Hvorslev, M.J. (1960) Physical components of the shear strength of saturated clays. *Proceedings of the Research Conference on Shear Strength of Cohesive Soils*, Boulder, CO, USA, pp. 169–273.

Ibrahim, A.A. and Kagawa, T. (1991) Microscopic measurement of sand fabric from cyclic tests causing liquefaction. *Geotechnical Testing Journal*, **14**(4), 371–382.

Ibsen, L.B. and Praastrup, U. (2002) The Danish rigid boundary true triaxial apparatus for soil testing. *Geotechnical Testing Journal*, **25**(3), 254–265.

Ishihara, K. (1993) Liquefaction and flow failure during earthquakes. *Geotechnique*, **43**(3), 351–415.

Iversen, K. and Moum, J. (1974) The paraffin method; triaxial testing without rubber membrane. *Geotechnique*, **24**(4), 665–670.

Iwasaki, T. and Tatsuoka, F. (1977) Effects of grain size and grading on dynamic shear moduli of sands. *Soils and Foundations*, **17**(3), 19–35.

Jacobsen, M. (1967) The undrained shear strength of preconsolidated boulder clay. *Proceedings of the Geotechnical Conference*, Oslo, Norway, vol. I, pp. 119–122.

Jacobsen, M. (1970) *New oedometer and new triaxial apparatus for firm soils*. Bulletin no. 27, The Danish Geotechnical Institute, Copenhagen, Denmark, pp. 7–20.

Jaky, J. (1948) Pressure in silos. *Proceedings of the 2nd International Conference on Soil Mechanics and Foundation Engineering*, Rotterdam, the Netherlands, vol. I, pp. 103–107.

Jang, D.-J. and Frost, J.D. (1998) Sand structure differences resulting from specimen preparation procedures. *Proceedings of the Specialty Conference on Geotechnical Earthquake Engineering and Soil Dynamics*, Seattle, WA, USA, vol. 1, pp. 234–245.

Jardine, R.J. (2014) Advanced laboratory testing in research and practice: the 2nd Bishop Lecture. *Geotechnical Research*, **1**(1), 2–31.

Jardine, R.J., Symes, M.J., and Burland, J.B. (1984) The measurement of soil stiffness in the triaxial apparatus. *Geotechnique*, **34**(3), 323–340.

Joer, H.A., Lanier, J., Desrues J., and Flavigny, E. (1992) 1γ2ε: A new shear apparatus to study the behaviour of granular materials. *Geotechnical Testing Journal*, **15**(2), 129–137.

Joer, H.A., Lanier, J., and Fahey, M. (1998) Deformation of granular materials due to rotation of principal axes. *Geotechnique*, **48**(5), 605–619.

Johnston, I.W. and Chiu, H.K. (1982) A simple device for automatic volume change measurements. *Geotechnical Engineering Journal of the SEAGS & AGSSEA*, **13**, 235–238.

Kang, X., Kang, G.-C., and Bate, B. (2014) Measurement of stiffness anisotropy in kaolinite using bender element tests in a floating wall consolidometer. *Geotechnical Testing Journal*, **37**(5), 869–883.

Karimpour, H. (2012) *Time effects and their relation to crushing in sand*." PhD thesis, The Catholic University of America.

Karlsrud, K. and Hernandez-Martinez, F.G. (2013) Strength and deformation properties of Norwegian clays from laboratory tests on high quality block samples. *Canadian Geotechnical Journal*, **50**(12), 1273–1293.

Kenney, T.C. and Chan, H.T. (1972) Use of radiographs in a geological and geotechnical investigation of varved soil. *Canadian Geotechnical Journal*, **10**(3), 195–205.

Khan, M.H. and Hoag, D.L. (1979) A noncontacting transducer for measurement of lateral strains. *Canadian Geotechnical Journal*, **16**(2), 409–411.

Kiekbusch, M. and Schuppener, B. (1977) Membrane penetration and its effects on pore pressures. *Journal of the Geotechnical Engineering Division*, **103**(GT11), 1267–1279.

Kim, M.K. and Lade, P.V. (1984) Modeling of rock strength in three dimensions. *International Journal of Rock Mechanics and Mining Sciences & Geomechanics Abstracts*, **21**(1), 21–33.

Kirkgard, M.M. and Lade, P.V. (1991) Anisotropy of normally consolidated San Francisco Bay Mud. *Geotechnical Testing Journal*, **14**(3), 231–246.

Kirkgard, M.M. and Lade, P.V. (1993) Anisotropic three-dimensional behavior of a normally consolidated clay. *Canadian Geotechnical Journal*, **30**(4), 848–858.

Kirkpatrick, W.M. and Belshaw, D.J. (1968) On the interpretation of the triaxial test. *Geotechnique*, **18**, 336–350.

Kirkpatrick, W.M. and Younger, J.S. (1970) Strain conditions in the compression cylinder. *Journal of the Soil Mechanics and Foundations Division*, **96**(SM5), 1683–1695.

Kjellman, W. (1951) Testing the shear strength of clay in Sweden. *Geotechnique*, **2**(3), 225–232.

Klementev, I. and Novak, J. (1978) Continuously water de-airing device. *Geotechnique*, **28**(3), 347–348.

Ko, H.-Y. and Scott, R.F. (1967) A new soil testing apparatus. *Geotechnique*, **17**(1), 40–57.

Kolbuszewski, J.J. (1948) An experimental study of the maximum and minimum porosities of sand. *Proceedings of the 2nd International Conference on Soil Mechanics and Foundation Engineering*, Rotterdam, the Netherlands, vol. 1, pp. 158–165.

Kolymbas, D. and Wu, W. (1989) A device for lateral strain measurement in triaxial tests with unsaturated specimens. *Geotechnical Testing Journal*, **12**(3), 227–229.

Korn, G.A. and Korn T.M. (1961) *Mathematical Handbook for Scientists and Engineers*. McGraw-Hill, New York, NY.

Kramer, S.L. and Sivaneswaran N. (1989) A nondestructive, specimen-specific method for measurement of membrane penetration in the triaxial tests. *Geotechnical Testing Journal*, **12**(1), 50–59.

Kramer, S.L., Sivaneswaran N., and Davis, R.O. (1990) Analysis of membrane penetration in triaxial test. *Journal of Engineering Mechanics*, **116**(4), 773–789.

Krizek, R.J., Edil, T.B., and Ozaydin, I.K. (1975) Preparation and identification of samples with controlled fabric. *Engineering Geology*, **9**(1), 13–38.

Kuerbis, R., Negussey, D., and Vaid, Y.P. (1988) Effect of gradation and fines content on the undrained response of sand. In: *Hydraulic Fill Structures, Geotechnical Special Publication No. 21*, pp. 330–345. ASCE, New York, NY.

Kuerbis, R. and Vaid, Y.P. (1988) Sand sample preparation – the slurry deposition method. *Soils and Foundations*, **28**(4), 107–118.

Kuwano, R., Connolly, T.M., and Jardine, R.J. (2000) Anisotropic stiffness measurements in a stress-path triaxial cell. *Geotechnical Testing Journal*, **23**(2), 141–157.

Ladd, C.C. and Foott, R. (1974) New design procedure for stability of soft clays. *Journal of the Geotechnical Engineering Division*, **100**(GT7), 763–785.

Ladd, C.C., Foott, R., Ishihara, K., Schlosser, F., and Poulos, H.G. (1977) State-of-the-art on stress deformation of soils. *Proceedings of the 9th Intenational Conference on Soil Mechanics and Foundation Engineering*, Tokyo, Japan, pp. 421–494.

Ladd, C.C. and Lambe, T.W. (1963) The strength of "undisturbed" clay determined from undrained tests. In: *Laboratory Shear Testing of Soils, ASTM STP 361*, pp. 342–371. ASTM, Philadelphia, PA.

Ladd, R.S. (1974) Specimen preparation and liquefaction of sands. *Journal of Geotechnical Engineering*, **100**(GT10), 1180–1184.

Ladd, R.S. (1978) Preparing test specimens using undercompaction. *Geotechnical Testing Journal*, **1**(1), 16–23.

Lade, P.V. (1972) *The stress-strain and strength characteristics of cohesionless soil*. PhD thesis, University of California, Berkeley.

Lade, P.V. (1975) Torsion shear tests on cohesionless soil. *Proceedings of the 5th Pan-American Conference on Soil Mechanics and Foundation Engineering*, Buenos Aires, Argentina, vol. I, pp. 117–127.

Lade, P.V. (1976) Interpretation of torsion shear tests on sand. *Proceedings of the 2nd International Conference on Numerical Methods in Geomechanics*, Blacksburg, VA, USA, vol. I, pp. 381–389.

Lade, P.V. (1977) Elasto-plastic stress-strain theory for cohesionless soil with curved yield surfaces. *International Journal of Solids and Structures*, **13**, 1019–1035.

Lade, P.V. (1978) Cubical triaxial apparatus for soil testing. *Geotechnical Testing Journal*, 1(2), 93–101.

Lade, P.V. (1981) Torsion shear apparatus for soil testing. In: *Laboratory Shear Strength of Soil, ASTM STP 740* (eds R.N. Yong and F.C. Townsend), pp. 145–163. ASTM, Philadelphia, PA.

Lade, P.V. (1982a) Localization effects in triaxial tests on sand. *Proceedings of the Symposium on Deformation and Failure of Granular Materials* (eds P.A. Vermeer and H.J. Luger), Delft, the Netherlands, pp. 461–471.

Lade, P.V. (1982b) Three-parameter failure criterion for concrete. *Journal of the Engineering Mechanics Division*, **108**(EM5), 850–863.

Lade, P.V. (1984) Failure criterion for frictional materials. In: *Mechanics of Engineering Materials* (eds C.S. Desai and R.H. Gallagher), pp. 385–402. John Wiley & Sons, Ltd, Chichester.

Lade, P.V. (1988a) Effects of voids and volume changes on the behaviour of frictional materials. *International Journal for Numerical and Analytical Methods in Geomechanics*, **12**(4), 351–370.

Lade, P.V. (1988b) Automatic volume change and pressure measurement devices for triaxial testing of soils. *Geotechnical Testing Journal*, **11**(4), 263–268.

Lade, P.V. (1989) Closure to discussion on triaxial testing by Tatsuoka and Oswell *et al. Geotechnical Testing Journal*, **12**(4), 327.

Lade, P.V. (1990) Single hardening model with application to NC clay. *Journal of Geotechnical Engineering*, **116**(3), 394–414.

Lade, P.V. (1993a) Initiation of static instability in the submarine Nerlerk berm. *Canadian Geotechnical Journal*, **30**(5), 895–904.

Lade, P.V. (1993b) Rock strength criteria: the theories and the evidence. In: *Comprehensive Rock Engineering, Principles, Practice & Projects* (ed. E.T. Brown), vol. 1, pp. 255–284. Pergamon Press, Oxford.

Lade, P.V. (2002) Instability, shear banding, and failure of granular materials. *International Journal of Solids and Structures*, **39**(13–14): 3337–3357.

Lade, P.V. (2003) Analysis and prediction of shear banding under 3D conditions in granular materials. *Soils and Foundations*, **43**(4), 161–172.

Lade, P.V. (2004) Shear banding in cross-anisotropic sand specimens. In: *Geotechnical Innovations* (eds R.B.J. Brinkgreve, H. Schad, H.F. Schweiger, and E. Willand), pp. 561–574. Verlag Glückauf, Essen.

Lade, P.V. (2006) Assessment of test data for selection of 3-D failure criterion for sand. *International Journal for Numerical and Analytical Methods in Geomechanics*, **30**(4), 307–333.

Lade, P.V. (2007) Modeling failure in cross-anisotropic frictional materials. *International Journal of Solids and Structures*, **44**(16), 5146–5162.

Lade, P.V. (2008) Failure criterion for cross-anisotropic soils. *Journal of Geotechnical and Geoenvironmental Engineering*, **134**(1), 117–124.

Lade, P.V. (2010) The mechanics of surficial failure in soil slopes. *Engineering Geology*, **136**(9), 1209–1219.

Lade, P.V. and Abelev, A.V. (2005) Characterization of cross-anisotropic soil deposits from isotropic compression tests. *Soils and Foundations*, **45**(5), 89–102.

Lade, P.V. and de Boer, R. (1997) The concept of effective stress for soil, concrete, and rock. *Geotechnique*, **47**(1), 61–78.

Lade, P.V. and Duncan, J.M. (1973) Cubical triaxial tests on cohesionless soil. *Journal of the Soil Mechanics and Foundations Division*, **99**(SM10), 793–812.

Lade, P.V. and Hernandez, S.B. (1977) Membrane penetration effects in undrained tests. *Journal of the Geotechnical Engineering Division*, **103**(GT2), 109–125.

Lade, P.V. and Karimpour, H. (2010) Static fatigue controls particle crushing and time effects in granular materials. *Soils and Foundations*, **50**(5), 573–583.

Lade, P.V. and Karimpour, H. (2015) Stress relaxation behavior in Virginia Beach sand. *Canadian Geotechnical Journal*, **52**(7), 813–835.

Lade, P.V. and Kirkgard, M.M. (1984) B-Value tests for soil specimens with anisotropic stress states. *Proceedings of the 5th ASCE-EMD Conference*, Laramie, WY, USA, pp. 1304–1307.

Lade P.V. and Kirkgard, M.M. (2000) Effects of stress rotation on cross-anisotropic behavior of natural K_0-consolidated soft clay. *Soils and Foundations*, **40**(6): 93–105.

Lade, P.V. and Liu, C.T. (1998) Experimental study of drained creep behavior of sand. *Journal of Engineering Mechanics*, **124**(8), 912–920.

Lade, P.V. and Musante, H.M. (1978) Three-dimensional behavior of remolded clay. *Journal of the Geotechnical Engineering Division*, **104**(GT2), 193–209.

Lade, P.V., Nam, J., and Hong, W.P. (2008) Shear banding and cross-anisotropic behavior observed in laboratory sand tests with stress rotation. *Canadian Geotechnical Journal*, **45**(1), 74–84.

Lade, P.V., Nam, J., and Hong, W.P. (2009) Interpretation of strains in torsion shear tests. *Computers and Geotechnics*, **36**(1–2), 211–225.

Lade, P.V., Rodriguez, N.M., and Van Dyck, E.J. (2014) Effects of Principal stress directions on 3D failure conditions in cross-anisotropic sand. *Journal of Geotechnical and Geoenvironmental Engineering*, **140**(2), 04013001-1-12.

Lade, P.V. and Tsai, J. (1985) Effects of localization in triaxial tests on clay. *Proceedings of the 11th International Conference on Soil Mechanics and Foundation Engineering*, San Francisco, CA, USA, vol. 1, pp. 549–552.

Lade, P.V. and Wang, Q. (2001) Analysis of shear banding in true triaxial tests on sand. *Journal of Engineering Mechanics*, **127**(8), 762–768.

Lade, P.V. and Wang, Q. (2012a) Method for uniform strain extension tests on sand. *Geotechnical Testing Journal*, **35**(4), 607–617.

Lade, P.V. and Wang, Q. (2012b) Effects of stiff and flexible boundary conditions in triaxial extension tests on cross-anisotropic sand behavior. *Geotechnical Testing Journal*, **35**(5), 715–727.

Lade, P.V. and Yamamuro, J.A. (1997) Effects of non-plastic fines on static liquefaction of sands. *Canadian Geotechnical Journal*, **34**(6), 918–928.

Lade, P.V., Yamamuro, J.A., and Skyers, B.D. (1996) Effects of shear band formation in triaxial extension tests. *Geotechnical Testing Journal*, **19**(4), 398–410.

Laloui, L., Peron, H., Geiser, F., Rifa'i, A., and Vulliet, L (2006) Advances in volume measurement in unsaturated soil triaxial tests. *Soils and Foundations*, **46**(3), 341–349.

Lam, W.-K. and Tatsuoka, F. (1988) Triaxial compressive and extension strength of sand affected by strength anisotropy and sample slenderness. In: *Advanced Triaxial Testing of Soil and Rock, ASTM STP 977* (eds R.T. Donaghe, R.C. Chaney, and M.L. Silver), pp. 655—666. ASTM, Philadelphia, PA.

Lambe, T.W. (1964) Methods of estimating settlements. *Journal of the Soil Mechanics and Foundations Division*, **90**(SM5), 43–67.

Lambe, T.W. and Whitman, R.V. (1979) *Soil Mechanics – SI Version*. John Wiley & Sons, Inc., New York, NY.

Landva, A.O. and Pheeney, P.E. (1980) Peat fabric and structure. *Canadian Geotechnical Journal*, **17**(3), 416–435.

Landva, A.O., Pheeney, P.E., and Mersereau D.E. (1983) Undisturbed sampling of peat. In: *Testing of*

Peats and Organic Soils, ASTM STP 820, pp. 141–156. ASTM, Philadelphia, PA.

La Rochelle, P. (1967) Membrane, drain and area correction in triaxial test on soil samples failing along a single shear plane. *Proceedings of the 3rd Pan-American Conference on Soil Mechanics*, Caracas, Venezuela, vol. 1, pp. 273–292.

La Rochelle, P. Leroueil, S., and Tavenas, F. (1986) A technique for long-term storage of clay samples. *Canadian Geotechnical Journal*, **23**(4), 602–605.

La Rochelle, P., Leroueil, S., Trak, B., Blais, L., and Tavenas, F. (1988) Observational approach to membrane and area corrections in triaxial tests. In: *Advanced Triaxial Testing of Soil and Rock, ASTM STP 977* (eds R.T Donaghe, R.C. Chaney, and M.L. Silver), pp. 715–731. ASTM, Philadelphia, PA.

Laudahn, A., Sosna, K., and Bohac, J. (2005) A simple method for air volume change measurements in triaxial tests. *Geotechnical Testing Journal*, **28**(3), 313–318.

Lee, I.K. (1966) Stress-dilatancy performance of feldspar. *Journal of the Soil Mechanics and Foundations Division*, **92**(SM2), 79–103.

Lee, J.S. and Santamarina, J.C. (2005) Bender elements: performance and signal interpretation. *Journal of Geotechnical and Geoenvironmental Engineering*, **131**(9), 1063–1070.

Lee, K.L. (1965) *Triaxial compressive strength of saturated sand under seismic loading conditions*. PhD thesis, University of California, Berkeley.

Lee, K.L. (1978) End restraint effects on undrained static triaxial strength of sand. *Journal of the Geotechnical Engineering Division*, **104**(GT6), 687–704.

Lee, K.L., Morrison, R.A., and Haley, S.C. (1969) A note on the pore pressure parameter B. *Proceedings of the 7th International Conference on Soil Mechanics and Foundation Engineering*, Mexico City, Mexico, vol. I, pp. 231–238.

Lee, K.L. and Seed, H.B. (1967) Drained strength characteristics of sands. *Journal of Soil Mechanics and Foundations Division*, **93**(SM6), 117–141.

Leong, E.C., Agus, S.S., and Rahardjo, H. (2004) Volume change measurement of soil specimen in triaxial test. *Geotechnical Testing Journal*, **27**(1), 47–56.

Leroueil, S., Tavenas, F., La Rochelle, P., and Tremblay, M. (1988) Influence of filter paper and leakage on triaxial testing. In: *Advanced Triaxial Testing of Soil and Rock, ASTM STP 977* (eds R.T. Donaghe, R.C. Chaney, and M. L. Silver), pp. 189–201. ASTM, Philadelphia, PA.

Lessard, G. and Mitchell, J.K. (1985) Causes and effects of aging in quick clays. *Canadian Geotechnical Journal*, **22**(3), 335–346.

Li, X.S., Chan, C.K., and Shen, C.K. (1988) An automated triaxial testing system. In: *Advanced Triaxial testing of Soil and Rock, ASTM STP 977* (eds R.T. Donaghe, R.C. Chaney, and M. L. Silver), pp. 95–106. ASTM, Philadelphia, PA.

Lo, S.-C. R., Chu, J., and Lee, I.K. (1989) A technique for reducing membrane penetration and bedding errors. *Geotechnical Testing Journal*, **12**(4), 311–316.

Lo Presti, D.C.F., Pallara, O., Constanzo, D., and Impavido, M. (1995a) Small strain measurements during triaxial tests: many problems, some solutions. *Proceedings of the 1st International Symposium on Pre-Failure Deformation Characteristics of Geomaterials*, Hokkaido, Japan, vol. 1, pp. 11–16.

Lo Presti, D.C.F., Pallara, O., and Puci, I. (1995b) A modified commercial triaxial testing system for small strain measurements: preliminary results on Pisa clay. *Geotechnical Testing Journal*, **18**(1), 15–31.

Lode, W. (1926) Versuche über den Einfluss der mittleren Hauptspannung auf das Fliessen der Metalle Eisen, Kupfer und Nickel. *Zeitschrift für Physik*, **36**, 913–939.

Lowe, J. and Johnson, T.C. (1960) Use of back pressure to increase degree of saturation of triaxial test specimen. *Proceedings of the Research Conference on Shear Strength of Cohesive Soils*, Boulder, CO, USA, pp. 819–836.

Lu, N. and Likos, W.J. (2004) *Unsaturated Soil Mechanics*. John Wiley & Sons, Inc., Hoboken, NJ.

Ludwik, P. (1909)*Elemente der technologishen Mechanik*. Springer, Berlin.

Lunne, T., Berre, T., and Strandvik, S. (1997) Sample disturbance effects in soft low plastic Norwegian clay. *Proceedings of the Conference on Recent Developments in Soil and Pavement Mechanics*, Rio de Janeiro, Brazil, pp. 81–102. A.A. Balkema, Rotterdam.

Macari, E.J., Parker, J.K., and Costes, N.C. (1997) Measurement of volume changes in triaxial tests using digital imaging techniques. *Geotechnical Testing Journal*, **20**(1), 103–109.

Macari-Pasqualino, E.J., Costes, N.C., and Parker, J.K. (1993) Digital image techniques for volume change measurement in triaxial tests. *Proceedings of the Engineering Foundation – National Science Foundation Conference on Digital Image Processing: Techniques and Applications in Civil Engineering*, Kona, HI, pp. 211–219. ASCE, New York, NY.

Maksimovic, M. (1989) Nonlinear failure envelope for soils. *Journal of Geotechnical Engineering*, **115**(4), 581–586.

Marachi, N.D., Chan, C.K., and Seed, H.B. (1972) Evaluation of properties of rockfill materials. *Journal of the Soil Mechanics and Foundations Division*, **98**(SM1), 95–114.

Marjanovic, J. and Germaine, J.T. (2013) Experimental study investigating the effects of setup conditions on bender element velocity results. *Geotechnical Testing Journal*, **36**(2), 187–197.

Martin, G.R., Liam Finn, W.D., and Seed, H.B. (1978) Effects of system compliance on liquefaction tests. *Journal of the Geotechnical Engineering Division*, **104**(GT4), 463–479.

Mayne, P.W. (1985) Stress anisotropy effects on clay strength. *Journal of Geotechnical Engineering*, **111**(3), 356–366.

Mayne, P.W. (1988) Determining OCR in clays from laboratory strength. *Journal of Geotechnical Engineering*, **114**(1), 76–92.

Mayne, P.W. and Kulhawy, F.H. (1982) K_0–OCR relationships in soil. *Journal of the Geotechnical Engineering Division*, **108**(GT6), 851–872.

Mayne, P.W. and Stewart, H. (1988) Pore pressure behavior of K_0-consolidated clays. *Journal of Geotechnical Engineering*, **114**(11), 1340–1346.

McKinlay, D.G. (1961) A laboratory study of consolidation in clays with particular reference to conditions of radial pore water drainage. *Proceedings of the 5th International Conference on Soil Mechanics and Foundation Engineering*, Paris, France, vol. 1, pp. 225–228.

Menzies, B.K. (1975) A device for measuring volume change. *Geotechnique*, **25**(1), 133–134.

Menzies, B.K. (1976) Design, manufacture and performance of a lateral strain device. *Geotechnique*, **26**(3), 542–544.

Menzies, B.K. (1988) A computer controlled hydraulic triaxial testing system. In: *Advanced Triaxial Testing of Soil and Rock, ASTM STP 977* (eds R.T. Donaghe, R.C. Chaney, and M.L. Silver), pp. 82–94. ASTM, Philadelphia, PA.

Mishu, L.P. (1966) *A study of stresses and strains in soil specimens in the triaxial test*. PhD thesis, Purdue University, West Lafayette, IN.

Mitachi, T., Kohata, Y., and Kudoh, Y. (1988) The influence of filter strip shape on consolidated undrained triaxial extension test results. In: *Advanced Triaxial Testing of Soil and Rock, ASTM STP 977* (eds R.T. Donaghe, R.C. Chaney, and M. L. Silver), pp. 667–678. ASTM, Philadelphia, PA.

Mitchell, J.K. (1956) The fabric of natural clays and its relation to engineering properties. *Proceedings of the 35th Annual Meeting of the Highway Research Board*, Washington, DC, USA, vol. 35, pp. 693–713.

Mitchell, J.K. (1960) Fundamental aspects of thixotropy in soils. *Journal of Soil Mechanics and Foundations Division*, **86**(SM3), 19–52.

Mitchell, J.K. (1976) *Fundamentals of Soil Behavior*. John Wiley & Sons, Inc., New York, NY.

Mitchell, R.J. and Burn, K.N. (1971) Electronic measurement of changes in the volume of pore water during testing of soil samples. *Canadian Geotechnical Journal*, **8**(2), 341–345.

Miura, S. and Toki, S. (1982) A sample preparation method and its effect on static and cyclic deformation-strength properties of sand. *Soils and Foundations*, **22**(1), 61–77.

Mohr, O. (1882) Uber die Darstellung des Spannungszustandes und des Deformation-Zustandes Lines Korper-Elements. *Civilingenieur*, **28**, 113–156.

Molenkamp, F. (1985) Comparison of frictional material models with respect to shear band initiation. *Geotechnique*, **35**(2), 127–143.

Molenkamp, F. and Luger, H.J. (1981) Modelling and minimization of membrane penetration effects in tests on grnular soils. *Geotechnique*, **31**(4), 471–486.

Molenkamp, F. and Tatsuoka, F. (1983) Discussion of 'Compression of "free ends" during triaxial testing.' *Journal of Geotechnial Engineering*, **109**(5), 766–771.

Morgan, J.R. and Moore, P.J. (1968) Experimental techniques. In: *Soil Mechanics – Selected Topics* (ed. I.K. Lee), pp. 295–340. American Elsevier Publishing Co., Inc., New York, NY.

Mualem, Y. (1976) A new model for predicting the hydraulic conductivity of unsaturated porous media. *Water Resources Research*, **12**, 513–522.

Mulilis, J.P., Seed, H.B., Chan, C.K., Mitchell, J.K., and Arulanandan, K. (1977) Effects of sample preparation on sand liquefaction. *Journal of Geotechnical Enginering*, **103**(GT2), 91–108.

Nakai, T. and Matsuoka, H. (1983) Shear behaviors of sand and clay under three-dimensional stres conditions. *Soils and Foundations*, **23**(2), 26–42.

Nataatmadja, A. and Parkin, A.K. (1990) Axial deformation measurement in repeated load triaxial testing. *Geotechnical Testing Journal*, **13**(1), 45–48.

Nayak, G.C. and Zienkiewicz, O.C. (1972) Convenient forms of stress invariants for plasticity. *Journal of the Structural Division*, **98**(ST4), 949–954.

Nazarian, S. and Baig, S.S. (1995) Evaluation of bender elements for use with coarse-grained soils. *Proceedings of the 3rd International Conference on Recent Advances in Geotechnical Earthquake Engineering and Soil Dynamics*, St Louis, MO, vol. 1, pp. 89–94.

Newland, P.L. and Allely, B.H. (1959) Volume changes during undrained triaxial tests on saturated dilatant granular materials. *Geotechnique*, **9**(4), 174–182.

Newson, T.A., Davies, M.C.R., and Bondok, A.R.A. (1997) Selecting the rate of loading for drained stress path triaxial tests. *Geotechnique*, **47**(5), 1063–1067.

Ng, C.W.W., Zhan, L.T., and Cui, Y.J. (2002) A new simple system for measuring volume changes in unsaturated soils. *Canadian Geotechnical Journal*, **39**(3), 757–764.

Nicholson, P.G., Seed, R.B., and Anwar, H.A. (1993a) Elimination of membrane compliance in undrained triaxial testing. I. Measurement and evaluation. *Canadian Geotechnical Journal*, **30**(5), 727–738.

Nicholson, P.G., Seed, R.B., and Anwar, H.A. (1993b) Elimination of membrane compliance in undrained triaxial testing. II. Mitigation by injection compensation. *Canadian Geotechnical Journal*, **30**(5), 739–746.

Nordal, S. (1994) Soil modelling: Chapters 1–6: A continuum mechanics based approach to elasto-plasticity for soils. Department of Geotechnical Engineering, The Norwegian Institute of Technology, Trondheim, Norway.

Norris, G.M. (1981) Effect of end membrane thickness on the strength of "frictionless" cap and base tests. In: *Laboratory Shear Strength of Soil, ASTM STP 740* (eds R.N. Yong and F.C. Townsend), pp. 303–314. ASTM, Philadelphia, PA.

Ochiai, H. and Lade, P.V. (1983) Three-dimensional behavior of sand with anisotropic fabric. *Journal of Geotechnical Engineering*, **109**(GT10), 1313–1328.

Oda, M. (1972a) Initial fabrics and their relations to mechanical properties of granular materials. *Soils and Foundations*, **12**(1), 17–36.

Oda, M. (1972b) The mechanism of fabric changes during compressional deformation of sand. *Soils and Foundations*, **12**(2), 1–18.

Oda, M. (1981) Anisotropic strength of cohesionless sands. *Journal of the Geotechnical Engineering Division*, **107**(GT9), 1219–1231.

Oda, M. and Koishikawa, I. (1977) Anisotropic fabric in sands. *Proceedings of the 9th International Conference on Soil Mechanics and Foundation Engineering*, Tokyo, Japan, vol. 1, pp. 235–238.

Oda, M., Koishikawa, I., and Higuchi, I. (1978) Experimental study of anisotropic shear strength of sand by plane strain tests. *Soils and Foundations*, **18**(1), 25–38.

Olson, R.E. and Daniel, D.E. (1981) Measurement of the hydraulic conductivity of fine-grained soils. In: *Permeability and Groundwater Containment Transport, ASTM STP 746*, pp. 18–64. ASTM, Philadelphia, PA.

Olson, R.E. and Kiefer, M.L. (1964) Effects of lateral filter drains on the triaxial shear characteristics of soils. In: *Laboratory Shear Testing of Soils, ASTM STP 361*, pp. 482–491. ASTM, Philadelphia, PA.

Oswell, J.M., Graham, J., Lingnau, B.E., and King, M.W. (1991) Use of side drains in triaxial testing at moderate to high pressures. *Geotechnical Testing Journal*, **14**(3), 315–319.

Oswell, J.M., Lingnau, B.E., Osiowy, K.B.P., and Graham, J. (1989) Discussion on "Automatic Volume Change and Pressure Measurement Devices for Triaxial Testing of Soils" by Poul V. Lade. *Geotechnical Testing Journal*, **12**(4), 325–326.

Pachakis, M.D. (1976) The influence of the membrane restraint on the measured strength of a soil sample failing along a single shear plane in the triaxial test. *Geotechnique*, **26**(1), 226–230.

Parry, R.H.G. (1960) Triaxial compression and extension tests on remoulded saturated clay. *Geotechnique*, **10**(4), 166–180.

Pearce, J.A. (1970) A truly triaxial machine for testing clays. *Veroffentlichungen des Institutes fur Bodenmechanik und Felsmechanik der Universitat Fredericiana in Karlsruhe*, **44**, 95–110.

Pearce, J.A. (1971) A new true triaxial apparatus. *Proceedings of Stress-Strain Behaviour of Soils, Proceeding, Roscoe Memorial Symposium*, Cambridge University, Cambridge, UK, pp. 330–339. G.T. Foulis and Co., Ltd, Henley-on-Thames.

Peters, J.F., Lade, P.V., and Bro, A. (1988) Shear band formation in triaxial and plane strain tests. In: *Advanced Triaxial Testing of Soil and Rock, ASTM STP 977* (eds R.T. Donaghe, R.C. Chaney, and M.L. Silver), pp. 604–627. ASTM, Philadelphia, PA.

Pitman, T.D., Robertson, P.K., and Sego, D.C. (1994) Influence of fines on the collapse of loose sands. *Canadian Geotechnical Journal*, **31**(5), 728–739.

Pollard, W.S., Sangrey, D.A., and Poulos, S.J. (1977) Air diffusion through membranes in triaxial tests.

Journal of the Geotechnical Engineering Division, **103**(GT10), 1169–1173.

Ponce, V.M. and Bell, J.M. (1971) Shear strength of sand at extremely low pressures. *Journal of the Soil Mechanics and Foundations Division*, **97**(SM4), 625–638.

Poulos, H.G. (1978) Normalized deformation parameters for kaolin. *Geotechnical Testing Journal*, **1**(2), 102–106.

Pradhan, T.B.S., Tatsuoka, F., and Molenkamp, F. (1986) Accuracy of automated volume change measurement by means of a differential pressure transducer. *Soils and Foundations*, **26**(4), 150–158.

Pradhan, T.B.S., Tatsuoka, F., and Sato, Y. (1989) Experimental stress-dilatancy relations of sand subjected to cyclic loading. *Soils and Foundations*, **29**(1), 45–64.

Prashant, A. and Penemadu, D. (2007) Effect of microfabric on mechanical behavior of kaolin clay using cubical true triaxial testing. *Journal of Geotechnical and Geoenvironmental Engineering*, **133**(4), 433–444.

Procter, D.C. and Barden, L. (1969) Correspondence on Green and Bishop: A note on the drained strength of sand under generalized strain conditions. *Geotechnique*, **19**(3), 424–426.

Rad, N.S. and Tumay, M.T. (1987) Factors affecting sand specimen preparation by raining. *Geotechnical Testing Journal*, **10**(1), 31–37.

Rad, S.R. and Clough, G.W. (1984) New procedure for saturating sand specimens. *Journal of Geotechnical Engineering*, **110**(9), 1205–1218.

Raju, V.S. and Deman, F. (1976) Homogeneous triaxial tests on sand – criterion for homogeneity and measurement of volume changes. *Canadian Geotechnical Journal*, **13**(1), 85–91.

Raju, V.S. and Sadasivan, S.K. (1974) Membrane penetration in triaxial tests on sands. *Journal of the Geotechnical Engineering Division*, **100**(GT4), 482–489.

Raju, V.S., Sadasivan, S.K., and Venkaraman, M. (1972) Use of lubricated and conventional end platens in triaxial tests on sands. *Soils and Foundations*, **12**(4), 35–43.

Raju, V.S. and Venkastaramana, K. (1980) Undrained triaxial tests to assess liquefaction potential of sands: effect of membrane penetration. *Proceedings of the International Symposium on Soils under Cyclic and Transient Loading*, Swansea, Wales, pp. 483–494.

Ramana, K.V. and Raju, V.S. (1982) Membrane penetration in triaxial tests. *Journal of the Geotechnical Engineering Division*, **108**(GT2), 305–310.

Ramanatha Iyer, T.S. (1973) Extension tests on soft clays without membranes. *Journal of the Soil Mechanics and Foundations Division*, **99**(SM6), 485–489.

Ramanatha Iyer, T.S. (1975) The behavior of Drammen plastic clay under low effective stresses. *Canadian Geotechnical Journal*, **12**(1), 70–83.

Raymond, G.P. and Soh, K.K. (1974) Making rubber membranes: discussion. *Canadian Geotechnical Journal*, **11**, 661.

Reades, D.W. and Green, G.E. (1976) Independent stress control and triaxial extension tests on sand. *Geotechnique*, **26**(4), 551–576.

Rifa'i, A., Laloui. L., and Vulliet, L. (2002) Volume measurement in unsaturated triaxial test using liquid variation and image processing. *Proceedings of the 3rd International Conference on Unsaturated Soils*, Recife, Brazil, vol. 2, pp. 441–445.

Rodriguez, N.M. and Lade, P.V. (2014) Non-coaxiality of strain increment and stress directions in cross-anisotropic sand. *International Journal of Solids and Structures*, **51**, 1103–1114.

Rolston, J.W. and Lade, P.V. (2009) Evaluation of practical procedure for compaction density and unit weight of rockfill material. *Geotechnical Testing Journal*, **32**(5), 410–417.

Romero, E., Facio, J.A., Lloret, A., Gens, A., and Alonso, E.E. (1997) A new suction and temperature controlled triaxial apparatus. *Proceedings of the 14th International Conference on Soil Mechanics and Foundation Engineering*, Hamburg, Germany, pp. 185–188.

Roscoe, K.H. (1953) An apparatus for the application of simple shear to soil samples. *Proceedings of the 3rd International Conference on Soil Mechanics ad Foundation Engineering*, Zurich, Switzerland, vol. I, pp. 186–191.

Roscoe, K.H. (1970) The influence of strains in soil mechanics. *Geotechnique*, **20**(2), 129–170.

Roscoe, K.H., Bassett, R.H., and Cole, E.R.L. (1967) Principal axes observed during simple shear of sand. *Proceedings of the Geotechnical Conference*, Oslo, Norway, vol. I, pp. 231–237.

Roscoe, K.H., Schofield, A.N., and Thurairajah, A. (1963) An Evaluation of test data for selecting a yield criterion for soils. In: Laboratory Shear Testing of Soils, ASTM STP 361, pp. 111–128. ASTM, Philadelphia, PA.

Rowe, P.W. (1959) Measurement of the coefficient of consolidation of lacustrine clay. *Geotechnique*, **9**(3), 107–118.

Rowe, P.W. (1970) Representative sampling in location, quality, and size. IN; *Sampling of Soil and Rock, ASTM STP 483*, pp. 77–106. ASTM, Philadelphia, PA.

Rowe, P.W. (1972) The relevance of soil fabric to site investigation practice. *Geotechnique*, **22**(2), 195–300.

Rowe, P.W. and Barden, L. (1964) Importance of free ends in triaxial testing. *Journal of the Soil Mechanics and Foundations Division*, **90**(SM1), 1–27.

Rowland, G.O. (1972) Apparatus for measuring volume change suitable for automatic logging. *Geotechnique*, **22**(3), 525–526.

Rutledge, P. (1947) *Cooperative triaxial shear research program of the Corps of Engineers, Triaxial shear research distribution studies on soils*, pp. 1–178. Waterways Experiment Station, Vicksburg, MS.

Saada, A.S. (1970) Testing of anisotropic clay soils. *Journal of the Soil Mechanics and Foundations Division*, **96**(SM5), 1847–1852.

Saada, A.S. and Bianchini, G.F. (1977) Closure to "Strength of one dimensionally consolidated clays." *Journal of the Geotechnical Engineering Division*, **103**(GT6), 655–660.

Saada, A.S., Bianchini, G.F., and Liang, L. (1994) Cracks, bifurcation and shear band propagation in saturated clays. *Geotechnique*, **44**(1), 35–64.

Safaqah, O.A. and Riemer, M.F. (2007) The elastomer gage for local strain measurement in monotonic and cyclic soil testing. *Geotechnical Testing Journal*, **30**(2), 164–172.

Santagata, M.C., Germaine, J.T., and Ladd, C.C. (1999) Initial stiffness of K_0-normally consolidated Boston Blue Clay measured in the triaxial apparatus. *Proceedings of the 2nd International Symposium on Pre-Failure Deformation Characteristics of Geomaterials*, Torino, Italy, vol. 1, pp. 27–34.

Sarsby, R.W., Kalteziotis, N., and Haddad, E.H. (1980) Bedding error in triaxial tests on granular media. *Geotechnique*, **30**(3), 302–309.

Sarsby, R.W., Kalteziotis, N., and Haddad, E.H. (1982) Compression of "free ends" during triaxial testing. *Journal of the Geotechnical Engineering Division*, **108**(GT1), 83–107.

Sayao, A. and Vaid, Y.P. (1991) A critical assessment of stress nonuniformities in hollow cylinder test specimens. *Soils and Foundations*, **31**(1), 60–72.

Schmidt, B. (1966) Discussion of "Earth pressures at rest related to stress history." *Canadian Geotechnical Journal*, **3**(4), 239–242.

Schofield, A. and Wroth, P. (1968) *Critical State Soil Mechanics*. McGraw-Hill, Maidenhead.

Scholey, G.K., Frost, J.D., Lo Presti, D.C.F., and Jamiolkowski, M. (1995) A review of instrumentation for measuring small strains during triaxial testing of soil specimens. *Geotechnical Testing Journal*, **18**(2), 137–156.

Schuurman, E. (1966) The compressibility of an air/water mixture and a theoretical relation between the air and water pressure. *Geotechnique*, **16**(4), 269–281.

Seed, H.B., Mitchell, J.K., and Chan, C.K. (1960) The strength of compacted cohesive soils. *Proceedings of the Research Conference on Shear Strength of Cohesive Soils*, Boulder, CO, USA, pp. 877–964.

Seed, H.B., Singh, S. Chan, C.K., and Vilela, T.F. (1982) Considerations in undisturbed sampling of sands. *Journal of Geotechnical Engineering*, **108**(GT2), 265–283.

Sharpe, P. (1978) A device for automatic measurement of volume change. *Geotechnique*, **28**(3), 348–350.

Sheahan, T.C. and Germaine, J.T. (1992) Computer automation of conventional triaxial equipment. *Geotechnical Testing Journal*, **15**(4), 311–322.

Sheahan, T.C., Germaine, J.T., and Ladd, C.C. (1990) Automated triaxial testing of soft clay: an upgraded commercial system. *Geotechnical Testing Journal*, **13**(3), 153–163.

Sheehan, D.E. and Krizek, R.J. (1971) Preparation of homogeneous soil samples by slurry consolidation. *Journal of Materials*, **6**(2), 356–373.

Shibata, T. and Karube, D. (1965) Influence of the variation of the intermediate principal stress on the mechanical properties of normally consolidated clays. *Proceedings of the 6th International Conference on Soil Mechanics and Foundation Engineering*, Montreal, Canada, vol. I, pp. 359–362.

Shibuya, S., Park, C.-S., Tatsuoka, F., *et al.* (1994) The significance of local lateral strain measurement of soil specimen for a wide range of strain. *Soils and Foundations*, **34**(2), 95–105.

Shirley, D.J. and Hampton, L.D. (1978) Shear wave measurements in laboratory sediments. *Journal of the Acoustical Society of America*, **63**(2), 607–613.

Silveira, I.D. (1953) Consolidation of a cylindrical clay sample with external radial flow of water. *Proceedings of the 3rd International Conference on Soil Mechanics and Foundation Engineering*, Zurich, Switzerland, vol. 1, pp. 55–56.

Silver, M.L. (1979) Automated data acquisition, transducers, and dynamic recording for the geotechnical testing laboratory. *Geotechnical Testing Journal*, **2**(4), 185–189.

Singh, S. Seed, H.B., and Chan, C.K. (1982) Undisturbed sampling of saturated sands by freezing. *Journal of Geotechnical Engineering*, **108**(GT2), 247–264.

Sivakumar, R., Sivakumar, V., Blatz, J., and Vimalan, L. (2006) Twin-cell stress path apparatus for testing unsaturated soils. *Geotechnical Testing Journal*, **29**(2), 175–179.

Sivakumar, V., Mackinnon, P., Zaini, J., and Cairns, P. (2010) Effectiveness of filters in reducing consolidation time in routine laboratory testing. *Geotechnique*, **60**(12), 949–956.

Skempton, A.W. (1954) The pore pressure coefficients A and B. *Geotechnique*, **4**(4), 143–147.

Skempton, A.W. and Bishop, A.W. (1954) Soils. In: *Building Materials, Their Elasticity and Inelasticity* (ed. M. Reiner), pp. 417–482. North Holland Publishing Company, Amsterdam.

Sladen, J.A. and Handford, G. (1987) A potential systematic error in laboratory testing of very loose sands. *Canadian Geotechnical Journal*, **24**(3), 462–466.

Steinbach, J. (1967) *Volume change due to membrane penetration in triaxial tests on granular materials*. MSc thesis, Cornell University, Ithaca, NY.

Stewart, W. and Wong, C.K. (1985) Temperature effects on volume measurements. *Journal of Geotechnical Engineering*, **111**(1), 140–144.

Sture, S., Costes, N.C., Baytiste, S.N., *et al.* (1998) Mechanics of granular materials at very low effective stresses. *Journal of Aerospace Engineering*, **11**(3), 67–72.

Sture, S. and Desai, C.S. (1979) Fluid cushion truly triaxial or multiaxial testing device. *Geotechnical Testing Journal*, **2**(1), 20–33.

Sture, S., Ko, H.-Y., Budiman, J.S., and Ontuna, A.K. (1985) Development and application of a directional shear cell. *Proceedings of the 11th International Conference on Soil Mechanics and Foundation Engineering*, San Francisco, CA, USA, vol. 2, pp. 1061–1064.

Symes, M.J. and Burland, J.B. (1984) Determination of local displacements on soil samples. *Geotechnical Testing Journal*, **7**(2), 49–59.

Symons, I.F. (1967) Discussion. *Proceedings of the Geotechnical Conference*, Oslo, Norway, vol. 2, pp. 175–177.

Tatsuoka, F. (1981) A simple method for automated measurements of volume change in laboratory tests. *Soils and Foundations*, **21**(3), 104–106.

Tatsuoka, F. (1988) Some recent developments in triaxial testing systems for cohesionless soils. In: *Advanced Triaxial Testing of Soil and Rock, ASTM STP 977* (eds R.T. Donaghe, R.C. Chaney, and M. L. Silver), pp. 7–67. ASTM, Philadelphia, PA.

Tatsuoka, F. (1989) Discussion on "Automatic volume change and pressure measurement devices for triaxial testing of soils" by Poul V. Lade. *Geotechnical Testing Journal*, **13**(4), 323–324.

Tatsuoka, F. and Haibara, O. (1985) Shear resistance between sand and smooth or lubricated surfaces. *Soils and Foundations*, **25**(1), 89–98.

Tatsuoka, F., Molenkamp, F. Torii, T., and Hino, T. (1984) Behavior of lubrication layers of platens in element tests. *Soils and Foundations*, **24**(1), 113–128.

Tatsuoka, F., Ochi, K., Fujii, S., and Okamoto, M. (1986) Cyclic undrained triaxial and torsional shear strength of sands for different sample preparation methods. *Soils and Foundations*, **26**(3), 23–41.

Tavenas, F., Jean, P., Leblond, P., and Leroueil, S. (1983) The permeability of natural clays. Part II: Permeability characteristics. *Canadian Geotechnical Journal*, **20**(4), 645–660.

Tavenas, F. and Leroueil, S. (1987) State-of-the-art on laboratory and in-situ stress-strain-time behavior of soft clays. *Proceedings of the International Symposium on Geotechnical Engineering of Soft Soils*, Mexico City, Mexico, pp. 1–46.

Taylor, D.W. (1948) *Fundamentals of Soil Mechanics*. John Wiley & Sons, Inc., New York, NY.

Taylor, J.R. (1997) An introduction to error analysis – the study of uncertainties in physical measurements. Oxford University Press, Oxford.

Terzaghi, K. (1925) *Erdbaumechanik auf Bodenphysikalischer Grundlage*. Franz Deuticke, Leipzig.

Thurairajah, A. and Roscoe, K.H. (1965) The correlation of triaxial compression tests data on cohesionless media. *Proceedings of the 6th International Conference on Soil Mechanics and Foundation Engineering*, London, UK, pp. 377–381.

Timoshenko, S. and Woinowski-Krieger, S. (1959) Theory of Plates and Shells. McGraw-Hill, New York, NY.

Tokimatsu, K. and Nakamura, K. (1986) A liquefaction test without membrane penetration effects. *Soils and Foundations*, **26**(4), 127–138.

Tsai, J.I. (1985) *Three-dimensional behavior of remolded overconsolidated clay*. PhD thesis, University of California, Los Angeles.

Tystovich, N.A. (1975) *The Mechanics of Frozen Ground*. McGraw-Hill, New York, NY.

Vaid, Y.P. and Campanella, R.G. (1973) Making rubber membranes. *Canadian Geotechnical Journal*, **10**(4), 643–644.

Vaid, Y.P. and Negussey, D. (1984a) A critical assessment of membrane penetration in the triaxial test. *Geotechnical Testing Journal*, **7**(2), 70–76.

Vaid, Y.P. and Negussey, D. (1984b) Relative density of air and water pluviated sand. *Soils and Foundations*, **24**(2), 101–105.

Vaid, Y.P. and Negussey, D. (1988) Preparation of reconstituted sand specimens. In: *Advanced Triaxial Testing of Soil and Rock, ASTM STP 977* (eds R.T. Donaghe, R.C. Chaney, and M.L. Silver), pp. 405–417. ASTM, Philadelphia, PA.

Vaid, Y.P. and Thomas, J. (1995) Liquefaction and postliquefaction behavior. *Journal of Geotechnical Engineering*, **121**(2), 163–173.

Vardoulakis, I. (1980) Shear band inclination and shear modulus of sand in biaxial tests. *International Journal of Numerical and Analytical Methods in Geomechanics*, **4**(2), 103–119.

Vesic, A.S. and Clough, G.W. (1968) Behavior of granular materials under high stresses. *Journal of the Soil Mechanics and Foundations Division*, **94**(SM3), 661–688.

Wang, Q. and Lade, P.V. (2001) Shear banding in true triaxial tests and its effect on failure in sand. *Journal of Engineering Mechanics*, **127**(8), 754–761.

Wheeler, S.J. (1988) The undrained shear strength of soils containing large gas bubbles. *Geotechnique*, **38**(3), 399–413.

Winter, H. and Goldscheider, M. (1978) Discussion of "Air diffusion through membranes in triaxial test" by W.S. Pollard, D.A. Sangrey, and S.J. Poulos. *Journal of the Geotechnical Engineering Division*, **104**(GT9), 1209–1211.

Wissa, A.E. (1969) Pore pressure measurement in saturated stiff soils. *Journal of the Soil Mechanics and Foundations Division*, **95**(SM4), 1063–1073.

Wissa, A.E. and Ladd, C.C. (1965) *Shear strength generation in stabilized soils*. Report no. R65-7, Massachusetts Institute of Technology, Cambridge, MA.

Wong, R.T., Seed, H.B., and Chan, C.K. (1975) Cyclic loading liquefaction of gravelly soils. *Journal of the Geotechnical Engineering Division*, **101**(GT6), 571–583.

Wood, D.M. (1990) *Soil Behaviour and Critical State Soil Mechanics*. Cambridge University Press, Cambridge.

Wood, D.M., Drescher, A., and Budhu, M. (1979) On the determination of stress state in the simple shear apparatus. *Geotechnical Testing Journal*, **2**(4), 211–221.

Wood, F.M., Yamamuro, J.A., and Lade, P.V. (2008) Effect of depositional method on the undrained response of silty sand. *Canadian Geotechnical Journal*, **45**(11), 1525–1537.

Wu, H.-C. and Chang, G.S. (1982) Stress analysis of dummy rod method for sand specimens. *Journal of the Geotechnical Engineering Division*, **108**(GT9), 1192–1197.

Wu, W. and Kolymbas, D. (1991) On some issues in triaxial extension tests. *Geotechnical Testing Journal*, **14**(3), 276–287.

Yamada, Y. and Ishihara, K. (1979) Anisotropic deformation characteristics of sand under three dimensional stress conditions. *Soils and Foundations*, **19**(2), 79–94.

Yamamuro, J.A., Abrantes, A.E., and Lade, P.V. (2011) Effect of strain rate on the stress-strain behavior of sand. *Journal of Geotechnical and Geoenvironmental Engineering*, **137**(12): 1169–1178.

Yamamuro, J.A. and Covert, K.M. (2001) Monotonic and cyclic liquefaction of very loose sands with high silt content. *Journal of Geotechnical and Geoenvironmental Engineering*, **127**(4), 314–324.

Yamamuro, J.A. and Lade, P.V. (1993a) B-value measurements for granular materials at high pressures. *Geotechnical Testing Journal*, **16**(2), 165–171.

Yamamuro, J.A. and Lade, P.V. (1993b) Effects of strain rate on instability of granular soils. *Geotechnical Testing Journal*, **16**(3), 304–313.

Yamamuro, J.A. and Lade, P.V. (1995) Strain localization in extension tests on granular materials. *Journal of Engineering Mechanics*, **121**(7), 828–836.

Yamamuro, J.A. and Lade, P.V. (1996) Drained sand behavior in axisymmetric tests at high pressures. *Journal of Geotechnical Engineering*, **122**(2), 109–119.

Yamamuro, J.A. and Lade, P.V. (1997) Static liquefaction of very loose sands. *Canadian Geotechnical Journal*, **34**(6), 905–917.

Yamamuro, J.A., Liu, Y., and Lade, P.V. (2012) Performance and suitability of radial drainage materials in axisymmetric testing of clayey soils at high confining pressures. *Geotechnical Testing Journal*, **35**(6), 1–10.

Yamamuro, J.A., Wood, F.M., and Lade, P.V. (2008) Effect of depositional method on the microstructure of silty sand. *Canadian Geotechnical Journal*, **45**(11), 1538–1555..

Yimsiri, S. (2001) *Pre-failure deformation characteristics of soils: anisotropy and soil fabric.* PhD thesis, Cambridge University.

Yimsiri, S. and Soga, K. (2002) A review of local strain measurement systems for triaxial testing of soils. *Journal of the Southeast Asian Geotechnical Society*, **33**(1), 41–52.

Yimsiri, S., Soga, K., and Chandler, S.G. (2005) Cantilever-type local deformation transducer for local axial strain measurement in triaxial test. *Geotechnical Testing Journal*, **28**(5), 1–7.

Yin, J.H. (2003) A double cell triaxial system for continuous measurement of volume changes of an unsaturated or saturated soil specimen in triaxial testing. *Geotechnical Testing Journal*, **26**(3), 353–358.

Yoshimi, Y., Hatanaka, M., and Oh-Oka, H. (1978) Undisturbed sampling of saturated sands by freezing. *Soils and Foundations*, **18**(3), 59–73.

Yuen, C.M.K., Lo, K.Y., Palmer, J.H.L., and Leonards, G.A. (1978) A New apparatus for measuring the principal strains in anisotropic clays. *Geotechnical Testing Journal*, **1**(1), 24–33.

Zhou, Y., Chen, Y., Asaka, Y., and Abe, T. (2008) Surface mounted bender elements for measuring horizontal shear wave velocity of soils. *Journal of Zhejiang University Science A*, **9**(11), 1490–1496.

Zlatovic, S. and Ishihara, K. (1997) Normalized behavior of very loose non-plastic soils: effects of fabric. *Soils and Foundations*, **37**(4), 47–56.

Zlatovic, S. and Szavits-Nossan, A. (1999) Local measurement of radial strain in triaxial apparatus: a new device. *Proceedings of the 2nd International Symposium on Pre-Failure Deformation Characteristics of Geomaterials*, Torino, Italy, vol. 1, pp. 245–248.

Index

Triaxial Testing of Soils, First Edition. Poul V. Lade.

Printed and bound by CPI Group (UK) Ltd, Croydon, CR0 4YY

17/09/2026

14964306-0002